T0332629

NONLINEAR VIBRATIONS AND STABILITY OF SHELLS AND PLATES

This unique book explores both theoretical and experimental aspects of nonlinear vibrations and stability of shells and plates. It is ideal for researchers, professionals, students and instructors. Expert researchers will find the most recent progresses in nonlinear vibrations and stability of shells and plates, including advanced problems of shells with fluid-structure interaction. Professionals will find many practical concepts, diagrams and numerical results, useful for the design of shells and plates made of traditional and advanced materials. They will be able to understand complex phenomena such as dynamic instability, bifurcations and chaos, without needing an extensive mathematical background. Graduate students will find (i) a complete text on nonlinear mechanics of shells and plates, collecting almost all the available theories in a simple form, (ii) an introduction to nonlinear dynamics and (iii) the state of the art on the nonlinear vibrations and stability of shells and plates, including fluid-structure interaction problems.

Marco Amabili is a professor and Director of the Laboratories in the Department of Industrial Engineering at the University of Parma. His main research is in vibrations of thin-walled structures and fluid-structure interaction. Professor Amabili is the winner of numerous awards in Italy and around the world, including the Bourse québécoise d'excellence from the Ministry of Education of Québec in 1999. He is the associate editor of the *Journal of Fluids and Structures*, a member of the editorial board of the *Journal of Sound and Vibration* and the editor of a special issue of the *Journal of Computers and Structures*. He is a co-organizer of 14 conferences or symposia, the secretary of the ASME Technical Committee on Dynamics and Control of Structures and Systems (AMD Division) and a member of the ASME Technical Committees on Vibration and Sound (DE Division) and Fluid-Structure Interaction (PVP Division). Professor Amabili is the author of more than 180 papers in vibrations and dynamics.

Nonlinear Vibrations and Stability of Shells and Plates

Marco Amabili
University of Parma, Italy

Shaftesbury Road, Cambridge CB2 8EA, United Kingdom

One Liberty Plaza, 20th Floor, New York, NY 10006, USA

477 Williamstown Road, Port Melbourne, VIC 3207, Australia

314–321, 3rd Floor, Plot 3, Splendor Forum, Jasola District Centre, New Delhi – 110025, India

103 Penang Road, #05–06/07, Visioncrest Commercial, Singapore 238467

Cambridge University Press is part of Cambridge University Press & Assessment, a department of the University of Cambridge.

We share the University's mission to contribute to society through the pursuit of education, learning and research at the highest international levels of excellence.

www.cambridge.org
Information on this title: www.cambridge.org/9780521883290

© Marco Amabili 2008

This publication is in copyright. Subject to statutory exception and to the provisions of relevant collective licensing agreements, no reproduction of any part may take place without the written permission of Cambridge University Press & Assessment.

First published 2008
First paperback edition 2015

A catalogue record for this publication is available from the British Library

Library of Congress Cataloging-in-Publication data
Amabili, M.
Nonlinear vibrations and stability of shells and plates / Marco Amabili.
p. cm.
Includes bibliographical references and index.
ISBN: 978-0-521-88329-0 (hardback)
1. Shells (Engineering) – Vibration – Mathematical models. 2. Plates (Engineering) – Vibration – Mathematical models. 3. Cylinders – Vibration – Mathematical models. 4. Nonlinear oscillations – Mathematical models. 5. Elastic plates and shells. I. Title.
TA660.S5A39 2008
624.1′776 – dc22 2007028688

ISBN 978-0-521-88329-0 Hardback
ISBN 978-1-107-43542-1 Paperback

Cambridge University Press & Assessment has no responsibility for the persistence or accuracy of URLs for external or third-party internet websites referred to in this publication and does not guarantee that any content on such websites is, or will remain, accurate or appropriate.

To my wonderful wife Olga

To my parents Vito and Antonietta

Contents

Preface

The present book is intended for (i) graduate students in engineering, (ii) researchers working on vibrations or on shell mechanics and (iii) engineers working on aircraft, missiles, launchers, cars, computer hard and optical disks, storage tanks, heat exchangers, nuclear plants, biomechanics, nano-resonators or thin-walled roofs and other structures in civil engineering. This book contains both theoretical and practical methods that could be of interest for a variety of readers. Expert researchers will find the most recent progress in nonlinear vibrations and the stability of shells and plates, including advanced problems in shells with fluid-structure interaction. Professionals will find many practical concepts, diagrams and numerical results, useful for the design of shells and plates made of traditional and advanced materials. They will be able to understand complex phenomena such as dynamic instability, bifurcations and chaos, without needing an extensive mathematical background. Graduate students will find (i) a complete text on the nonlinear mechanics of shells and plates, collecting almost all the available theories in a simple form, (ii) an introduction to nonlinear dynamics and (iii) the state of art on nonlinear vibrations and the stability of shells and plates, including fluid-structure interaction problems.

My interest in vibrations of shells and plates started with my Master's thesis, more than 16 years ago. However, the turning of my research toward nonlinear vibrations is due to the lucky meeting of two friends: Michael P. Païdoussis and Francesco Pellicano. The three of us have worked together as a team since 1997 on challenging problems of nonlinear vibrations and the stability of circular cylindrical shells with fluid-structure interaction. This book would have never been written without their precious contribution during all these years.

Several other eminent researchers worked with me on nonlinear vibration problems (in chronological order): Alexander Vakakis, Arun K. Misra, Konstantinos Karagiozis, Eugeni Grinevitch, Rinaldo Garziera, Abijit Sarkar, Cyril Touzé, Olivier Thomas, Roger Ohayon, Jean-Sébastien Schotté, Yuri Mikhlin, Konstantin Avramov, Yevgeniy Kurilov, Lidia Kurpa, Galina Pilgun, Silvia Carra and Korosh Khorshidi. My thanks go to all of them.

Among my coauthors of other studies on shells and plates, I cannot forget Giovanni Frosali, Moon K. Kwak and Kyeong-Hoon Jeong.

During the writing of the present book, I had many helpers. In particular, I would like to thank Michael P. Païdoussis, who fully corrected the first three chapters of the book; my colleague Rinaldo Garziera, who reviewed Chapter 3; Michael P. Nemeth who authorized the use of six figures used in Chapter 10; Arvind Raman, Olivier Thomas and Cyril Touzé, who corrected Chapter 12, which is based on their work, and authorized the use of their figures; Francesco Pellicano, who obtained the results presented in Chapter 13; Konstantinos Karagiozis, who drew Figure 14.6, merging my calculations with his results; the students of my course Meccanica delle Vibrazioni *B* at the University of Parma, year 2006, who helped to find errors and drew Figures 3.3 and 3.9(b); my editor Peter Gordon and his assistant Erin Fae of Cambridge University Press, who have found the photographs that I needed; and finally, my wife Olga, for her patience.

During the past ten years, many students of mechanical engineering at the Dipartimento di Ingegneria Industriale of the University of Parma helped me perform experiments under my supervision. In particular, the Ph.D. students: Silvia Carra, Michele Pellegrini, Carlo Augenti and Fabrizio Pagnanelli; the Master's students: Angelo Negri, Dario Rabotti, Andrea Bellingeri, Salvatore Falanga, Marco Tommesani, Valentina Pinelli, Carlotta Truzzi, Paolo Evangelista, Simone Varesi, Antonio Segalini, Paolo Campioli, Margherita Mangia, Francesco Boccia, Simone Sabaini, Fabio Magnani, Marco Venturini, Francesca Righi, Federico Colla, Filippo Vinci, Marco Codeghini, Innocente Granelli, Giovanni Zannoni, Giangiacomo Clausi, Filippo Huller and Pier Paolo Lanzi.

Thanks go to my friend Yaroslav Bysaga for his warm hospitality during the proofreading and to Giovanna Magazzù for preparing the multimedia presentation of the book.

Warm encouragement to complete this book was given to me during the 2006 ASME winter annual meeting by Isaac Elishakoff, Roger Ohayon and Liviu Librescu. On April 16, 2007, Liviu Librescu was murdered as a hero during the terrible massacre at Virginia Tech, so I take this opportunity to celebrate this extraordinary and indefatigable researcher who helped colleagues and students up to the end.

I had the privilege to be at McGill University when Michael P. Païdoussis was writing the second volume of his fundamental monograph *Fluid-Structure Interactions: Slender Structures and Axial Flow*; this has been an extraordinary school.

Marco Amabili

Introduction

Plates are structural elements given by a flat surface with a given thickness h. The flat surface is the middle surface of the plate; the upper and lower surfaces delimiting the plate are at distance $h/2$ from the middle surface. The thickness is small compared with the in-plane dimensions and can be either constant or variable. Thin plates are very stiff for in-plane loads, but they are quite flexible in bending. Many applications of plates, made of extremely different materials, can be found in engineering. For example, very thin circular plates are used in computer hard-disk drives; rectangular and trapezoidal plates can be found in the wing skin, horizontal tail surfaces, flaps and vertical fins of aircraft; cantilever rectangular plates are used as nano-resonators for drug detection; and flat rectangular panels are largely used in civil buildings.

If the middle surface describing the structural elements is folded, shells are obtained. Such structures are abundantly present in nature. In fact, because of the curvature of the middle surface, shells are very stiff for both in-plane and bending loads; therefore, they can span over large areas by using a minimum amount of material.

Shells are largely used in engineering; some shell structures are impressive and beautiful. In automotive engineering, the bodies of cars are shells; in biomechanics, arteries are shells conveying flow. Shell structural elements are largely present in the NASA space shuttle in Figure 1, where the solid rocket boosters and the big orange tank for liquid fuels are large shell bodies. Actually, the super lightweight orange tank is composed of a smaller top tank of ogival shape containing liquid oxygen and a bigger bottom tank of circular cylindrical shape containing liquid hydrogen joined by an intertank element. Shells are: the hull of the *Queen Mary 2* transatlantic boat shown in Figure 2 and of the Los Angeles-class fast attack submarine in Figure 3; the roof of *L'Hemisféric* designed by Santiago Calatrava in the Ciutat del les Arts i les Ciéncies, Valencia, Spain (Figure 4); and the fuselage and wing panels of the huge Airbus A380 civil aircraft in Figure 5.

One of the main targets in the design of shell structural elements is to make the thickness as small as possible to spare material and to make the structure light. The analysis of shells has difficulty related to the curvature, which is also the reason for the carrying load capacity of these structures. In fact, a change of the curvature can give a totally different strength (Chapelle and Bathe 2003). Moreover, because of the optimal distribution of material, shells collapse for buckling much before the failure strength of the material is reached. For their thin nature, they can present large displacements, with respect to the shell thickness, associated to small strains

Figure 1. NASA space shuttle: orbiter *Discovery* with the two solid rocket boosters and the large orange external tank. Courtesy of NASA.

before collapse. This is the rationale for using a nonlinear shell and plate theory for studying shell stability.

Shells are often subjected to dynamic loads that cause vibrations; vibration amplitudes of the order of the shell thickness can be easily reached in many applications. Therefore, a nonlinear shell theory should be applied.

Figure 2. The *Queen Mary 2* transatlantic boat. Courtesy of Carnival Corporation & PLC.

Figure 3. The Los Angeles-class fast attack submarine USS *Asheville*. Courtesy of the U.S. Navy.

The book is organized into 15 chapters. Chapter 1 obtains classical nonlinear theories for rectangular and circular plates, circular cylindrical and spherical shells. Classical shell theories for doubly curved shallow shells and for shells of arbitrary shape are obtained in Chapter 2. Composite and innovative functionally graded materials

Figure 4. *L'Hemisféric* designed by Santiago Calatrava in the Ciutat del les Arts i les Ciéncies, Valencia, Spain. Alan Copson/age fotostock.

Figure 5. Airbus A380 aircraft. Plath/age fotostock.

are introduced and nonclassical nonlinear theories, including shear deformation and rotary inertia, for laminated and functionally graded shells are developed; thermal stresses are also introduced. The first two chapters are self-contained with the full development of the theories under clear hypotheses and limitations. They present material that is usually spread in several articles and books with different approaches and symbols. The shell theories are expressed in lines-of-curvature coordinates, which is the form suitable for applications and computer implementation. In some cases, slightly improved formulations, suitable for moderately thick shells and large rotations, have been developed.

The nonlinear dynamics, stability, bifurcation analysis and modern computational tools are introduced in Chapter 3. The Galerkin method and the energy approach that leads to the Lagrange equations of motion are introduced here.

Linear and nonlinear vibrations of rectangular plates and simply supported circular cylindrical shells (closed around the circumference) are throughly studied in Chapters 4 and 5, respectively. In particular, the Lagrange equations are used for discretizing plates and the Galerkin method is used for discretizing circular shells to show both methods. Numerical and experimental results are presented and compared. The effect of geometric imperfections is also addressed in both chapters. The problem of inertial coupling in the equations of motion is analyzed in Chapter 4. Fluid-filled circular shells are investigated with great care in Chapter 5 for their important applications. Chaotic vibrations are analyzed for large harmonic excitation.

Modern numerical techniques, specifically proper orthogonal decomposition (POD) and nonlinear normal modes (NNM) methods, that reduce the number of degrees of freedom in nonlinear shell models are presented in Chapter 6. The comparison of different classical nonlinear shell theories to study large-amplitude vibrations of simply supported circular cylindrical shells is performed in Chapter 7, where the Lagrange equations are used; the nonlinear stability of pressurized shells is also investigated. Nonlinear vibrations of circular cylindrical shells with different boundary conditions are addressed in Chapter 8. Linear and nonlinear vibrations of circular cylindrical panels (open shells) with different boundary conditions are studied in Chapter 9. Also in this case numerical and experimental results are presented and successfully compared. Chaotic vibrations are detected for large harmonic excitations. Nonlinear vibrations of doubly curved shallow (i.e. with small rise) shells with rectangular base, including spherical and hyperbolic paraboloidal shells, are investigated in Chapter 10. Both classical and first-order shear deformation theories are used to study nonlinear vibrations of laminated composite shells. Static buckling, including the example of the external tank of the NASA space shuttle, is also addressed.

The R-function method for meshless discretization of shells and plates of complex shape is introduced in Chapter 11. This is a method with great potential to develop commercial software, but very little is still known about it outside of the Ukraine and Russia. Linear and nonlinear vibrations of circular plates and rotating disks are investigated in Chapter 12. They have an important application in engineering; for example, in hard-disk, CD and DVD drives of computers.

Nonlinear stability of circular cylindrical shells under static and periodic axial loads are studied in Chapter 13. The fundamental effects of geometric imperfections on the buckling load are investigated, and the period-doubling bifurcation giving dynamic instability in periodic loads is deeply analyzed. The problem of stability of circular cylindrical shells conveying a subsonic flow is addressed in Chapter 14, where a strongly subcritical divergence is detected. The fluid-structure interaction problem for inviscid and incompressible flow is fully studied and the numerical and experimental results obtained in a water tunnel are compared. Nonlinear forced vibrations of circular cylindrical shells conveying water flow are studied for both moderate and large excitations, giving periodic response and complex dynamics, respectively. Flutter instability of circular cylindrical shells inserted in axial supersonic airflow is finally investigated in Chapter 15 by using either linear or nonlinear piston theory to model the aerodynamic loads, and a nonlinear shell model taking into account geometric imperfections. Numerical results are compared with the results of experiments performed by the National Aeronautics and Space Administration (NASA).

REFERENCE

D. Chapelle and K. J. Bathe 2003 *The Finite Element Analysis of Shells – Fundamentals.* Springer, Berlin, Germany.

1 Nonlinear Theories of Elasticity of Plates and Shells

1.1 Introduction

It is well known that certain elastic bodies may undergo large displacements while the strain at each point remains small. The classical theory of elasticity treats only problems in which displacements and their derivatives are small. Therefore, to treat such cases, it is necessary to introduce a theory of nonlinear elasticity with small strains. If the strains are small, the deformation in the neighborhood of each point can be identified with a deformation to which the linear theory is applicable. This gives a rationale for adopting Hooke's stress-strain relations, and in the resulting nonlinear theory large parts of the classical theory are preserved (Stoker 1968). However, the original and deformed configuration of a solid now cannot be assumed to be coincident, and the strains and stresses can be evaluated in the original undeformed configuration by using Lagrangian description, or in the deformed configuration by using Eulerian description (Fung 1965).

In this chapter, the classical geometrically nonlinear theories for rectangular plates, circular cylindrical shells, circular plates and spherical shells are derived, classical theories being those that neglect the shear deformation. Results are obtained in Lagrangian description, the effect of geometric imperfections is considered and the formulation of the elastic strain energy is also given. Classical theories for shells of any shape, as well as theories including shear deformation, are addressed in Chapter 2.

1.1.1 Literature Review

A short overview of some theories for geometrically nonlinear shells and plates will now be given. Some information is taken from the review by Amabili and Païdoussis (2003).

In the classical linear theory of plates, there are two fundamental methods for the solution of the problem. The first method was proposed by Cauchy (1828) and Poisson (1829) and the second by Kirchhoff (1850). The method of Cauchy and Poisson is based on the expansion of displacements and stresses in the plate in power series of the distance z from the middle surface. Disputes concerning the convergence of these series and about the necessary boundary conditions made this method unpopular. Moreover, the method proposed by Kirchhoff has the advantage of introducing physical meaning into the theory of plates. Von Kármán (1910) extended this method to study finite deformation of plates, taking into account nonlinear terms. The nonlinear dynamic case was studied by Chu and Herrmann (1956), who were the pioneers in studying nonlinear vibrations of rectangular plates. In order to deal with thicker and laminated composite plates, the Reissner-Mindlin theory of plates (first-order shear deformation theory) was introduced to take into account transverse shear strains. Five variables are used in this theory to describe the deformation: three displacements of the middle surface and two rotations. The Reissner-Mindlin approach does not satisfy the transverse shear boundary conditions at the top and bottom surfaces of the plate, because a constant shear angle through the thickness is assumed, and plane sections remain plane after deformation. As a consequence of this approximation, the Reissner-Mindlin theory of plates requires shear correction factors for equilibrium considerations. For this reason, Reddy (1990) has developed a nonlinear plate theory that includes cubic terms (in the distance from the middle surface of the plate) in the in-plane displacement kinematics. This higher-order shear deformation theory satisfies zero transverse shear stresses at the top and bottom surfaces of the plate; up to cubic terms are retained in the expression of the shear, giving a parabolic shear strain distribution through the thickness, resembling with good approximation the results of three-dimensional elasticity. The same five variables of the Reissner-Mindlin theory are used to describe the kinematics in this higher-order shear deformation theory, but shear correction factors are not required.

Donnell (1934) established the nonlinear theory of circular cylindrical shells under the simplifying shallow-shell hypothesis. Because of its relative simplicity and practical accuracy, this theory has been widely used. The most frequently used form of Donnell's nonlinear shallow-shell theory (also referred to as Donnell-Mushtari-Vlasov theory) introduces a stress function in order to combine the three equations of equilibrium involving the shell displacements in the radial, circumferential and axial directions into two equations involving only the radial displacement w and the stress function F. This theory is accurate only for modes with circumferential wavenumber n that are not small: specifically, $1/n^2 \ll 1$ must be satisfied, so that $n \geq 4$ or 5 is required in order to have fairly good accuracy. Donnell's nonlinear shallow-shell equations are obtained by neglecting the in-plane inertia, transverse shear deformation and rotary inertia, giving accurate results only for very thin shells. The predominant nonlinear terms are retained, but other secondary effects, such as the nonlinearities in curvature strains, are neglected; specifically, the curvature changes are expressed by linear functions of w only.

Von Kármán and Tsien (1941) performed a seminal study on the stability of axially loaded circular cylindrical shells, based on Donnell's nonlinear shallow-shell theory. In their book, Mushtari and Galimov (1957) presented nonlinear theories for moderate and large deformations of thin elastic shells. The nonlinear theory of shallow-shells is also discussed in the book by Vorovich (1999), where the classical Russian studies, for example, due to Mushtari and Vlasov, are presented.

Sanders (1963) developed a more refined nonlinear theory of shells, expressed in tensorial form. The same equations were obtained by Koiter (1966) around the same period, leading to the designation of these equations as the Sanders-Koiter equations. Later, this theory was reformulated in lines-of-curvature coordinates, that is, in a form that can be more suitable for applications; see, for example, Budiansky (1968), where only linear terms are given. According to the Sanders-Koiter theory, all three displacements are used in the equations of motion. Changes in curvature and torsion are linear according to both the Donnell and the Sanders-Koiter nonlinear theories (Yamaki 1984). The Sanders-Koiter theory gives accurate results for vibration amplitudes significantly larger than the shell thickness for thin shells (Amabili 2003).

Details on the above-mentioned nonlinear shell theories may be found in Yamaki (1984) and Amabili (2003), with an introduction to another accurate theory called the modified Flügge nonlinear theory of shells, also referred to as the Flügge-Lur'e-Byrne nonlinear shell theory (Ginsberg 1973). The Flügge-Lur'e-Byrne theory is close to the general large deflection theory of thin shells developed by Novozhilov (1953) and differs only in terms for change in curvature and torsion.

Additional nonlinear shell theories were formulated by Naghdi and Nordgren (1963), using the Kirchhoff hypotheses, and by Libai and Simmonds (1988).

In order to treat moderately thick laminated shells, the nonlinear first-order shear deformation theory of shells was introduced by Reddy and Chandrashekhara (1985), which is based on the linear first-order shear deformation theory introduced by Reddy (1984). Five independent variables, three displacements and two rotations, are used to describe the shell deformation. This theory may be regarded as the thick-shell version of the Sanders theory for linear terms and of the Donnell nonlinear shell theory for nonlinear terms. A linear higher-order shear deformation theory of shells has been introduced by Reddy and Liu (1985); see also Reddy (2003). Dennis and Palazotto have extended this theory to nonlinear deformations (1990); see also Soldatos (1992).

The nonlinear mechanics of composite laminated shells has also been investigated by many authors. Librescu (1987) developed refined nonlinear theories for anisotropic laminated shells. Other theories applied to the dynamics of laminated shells have been developed, for example, by Tsai and Palazotto (1991), Kobayashi and Leissa (1995), Sansour et al. (1997), Gummadi and Palazotto (1999) and Pai and Nayfeh (1994). Nonlinear electromechanics of piezoelectric laminated shallow spherical shells was developed by Zhou and Tzou (2000).

1.2 Large Deflection of Rectangular Plates

1.2.1 Green's and Almansi Strain Tensors for Finite Deformation

It is assumed that a continuous body changes its configuration under physical actions and the change is continuous (no fractures are considered). A system of coordinates x_1, x_2, x_3 is chosen so that a point P of a body at a certain instant of time is described by the coordinates x_i ($i = 1, 2, 3$). At a later instant of time, the body has moved and deformed to a new configuration; the point P has moved to Q with coordinates $a_i (i = 1, 2, 3)$ with respect to a new coordinate system a_1, a_2, a_3 (see Figure 1.1). Both coordinate systems are assumed to be the same rectangular Cartesian (rectilinear and orthogonal) coordinates for simplicity. The point transformation from P to Q is considered to be one-to-one, so that there is a unique inverse of the transformation. The functions x_i and a_i, describing the coordinates, are assumed to be continuous and differentiable, and the Jacobian determinant of the transformation is positive (i.e. a right-hand set of coordinates is transformed into another right-hand set) and does not vanish at any point. The displacement vector $\mathbf{u}$ is introduced having the following components:

$$u_i = a_i - x_i \quad \text{for } i = 1, \ldots, 3. \tag{1.1}$$

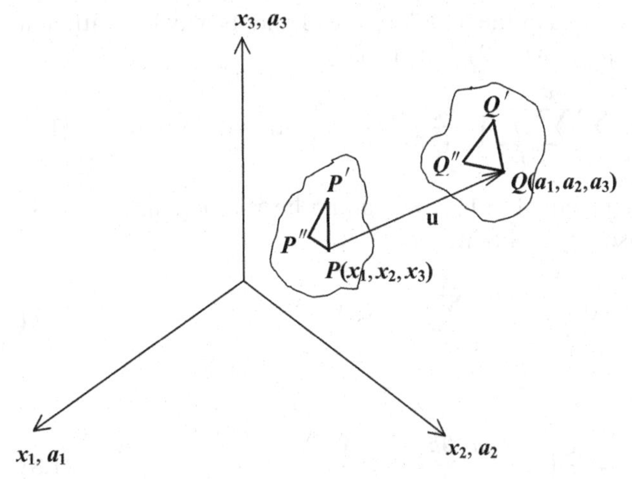

Figure 1.1. Body in the original configuration and after displacement **u**, which moves point P to Q.

In the present book, the Lagrangian description of the continuous systems is used for convenience; therefore u_i are considered to be functions of x_i in order to evaluate the Lagrangian (or Green's) strain tensor, that is, the strains are referred to the original undeformed configuration.

If P, P', P'' are three neighboring points forming a triangle in the original configuration, and if they are transformed to points Q, Q', Q'' in the deformed configuration, the change in the area and angles of the triangle is completely determined if the change in the length of the sides is known. However, the location of the triangle in space is not determined by the change of the sides. Similarly, if the change in length between any two arbitrary points of the body is known, the new configuration of the body is completely defined except for the location of the body in the space. Because interest here is on strains, and these are related to stresses, attention is now focused on the change in distance between any two points of the body.

An infinitesimal line connecting point P (x_1, x_2, x_3) to a neighborhood point P' $(x_1 + \mathrm{d}x_1, x_2 + \mathrm{d}x_2, x_3 + \mathrm{d}x_3)$ is considered; the square of its length in the original configuration is given by

$$\overline{PP'}^2 = \mathrm{d}s_0^2 = \mathrm{d}x_1^2 + \mathrm{d}x_2^2 + \mathrm{d}x_3^2. \tag{1.2}$$

When, due to deformation, P and P' become $Q(a_1, a_2, a_3)$ and $Q'(a_1 + \mathrm{d}a_1, a_2 + \mathrm{d}a_2, a_3 + \mathrm{d}a_3)$, respectively, the square of the distance is

$$\overline{QQ'}^2 = \mathrm{d}s^2 = \mathrm{d}a_1^2 + \mathrm{d}a_2^2 + \mathrm{d}a_3^2, \tag{1.3}$$

in the coordinate system a_i. The differentials $\mathrm{d}a_i$ can be transformed in the original coordinate system x_i:

$$\mathrm{d}a_i = \frac{\partial a_i}{\partial x_1}\mathrm{d}x_1 + \frac{\partial a_i}{\partial x_2}\mathrm{d}x_2 + \frac{\partial a_i}{\partial x_3}\mathrm{d}x_3. \tag{1.4}$$

Therefore, by using equation (1.4), equation (1.3) is transformed into

$$\mathrm{d}s^2 = \sum_{k=1}^{3}\sum_{i=1}^{3}\sum_{j=1}^{3} \frac{\partial a_k}{\partial x_i}\frac{\partial a_k}{\partial x_j}\,\mathrm{d}x_i\,\mathrm{d}x_j. \tag{1.5}$$

The difference between the squares of the length of the elements may be written in the following form by using equations (1.2) and (1.5):

$$\mathrm{d}s^2 - \mathrm{d}s_0^2 = \sum_{k=1}^{3}\sum_{i=1}^{3}\sum_{j=1}^{3}\left(\frac{\partial a_k}{\partial x_i}\frac{\partial a_k}{\partial x_j} - \delta_{ij}\right)\mathrm{d}x_i\,\mathrm{d}x_j, \tag{1.6}$$

where δ_{ij} is the Kronecker delta, equal to 1 if $i = j$ and otherwise equal to zero. By definition, Green's strain tensor ε_{ij} is obtained as

$$\mathrm{d}s^2 - \mathrm{d}s_0^2 = 2\sum_{i=1}^{3}\sum_{j=1}^{3}\varepsilon_{ij}\,\mathrm{d}x_i\,\mathrm{d}x_j, \tag{1.7}$$

and therefore is given by

$$\varepsilon_{ij} = \frac{1}{2}\left(\sum_{k=1}^{3}\frac{\partial a_k}{\partial x_i}\frac{\partial a_k}{\partial x_j} - \delta_{ij}\right). \tag{1.8}$$

By using equation (1.1), the following expression is obtained:

$$\frac{\partial a_k}{\partial x_i} = \frac{\partial u_k}{\partial x_i} + \delta_{ki}. \tag{1.9}$$

Finally, by substituting equation (1.9) into equation (1.8), Green's strain tensor is expressed as

$$\begin{aligned}\varepsilon_{ij} &= \frac{1}{2}\left[\left(\sum_{k=1}^{3}\frac{\partial u_k}{\partial x_i} + \delta_{ki}\right)\left(\sum_{k=1}^{3}\frac{\partial u_k}{\partial x_j} + \delta_{kj}\right) - \delta_{ij}\right]\\ &= \frac{1}{2}\left(\frac{\partial u_i}{\partial x_j} + \frac{\partial u_j}{\partial x_i} + \sum_{k=1}^{3}\frac{\partial u_k}{\partial x_i}\frac{\partial u_k}{\partial x_j}\right).\end{aligned} \tag{1.10}$$

In unabridged notation (x, y, z for x_1, x_2, x_3), the typical formulations are obtained:

$$\varepsilon_{xx} = \frac{\partial u_1}{\partial x} + \frac{1}{2}\left[\left(\frac{\partial u_1}{\partial x}\right)^2 + \left(\frac{\partial u_2}{\partial x}\right)^2 + \left(\frac{\partial u_3}{\partial x}\right)^2\right], \tag{1.11a}$$

$$\gamma_{xy} = \frac{1}{2}\left[\frac{\partial u_1}{\partial y} + \frac{\partial u_2}{\partial x} + \left(\frac{\partial u_1}{\partial x}\frac{\partial u_1}{\partial y} + \frac{\partial u_2}{\partial x}\frac{\partial u_2}{\partial y} + \frac{\partial u_3}{\partial x}\frac{\partial u_3}{\partial y}\right)\right], \tag{1.11b}$$

where the usual symbol γ_{xy} has been used instead of ε_{xy} in equation (1.11b). Equation (1.10) shows that Green's strain tensor is symmetric.

If the Eulerian description of the continuous systems is used, u_i are considered functions of a_i in order to evaluate the Eulerian (usually referred to as the Almansi) strain tensor (i.e. the strains are referred to the deformed configuration). With analogous mathematical development, the Almansi strain tensor is given by

$$\begin{aligned}\varepsilon_{ij}^{(E)} &= \frac{1}{2}\left[\delta_{ij} - \left(-\sum_{k=1}^{3}\frac{\partial u_k}{\partial a_i} + \delta_{ki}\right)\left(-\sum_{k=1}^{3}\frac{\partial u_k}{\partial a_j} + \delta_{kj}\right)\right]\\ &= \frac{1}{2}\left(\frac{\partial u_i}{\partial a_j} + \frac{\partial u_j}{\partial a_i} - \sum_{k=1}^{3}\frac{\partial u_k}{\partial a_i}\frac{\partial u_k}{\partial a_j}\right),\end{aligned} \tag{1.12}$$

which is also symmetric; the superscript denotes "Eulerian."

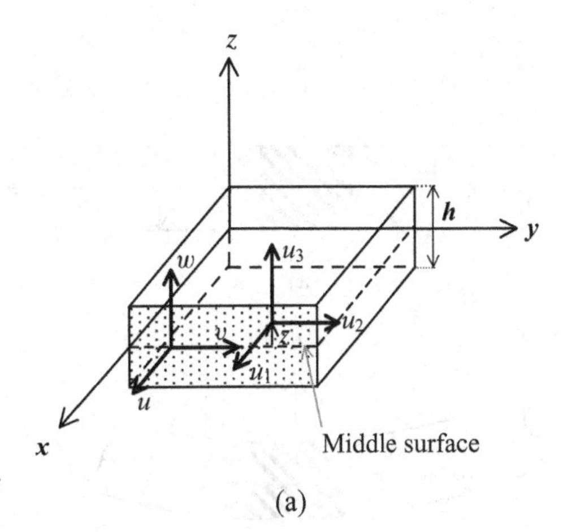

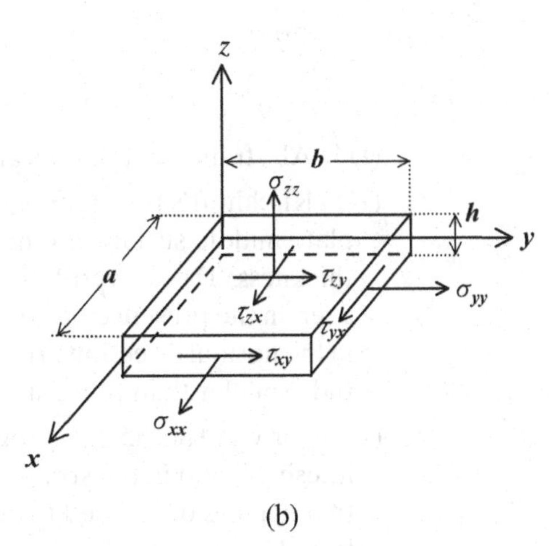

Figure 1.2. Rectangular plate. (a) Symbols used for displacements of middle surface and generic point. (b) Symbols used for dimensions and Kirchhoff stresses.

1.2.2 Strains for Finite Deflection of Rectangular Plates: Von Kármán Theory

The displacements of a generic point of the middle surface of the plate are indicated by u, v and w in x, y and z direction, respectively; the corresponding displacements of a generic point of the plate at distance z from the middle surface are denoted by u_1, u_2 and u_3, as shown in Figure 1.2(a).

When the plate deflection w is of the same order of magnitude as the plate thickness h, results obtained by using linear theories become quite inaccurate. Here a theory for large deflections (large refers to the fact that w is not small compared to h, so that the original and deformed configurations are different) of plates is developed in rectangular coordinates (suitable for rectangular plates of in-plane dimensions a and b, see Figure 1.2[b]). In the theory, the following hypotheses are made (Fung 1965):

(H1) The plate is thin: $h \ll a, h \ll b$.

(H2) The magnitude of deflection w is of the same order as the thickness h of the plate, therefore small compared to the plate dimensions a and b for (H1): $|w| = O(h)$.

(H3) The slope is small at each point: $|\partial w/\partial x| \ll 1$, $|\partial w/\partial y| \ll 1$.

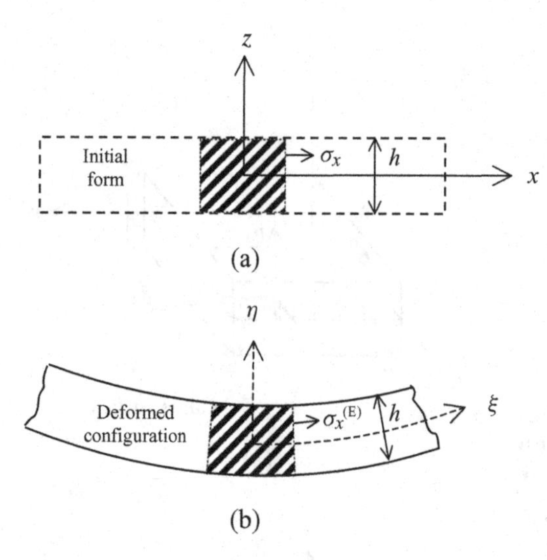

Figure 1.3. Cross-section of the rectangular plate. (a) Initial configuration; σ_x is the normal Kirchhoff stress. (b) Deformed configuration; $\sigma_x^{(E)}$ is the normal Eulerian stress.

(H4) All strain components are small so that linear elasticity can be applied.

(H5) Kirchhoff's hypotheses hold, that is, stresses in the direction normal to the plate middle surface are negligible, and strains vary linearly within the plate thickness. These hypotheses are a good approximation for thin plates. However, in the presence of external loads orthogonal to the plate surface, stresses in the normal direction arise, even if they are generally of some order of magnitude smaller than other stresses.

(H6) For von Kármán's hypothesis, the in-plane displacements u and v are infinitesimal, and in the strain-displacement relations only those nonlinear terms that depend on w need to be retained. All other nonlinear terms may be neglected.

Hypothesis (H6) can be removed in order to get more accurate nonlinear plate theories.

Figure 1.3 shows that the deformed configuration of the plate differs significantly from the original one. The Lagrangian description of the plate is used, so that the plate surfaces are always designated as $z = \pm h/2$; a right-handed rectangular Cartesian reference system (O; x, y, z) is used, with the x, y plane coinciding with the middle surface of the plate in its initial, undeformed configuration and the z axis normal to it. In the Lagrangian description, Green's strain tensor, referred to in the initial configuration, is used; it is given by equations (1.10) and (1.11). By using hypothesis (H5), we have

$$u_1 = u(x, y) - z\frac{\partial w}{\partial x}, \tag{1.13}$$

$$u_2 = v(x, y) - z\frac{\partial w}{\partial y}, \tag{1.14}$$

$$u_3 = w(x, y). \tag{1.15}$$

Equations (1.13–1.15) are linear expressions. In particular, these equations are obtained as follows: hypothesis (H5) requires that strains vary linearly within the plate thickness and Kirchhoff stresses in the direction normal to the plate middle surface are negligible (i.e. $\sigma_{zz} = \tau_{zx} = \tau_{zy} = 0$, where σ_{ij} is the normal stress acting on the surface normal to i in direction j, and τ_{ij} is the tangential stress as shown in Figure 1.2[b]); therefore, by using linear elasticity,

$$\sigma_{xx} = a_1(x, y) + b_1(x, y)\, z, \tag{1.16}$$

$$\sigma_{yy} = a_2(x, y) + b_2(x, y)\, z, \tag{1.17}$$

$$\varepsilon_{zz} \simeq \frac{\partial u_3}{\partial z} = -\frac{\nu}{E}\left(\sigma_{xx} + \sigma_{yy}\right), \tag{1.18}$$

where E is Young's modulus and ν is Poisson's ratio. In the last equation, the linearized expression of ε_{zz} is used (this is the reason for using $\simeq$ instead of $=$) as a consequence of equations (1.13–1.15) being linear. Integration of equation (1.18) gives

$$u_3 = w(x, y) - \frac{\nu}{E}\left[a_1(x, y) + a_2(x, y)\right] z - \frac{\nu}{E}\left[b_1(x, y) + b_2(x, y)\right] \frac{z^2}{2}, \tag{1.19}$$

where the term w is the integration constant. The last two terms on the right-hand side are in general very small for thin plates, as a consequence of ν/E being a very small number, and they can be neglected; therefore, equation (1.15) is verified, and $\varepsilon_{zz} \simeq 0$. Also for hypothesis (H5), the following expressions are obtained:

$$\gamma_{zx} \simeq \frac{1}{2}\left(\frac{\partial u_1}{\partial z} + \frac{\partial u_3}{\partial x}\right) = \frac{\tau_{zx}}{E} - \frac{\nu}{E}(\tau_{zy} + \sigma_{zz}) = 0, \tag{1.20}$$

$$\gamma_{zy} \simeq \frac{1}{2}\left(\frac{\partial u_2}{\partial z} + \frac{\partial u_3}{\partial y}\right) = 0. \tag{1.21}$$

Inserting $u_3 = w(x, y)$ from equation (1.15) into equations (1.20) and (1.21) and integrating, equations (1.13) and (1.14) are obtained.

Equations (1.13–1.15) are inserted into Green's strain tensor, equations (1.10) and (1.11), in order to obtain the strain-displacement relations for a plate in rectangular coordinates. The strain components ε_{xx}, ε_{yy} and γ_{xy} at an arbitrary point of the plate are related to the middle surface strains $\varepsilon_{x,0}$, $\varepsilon_{y,0}$ and $\gamma_{xy,0}$ and to the changes in the curvature and torsion of the middle surface k_x, k_y and k_{xy} by the following three relations:

$$\varepsilon_{xx} = \varepsilon_{x,0} + z\, k_x, \tag{1.22}$$

$$\varepsilon_{yy} = \varepsilon_{y,0} + z\, k_y, \tag{1.23}$$

$$\gamma_{xy} = \gamma_{xy,0} + z\, k_{xy}, \tag{1.24}$$

where z is, as usual, the distance of the arbitrary point of the plate from the middle surface. If von Kármán hypothesis (H6) is used, the following expressions for the middle surface strains and the changes in the curvature and torsion of the middle surface are obtained, namely

$$\varepsilon_{x,0} = \frac{\partial u}{\partial x} + \frac{1}{2}\left(\frac{\partial w}{\partial x}\right)^2, \tag{1.25}$$

$$\varepsilon_{y,0} = \frac{\partial v}{\partial y} + \frac{1}{2}\left(\frac{\partial w}{\partial y}\right)^2, \tag{1.26}$$

$$\gamma_{xy,0} = \frac{\partial u}{\partial y} + \frac{\partial v}{\partial x} + \frac{\partial w}{\partial x}\frac{\partial w}{\partial y}, \tag{1.27}$$

$$k_x = -\frac{\partial^2 w}{\partial x^2}, \tag{1.28}$$

$$k_y = -\frac{\partial^2 w}{\partial y^2}, \tag{1.29}$$

$$k_{xy} = -2\frac{\partial^2 w}{\partial x\,\partial y}. \tag{1.30}$$

These expressions are in general accurate enough for moderately large vibrations of plates. If more accurate expressions are needed, hypothesis (H6) can be removed and equations (1.25–1.30) can be computed, retaining all the nonlinear terms.

1.2.3 Geometric Imperfections

Initial geometric imperfections of the rectangular plate associated with zero initial stress are denoted by normal displacement w_0; in-plane initial imperfections are neglected. Therefore, equation (1.15) is replaced by

$$u_3 = w(x, y) + w_0(x, y). \tag{1.31}$$

Substituting equation (1.31) into Green's strain tensor and neglecting all terms that depend on w_0 only (as they are associated with initial deformation and therefore with zero stress), the following expressions are obtained, replacing equations (1.25–1.27):

$$\varepsilon_{x,0} = \frac{\partial u}{\partial x} + \frac{1}{2}\left(\frac{\partial w}{\partial x}\right)^2 + \frac{\partial w}{\partial x}\frac{\partial w_0}{\partial x}, \tag{1.32}$$

$$\varepsilon_{y,0} = \frac{\partial v}{\partial y} + \frac{1}{2}\left(\frac{\partial w}{\partial y}\right)^2 + \frac{\partial w}{\partial y}\frac{\partial w_0}{\partial y}, \tag{1.33}$$

$$\gamma_{xy,0} = \frac{\partial u}{\partial y} + \frac{\partial v}{\partial x} + \frac{\partial w}{\partial x}\frac{\partial w}{\partial y} + \frac{\partial w}{\partial x}\frac{\partial w_0}{\partial y} + \frac{\partial w_0}{\partial x}\frac{\partial w}{\partial y}. \tag{1.34}$$

Equations (1.28–1.30) are unchanged by the presence of the imperfection w_0. It can be observed that, in the presence of geometric imperfections, the plate is no longer perfectly flat.

1.2.4 Eulerian, Lagrangian and Kirchhoff Stress Tensors

In the deformed configuration of a body, the equations of equilibrium of an infinitesimal parallelepiped with surfaces parallel to the coordinate planes (see Figure 1.4) are given by

$$\sum_{j=1}^{3} \frac{\partial \sigma_{ji}^{(E)}}{\partial a_j} + X_i = 0 \quad \text{for } i = 1, \ldots, 3, \tag{1.35}$$

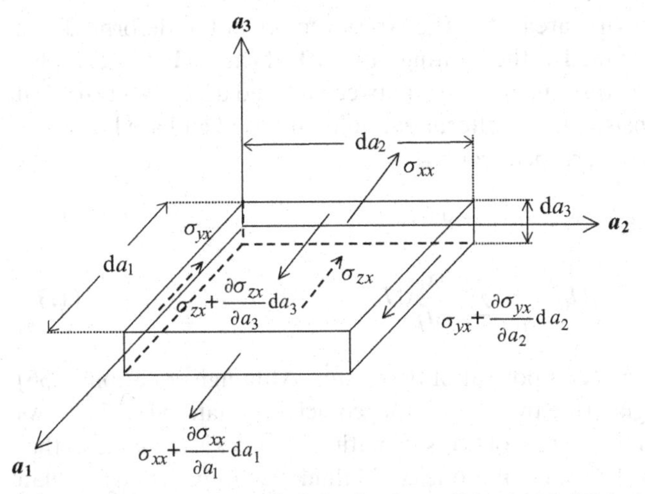

Figure 1.4. Stresses on an infinitesimal parallelepiped of the plate.

where $\sigma_{ji}^{(E)}$ is the Eulerian stress tensor (for simplicity, the symbol σ is used here also for the tangential stresses instead of τ), that is, the stresses referred to the deformed configuration a_j are the coordinates describing the deformed configuration and X_i are the body forces, including inertia, per unit volume. The Eulerian stresses are symmetric, that is, $\sigma_{ji}^{(E)} = \sigma_{ij}^{(E)}$. Equation (1.35) shows that in the Eulerian description the dynamic equations are simple, but the kinematics is complex; for this reason it is convenient to find the relationships to transform the Eulerian stresses in the original undeformed configuration in order to use the Lagrangian description. True loading and stresses in a body exist only in the deformed configuration; therefore, the loading and stresses in the initial undeformed configuration are somehow fictitious.

In Figure 1.5, a solid body with initial configuration drawn on the left-hand side (described in the coordinate system x_i, with $i = 1, \ldots, 3$) is transformed into the parallelepiped shown on the right-hand side (described in the deformed coordinate system a_i, with $i = 1, \ldots, 3$). The force vector $\mathbf{dT}$ acting on the deformed surface PQRS of area dS corresponds to the force $\mathbf{dT}_0$ in the original configuration, which acts

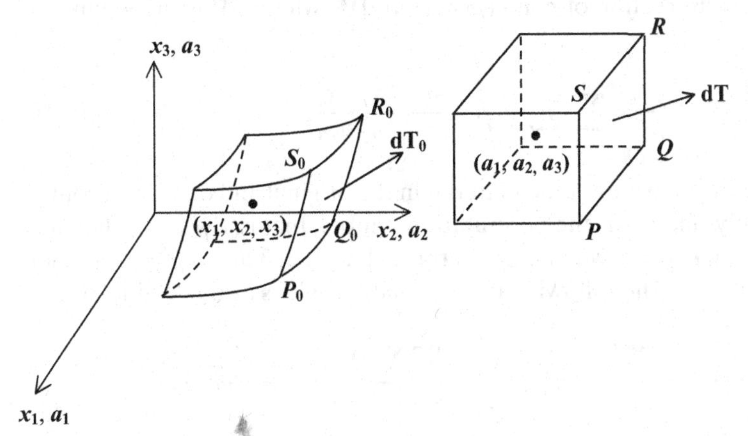

Figure 1.5. Infinitesimal parallelepiped in deformed configuration and corresponding body in the initial configuration.

on the surface $P_0Q_0R_0S_0$ having area $\mathrm{d}S_0$. The stress vectors in the deformed and original configurations are defined as the limiting ratios $\mathbf{dT}/\mathrm{d}S$ and $\mathbf{dT}_0/\mathrm{d}S_0$, respectively. The assignment of a correspondence rule between $\mathbf{dT}$ and $\mathbf{dT}_0$ is arbitrary, but must be mathematically consistent; two alternative rules are used and are known as Lagrangian and Kirchhoff rules, respectively:

$$\mathrm{d}T_{0_i}^{(L)} = \mathrm{d}T_i, \tag{1.36}$$

$$\mathrm{d}T_{0_i}^{(K)} = \sum_{j=1}^{3} \frac{\partial x_i}{\partial a_j} \mathrm{d}T_j, \tag{1.37}$$

where $i = 1, 2, 3$ indicates each component of the vector. Although equation (1.36) shows that with the Lagrangian rule the two vectors coincide, equation (1.37) shows that with the Kirchhoff formula the rule of transformation of forces is the same as that obtained for $\mathrm{d}a_i$ in equation (1.4). Actually, one could think that only the Lagrangian rule could be introduced; the reason for introducing the Kirchhoff rule is that it gives useful properties to the related stress tensor, as will be shown soon.

By using the Cauchy formula, it is possible to relate the force $\mathbf{dT}$ to the Eulerian stresses

$$\mathrm{d}T_i = \left(\sum_{j=1}^{3} \sigma_{ji}^{(E)} n_j \right) \mathrm{d}S, \tag{1.38}$$

where n_j are the components of the unit vector $\mathbf{n}$ normal to the surface $\mathrm{d}S$. Similarly, it is possible to write

$$\mathrm{d}T_{0_i}^{(L)} = \sum_{j=1}^{3} S_{ji} n_{0_i} \mathrm{d}S_0 = \mathrm{d}T_i, \tag{1.39}$$

$$\mathrm{d}T_{0_i}^{(K)} = \sum_{j=1}^{3} \sigma_{ji} n_{0_i} \mathrm{d}S_0 = \sum_{j=1}^{3} \frac{\partial x_i}{\partial a_j} \mathrm{d}T_j, \tag{1.40}$$

where S_{ij} and σ_{ij} are the Lagrangian and Kirchhoff stresses, respectively, and n_{0i} are the components of the unit vector $\mathbf{n}_0$ normal to the surface $\mathrm{d}S_0$.

For the law of conservation of mass ($\rho \mathrm{d}V = \rho_0 \mathrm{d}V_0$, where $\mathrm{d}V$ is the volume of the solid element),

$$\frac{\rho}{\rho_0} = \sum_{i=1}^{3} \sum_{j=1}^{3} \sum_{k=1}^{3} e_{ijk} \frac{\partial x_i}{\partial a_1} \frac{\partial x_j}{\partial a_2} \frac{\partial x_k}{\partial a_3}, \tag{1.41}$$

where ρ_0 and ρ are the densities of the material in the original and deformed configurations, respectively, and e_{ijk} is the permutation symbol, that is, $e_{ijk} = 0$ when any two indices are equal, $e_{ijk} = 1$ when i, j, k permute like 1, 2, 3 and $e_{ijk} = -1$ when i, j, k permute like 1, 3, 2. The following transformation rule is also introduced:

$$\sum_{i=1}^{3} \sum_{j=1}^{3} \sum_{k=1}^{3} e_{ijk} \frac{\partial x_i}{\partial a_\gamma} \frac{\partial x_j}{\partial a_\alpha} \frac{\partial x_k}{\partial a_\beta} = e_{\gamma\alpha\beta} \sum_{i=1}^{3} \sum_{j=1}^{3} \sum_{k=1}^{3} e_{ijk} \frac{\partial x_l}{\partial a_1} \frac{\partial x_j}{\partial a_2} \frac{\partial x_k}{\partial a_3}.$$

Two lines $\mathrm{d}\boldsymbol{a}$, with components ($\mathrm{d}a_1$, $\mathrm{d}a_2$, $\mathrm{d}a_3$), and $\delta\boldsymbol{a}$, with components (δa_1, δa_2, δa_3), are considered in the deformed body, which correspond to $\mathrm{d}\boldsymbol{x}$ and $\delta\boldsymbol{x}$, respectively, in the undeformed body. The area $\mathrm{d}S$ of the parallelogram with $\mathrm{d}\boldsymbol{a}$ and $\delta\boldsymbol{a}$ as sides is given by the vector product $\mathbf{da} \wedge \delta\mathbf{a}$, which can be rewritten as

$$n_i\,\mathrm{d}S = \sum_{j=1}^{3}\sum_{k=1}^{3} e_{ijk}\,\mathrm{d}a_j\,\delta a_k.$$

Similarly,

$$n_{0_i}\mathrm{d}S_0 = \sum_{j=1}^{3}\sum_{k=1}^{3} e_{ijk}\,\mathrm{d}x_j\,\delta x_k = \sum_{\alpha=1}^{3}\sum_{\beta=1}^{3}\sum_{j=1}^{3}\sum_{k=1}^{3} e_{ijk}\frac{\partial x_j}{\partial a_\alpha}\frac{\partial x_k}{\partial a_\beta}\mathrm{d}a_\alpha\delta a_\beta.$$

Multiplying both the right-hand and left-hand sides by $\partial x_i/\partial a_\gamma$, summing up on i and substituting equation (1.41) in order to make the notation more compact, the relationship involving the areas is obtained:

$$\sum_{i=1}^{3}\frac{\partial x_i}{\partial a_\gamma}n_{0_i}\mathrm{d}S_0 = \frac{\rho}{\rho_0}\sum_{\alpha=1}^{3}\sum_{\beta=1}^{3} e_{\gamma\alpha\beta}\,\mathrm{d}a_\alpha\delta a_\beta = \frac{\rho}{\rho_0}n_\gamma\mathrm{d}S. \tag{1.42}$$

Hence, from equations (1.38), (1.39) and (1.42),

$$\sum_{j=1}^{3} S_{ji}n_{0_j}\mathrm{d}S_0 = \frac{\rho_0}{\rho}\sum_{\alpha=1}^{3}\sigma_{\alpha i}^{(E)}\sum_{j=1}^{3}\frac{\partial x_j}{\partial a_\alpha}n_{0_j}\mathrm{d}S_0, \tag{1.43}$$

which, eliminating the two sums on j on both the left-hand and right-hand sides and identifying term to term, gives

$$S_{ji} = \frac{\rho_0}{\rho}\sum_{\alpha=1}^{3}\frac{\partial x_j}{\partial a_\alpha}\sigma_{\alpha i}^{(E)}. \tag{1.44}$$

Similarly, from equations (1.38), (1.40) and (1.42),

$$\sigma_{ji} = \frac{\rho_0}{\rho}\sum_{\alpha=1}^{3}\sum_{\beta=1}^{3}\frac{\partial x_i}{\partial a_\alpha}\frac{\partial x_j}{\partial a_\beta}\sigma_{\beta\alpha}^{(E)}. \tag{1.45}$$

Equation (1.44) shows that the Lagrangian stress tensor S_{ji} is not symmetric; on the other hand, the Kirchhoff stress tensor (1.45) is symmetric ($\sigma_{ij} = \sigma_{ji}$) and is more suitable for use in a stress-strain law where the strain tensor is symmetric. By substituting equation (1.44) into (1.45), it is finally possible to obtain the relation linking Kirchhoff to Lagrangian stresses

$$\sigma_{ji} = \sum_{k=1}^{3}\frac{\partial x_i}{\partial a_k}S_{jk}, \tag{1.46a}$$

which makes the pair with the inverse transformation

$$S_{ij} = \sum_{k=1}^{3}\sigma_{ik}\frac{\partial a_j}{\partial x_k}. \tag{1.46b}$$

For an elastic material, there is a one-to-one correspondence between the Eulerian stress tensor $\sigma_{ij}^{(E)}$ and the strain tensor in an Eulerian description (Almansi strain tensor), which is given by the Hooke's stress-strain relations. Because the Eulerian stress tensor and the Almansi strain tensor are related uniquely to the Kirchhoff stress tensor and Green's strain tensor (which is the strain tensor in a Lagrangian description), respectively, an elastic material has a single-valued correspondence between stress and strain also in Lagrangian description and the Hooke's law still applies.

1.2.5 Equations of Motion in Lagrangian Description

The equations of motion (1.35) can now be rewritten in the original configuration as

$$\sum_{j=1}^{3} \frac{\partial S_{ji}}{\partial x_j} + \frac{\rho_0}{\rho} X_i = 0, \quad \text{for } i = 1, \ldots, 3, \tag{1.47}$$

where the Lagrangian description is used. However, in order to use the more useful Kirchhoff stresses, it is necessary to transform equation (1.47) into

$$\sum_{j=1}^{3} \frac{\partial}{\partial x_j} \left(\sum_{k=1}^{3} \sigma_{jk} \frac{\partial a_i}{\partial x_k} \right) + \frac{\rho_0}{\rho} X_i = 0, \quad \text{for } i = 1, \ldots, 3. \tag{1.48}$$

Transformation (1.46b) has been used in order to obtain equation (1.48). It is known from equation (1.1) that $a_i = x_i + u_i$, and substituting into equation (1.48), the final equations of equilibrium in Lagrangian description with Kirchhoff stresses become

$$\sum_{j=1}^{3} \frac{\partial}{\partial x_j} \left[\sum_{k=1}^{3} \sigma_{jk} \left(\delta_{ik} + \frac{\partial u_i}{\partial x_k} \right) \right] + \frac{\rho_0}{\rho} X_i = 0, \quad \text{for } i = 1, \ldots, 3. \tag{1.49}$$

1.2.6 Elastic Strain Energy

The elastic strain energy U_P of a rectangular plate, under Kirchhoff's hypothesis $\sigma_{zz} = \tau_{zx} = \tau_{zy} = 0$, is given by

$$U_P = \frac{1}{2} \int_0^a \int_0^b \int_{-h/2}^{h/2} (\sigma_{xx}\, \varepsilon_{xx} + \sigma_{yy}\, \varepsilon_{yy} + \tau_{xy}\, \gamma_{xy})\, \mathrm{d}x\, \mathrm{d}y\, \mathrm{d}z, \tag{1.50}$$

where h is the plate thickness; a and b are the in-plane dimensions in x and y directions, respectively; σ_{xx}, σ_{yy} and τ_{xy} are Kirchhoff stresses; and ε_{xx}, ε_{yy} and γ_{xy} are Green's strains. The simplicity of equation (1.50) is due to the Lagrangian description of the plate, which allows integration over the plate in the original undeformed configuration. The Kirchhoff stresses for a homogeneous and isotropic material ($\sigma_{zz} = 0$, case of plane stress) are given by

$$\sigma_{xx} = \frac{E}{1-\nu^2} (\varepsilon_{xx} + \nu\, \varepsilon_{yy}), \quad \sigma_{yy} = \frac{E}{1-\nu^2} (\varepsilon_{yy} + \nu\, \varepsilon_{xx}), \quad \tau_{xy} = \frac{E}{2(1+\nu)} \gamma_{xy}. \tag{1.51}$$

Analogous expressions can be used in non-homogeneous or nonisotropic material. By using equations (1.22–1.24, 1.50, 1.51), the following expression is obtained for uniform thickness:

$$U_P = \frac{1}{2}\frac{Eh}{1-\nu^2}\int_0^a\int_0^b \left(\varepsilon_{x,0}^2 + \varepsilon_{y,0}^2 + 2\nu\varepsilon_{x,0}\,\varepsilon_{y,0} + \frac{1-\nu}{2}\gamma_{xy,0}^2\right) \mathrm{d}x\,\mathrm{d}y$$

$$+\frac{1}{2}\frac{Eh^3}{12(1-\nu^2)}\int_0^a\int_0^b \left(k_x^2 + k_y^2 + 2\nu k_x k_y + \frac{1-\nu}{2}k_{xy}^2\right) \mathrm{d}x\,\mathrm{d}y, \quad (1.52)$$

where the first term, on the right-hand side, is the membrane (also referred to as stretching) energy and the second one is the bending energy. Equation (1.52) can be written as a function of the displacements u, v, w by using equations (1.25–1.30) in the case of a perfect plate and equations (1.32–1.34) and (1.28–1.30) in the case of a plate with initial geometric imperfections associated with zero stress.

1.2.7 Von Kármán Equation of Motion

If the in-plane inertia of the plate is neglected, a simplified version of the equations of motion can be obtained, in the form originally presented by von Kármán, without proof, in 1910. Here the elegant development in Lagrangian coordinates by Fung (1965) is used.

The stress resultants, which are forces per unit length, and the stress moments, which are moments per unit length (see Figure 1.6(a,b)) are defined as

$$N_x = \int_{-h/2}^{h/2} \sigma_{xx}\,\mathrm{d}z, \quad N_y = \int_{-h/2}^{h/2} \sigma_{yy}\,\mathrm{d}z, \quad N_{xy} = \int_{-h/2}^{h/2} \sigma_{xy}\,\mathrm{d}z, \quad (1.53\text{a–c})$$

$$M_x = \int_{-h/2}^{h/2} \sigma_{xx}\,z\,\mathrm{d}z, \quad M_y = \int_{-h/2}^{h/2} \sigma_{yy}\,z\,\mathrm{d}z, \quad M_{xy} = \int_{-h/2}^{h/2} \sigma_{xy}\,z\,\mathrm{d}z, \quad (1.53\text{d–f})$$

$$Q_x = \int_{-h/2}^{h/2} \sigma_{xz}\,\mathrm{d}z, \quad Q_y = \int_{-h/2}^{h/2} \sigma_{yz}\,\mathrm{d}z. \quad (1.53\text{g,h})$$

According to hypothesis (H5), the Kirchhoff stress in the direction normal to the plate middle surface is negligible so that Q_x and Q_y should be zero; however, because (H5) is only an approximation, for equilibrium, Q_x and Q_y are not zero and must be introduced. The stress resultants and stress moments are based on Kirchhoff stresses σ_{ij} and are referred to the initial, undeformed configuration of the plate.

By using the strain-displacement relations (1.24–1.30) and the stress-strain relations (1.51), the stress and moment resultants can be written in the following form:

$$N_x = \frac{Eh}{1-\nu^2}\left[\frac{\partial u}{\partial x} + \nu\frac{\partial v}{\partial y} + \frac{1}{2}\left(\frac{\partial w}{\partial x}\right)^2 + \frac{\nu}{2}\left(\frac{\partial w}{\partial y}\right)^2\right], \quad (1.54\text{a})$$

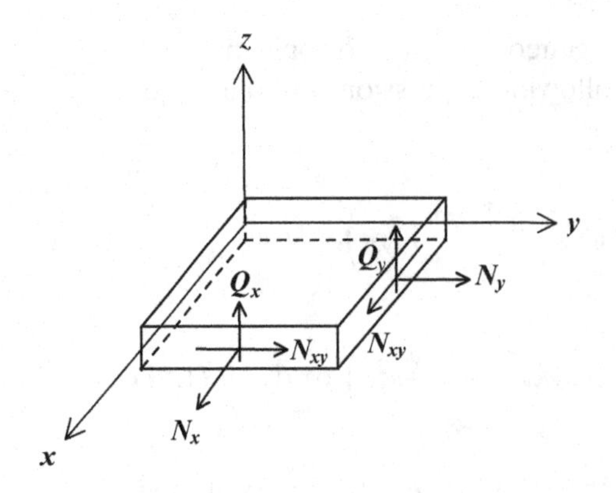

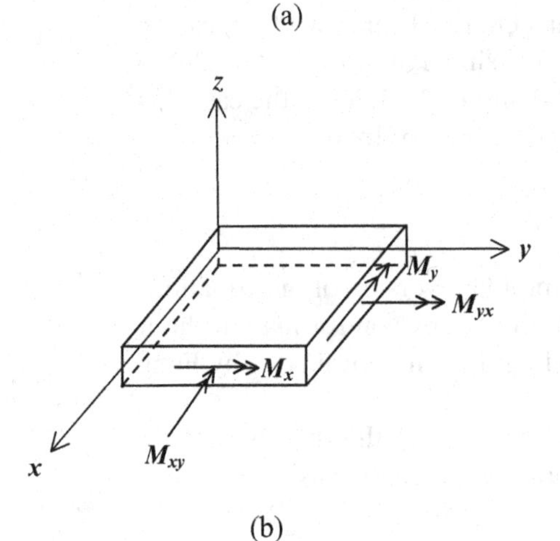

Figure 1.6. Forces and moments per unit length acting on a plate element. (a) Stress resultants. (b) Stress moments.

$$N_y = \frac{Eh}{1-\nu^2}\left[\frac{\partial v}{\partial y} + \nu\frac{\partial u}{\partial x} + \frac{1}{2}\left(\frac{\partial w}{\partial y}\right)^2 + \frac{\nu}{2}\left(\frac{\partial w}{\partial x}\right)^2\right], \tag{1.54b}$$

$$N_{xy} = \frac{Eh}{2(1+\nu)}\left[\frac{\partial u}{\partial y} + \frac{\partial v}{\partial x} + \frac{\partial w}{\partial x}\frac{\partial w}{\partial y}\right], \tag{1.54c}$$

$$M_x = -D\left(\frac{\partial^2 w}{\partial x^2} + \nu\frac{\partial^2 w}{\partial y^2}\right), \quad M_y = -D\left(\frac{\partial^2 w}{\partial y^2} + \nu\frac{\partial^2 w}{\partial x^2}\right), \quad M_{xy} = -(1-\nu)D\frac{\partial^2 w}{\partial x\,\partial y}, \tag{1.54d–f}$$

where $D = Eh^3/[12(1-\nu^2)]$.

The three-dimensional equations of equilibrium of an infinitesimal plate element are given by equation (1.49) in Lagrangian description; in particular, the following notation is introduced into equation (1.49): $x_1 = x, x_2 = y, x_3 = z, a_1 = x + u_1, a_2 = y + u_2, a_3 = z + u_3$. Moreover, by using hypothesis (H6), the following approximations are used: $\partial u_1/\partial x_i = \partial u_2/\partial x_i = \partial u_3/\partial z = 0$. Therefore, the three equations of motion can be written as

$$\frac{\partial \sigma_{xx}}{\partial x} + \frac{\partial \sigma_{yx}}{\partial y} + \frac{\partial \sigma_{zx}}{\partial z} + X = 0, \qquad \frac{\partial \sigma_{xy}}{\partial x} + \frac{\partial \sigma_{yy}}{\partial y} + \frac{\partial \sigma_{zy}}{\partial z} + Y = 0, \tag{1.55a,b}$$

$$\frac{\partial}{\partial x}\left(\sigma_{xx}\frac{\partial w}{\partial x}+\sigma_{xy}\frac{\partial w}{\partial y}+\sigma_{xz}\right)+\frac{\partial}{\partial y}\left(\sigma_{yx}\frac{\partial w}{\partial x}+\sigma_{yy}\frac{\partial w}{\partial y}+\sigma_{yz}\right)$$
$$+\frac{\partial}{\partial z}\left(\sigma_{zx}\frac{\partial w}{\partial x}+\sigma_{zy}\frac{\partial w}{\partial y}+\sigma_{zz}\right)+Z=0, \quad (1.56)$$

where it has been assumed that the density of the body before and after deformation is the same, that is, $\rho_0=\rho$, and X, Y and Z are the body forces per unit volume, including inertia.

Multiplying equations (1.55a,b) by dz and integrating from $-h/2$ to $h/2$, and by using definitions (1.53a–c), the following expressions are obtained:

$$\frac{\partial N_x}{\partial x}+\frac{\partial N_{xy}}{\partial y}+f_x=0, \quad \frac{\partial N_{xy}}{\partial x}+\frac{\partial N_y}{\partial y}+f_y=0, \quad (1.57a,b)$$

where $f_x=\sigma_{zx}|_{h/2}-\sigma_{zx}|_{-h/2}+\int_{-h/2}^{h/2}X\mathrm{d}z$ and $f_y=\sigma_{zy}|_{h/2}-\sigma_{zy}|_{-h/2}+\int_{-h/2}^{h/2}Y\mathrm{d}z$ are the external loading tangential to the plate per unit area, usually equal to zero if in-plane inertia is neglected.

Similarly, by multiplying equations (1.55a,b) by $z\,\mathrm{d}z$ and integrating from $-h/2$ to $h/2$,

$$\frac{\partial M_x}{\partial x}+\frac{\partial M_{xy}}{\partial y}-Q_x+m_x=0, \quad \frac{\partial M_{xy}}{\partial x}+\frac{\partial M_y}{\partial y}-Q_y+m_y=0, \quad (1.58a,b)$$

where $m_x=\frac{h}{2}(\sigma_{zx}|_{h/2}-\sigma_{zx}|_{-h/2})+\int_{-h/2}^{h/2}Xz\,\mathrm{d}z$ and $m_y=\frac{h}{2}(\sigma_{zy}|_{h/2}-\sigma_{zy}|_{-h/2})+\int_{-h/2}^{h/2}Yz\,\mathrm{d}z$ are the external moments per unit area, in general, equal to zero.

The integration of the third equation of motion (1.56) with respect to z gives

$$\frac{\partial}{\partial x}\left(Q_x+N_x\frac{\partial w}{\partial x}+N_{xy}\frac{\partial w}{\partial y}\right)+\frac{\partial}{\partial y}\left(Q_y+N_{xy}\frac{\partial w}{\partial x}+N_y\frac{\partial w}{\partial y}\right)+q=0, \quad (1.59)$$

where q is the transverse load (including inertia) per unit area of the undeformed middle plane, which is given by

$$q=\left(\sigma_{zz}+\sigma_{zx}\frac{\partial w}{\partial x}+\sigma_{zy}\frac{\partial w}{\partial y}\right)\Bigg|_{-h/2}^{h/2}+\int_{-h/2}^{h/2}Z\,\mathrm{d}z. \quad (1.60)$$

The transverse load can be rewritten as

$$q=f+p-\rho h\ddot{w}-ch\dot{w}, \quad (1.61)$$

where f is the external pressure load, p is the fluid pressure in case of fluid-structure interaction, $\rho h\ddot{w}$ is the inertia force per unit area and $ch\dot{w}$ is the viscous damping force per unit area; expression (1.61) covers most applications.

By using equations (1.57a,b), equation (1.59) can be simplified into

$$\frac{\partial Q_x}{\partial x}+\frac{\partial Q_y}{\partial y}+q+N_x\frac{\partial^2 w}{\partial x^2}+2N_{xy}\frac{\partial^2 w}{\partial x\,\partial y}+N_y\frac{\partial^2 w}{\partial y^2}-f_x\frac{\partial w}{\partial x}-f_y\frac{\partial w}{\partial y}=0. \quad (1.62)$$

Eliminating Q_x and Q_y by using (1.58a,b),

$$\frac{\partial^2 M_x}{\partial x^2} + 2\frac{\partial^2 M_{xy}}{\partial x\,\partial y} + \frac{\partial^2 M_y}{\partial y^2} = -q - \frac{\partial m_x}{\partial x} - \frac{\partial m_y}{\partial y} - N_x \frac{\partial^2 w}{\partial x^2} - 2N_{xy}\frac{\partial^2 w}{\partial x\,\partial y} - N_y\frac{\partial^2 w}{\partial y^2} + f_x\frac{\partial w}{\partial x} + f_y\frac{\partial w}{\partial y}. \quad (1.63)$$

Finally, substituting equations (1.54d–f), the third equation of equilibrium takes the form

$$D\left(\frac{\partial^4 w}{\partial x^4} + 2\frac{\partial^4 w}{\partial x^2 \partial y^2} + \frac{\partial^4 w}{\partial y^4}\right) = q + \frac{\partial m_x}{\partial x} + \frac{\partial m_y}{\partial y} + N_x\frac{\partial^2 w}{\partial x^2} + 2N_{xy}\frac{\partial^2 w}{\partial x\,\partial y} + N_y\frac{\partial^2 w}{\partial y^2} - f_x\frac{\partial w}{\partial x} - f_y\frac{\partial w}{\partial y}. \quad (1.64)$$

Coming back to the other two equations of equilibrium, equations (1.57a,b), and substituting in them equations (1.54a–c), the following two expressions are obtained:

$$\frac{\partial}{\partial x}\left[\frac{\partial u}{\partial x} + \nu\frac{\partial v}{\partial y} + \frac{1}{2}\left(\frac{\partial w}{\partial x}\right)^2 + \frac{\nu}{2}\left(\frac{\partial w}{\partial y}\right)^2\right] + \frac{1-\nu}{2}\frac{\partial}{\partial y}\left[\frac{\partial u}{\partial y} + \frac{\partial v}{\partial x} + \frac{\partial w}{\partial x}\frac{\partial w}{\partial y}\right] + \frac{1-\nu^2}{Eh} f_x = 0, \quad (1.65)$$

$$\frac{\partial}{\partial y}\left[\frac{\partial v}{\partial y} + \nu\frac{\partial u}{\partial x} + \frac{1}{2}\left(\frac{\partial w}{\partial y}\right)^2 + \frac{\nu}{2}\left(\frac{\partial w}{\partial x}\right)^2\right] + \frac{1-\nu}{2}\frac{\partial}{\partial x}\left[\frac{\partial u}{\partial y} + \frac{\partial v}{\partial x} + \frac{\partial w}{\partial x}\frac{\partial w}{\partial y}\right] + \frac{1-\nu^2}{Eh} f_y = 0. \quad (1.66)$$

Equations (1.64–1.66) are the equations of motion of the plate in the three displacements u, v, w. However, it is possible to reduce these equations to a single equation of motion in w and a compatibility equation by introducing a stress function $F(x, y)$ and neglecting in-plane inertia, that is, $\ddot{u}$, $\ddot{v}$.

The in-plane stress function F is introduced in order to satisfy identically equations (1.57a,b) by introducing the following expressions for the stress resultants:

$$N_x = \frac{\partial^2 F}{\partial y^2} - \int f_x\,\mathrm{d}x, \quad N_y = \frac{\partial^2 F}{\partial x^2} - \int f_y\,\mathrm{d}y, \quad N_{xy} = -\frac{\partial^2 F}{\partial x\,\partial y}, \quad (1.67a–c)$$

where the indefinite integrals give the potentials of tangential loadings. Subtracting equation (1.54b) multiplied by ν from equation (1.54a) yields

$$\frac{\partial u}{\partial x} = \frac{N_x - \nu N_y}{Eh} - \frac{1}{2}\left(\frac{\partial w}{\partial x}\right)^2; \quad (1.68a)$$

similarly, subtracting equation (1.54a) multiplied by ν from equation (1.54b) yields

$$\frac{\partial v}{\partial y} = \frac{N_y - \nu N_x}{Eh} - \frac{1}{2}\left(\frac{\partial w}{\partial y}\right)^2. \quad (1.68b)$$

The compatibility equation is obtained by applying the differential operator $\partial^2/\partial x\,\partial y$ to equation (1.54c),

$$\frac{\partial^2 N_{xy}}{\partial x\,\partial y} = \frac{Eh}{2(1+\nu)}\left[\frac{\partial^3 u}{\partial x\,\partial y^2} + \frac{\partial^3 v}{\partial x^2\,\partial y} + \frac{\partial^3 w}{\partial x^2\,\partial y}\frac{\partial w}{\partial y} + \frac{\partial^3 w}{\partial x\,\partial y^2}\frac{\partial w}{\partial x} + \left(\frac{\partial^2 w}{\partial x\,\partial y}\right) + \frac{\partial^2 w}{\partial x^2}\frac{\partial^2 w}{\partial y^2}\right]; \tag{1.69}$$

substituting in it equations (1.68a,b) after applying operators $\partial^2/\partial y^2$ and $\partial^2/\partial x^2$, respectively, and finally using equations (1.67a–c) in order to have an expressions involving only F and w, the resulting equation is

$$\nabla^4 F = Eh\left[\left(\frac{\partial^2 w}{\partial x\,\partial y}\right)^2 - \frac{\partial^2 w}{\partial x^2}\frac{\partial^2 w}{\partial y^2}\right] + \frac{\partial^2}{\partial y^2}\left(\int f_x\,\mathrm{d}x - \nu\int f_y\,\mathrm{d}y\right) + \frac{\partial^2}{\partial x^2}\left(\int f_y\,\mathrm{d}y - \nu\int f_x\,\mathrm{d}x\right), \tag{1.70}$$

where the biharmonic operator is defined as $\nabla^4 = [\partial^2/\partial x^2 + \partial^2/(\partial y^2)]^2$.

The compatibility equation makes a pair with the following equation of motion, obtained by inserting equations (1.67a–c) into (1.64):

$$D\nabla^4 w = q + \left(\frac{\partial^2 F}{\partial y^2}\frac{\partial^2 w}{\partial x^2} - 2\frac{\partial^2 F}{\partial x\,\partial y}\frac{\partial^2 w}{\partial x\,\partial y} + \frac{\partial^2 F}{\partial x^2}\frac{\partial^2 w}{\partial y^2}\right) + \frac{\partial m_x}{\partial x} + \frac{\partial m_y}{\partial y} - f_x\frac{\partial w}{\partial x} - f_y\frac{\partial w}{\partial y} - \frac{\partial^2 w}{\partial x^2}\int f_x\,\mathrm{d}x - \frac{\partial^2 w}{\partial y^2}\int f_y\,\mathrm{d}y. \tag{1.71}$$

In most of applications, $m_x, m_y, f_x, f_y, \int f_x\,\mathrm{d}x$ and $\int f_y\,\mathrm{d}y$ are negligible (including tangential inertia and tangential dissipation, which appears in f_x, f_y) and the equation of motion can be written in the classical form

$$D\nabla^4 w + ch\dot{w} + \rho h\ddot{w} = f + p + \left(\frac{\partial^2 F}{\partial y^2}\frac{\partial^2 w}{\partial x^2} - 2\frac{\partial^2 F}{\partial x\,\partial y}\frac{\partial^2 w}{\partial x\,\partial y} + \frac{\partial^2 F}{\partial x^2}\frac{\partial^2 w}{\partial y^2}\right), \tag{1.72}$$

which makes a pair with the compatibility equation

$$\frac{1}{Eh}\nabla^4 F = \left[\left(\frac{\partial^2 w}{\partial x\,\partial y}\right)^2 - \frac{\partial^2 w}{\partial x^2}\frac{\partial^2 w}{\partial y^2}\right]. \tag{1.73}$$

1.2.8 Von Kármán Equation of Motion Including Geometric Imperfections

In the presence of initial geometric imperfections of the rectangular plate associated with zero initial stress, denoted by normal displacement w_0 (see equation [1.31]), the von Kármán equation of motion becomes

$$D\nabla^4 w + ch\dot{w} + \rho h\ddot{w} = f + p + \left[\frac{\partial^2 F}{\partial y^2}\frac{\partial^2 (w+w_0)}{\partial x^2} - 2\frac{\partial^2 F}{\partial x\,\partial y}\frac{\partial^2 (w+w_0)}{\partial x\,\partial y} + \frac{\partial^2 F}{\partial x^2}\frac{\partial^2 (w+w_0)}{\partial y^2}\right], \tag{1.74}$$

which makes a pair with the compatibility equation

$$\frac{1}{Eh}\nabla^4 F = \left[\left(\frac{\partial^2 w}{\partial x\,\partial y}\right)^2 + 2\frac{\partial^2 w}{\partial x\,\partial y}\frac{\partial^2 w_0}{\partial x\,\partial y} - \left(\frac{\partial^2 w}{\partial x^2} + \frac{\partial^2 w_0}{\partial x^2}\right)\frac{\partial^2 w}{\partial y^2} - \frac{\partial^2 w}{\partial x^2}\frac{\partial^2 w_0}{\partial y^2}\right]. \tag{1.75}$$

Equations (1.74) and (1.75) are obtained by substituting equation (1.31) into the von Kármán equation of motion and neglecting all terms that depend on w_0 only (as they are associated with initial deformation and therefore with zero stress).

1.3 Large Deflection of Circular Cylindrical Shells

1.3.1 Euclidean Metric Tensor

The metric tensor gives information on how to measure length in a reference system different from the rectangular Cartesian one, here denoted by (x_1, x_2, x_3) in a three-dimensional Euclidean space. A new coordinate system $(\theta_1, \theta_2, \theta_3)$ is introduced, with the following one-to-one transformation of coordinates:

$$\theta_i = \theta_i(x_1, x_2, x_3), \qquad \text{for } i = 1, \ldots 3. \tag{1.76}$$

The inverse transformation,

$$x_i = x_i(\theta_1, \theta_2, \theta_3), \tag{1.77}$$

is assumed to exist with one-to-one correspondence. A line element given by differentials $\mathrm{d}x_1$, $\mathrm{d}x_2$, $\mathrm{d}x_3$ has length given by Pythagoras' theorem in the rectangular Cartesian coordinate system

$$\mathrm{d}s^2 = \sum_{i=1}^{3} (\mathrm{d}x_i)^2 = \sum_{i=1}^{3}\sum_{j=1}^{3} \delta_{ij}\,\mathrm{d}x_i\,\mathrm{d}x_j. \tag{1.78}$$

By using the transformation of coordinates (1.77), the following differentiation rule is obtained:

$$\mathrm{d}x_i = \sum_{k=1}^{3} \frac{\partial x_i}{\partial \theta_k}\,\mathrm{d}\theta_k. \tag{1.79}$$

By inserting equation (1.79) into (1.78), the following expression is obtained:

$$\mathrm{d}s^2 = \sum_{k=1}^{3}\sum_{m=1}^{3}\sum_{i=1}^{3} \frac{\partial x_i}{\partial \theta_k}\frac{\partial x_i}{\partial \theta_m}\,\mathrm{d}\theta_k\,\mathrm{d}\theta_m. \tag{1.80}$$

It is useful to define the function $g_{km}(\theta_1, \theta_2, \theta_3)$ given by

$$g_{km} = \sum_{i=1}^{3} \frac{\partial x_i}{\partial \theta_k}\frac{\partial x_i}{\partial \theta_m}, \tag{1.81}$$

so that equation (1.80) can be rewritten as follows:

$$\mathrm{d}s^2 = \sum_{k=1}^{3}\sum_{m=1}^{3} g_{km}\,\mathrm{d}\theta_k\,\mathrm{d}\theta_m. \tag{1.82}$$

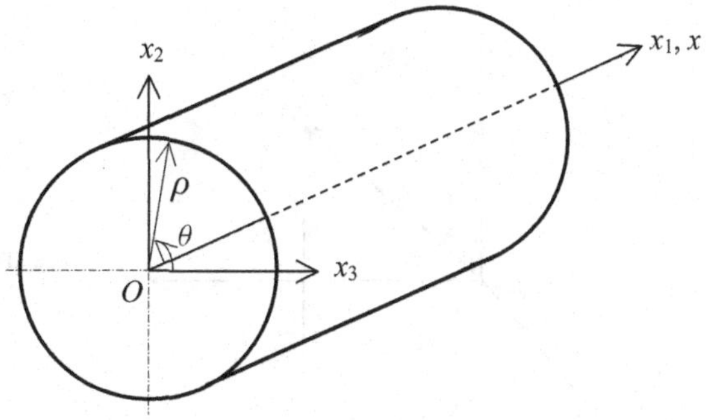

Figure 1.7. Cylindrical coordinate system.

Equation (1.80) is only formally similar to (1.5), which gives the length of a deformed line element in the new coordinate system. The functions g_{km} are symmetric so that $g_{km} = g_{mk}$ and are the components of the Euclidean metric tensor in the coordinate system $(\theta_1, \theta_2, \theta_3)$. As a consequence, the line element of components $(\mathrm{d}\theta_1, 0, 0)$ has length $\sqrt{g_{11}}\,|\mathrm{d}\theta_1|$. If the coordinate system $(\theta_1, \theta_2, \theta_3)$ is orthogonal, then $g_{ij} = \delta_{ij} g_{ii}$.

1.3.1.1 Example: Cylindrical Coordinates

The transformation from the cylindrical coordinate system (x, θ, ρ) (see Figure 1.7) to the rectangular Cartesian system (x_1, x_2, x_3) is given by

$$x_1 = x, \quad x_2 = \rho \sin\theta, \quad x_3 = \rho\cos\theta. \tag{1.83}$$

The components of the Euclidean metric tensor in the cylindrical coordinate system are

$$g_{11} = \left(\frac{\partial x}{\partial x}\right)^2 + \left(\frac{\partial \rho \sin\theta}{\partial x}\right)^2 + \left(\frac{\partial \rho\cos\theta}{\partial x}\right)^2 = 1, \tag{1.84a}$$

$$g_{22} = \left(\frac{\partial x}{\partial \theta}\right)^2 + \left(\frac{\partial \rho \sin\theta}{\partial \theta}\right)^2 + \left(\frac{\partial \rho\cos\theta}{\partial \theta}\right)^2 = \rho^2\sin^2\theta + \rho^2\cos^2\theta = \rho^2, \tag{1.84b}$$

$$g_{33} = \left(\frac{\partial x}{\partial \rho}\right)^2 + \left(\frac{\partial \rho \sin\theta}{\partial \rho}\right)^2 + \left(\frac{\partial \rho\cos\theta}{\partial \rho}\right)^2 = \sin^2\theta + \cos^2\theta = 1. \tag{1.84c}$$

1.3.1.2 Example: Spherical Coordinates

The transformation from the spherical coordinate system (φ, θ, ρ) (see Figure 1.8) to the Cartesian system (x_1, x_2, x_3) is given by

$$x_1 = \rho\sin\varphi\cos\theta, \quad x_2 = \rho\sin\varphi\sin\theta, \quad x_3 = \rho\cos\varphi, \tag{1.85}$$

where $0 \le \varphi \le \pi$ and $0 \le \theta \le 2\pi$.

The components of the Euclidean metric tensor in the spherical coordinate system are

$$g_{11} = \left(\frac{\partial \rho\sin\varphi\cos\theta}{\partial \rho}\right)^2 + \left(\frac{\partial \rho\sin\varphi\sin\theta}{\partial \rho}\right)^2 + \left(\frac{\partial \rho\cos\varphi}{\partial \rho}\right)^2$$
$$= \sin^2\varphi(\cos\theta^2 + \sin\theta^2) + \cos^2\varphi = 1, \tag{1.86a}$$

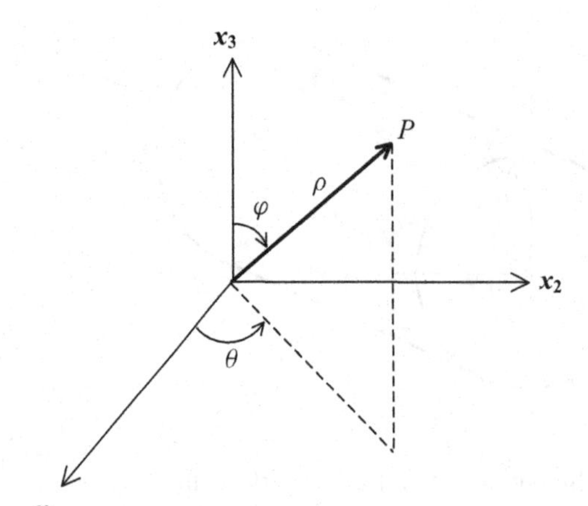

Figure 1.8. Spherical coordinate system.

$$g_{22} = \left(\frac{\partial \rho \sin\varphi \cos\theta}{\partial \varphi}\right)^2 + \left(\frac{\partial \rho \sin\varphi \sin\theta}{\partial \varphi}\right)^2 + \left(\frac{\partial \rho \cos\varphi}{\partial \varphi}\right)^2$$
$$= \rho^2 \cos^2\varphi(\cos\theta^2 + \sin\theta^2) + \rho^2 \sin^2\varphi = \rho^2, \tag{1.86b}$$

$$g_{33} = \left(\frac{\partial \rho \sin\varphi \cos\theta}{\partial \theta}\right)^2 + \left(\frac{\partial \rho \sin\varphi \sin\theta}{\partial \theta}\right)^2 + \left(\frac{\partial \rho \cos\varphi}{\partial \theta}\right)^2$$
$$= \rho^2 \sin^2\varphi(\sin\theta^2 + \cos\theta^2) = \rho^2 \sin^2\varphi \tag{1.86c}$$

1.3.2 Green's Strain Tensor in a Generic Coordinate System

In Section 1.2.1, the Green's strain tensor was obtained in rectangular Cartesian coordinate systems. Here this limitation is removed and Section 1.2.1 is extended to generic coordinate systems.

In the original generic coordinate system (x_1, x_2, x_3), an infinitesimal line connecting point P (x_1, x_2, x_3) to a neighborhood point P' $(x_1 + \mathrm{d}x_1, x_2 + \mathrm{d}x_2, x_3 + \mathrm{d}x_3)$ is considered; the square of its length, in the original undeformed configuration, is given by

$$\mathrm{d}s_0^2 = \sum_{i=1}^{3}\sum_{j=1}^{3} g_{ij}\, \mathrm{d}x_i\, \mathrm{d}x_j. \tag{1.87a}$$

When the system is deformed so that points P, P' migrate to Q (a_1, a_2, a_3), Q' $(a_1 + \mathrm{d}a_1,\ a_2 + \mathrm{d}a_2,\ a_3 + \mathrm{d}a_3)$ in the new generic coordinate system a_1, a_2, a_3, the square of the distance is

$$\mathrm{d}s^2 = \sum_{k=1}^{3}\sum_{m=1}^{3} h_{km}\, \mathrm{d}a_k\, \mathrm{d}a_m. \tag{1.87b}$$

In equations (1.87a,b), g_{ij} and h_{km} are the Euclidean metric tensors in the coordinate systems x_i and a_k, respectively. By using a Lagrangian description,

$$\mathrm{d}s^2 = \sum_{k=1}^{3}\sum_{m=1}^{3}\sum_{i=1}^{3}\sum_{j=1}^{3} h_{km} \frac{\partial a_k}{\partial x_i}\frac{\partial a_m}{\partial x_j} \mathrm{d}x_i\, \mathrm{d}x_j. \tag{1.88}$$

The difference between the squares of the length of the elements may be written in the following form by using equations (1.87a) and (1.88):

$$\mathrm{d}s^2 - \mathrm{d}s_0^2 = \left(\sum_{k=1}^{3} \sum_{m=1}^{3} \sum_{i=1}^{3} \sum_{j=1}^{3} h_{km} \frac{\partial a_k}{\partial x_i} \frac{\partial a_m}{\partial x_j} - g_{ij} \right) \mathrm{d}x_i \, \mathrm{d}x_j. \tag{1.89}$$

By definition, the Green's strain tensor ε_{ij} is obtained as

$$\mathrm{d}s^2 - \mathrm{d}s_0^2 = 2 \sum_{i=1}^{3} \sum_{j=1}^{3} \varepsilon_{ij} \sqrt{g_{ii}} \, \mathrm{d}x_i \sqrt{g_{jj}} \, \mathrm{d}x_j, \tag{1.90}$$

where the only difference with respect to equation (1.7), valid for the rectangular Cartesian system, is the presence of the factor $\sqrt{g_{ii}}\,\mathrm{d}x_i$ instead of $\mathrm{d}x_i$; in fact, $\sqrt{g_{ii}}\,\mathrm{d}x_i$ is the length of a line element in a generic coordinate system, as discussed in Section 1.3.1. By using equations (1.89) and (1.90), the Green's strain tensor ε_{ij} in generic coordinate systems is

$$\varepsilon_{ij} = \frac{1}{2} \sum_{k=1}^{3} \sum_{m=1}^{3} \frac{1}{\sqrt{g_{ii}}} \frac{1}{\sqrt{g_{jj}}} \left(h_{km} \frac{\partial a_k}{\partial x_i} \frac{\partial a_m}{\partial x_j} - g_{ij} \right). \tag{1.91}$$

Equation (1.91) is the generalization of equation (1.8). If the coordinate system (x_1, x_2, x_3) is orthogonal, $g_{ij} = \delta_{ij} g_{ii}$; similarly, if ($a_1, a_2, a_3$) is orthogonal, $h_{km} = \delta_{km} h_{kk}$. If ($a_1, a_2, a_3$) is a rectangular Cartesian coordinate system, that is, $h_{km} = \delta_{km}$, and (x_1, x_2, x_3) is orthogonal, the Green's strain tensor can be simplified into

$$\varepsilon_{ij} = \frac{1}{2} \sum_{k=1}^{3} \frac{1}{\sqrt{g_{ii}}} \frac{1}{\sqrt{g_{jj}}} \left(\frac{\partial a_k}{\partial x_i} \frac{\partial a_k}{\partial x_j} - \delta_{ij} g_{ii} \right). \tag{1.92}$$

1.3.3 Green's Strain Tensor in Cylindrical Coordinates

A point P of coordinates (x, θ, ρ) in a cylindrical coordinate system is considered, as shown in Figure 1.9(a,b). The coordinates of P in a rectangular Cartesian coordinate system (a_1, a_2, a_3) are given by equation (1.83). Now, three orthogonal displacements u_1, u_2, u_3, in the axial, circumferential and radial direction, respectively, are given to the point P as shown in Figure 1.9(b), so that a new point P' is obtained. The coordinates of P' in the rectangular Cartesian system (a_1, a_2, a_3) are

$$a_1 = x + u_1, \quad a_2 = \rho \sin\theta + u_3 \sin\theta + u_2 \cos\theta, \quad a_3 = \rho\cos\theta + u_3\cos\theta - u_2 \sin\theta. \tag{1.93}$$

The Green's strain tensor in cylindrical coordinates is obtained by using equations (1.92), in which (1.84) and (1.93) must be substituted. In particular,

$$\begin{aligned} \left(\frac{\partial a_1}{\partial x}\right)^2 + \left(\frac{\partial a_2}{\partial x}\right)^2 + \left(\frac{\partial a_3}{\partial x}\right)^2 &= \left(1 + \frac{\partial u_1}{\partial x}\right)^2 + \left(\frac{\partial u_3}{\partial x}\sin\theta + \frac{\partial u_2}{\partial x}\cos\theta\right)^2 \\ &\quad + \left(\frac{\partial u_3}{\partial x}\cos\theta - \frac{\partial u_2}{\partial x}\sin\theta\right)^2 \\ &= 1 + 2\frac{\partial u_1}{\partial x} + \left(\frac{\partial u_1}{\partial x}\right)^2 + \left(\frac{\partial u_2}{\partial x}\right)^2 + \left(\frac{\partial u_3}{\partial x}\right)^2, \end{aligned}$$

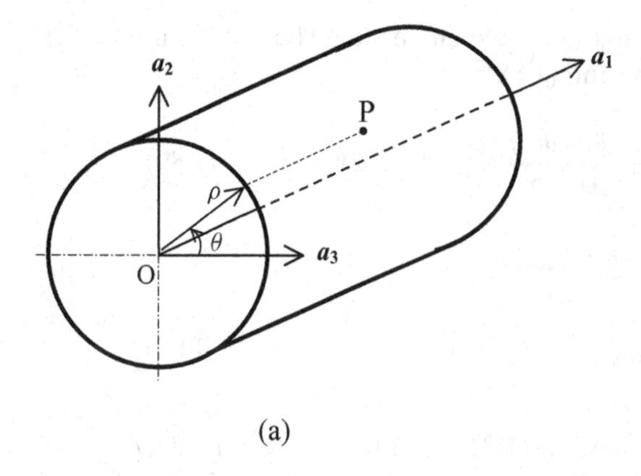

(a)

Figure 1.9. Point P in initial position and after displacement in rectangular Cartesian and cylindrical coordinate systems. (a) Initial position. (b) Initial position P and position P' after displacement.

(b)

which gives

$$\varepsilon_{xx} = \frac{\partial u_1}{\partial x} + \frac{1}{2}\left[\left(\frac{\partial u_1}{\partial x}\right)^2 + \left(\frac{\partial u_2}{\partial x}\right)^2 + \left(\frac{\partial u_3}{\partial x}\right)^2\right], \tag{1.94}$$

which is unchanged with respect to equation (1.11a) obtained in rectangular Cartesian coordinates.

Similarly,

$$\begin{aligned}\left(\frac{\partial a_1}{\partial \theta}\right)^2 + \left(\frac{\partial a_2}{\partial \theta}\right)^2 + \left(\frac{\partial a_3}{\partial \theta}\right)^2 &= \left(\frac{\partial u_1}{\partial \theta}\right)^2 \\ &+ \left(\rho \cos\theta + u_3 \cos\theta - u_2 \sin\theta + \frac{\partial u_3}{\partial \theta}\sin\theta + \frac{\partial u_2}{\partial \theta}\cos\theta\right)^2 \\ &+ \left(-\rho \sin\theta - u_3 \sin\theta - u_2 \cos\theta + \frac{\partial u_3}{\partial \theta}\cos\theta - \frac{\partial u_2}{\partial \theta}\sin\theta\right)^2,\end{aligned}$$

which, after some manipulation, gives

$$\varepsilon_{\theta\theta} = \frac{1}{\rho}\left(\frac{\partial u_2}{\partial \theta} + u_3\right) + \frac{1}{2\rho^2}\left[\left(\frac{\partial u_1}{\partial \theta}\right)^2 + \left(\frac{\partial u_3}{\partial \theta} - u_2\right)^2 + \left(\frac{\partial u_2}{\partial \theta} + u_3\right)^2\right]; \tag{1.95}$$

this is a more complex expression than equation (1.94), as a consequence of the curvature of the cylindrical coordinates in θ direction. In equation (1.95), the quadratic term $(\partial u_2/\partial\theta + u_3)^2$ is neglected in some shell theories with respect to the analogous linear term $(\partial u_2/\partial\theta + u_3)$ for small displacements u_1, u_2, u_3.

Finally,

$$\frac{\partial a_1}{\partial x}\frac{\partial a_1}{\partial \theta}+\frac{\partial a_2}{\partial x}\frac{\partial a_2}{\partial \theta}+\frac{\partial a_3}{\partial x}\frac{\partial a_3}{\partial \theta}=\left(1+\frac{\partial u_1}{\partial x}\right)\frac{\partial u_1}{\partial \theta}+\left(\frac{\partial u_3}{\partial x}\sin\theta+\frac{\partial u_2}{\partial x}\cos\theta\right)$$
$$\times\left(\rho\cos\theta+u_3\cos\theta-u_2\sin\theta+\frac{\partial u_3}{\partial \theta}\sin\theta+\frac{\partial u_2}{\partial \theta}\cos\theta\right)+\left(\frac{\partial u_3}{\partial x}\cos\theta-\frac{\partial u_2}{\partial x}\sin\theta\right)$$
$$\times\left(-\rho\sin\theta-u_3\sin\theta-u_2\cos\theta+\frac{\partial u_3}{\partial \theta}\cos\theta-\frac{\partial u_2}{\partial \theta}\sin\theta\right),$$

which, after some manipulation, gives

$$\gamma_{x\theta}=\left(\frac{\partial u_1}{\rho\,\partial \theta}+\frac{\partial u_2}{\partial x}\right)+\frac{1}{\rho}\left[\left(\frac{\partial u_1}{\partial x}\frac{\partial u_1}{\partial \theta}\right)+\frac{\partial u_2}{\partial x}\left(\frac{\partial u_2}{\partial \theta}+u_3\right)+\frac{\partial u_3}{\partial x}\left(\frac{\partial u_3}{\partial \theta}-u_2\right)\right]. \tag{1.96}$$

This differs from the analogous equation (1.11b) obtained in rectangular Cartesian coordinates; in equation (1.96), the symbol $\gamma_{x\theta}$ is traditionally used instead of $\varepsilon_{x\theta}$. Equations (1.94–1.96) give the Green's strains in cylindrical coordinates.

1.3.4 Strains for Finite Deflection of Circular Cylindrical Shells: Donnell's Nonlinear Theory

A shell is a body bounded by two curved surfaces, the distance between the surfaces being small in comparison with other dimensions. The middle surface of the shell is the locus of points that lie at equal distance from these two surfaces.

A circular cylindrical shell of mean radius R, thickness h and length L is considered. A theory for large deflection w (in radial direction) of circular cylindrical shells was developed in cylindrical coordinates. In the theory, the following hypotheses are made (see Figure 1.10):

(H1) The shell is thin: $h \ll R$, $h \ll L$ (in practical applications $h/R \le 1/20$ in order to have a thin shell).

(H2) The magnitude of deflection w is of the same order as the thickness h of the shell, therefore small compared to the shell dimensions R and L for hypothesis (H1): $|w| = O(h)$.

(H3) The slope is small at each point: $|\partial w/\partial x| \ll 1$, $|\partial w/(R\partial\theta)| \ll 1$.

(H4) All strain components are small, so that linear elasticity can be applied.

(H5) The Kirchhoff-Love hypotheses hold; that is, stresses in the direction normal to the shell middle surface are negligible, and strains vary linearly within the shell thickness. These hypotheses are a good approximation for thin shells. However, in the presence of external loads normal to the shell surface, stresses in the normal direction arise, even if they are generally of some order of magnitude smaller than other stresses, except in proximity to a concentrated load.

(H6) In Donnell's hypothesis, the tangential displacements u and v are infinitesimal, and in the strain-displacement relations only those nonlinear terms that depend on w are to be retained. All other nonlinear terms are to be neglected.

Hypothesis (H6) will be removed in the next sections of this chapter in order to get more accurate nonlinear shell theories.

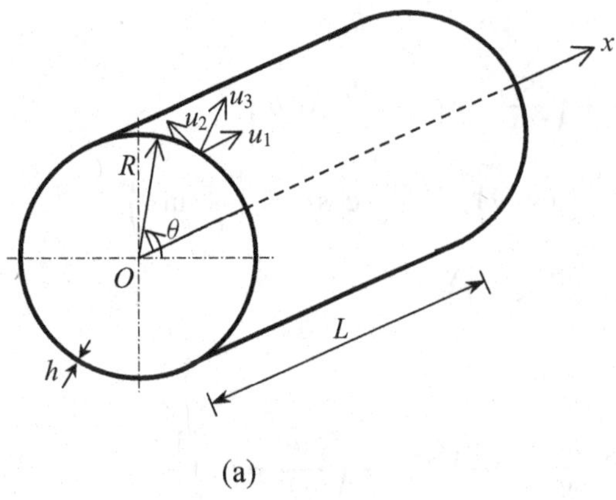

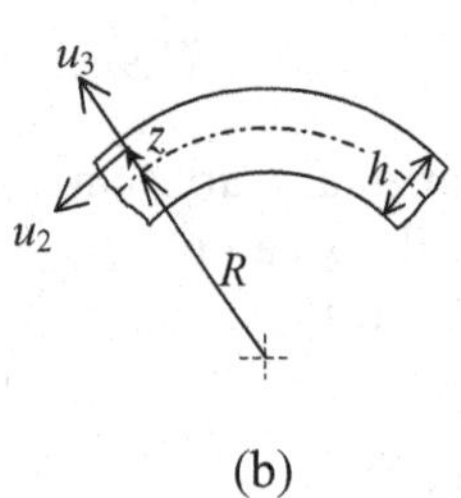

Figure 1.10. Circular cylindrical shell. (a) Symbols used for dimensions and displacements of a generic point. (b) Cross-section of the shell with close-up view.

In analogy with the formulation for plates, the Lagrangian description is used. A cylindrical coordinate system (O; x, θ, ρ) is used, with the surface $\rho = R$ coinciding with the middle surface of the shell in its initial, undeformed configuration, and the ρ axis normal to it; the shell external and internal surfaces are designated as $\rho = R \pm h/2$.

The orthogonal displacements of a generic point of the shell at coordinates (x, θ, $\rho = R + z$) are denoted by u_1, u_2, u_3, in the axial, circumferential and radial directions, respectively, whereas u, v, w are the displacements of points on the middle surface of the shell (the middle surface is obtained for $z = 0$, where z is the coordinate along the shell thickness; see Figure 1.10). Note that the radial displacements u_3 and w are taken positive outward with respect to the shell surface. In the Lagrangian description, Green's strain tensor, referred to the initial configuration, is used; it is given in equations (1.94–1.96). By using hypothesis (H5), we obtain

$$u_1 = u(x, \theta) - z\frac{\partial w}{\partial x}, \tag{1.97}$$

$$u_2 = v(x, \theta) - z\frac{\partial w}{R\partial \theta}, \tag{1.98}$$

$$u_3 = w(x, \theta). \tag{1.99}$$

Equations (1.97–1.99) are linear expressions that are obtained exactly in the same way as for rectangular plates in Section 1.2.2, via substitution of y by $R\theta$; assuming that $R + z \simeq R$, Figures 1.11(a,b) give the geometrical interpretation of equations (1.97) and (1.98), respectively. These equations are inserted into Green's strain tensor, equations (1.94–1.96), in order to obtain the strain-displacement relations for a circular cylindrical shell; moreover, $\varepsilon_{\rho\rho} = \varepsilon_{zz} \simeq 0$, $\gamma_{x\rho} \simeq 0$, $\gamma_{\theta\rho} \simeq 0$. The strain components ε_{xx}, $\varepsilon_{\theta\theta}$ and $\gamma_{x\theta}$ at an arbitrary point of the shell are related to the middle surface strains $\varepsilon_{x,0}$, $\varepsilon_{\theta,0}$ and $\gamma_{x\theta,0}$ and to the changes in the curvature and torsion of

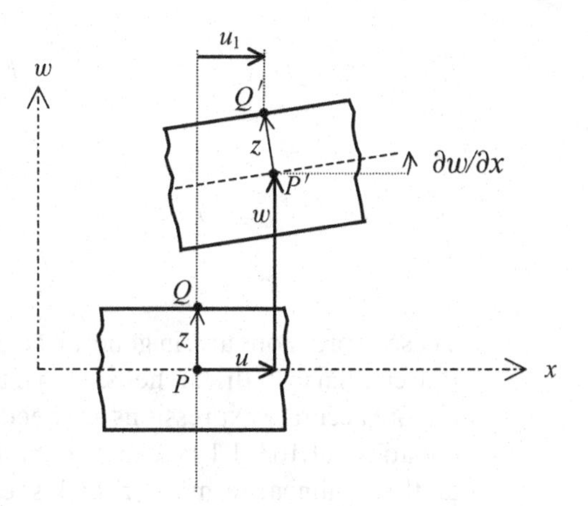

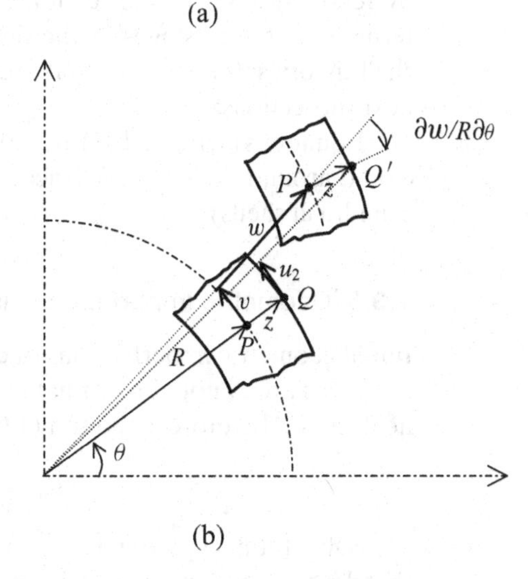

Figure 1.11. Displacements of a middle surface point P and of a generic point Q of a circular cylindrical shell. (a) Longitudinal section. (b) Cross-section.

the middle surface k_x, k_θ and $k_{x\theta}$ by the following three relations, which are the same as those used for plates:

$$\varepsilon_{xx} = \varepsilon_{x,0} + z\,k_x, \tag{1.100}$$

$$\varepsilon_{\theta\theta} = \varepsilon_{\theta,0} + z\,k_\theta, \tag{1.101}$$

$$\gamma_{x\theta} = \gamma_{x\theta,0} + z\,k_{x\theta}, \tag{1.102}$$

where z is, as usual, the distance of the arbitrary point of the shell from the middle surface. If Donnell's hypothesis (H6) is used, and z $(-h/2 \le z \le h/2)$ is neglected with respect to R, the following expressions for the middle surface strains and the changes in the curvature and torsion of the middle surface are obtained:

$$\varepsilon_{x,0} = \frac{\partial u}{\partial x} + \frac{1}{2}\left(\frac{\partial w}{\partial x}\right)^2, \tag{1.103}$$

$$\varepsilon_{\theta,0} = \frac{\partial v}{R\partial\theta} + \frac{w}{R} + \frac{1}{2}\left(\frac{\partial w}{R\partial\theta}\right)^2, \tag{1.104}$$

$$\gamma_{x\theta,0} = \frac{\partial u}{R\partial\theta} + \frac{\partial v}{\partial x} + \frac{\partial w}{\partial x}\frac{\partial w}{R\partial\theta}, \tag{1.105}$$

$$k_x = -\frac{\partial^2 w}{\partial x^2}, \tag{1.106}$$

$$k_\theta = -\frac{\partial^2 w}{R^2\, \partial \theta^2}, \tag{1.107}$$

$$k_{x\theta} = -2\frac{\partial^2 w}{R\, \partial x\, \partial \theta}. \tag{1.108}$$

These expressions are, in general, accurate enough for moderately large vibrations of thin circular cylindrical shells, even if the linear terms can give rise to some inaccuracy; if more accurate expressions are needed, equations (1.97–1.99) can be improved, and equations (1.103–1.108) can be computed with more accurate linear parts and retain all the nonlinear terms. For thick shells, more accurate expressions can be obtained by retaining z in the expression $z + R$. The possibility of retaining or neglecting terms at different stages of the derivation gives rise to several different nonlinear shell theories; the theories that are applied most frequently will be discussed in the next subsections.

Equations (1.103–1.108) are valid for both complete circular cylindrical shells (i.e. closed around the circumference) and circular cylindrical panels (i.e. open circular cylindrical shells).

1.3.5 Geometric Imperfections in Donnell's Nonlinear Shell Theory

Initial geometric imperfections of circular cylindrical shells associated with zero initial stress are denoted by radial displacement w_0; in-plane initial imperfections are neglected. Therefore, equation (1.99) is replaced by

$$u_3 = w(x, y) + w_0(x, y). \tag{1.109}$$

Substituting equation (1.109) into Green's strain tensor and neglecting all terms that depend only on w_0 (as they are associated with initial deformation and therefore with zero stress), the following expressions are obtained, replacing equations (1.103–1.105):

$$\varepsilon_{x,0} = \frac{\partial u}{\partial x} + \frac{1}{2}\left(\frac{\partial w}{\partial x}\right)^2 + \frac{\partial w}{\partial x}\frac{\partial w_0}{\partial x}, \tag{1.110}$$

$$\varepsilon_{\theta,0} = \frac{\partial v}{R\partial \theta} + \frac{w}{R} + \frac{1}{2}\left(\frac{\partial w}{R\partial \theta}\right)^2 + \frac{\partial w}{R\partial \theta}\frac{\partial w_0}{R\partial \theta}, \tag{1.111}$$

$$\gamma_{x\theta,0} = \frac{\partial u}{R\partial \theta} + \frac{\partial v}{\partial x} + \frac{\partial w}{\partial x}\frac{\partial w}{R\partial \theta} + \frac{\partial w}{\partial x}\frac{\partial w_0}{R\partial \theta} + \frac{\partial w_0}{\partial x}\frac{\partial w}{R\partial \theta}. \tag{1.112}$$

Equations (1.106–1.108) are unchanged by the presence of the imperfection w_0. It can be observed that, in the presence of geometric imperfections, the shell is no longer perfectly circular cylindrical.

1.3.6 The Flügge-Lur'e-Byrne Nonlinear Shell Theory

According to the Flügge-Lur'e-Byrne nonlinear theory (Ginsberg 1973; Yamaki 1984; Amabili 2003), the thinness assumption is delayed in the derivations; moreover, hypothesis (H6) is eliminated. For this reason, displacements u_1, u_2, u_3 of points at distance z from the middle surface are related to displacements u, v, w on the middle surface by the relationships

$$u_1 = u - z\frac{\partial (w + w_0)}{\partial x}, \tag{1.113}$$

$$u_2 = v - \frac{z}{R}\left(\frac{\partial (w + w_0)}{\partial \theta} - v\right), \tag{1.114}$$

$$u_3 = w + w_0. \tag{1.115}$$

Equation (1.113) is easily explained by using equation (1.115) and observing Figure 1.11(a), where the displacement of a generic point Q of the shell is related to the displacements of the corresponding point P on the middle surface. Similarly, equation (1.114) is justified by the diagram in Figure 1.11(b). The differences between equations (1.113–1.115) and (1.97–1.99) are (i) the addition of the geometric imperfection w_0 and (ii) the substitution of the term v in equation (1.98) with $v(1 + z/R)$ as a consequence of the curvature of the shell in u_2 direction, so that the middle surface tangential displacement v gives a contribution linearly increasing with z (in particular, with the distance from the center of curvature) to u_2. This is a better approximation than keeping the contribution of v constant as for a flat surface, as in equation (1.98).

The strain-displacement relationships for a generic point of the shell are given by equations (1.94–1.96), where $\rho = R + z$. The following thinness approximations are now introduced:

$$\frac{1}{R+z} = \frac{1}{R}\left[1 - \frac{z}{R} + O(z/R)^2\right], \tag{1.116}$$

$$\frac{1}{(R+z)^2} = \frac{1}{R^2}\left[1 - \frac{2z}{R} + O(z/R)^2\right], \tag{1.117}$$

where $O(z/R)^2$ is a small quantity of order $(z/R)^2$, which is neglected. By introducing equations (1.113–1.115) into equations (1.94–1.96) and using approximations (1.116) and (1.117), the middle surface strain-displacement relationships, changes in the curvature and torsion are obtained for the Flügge-Lur'e-Byrne nonlinear shell theory of circular cylindrical shells, as follows:

$$\varepsilon_{x,0} = \frac{\partial u}{\partial x} + \frac{1}{2}\left[\left(\frac{\partial u}{\partial x}\right)^2 + \left(\frac{\partial v}{\partial x}\right)^2 + \left(\frac{\partial w}{\partial x}\right)^2\right] + \frac{\partial w}{\partial x}\frac{\partial w_0}{\partial x}, \tag{1.118}$$

$$\begin{aligned}\varepsilon_{\theta,0} = {} & \frac{\partial v}{R\partial \theta} + \frac{w}{R} + \frac{1}{2R^2}\left[\left(\frac{\partial u}{\partial \theta}\right)^2 + \left(\frac{\partial v}{\partial \theta} + w\right)^2 + \left(\frac{\partial w}{\partial \theta} - v\right)^2\right] \\ & + \frac{1}{R^2}\left[\frac{\partial w_0}{\partial \theta}\left(\frac{\partial w}{\partial \theta} - v\right) + w_0\left(w + \frac{\partial v}{\partial \theta}\right)\right],\end{aligned} \tag{1.119}$$

$$\begin{aligned}\gamma_{x\theta,0} = {} & \frac{\partial v}{\partial x} + \frac{\partial u}{R\partial \theta} + \frac{1}{R}\left[\frac{\partial u}{\partial x}\frac{\partial u}{\partial \theta} + \frac{\partial v}{\partial x}\left(\frac{\partial v}{\partial \theta} + w\right) + \frac{\partial w}{\partial x}\right. \\ & \left. \times\left(\frac{\partial w}{\partial \theta} - v\right) + \frac{\partial w_0}{\partial x}\left(\frac{\partial w}{\partial \theta} - v\right) + \frac{\partial w}{\partial x}\frac{\partial w_0}{\partial \theta} + \frac{\partial v}{\partial x}w_0\right].\end{aligned} \tag{1.120}$$

$$k_x = -\frac{\partial^2 w}{\partial x^2} - \frac{\partial u}{\partial x}\frac{\partial^2 (w+w_0)}{\partial x^2} - \frac{\partial v}{\partial x}\frac{\partial^2 (w+w_0)}{R\,\partial x\,\partial \theta} + \frac{1}{R}\left(\frac{\partial v}{\partial x}\right)^2, \tag{1.121}$$

$$k_\theta = -\frac{\partial^2 w}{R^2\,\partial \theta^2} - \frac{w}{R^2} - \frac{(w+w_0)}{R^3}\left(w + \frac{\partial^2 w}{\partial \theta^2} + \frac{\partial v}{\partial \theta}\right) - \frac{w}{R^3}\left(w_0 + \frac{\partial^2 w_0}{\partial \theta^2}\right)$$
$$- \frac{\partial u}{R^2\,\partial \theta}\left(\frac{\partial u}{R\partial \theta} + \frac{\partial^2 (w+w_0)}{\partial x\,\partial \theta}\right) - \frac{\partial v}{R^3\,\partial \theta}\frac{\partial^2 (w+w_0)}{\partial \theta^2}, \tag{1.122}$$

$$k_{x\theta} = -2\frac{\partial^2 w}{R\,\partial x\,\partial \theta} + \frac{\partial v}{R\,\partial x} - \frac{\partial u}{R^2\,\partial \theta} - \frac{\partial u}{R\,\partial x}\left(\frac{\partial u}{R\,\partial \theta} + \frac{\partial^2 (w+w_0)}{\partial x\,\partial \theta}\right)$$
$$+ \frac{\partial v}{R^2\,\partial x}\left(\frac{\partial v}{\partial \theta} - \frac{\partial^2 (w+w_0)}{\partial \theta^2}\right) - \frac{\partial^2 (w+w_0)}{R^2\,\partial x\,\partial \theta}\left(w + \frac{\partial v}{\partial \theta}\right)$$
$$- \frac{\partial^2 w}{R^2\,\partial x\,\partial \theta}w_0 - \frac{\partial^2 (w+w_0)}{\partial x^2}\frac{\partial u}{R\,\partial \theta} - \frac{\partial v}{\partial x}\frac{\partial^2 (w+w_0)}{R^2\,\partial \theta^2}. \tag{1.123}$$

It is important to note that in the present formulation, differently from Ginsberg (1973) and Yamaki (1984), nonlinear terms in changes in curvature and torsion are retained in equations (1.121–1.123). Equations (1.118–1.123) have been obtained by neglecting higher-order terms in z, in order to reduce equations (1.94–1.96) to the form given by equations (1.100–1.102).

1.3.7 The Novozhilov Nonlinear Shell Theory

According to the Novozhilov (1953) nonlinear shell theory, equations (1.113–1.115) are replaced by the following more complex nonlinear kinematic relationships:

$$u_1 = u + z\Theta, \tag{1.124}$$

$$u_2 = v + z\Psi, \tag{1.125}$$

$$u_3 = w + w_0 + z\chi, \tag{1.126}$$

where

$$\Theta = -\frac{\partial\,(w+w_0)}{\partial x}\left(1 + \frac{\partial v}{R\partial \theta} + \frac{w}{R}\right) + \left(\frac{\partial\,(w+w_0)}{R\partial \theta} - \frac{v}{R}\right)\frac{\partial u}{R\partial \theta} - \frac{\partial w}{\partial x}\frac{w_0}{R}, \tag{1.127}$$

$$\Psi = -\left(\frac{\partial\,(w+w_0)}{R\partial \theta} - \frac{v}{R}\right)\left(1 + \frac{\partial u}{\partial x}\right) + \frac{\partial\,(w+w_0)}{\partial x}\frac{\partial v}{\partial x}, \tag{1.128}$$

$$\chi \cong \frac{\partial u}{\partial x} + \frac{\partial v}{R\partial \theta} + \frac{w+w_0}{R} + \frac{\partial u}{\partial x}\left(\frac{\partial v}{R\partial \theta} + \frac{w+w_0}{R}\right) - \frac{\partial u}{R\partial \theta}\frac{\partial v}{\partial x}. \tag{1.129}$$

Equations (1.124–1.129) were obtained by Novozhilov (1953) by imposing analytically that the straight fibers that are normal to the middle surface of the shell remain straight and normal to the middle surface after deformation, and are not elongated. This assumption replaces the second part of hypothesis (H5) in Donnell's theory (i.e.

strains vary linearly within the shell thickness); hypothesis (H6) is removed. More details on this theory and a complete derivation are given in Chapter 2.

Substituting equations (1.124–1.129) into equations (1.94–1.96) and using approximations (1.116) and (1.117), the following expressions for changes in the curvature and torsion are obtained:

$$\begin{aligned} k_x = &-\frac{\partial^2 w}{\partial x^2} + \frac{\partial v}{R\partial x}\left(-\frac{\partial u}{R\partial \theta} + \frac{\partial v}{\partial x} - \frac{\partial^2 (w+w_0)}{\partial x\,\partial \theta}\right) + \frac{\partial\,(w+w_0)}{\partial x}\frac{\partial^2 u}{\partial x^2} \\ &+ \frac{\partial^2 u}{R\partial x\,\partial \theta}\left(-\frac{v}{R} + \frac{\partial\,(w+w_0)}{R\partial \theta}\right) + \frac{\partial^2 (w+w_0)}{R\partial x\,\partial \theta}\frac{\partial u}{R\partial \theta} \\ &- \frac{\partial^2 (w+w_0)}{\partial x^2}\left(\frac{w}{R} + \frac{\partial v}{R\partial \theta} + \frac{\partial u}{\partial x}\right) - \frac{\partial^2 w}{\partial x^2}\frac{w_0}{R}, \end{aligned} \tag{1.130}$$

$$\begin{aligned} k_\theta = &-\frac{\partial^2 w}{R^2\,\partial \theta^2} + \frac{\partial u}{R\partial x} + \frac{\partial v}{R^2\,\partial \theta} - \frac{(w+w_0)}{R}\left(\frac{\partial^2 w}{R^2\,\partial \theta^2} - \frac{\partial v}{R^2\,\partial \theta} - 2\frac{\partial u}{R\partial x}\right) \\ &- \frac{w}{R}\frac{\partial^2 w_0}{R^2\,\partial \theta^2} - \frac{\partial u}{R^2\,\partial \theta}\left(\frac{\partial u}{R\partial \theta} + \frac{\partial v}{\partial x} + \frac{\partial^2\,(w+w_0)}{\partial x\,\partial \theta}\right) + \frac{\partial v}{R^2\,\partial \theta} \\ &\times\left(\frac{\partial v}{R\partial \theta} + 3\frac{\partial u}{\partial x} - \frac{\partial^2\,(w+w_0)}{R\partial \theta^2}\right) + \frac{\partial^2 v}{R^3\,\partial \theta^2}\left(\frac{\partial (w+w_0)}{\partial \theta} - v\right) \\ &+ \frac{\partial (w+w_0)}{R^2\,\partial \theta}\left(-\frac{v}{R} + \frac{\partial w}{R\partial \theta}\right) - \frac{\partial^2\,(w+w_0)}{R^2\,\partial \theta^2}\frac{\partial u}{\partial x} + \frac{\partial w}{R^2\,\partial \theta}\frac{\partial w_0}{R\partial \theta} \\ &+ \frac{\partial\,(w+w_0)}{\partial x}\frac{\partial^2 v}{R\partial x\,\partial \theta} + \frac{\partial v}{\partial x}\frac{\partial^2 (w+w_0)}{R\partial x\,\partial \theta}, \end{aligned} \tag{1.131}$$

$$\begin{aligned} k_{x\theta} = &-2\frac{\partial^2 w}{R\partial x\,\partial \theta} - \frac{\partial u}{R^2\,\partial \theta} + \frac{\partial v}{R\partial x} + \frac{\partial u}{R^2\,\partial \theta}\left(-\frac{\partial u}{\partial x} + \frac{\partial^2 (w+w_0)}{R\partial \theta^2} - \frac{\partial v}{R\partial \theta}\right) \\ &+ \frac{\partial^2 (w+w_0)}{\partial x^2}\frac{\partial v}{\partial x} - \frac{v}{R^2}\left(\frac{\partial^2 u}{R\partial \theta^2} + \frac{\partial^2 v}{\partial x\,\partial \theta} + \frac{\partial\,(w+w_0)}{\partial x}\right) \\ &+ \frac{\partial\,(w+w_0)}{R^2\,\partial \theta}\left(\frac{\partial^2 u}{R\partial \theta^2} + \frac{\partial w}{\partial x} + \frac{\partial^2 v}{\partial x\,\partial \theta}\right) + \frac{\partial w}{R^2\,\partial \theta}\frac{\partial w_0}{\partial x} + \frac{\partial v}{R\partial x} \\ &\times\left(\frac{w+w_0}{R} + 2\frac{\partial v}{R\partial \theta} - \frac{\partial^2 (w+w_0)}{R\partial \theta^2} + 2\frac{\partial u}{\partial x}\right) + \frac{\partial\,(w+w_0)}{\partial x}\left(\frac{\partial^2 u}{R\partial x\,\partial \theta} + \frac{\partial^2 v}{\partial x^2}\right) \\ &- 2\frac{\partial^2 (w+w_0)}{R\partial x\,\partial \theta}\left(\frac{w}{R} + \frac{\partial v}{R\partial \theta} + \frac{\partial u}{\partial x}\right) - 2\frac{\partial^2 w}{R^2\,\partial x\,\partial \theta}w_0. \end{aligned} \tag{1.132}$$

The middle surface strain-displacement relationships are still given by equations (1.118–1.120); therefore, the only difference resulting from using such complex kinematic relationships as compared to the Flügge-Lur'e-Byrne nonlinear theory is in the changes in curvature and torsion. Equations (1.118–1.120) and (1.130–1.132) together are an improved version of the classical Novozhilov nonlinear shell theory, as approximations (1.116) and (1.117) have been used instead of neglecting terms in $(z/R)^2$ in the derivation, as originally done by Novozhilov (1953).

1.3.8 The Sanders-Koiter Nonlinear Shell Theory

Sanders (1963) developed a refined nonlinear theory of shells, originally expressed in tensorial form; the same equations were obtained by Koiter (1966) around the same period, leading to the designation of the equations as the Sanders-Koiter equations. These equations were obtained by using consistent simplification in order to obtain mathematical expressions that respect the mechanics of shells and keep only terms of the same magnitude. They are suitable for finite deformations with small strains and moderately small rotations; therefore, hypothesis (H6) is removed. Transverse shear strains are neglected. The middle surface strain-displacement relationships, changes in the curvature and torsion for circular cylindrical shells are (Yamaki 1984)

$$\varepsilon_{x,0} = \frac{\partial u}{\partial x} + \frac{1}{2}\left(\frac{\partial w}{\partial x}\right)^2 + \frac{1}{8}\left(\frac{\partial v}{\partial x} - \frac{\partial u}{R\partial\theta}\right)^2 + \frac{\partial w}{\partial x}\frac{\partial w_0}{\partial x}, \tag{1.133}$$

$$\varepsilon_{\theta,0} = \frac{\partial v}{R\partial\theta} + \frac{w}{R} + \frac{1}{2}\left(\frac{\partial w}{R\partial\theta} - \frac{v}{R}\right)^2 + \frac{1}{8}\left(\frac{\partial u}{R\partial\theta} - \frac{\partial v}{\partial x}\right)^2 + \frac{\partial w_0}{R\partial\theta}\left(\frac{\partial w}{R\partial\theta} - \frac{v}{R}\right), \tag{1.134}$$

$$\gamma_{x\theta,0} = \frac{\partial u}{R\partial\theta} + \frac{\partial v}{\partial x} + \frac{\partial w}{\partial x}\left(\frac{\partial w}{R\partial\theta} - \frac{v}{R}\right) + \frac{\partial w_0}{\partial x}\left(\frac{\partial w}{R\partial\theta} - \frac{v}{R}\right) + \frac{\partial w}{\partial x}\frac{\partial w_0}{R\partial\theta}, \tag{1.135}$$

$$k_x = -\frac{\partial^2 w}{\partial x^2}, \tag{1.136}$$

$$k_\theta = \frac{\partial v}{R^2\,\partial\theta} - \frac{\partial^2 w}{R^2\,\partial\theta^2}, \tag{1.137}$$

$$k_{x\theta} = -2\frac{\partial^2 w}{R\partial x\,\partial\theta} + \frac{1}{2\,R}\left(3\frac{\partial v}{\partial x} - \frac{\partial u}{R\partial\theta}\right). \tag{1.138}$$

Changes in curvature and torsion are linear according to the Sanders-Koiter nonlinear shell theory.

1.3.9 Elastic Strain Energy

The elastic strain energy U_S of a circular cylindrical shell, under Love's hypothesis $\sigma_{zz} = \tau_{zx} = \tau_{z\theta} = 0$, is given by

$$U_S = \frac{1}{2}\int_0^L\int_0^{2\pi}\int_{-h/2}^{h/2}(\sigma_{xx}\,\varepsilon_{xx} + \sigma_{\theta\theta}\,\varepsilon_{\theta\theta} + \tau_{x\theta}\,\gamma_{x\theta})\,\mathrm{d}x\,R\mathrm{d}\theta\,\mathrm{d}z, \tag{1.139}$$

where h is the shell thickness; L is the length; R the mean radius; σ_{xx}, σ_{yy} and τ_{xy} are Kirchhoff stresses and ε_{xx}, $\varepsilon_{\theta\theta}$ and $\gamma_{x\theta}$ are Green's strains. The simplicity of equation (1.139) is due to the Lagrangian description of the shell, which allows integration over the shell in the original undeformed configuration. The Kirchhoff stresses for a homogeneous and isotropic material ($\sigma_{zz} = 0$, case of plane stress) are given by

$$\sigma_{xx} = \frac{E}{1-\nu^2}\left(\varepsilon_{xx} + \nu\,\varepsilon_{\theta\theta}\right), \quad \sigma_{\theta\theta} = \frac{E}{1-\nu^2}\left(\varepsilon_{\theta\theta} + \nu\,\varepsilon_{xx}\right), \quad \tau_{x\theta} = \frac{E}{2\,(1+\nu)}\gamma_{x\theta}. \tag{1.140}$$

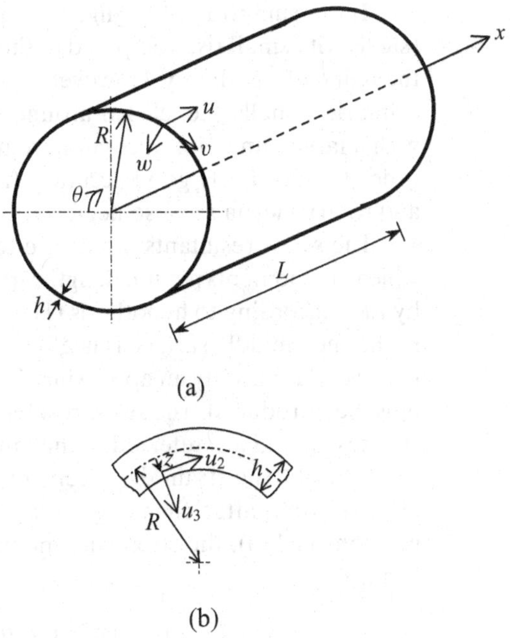

Figure 1.12. Circular cylindrical shell and notation in case of Donnell's nonlinear shallow-shell theory. (a) Symbols used for dimensions and displacements of a generic point. (b) Cross-section of the shell with close-up view.

By using equations (1.100–1.102), (1.139) and (1.140), the following expression is obtained in the case of uniform thickness:

$$U_S = \frac{1}{2}\frac{Eh}{1-\nu^2}\int_0^{2\pi}\int_0^L \left(\varepsilon_{x,0}^2 + \varepsilon_{\theta,0}^2 + 2\nu\,\varepsilon_{x,0}\,\varepsilon_{\theta,0} + \frac{1-\nu}{2}\gamma_{x\theta,0}^2\right) \mathrm{d}x\, R\mathrm{d}\theta$$
$$+\frac{1}{2}\frac{Eh^3}{12(1-\nu^2)}\int_0^{2\pi}\int_0^L \left(k_x^2 + k_\theta^2 + 2\nu\, k_x\, k_\theta + \frac{1-\nu}{2}k_{x\theta}^2\right) \mathrm{d}x\, R\mathrm{d}\theta$$
$$+\frac{1}{2}\frac{Eh^3}{6R(1-\nu^2)}\int_0^{2\pi}\int_0^L \left(\varepsilon_{x,0}\,k_x + \varepsilon_{\theta,0}\,k_\theta + \nu\,\varepsilon_{x,0}\,k_\theta + \nu\,\varepsilon_{\theta,0}\,k_x + \frac{1-\nu}{2}\,\gamma_{x\theta,0}k_{x\theta}\right)$$
$$\times\, \mathrm{d}x\, R\mathrm{d}\theta + O(h^4), \tag{1.141}$$

where the first term, on the right-hand side, is the membrane (also referred to as stretching) energy and the second one is the bending energy. If the last term is retained, membrane and bending energies are coupled. Equation (1.141) can be written as function of the middle-surface displacements u, v, w by using the appropriate strain-displacement relationships according with the selected shell theory.

1.3.10 Donnell's Nonlinear Shallow-Shell Theory

Neglecting in-plane inertia, a simplified version of the shell equations of motion can be obtained, in the form originally obtained by Donnell (1934). Here a development similar to the one in Section 1.2.7 for rectangular plates is used. Note that, for this theory only, the radial displacements u_3 (of a generic point) and w (of points on the middle surface) are taken positive inward of the shell surface (see Figure 1.12) in order to conform with the commonly used notation.

The assumption of negligible in-plane inertia is well verified by shallow-shells (shells with small rise compared to the radius of curvature), and this is the reason for the name of the theory. However, the theory is also applicable to complete circular cylindrical shells (i.e. closed around the circumference) to study vibration modes with a large number n of circumferential waves; specifically, $1/n^2 \ll 1$ is required in order to have fairly good accuracy (i.e. $n \geq 4$ or 5). Transverse shear deformation and rotary inertia are also neglected.

The stress resultants, which are forces per unit length, and the stress moments, which are moments per unit length, are defined in equation (1.53), where y is replaced by $R\theta$. According to hypothesis (H5), the Kirchhoff stresses in the direction normal to the shell middle surface is negligible, so that Q_x and Q_θ should be zero. However, because (H5) is only an approximation, Q_x and Q_θ are not zero for equilibrium and must be introduced. The stress resultants and stress moments are based on Kirchhoff stresses σ_{ij} and are referred to the initial, undeformed, configuration of the shell.

By using the strain-displacement relations for Donnell's nonlinear shell theory (1.100–1.108), after changing the sign to w in the equations, and the stress-strain relations (1.140), the stress and moment resultants can be written in the following form:

$$N_x = \frac{Eh}{1-\nu^2}\left[-\frac{\nu w}{R} + \frac{1}{2}\left(\frac{\partial w}{\partial x}\right)^2 + \frac{\nu}{2}\left(\frac{\partial w}{R\partial\theta}\right)^2 + \frac{\partial u}{\partial x} + \frac{\nu}{R}\frac{\partial v}{\partial\theta}\right], \quad (1.142a)$$

$$N_\theta = \frac{Eh}{1-\nu^2}\left[-\frac{w}{R} + \frac{\nu}{2}\left(\frac{\partial w}{\partial x}\right)^2 + \frac{1}{2}\left(\frac{\partial w}{R\partial\theta}\right)^2 + \nu\frac{\partial u}{\partial x} + \frac{1}{R}\frac{\partial v}{\partial\theta}\right], \quad (1.142b)$$

$$N_{x\theta} = \frac{Eh}{2(1+\nu)}\left[\frac{1}{R}\frac{\partial w}{\partial x}\frac{\partial w}{\partial\theta} + \frac{1}{R}\frac{\partial u}{\partial\theta} + \frac{\partial v}{\partial x}\right], \quad (1.142c)$$

$$M_x = -D\left(\frac{\partial^2 w}{\partial x^2} + \nu\frac{\partial^2 w}{R^2\,\partial\theta^2}\right), \quad M_\theta = -D\left(\frac{\partial^2 w}{R^2\,\partial\theta^2} + \nu\frac{\partial^2 w}{\partial x^2}\right),$$

$$M_{x\theta} = -(1-\nu)D\frac{\partial^2 w}{\partial x\,R\partial\theta}, \quad (1.142d–f)$$

where $D = Eh^3/[12(1-\nu^2)]$.

The equations of motion (1.47) can now be rewritten for an infinitesimal cylindrical element in cylindrical coordinates and in the original configuration (see Figure 1.13):

$$\frac{\partial S_{xx}}{\partial x} + \frac{\partial S_{\theta x}}{R\partial\theta} + \frac{\partial S_{zx}}{\partial z} - \frac{S_{zx}}{R} + \frac{\rho_0}{\rho}X = 0,$$

$$\frac{\partial S_{x\theta}}{\partial x} + \frac{\partial S_{\theta\theta}}{R\partial\theta} + \frac{\partial S_{z\theta}}{\partial z} - \frac{S_{z\theta}}{R} + \frac{\rho_0}{\rho}Y = 0,$$

$$\frac{\partial S_{xz}}{\partial x} + \frac{\partial S_{\theta z}}{R\partial\theta} + \frac{\partial S_{zz}}{\partial z} + \frac{S_{\theta\theta} - S_{zz}}{R} + \frac{\rho_0}{\rho}Z = 0,$$

where the Lagrangian description is used and S_{ij} are the Lagrangian stresses. However, in order to use Kirchhoff stresses, it is necessary to use the transformation (1.46b)

$$S_{ij} = \sum_{k=1}^{3}\sigma_{ik}\frac{\partial a_j}{\partial x_k},$$

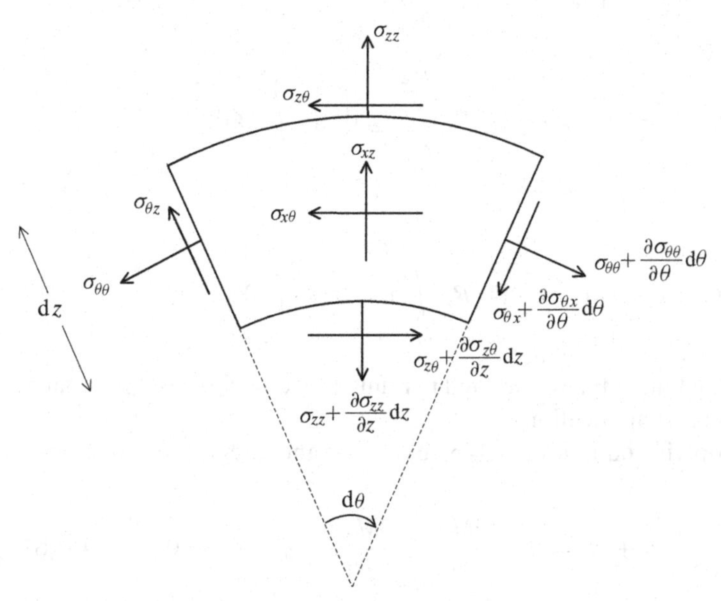

Figure 1.13. Stresses on an infinitesimal element of a circular cylindrical shell.

where $x_1 = x, x_2 = R\theta, x_3 = z, a_1 = x + u_1, a_2 = R\theta + u_2, a_3 = z + u_3$. Moreover, by using hypothesis (H6), the following approximations are used: $\partial u_1/\partial x_i = \partial u_2/\partial x_i = \partial u_3/\partial z = 0$. After manipulation, the transformations for the stresses are

$$S_{ij} = \sigma_{ij}, \text{ for } i = 1, 2, 3 \text{ and } j = 1, 2 \text{ (i.e. excluding } z \text{ axis for } j)$$

$$S_{iz} = \sigma_{ix}\frac{\partial w}{\partial x} + \sigma_{i\theta}\frac{\partial w}{R\partial\theta} + \sigma_{iz}, \text{ for } i = 1, 2, 3.$$

Therefore, the three equations of motion can be written as

$$\frac{\partial\sigma_{xx}}{\partial x} + \frac{\partial\sigma_{\theta x}}{R\partial\theta} + \frac{\partial\sigma_{zx}}{\partial z} - \frac{\sigma_{zx}}{R} + X = 0, \tag{1.143a}$$

$$\frac{\partial\sigma_{x\theta}}{\partial x} + \frac{\partial\sigma_{\theta\theta}}{R\partial\theta} + \frac{\partial\sigma_{z\theta}}{\partial z} - \frac{\sigma_{z\theta}}{R} + Y = 0, \tag{1.143b}$$

$$\frac{\partial}{\partial x}\left(\sigma_{xx}\frac{\partial w}{\partial x} + \sigma_{x\theta}\frac{\partial w}{R\partial\theta} + \sigma_{xz}\right) + \frac{\partial}{R\partial\theta}\left(\sigma_{\theta x}\frac{\partial w}{\partial x} + \sigma_{\theta\theta}\frac{\partial w}{R\partial\theta} + \sigma_{\theta z}\right)$$
$$+ \frac{\partial}{\partial z}\left(\sigma_{zx}\frac{\partial w}{\partial x} + \sigma_{z\theta}\frac{\partial w}{R\partial\theta} + \sigma_{zz}\right) + \frac{1}{R}\left(\sigma_{\theta\theta} - \sigma_{zx}\frac{\partial w}{\partial x} - \sigma_{z\theta}\frac{\partial w}{R\partial\theta} - \sigma_{zz}\right) + Z = 0, \tag{1.144}$$

where it has been assumed that the density of the body before and after deformation is the same, that is, $\rho_0 = \rho$, and X, Y and Z are the body forces per unit volume, including inertia.

Multiplying equations (1.143a,b) by dz and integrating from $-h/2$ to $h/2$, and by using definitions (1.53a–c), the following expressions are obtained:

$$\frac{\partial N_x}{\partial x} + \frac{\partial N_{x\theta}}{R\partial\theta} + f_x = 0, \qquad \frac{\partial N_{x\theta}}{\partial x} + \frac{\partial N_\theta}{R\partial\theta} + f_\theta = 0, \tag{1.145a,b}$$

where

$$f_x = \sigma_{zx}|_{h/2} - \sigma_{zx}|_{-h/2} - (1/R)\int_{-h/2}^{h/2} \sigma_{zx}\,\mathrm{d}z + \int_{-h/2}^{h/2} X\mathrm{d}z$$

and

$$f_\theta = \sigma_{z\theta}|_{h/2} - \sigma_{z\theta}|_{-h/2} - (1/R)\int_{-h/2}^{h/2} \sigma_{z\theta}\,\mathrm{d}z + \int_{-h/2}^{h/2} Y\mathrm{d}z$$

are the external loading tangential to the shell per unit area equal to zero if in-plane inertia is neglected in most applications.

Similarly, by multiplying equations (1.143a,b) by $z\,\mathrm{d}z$ and integrating from $-h/2$ to $h/2$,

$$\frac{\partial M_x}{\partial x} + \frac{\partial M_{x\theta}}{R\partial\theta} - Q_x + m_x = 0, \quad \frac{\partial M_{x\theta}}{\partial x} + \frac{\partial M_\theta}{R\partial\theta} - Q_\theta + m_\theta = 0, \quad (1.146\text{a,b})$$

where

$$m_x = \frac{h}{2}(\sigma_{zx}|_{h/2} - \sigma_{zx}|_{-h/2}) - (1/R)\int_{-h/2}^{h/2} \sigma_{zx} z\,\mathrm{d}z + \int_{-h/2}^{h/2} X z\,\mathrm{d}z$$

and

$$m_\theta = \frac{h}{2}(\sigma_{z\theta}|_{h/2} - \sigma_{z\theta}|_{-h/2}) - (1/R)\int_{-h/2}^{h/2} \sigma_{z\theta} z\,\mathrm{d}z + \int_{-h/2}^{h/2} Y z\,\mathrm{d}z$$

are the external moments per unit area, in general, equal to zero.

The integration with respect to z of the third equation of motion (1.144) gives

$$\frac{\partial}{\partial x}\left(Q_x + N_x\frac{\partial w}{\partial x} + N_{x\theta}\frac{\partial w}{R\partial\theta}\right) + \frac{\partial}{R\partial\theta}\left(Q_\theta + N_{x\theta}\frac{\partial w}{\partial x} + N_\theta\frac{\partial w}{R\partial\theta}\right) + \frac{N_\theta}{R} + q = 0, \quad (1.147)$$

where q is the transverse load (including inertia) per unit area of the undeformed middle surface, which is given by

$$q = \left(\sigma_{zz} + \sigma_{zx}\frac{\partial w}{\partial x} + \sigma_{z\theta}\frac{\partial w}{R\partial\theta}\right)\Bigg|_{-h/2}^{h/2} - \frac{1}{R}\int_{-h/2}^{h/2}\left(\sigma_{zz} + \sigma_{zx}\frac{\partial w}{\partial x} + \sigma_{z\theta}\frac{\partial w}{R\partial\theta}\right)\mathrm{d}z + \int_{-h/2}^{h/2} Z\mathrm{d}z. \quad (1.148)$$

The transverse load can be rewritten as

$$q = f - p - \rho h\ddot{w} - ch\dot{w}, \quad (1.149)$$

where f is the external pressure load (assumed positive with w), p is the internal fluid pressure in fluid-structure interaction (positive with $-w$), $\rho h\ddot{w}$ is the inertia force per unit area and $ch\dot{w}$ is the viscous damping force per unit area. Expression (1.149) takes most of applications into account.

By using equations (1.145a,b), equation (1.147) can be simplified into

$$\frac{\partial Q_x}{\partial x} + \frac{\partial Q_\theta}{R\partial\theta} + q + N_x\frac{\partial^2 w}{\partial x^2} + 2N_{x\theta}\frac{\partial^2 w}{\partial x\, R\partial\theta} + N_\theta\frac{\partial^2 w}{R^2\,\partial\theta^2} + \frac{N_\theta}{R} - f_x\frac{\partial w}{\partial x} - f_\theta\frac{\partial w}{R\partial\theta} = 0. \tag{1.150}$$

Eliminating Q_x and Q_θ by using (1.146a,b),

$$\frac{\partial^2 M_x}{\partial x^2} + 2\frac{\partial^2 M_{x\theta}}{\partial x\, R\partial\theta} + \frac{\partial^2 M_\theta}{R^2\,\partial\theta^2} = -q - \frac{\partial m_x}{\partial x} - \frac{\partial m_\theta}{R\partial\theta} - N_x\frac{\partial^2 w}{\partial x^2} - 2N_{x\theta}\frac{\partial^2 w}{\partial x\, R\partial\theta} - N_\theta\frac{\partial^2 w}{R^2\,\partial\theta^2} - \frac{N_\theta}{R} + f_x\frac{\partial w}{\partial x} + f_\theta\frac{\partial w}{R\partial\theta}. \tag{1.151}$$

Finally, substituting equations (1.142d–f), the third equation of equilibrium takes the form

$$D\left(\frac{\partial^4 w}{\partial x^4} + 2\frac{\partial^4 w}{\partial x^2 R^2\,\partial\theta^2} + \frac{\partial^4 w}{R^4\partial\theta^4}\right) = q + \frac{\partial m_x}{\partial x} + \frac{\partial m_\theta}{R\partial\theta} + N_x\frac{\partial^2 w}{\partial x^2} + 2N_{x\theta}\frac{\partial^2 w}{\partial x\, R\partial\theta} + N_\theta\frac{\partial^2 w}{R^2\,\partial\theta^2} + \frac{N_\theta}{R} - f_x\frac{\partial w}{\partial x} - f_\theta\frac{\partial w}{R\partial\theta}. \tag{1.152}$$

Coming back to the other two equations of equilibrium, equations (1.145a,b), and substituting equations (1.142a–c) into them, the following two expressions are obtained:

$$\frac{\partial}{\partial x}\left[-\frac{\nu w}{R} + \frac{1}{2}\left(\frac{\partial w}{\partial x}\right)^2 + \frac{\nu}{2}\left(\frac{\partial w}{R\partial\theta}\right)^2 + \frac{\partial u}{\partial x} + \frac{\nu\,\partial v}{R\,\partial\theta}\right] + \frac{1-\nu}{2}\frac{\partial}{R\partial\theta}\left[\frac{1}{R}\frac{\partial w}{\partial x}\frac{\partial w}{\partial\theta} + \frac{1}{R}\frac{\partial u}{\partial\theta} + \frac{\partial v}{\partial x}\right] + \frac{1-\nu^2}{Eh}f_x = 0, \tag{1.153}$$

$$\frac{\partial}{R\partial\theta}\left[-\frac{w}{R} + \frac{\nu}{2}\left(\frac{\partial w}{\partial x}\right)^2 + \frac{1}{2}\left(\frac{\partial w}{R\partial\theta}\right)^2 + \nu\frac{\partial u}{\partial x} + \frac{1}{R}\frac{\partial v}{\partial\theta}\right] + \frac{1-\nu}{2}\frac{\partial}{\partial x}\left[\frac{1}{R}\frac{\partial w}{\partial x}\frac{\partial w}{\partial\theta} + \frac{1}{R}\frac{\partial u}{\partial\theta} + \frac{\partial v}{\partial x}\right] + \frac{1-\nu^2}{Eh}f_\theta = 0. \tag{1.154}$$

Equations (1.152–1.154) are the equations of motion of the circular cylindrical shell in the three displacements u, v, w. However, it is possible to reduce these equations to a single equation of motion in w and a compatibility equation by introducing a stress function $F(x,\theta)$ and neglecting in-plane inertia, i.e. $\ddot{u}$, $\ddot{v}$.

The in-plane stress function F is introduced in order to satisfy identically equations (1.146a,b), by introducing the following expressions for the stress resultants:

$$N_x = \frac{\partial^2 F}{R^2\,\partial\theta^2} - \int f_x\,\mathrm{d}x, \quad N_\theta = \frac{\partial^2 F}{\partial x^2} - \int f_\theta\, R\,\mathrm{d}\theta, \quad N_{x\theta} = -\frac{\partial^2 F}{\partial x\, R\partial\theta}, \tag{1.155a–c}$$

where the indefinite integrals give the potentials of tangential loadings. Subtracting equation (1.142b) multiplied by ν from equation (1.142a),

$$\frac{\partial u}{\partial x} = \frac{N_x - \nu N_\theta}{Eh} - \frac{1}{2}\left(\frac{\partial w}{\partial x}\right)^2. \tag{1.156a}$$

Similarly, subtracting equation (1.142a) multiplied by ν from equation (1.142b),

$$\frac{\partial v}{R\,\partial\theta} = \frac{N_\theta - \nu N_x}{Eh} - \frac{1}{2}\left(\frac{\partial w}{R\,\partial\theta}\right)^2 + \frac{w}{R}. \tag{1.156b}$$

The compatibility equation is obtained by applying the differential operator $\partial^2/\partial x\,R\,\partial\theta$ to equation (1.142c),

$$\frac{\partial^2 N_{x\theta}}{\partial x\,R\,\partial\theta} = \frac{Eh}{2(1+\nu)}\left[\frac{\partial^3 u}{\partial x\,R^2\,\partial\theta^2} + \frac{\partial^3 v}{\partial x^2 R\,\partial\theta} + \frac{\partial^3 w}{\partial x^2 R\,\partial\theta}\frac{\partial w}{R\,\partial\theta} + \frac{\partial^3 w}{\partial x\,R^2\,\partial\theta^2}\frac{\partial w}{\partial x} + \left(\frac{\partial^2 w}{\partial x\,R\,\partial\theta}\right) + \frac{\partial^2 w}{\partial x^2}\frac{\partial^2 w}{R^2\,\partial\theta^2}\right], \tag{1.157}$$

substituting in it equations (1.156a,1.156b) after applying operators $\partial^2/R^2\,\partial\theta^2$ and $\partial^2/\partial x^2$, respectively, and finally using equations (1.155a–c) in order to obtain an expression involving only F and w; the resulting equation is

$$\nabla^4 F = Eh\left[-\frac{1}{R}\frac{\partial^2 w}{\partial x^2} + \left(\frac{\partial^2 w}{\partial x\,R\,\partial\theta}\right)^2 - \frac{\partial^2 w}{\partial x^2}\frac{\partial^2 w}{R^2\,\partial\theta^2}\right] + \frac{\partial^2}{R^2\,\partial\theta^2}\left(\int f_x\,\mathrm{d}x - \nu\int f_\theta\,R\,\mathrm{d}\theta\right) + \frac{\partial^2}{\partial x^2}\left(\int f_\theta\,R\,\mathrm{d}\theta - \nu\int f_x\,\mathrm{d}x\right), \tag{1.158}$$

where the biharmonic operator in cylindrical coordinates is defined as $\nabla^4 = [\partial^2/\partial x^2 + \partial^2/(R^2\,\partial\theta^2)]^2$.

The compatibility equation makes a pair with the following equation of motion, obtained by inserting equations (1.155a–c) into (1.152):

$$D\nabla^4 w = q + \frac{1}{R}\frac{\partial^2 F}{\partial x^2} + \left(\frac{\partial^2 F}{R^2\,\partial\theta^2}\frac{\partial^2 w}{\partial x^2} - 2\frac{\partial^2 F}{\partial x\,R\,\partial\theta}\frac{\partial^2 w}{\partial x\,R\,\partial\theta} + \frac{\partial^2 F}{\partial x^2}\frac{\partial^2 w}{R^2\,\partial\theta^2}\right) + \frac{\partial m_x}{\partial x} + \frac{\partial m_\theta}{R\,\partial\theta} - f_x\frac{\partial w}{\partial x} - f_\theta\frac{\partial w}{R\,\partial\theta} - \frac{\partial^2 w}{\partial x^2}\int f_x\,\mathrm{d}x - \frac{\partial^2 w}{R^2\,\partial\theta^2}\int f_\theta\,R\,\mathrm{d}\theta. \tag{1.159}$$

In most of applications, $m_x, m_\theta, f_x, f_\theta, \int f_x\,\mathrm{d}x$ and $\int f_\theta\,R\,\mathrm{d}\theta$ are negligible (including tangential inertia and tangential dissipation, which appears in f_x, f_θ) and the equation of motion can be written in the classical form

$$D\nabla^4 w + ch\dot{w} + \rho h\ddot{w} = f - p + \frac{1}{R}\frac{\partial^2 F}{\partial x^2} + \left(\frac{\partial^2 F}{R^2\,\partial\theta^2}\frac{\partial^2 w}{\partial x^2} - 2\frac{\partial^2 F}{R\,\partial x\,\partial\theta}\frac{\partial^2 w}{R\,\partial x\,\partial\theta} + \frac{\partial^2 F}{\partial x^2}\frac{\partial^2 w}{R^2\,\partial\theta^2}\right), \tag{1.160}$$

which makes a pair with the compatibility equation

$$\frac{1}{Eh}\nabla^4 F = -\frac{1}{R}\frac{\partial^2 w}{\partial x^2} + \left[\left(\frac{\partial^2 w}{R\,\partial x\,\partial\theta}\right)^2 - \frac{\partial^2 w}{\partial x^2}\frac{\partial^2 w}{R^2\,\partial\theta^2}\right]. \tag{1.161}$$

Donnell's nonlinear shallow-shell equations (1.160) and (1.161) can be compared to the von Kármán equation of motion for a rectangular plate (1.72) and (1.73). By using the notation $y = R\theta$ and verifying that the different sign of the fluid pressure p is only due to notation (the fluid pressure is here assumed to be opposite to f), it is immediately verified that both the equation of motion and the compatibility equation here involve an additional term due to the curvature of the shell in θ direction.

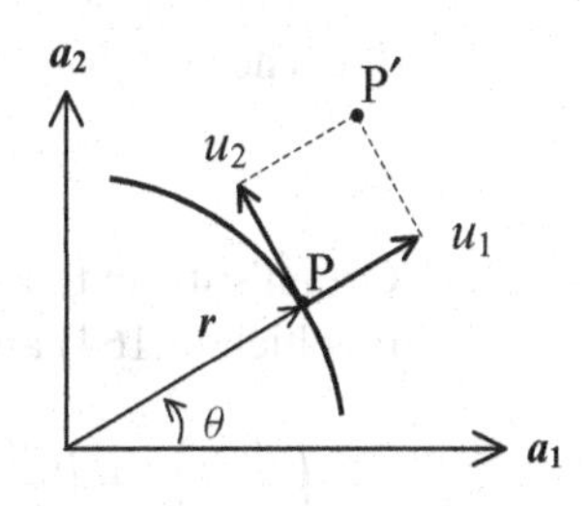

Figure 1.14. Point P in initial position and P' after displacement in rectangular Cartesian and polar coordinate systems.

1.3.11 Donnell's Nonlinear Shallow-Shell Theory Including Geometric Imperfections

In the presence of initial geometric imperfections of circular cylindrical shells associated with zero initial stress and denoted by the radial displacement w_0 (assumed positive inward) (see equation [1.109]), Donnell's nonlinear shallow-shell equation of motion becomes

$$D\nabla^4 w + ch\dot{w} + \rho h\ddot{w} = f - p + \frac{1}{R}\frac{\partial^2 F}{\partial x^2} + \frac{1}{R^2}\left[\frac{\partial^2 F}{\partial \theta^2}\left(\frac{\partial^2 w}{\partial x^2} + \frac{\partial^2 w_0}{\partial x^2}\right)\right.$$
$$\left. -2\frac{\partial^2 F}{\partial x\,\partial \theta}\left(\frac{\partial^2 w}{\partial x\,\partial \theta} + \frac{\partial^2 w_0}{\partial x\,\partial \theta}\right) + \frac{\partial^2 F}{\partial x^2}\left(\frac{\partial^2 w}{\partial \theta^2} + \frac{\partial^2 w_0}{\partial \theta^2}\right)\right], \tag{1.162}$$

which makes a pair with the compatibility equation

$$\frac{1}{Eh}\nabla^4 F = -\frac{1}{R}\frac{\partial^2 w}{\partial x^2} + \frac{1}{R^2}\left[\left(\frac{\partial^2 w}{\partial x\,\partial \theta}\right)^2 + 2\frac{\partial^2 w}{\partial x\,\partial \theta}\frac{\partial^2 w_0}{\partial x\,\partial \theta}\right.$$
$$\left. -\left(\frac{\partial^2 w}{\partial x^2} + \frac{\partial^2 w_0}{\partial x^2}\right)\frac{\partial^2 w}{\partial \theta^2} - \frac{\partial^2 w}{\partial x^2}\frac{\partial^2 w_0}{\partial \theta^2}\right]. \tag{1.163}$$

Equations (1.162) and (1.163) are obtained by substituting equation (1.109) into Donnell's nonlinear shallow-shell equation of motion and neglecting all terms which depend on w_0 only (as they are associated with initial deformation and therefore with zero stress).

1.4 Large Deflection of Circular Plates

Circular plates are a common structural element in engineering. By using the results obtained in the previous sections, the strain-displacement relationships and the von Kármán equation of motion are presented for circular plates.

1.4.1 Green's Strain Tensor for Circular Plates

A point P of coordinates (r, θ, z) in a polar coordinate system, with the additional z axis orthogonal to the plate plane, is considered, as shown in Figure 1.14. Three orthogonal displacements u_1, u_2, u_3, in the radial, circumferential and z direction, respectively, are given to the point P as shown in Figure 1.14, so that a new point P' is obtained. The coordinates of P' in the rectangular Cartesian system (a_1, a_2, a_3) are

$$a_1 = r\cos\theta + u_1\cos\theta - u_2\sin\theta, \quad a_2 = r\sin\theta + u_1\sin\theta + u_2\cos\theta, \quad a_3 = z + u_3. \tag{1.164}$$

The components of the Euclidean metric tensor are

$$g_{11} = 1, \quad g_{22} = r^2, \quad g_{33} = 1. \tag{1.165}$$

Green's strain tensor in the (r, θ, z) coordinates is obtained by using equations (1.92), in which (1.164) and (1.165) must be substituted. In particular,

$$\left(\frac{\partial a_1}{\partial r}\right)^2 + \left(\frac{\partial a_2}{\partial r}\right)^2 + \left(\frac{\partial a_3}{\partial r}\right)^2 = 1 + 2\frac{\partial u_1}{\partial r} + \left(\frac{\partial u_1}{\partial r}\right)^2 + \left(\frac{\partial u_2}{\partial r}\right)^2 + \left(\frac{\partial u_3}{\partial r}\right)^2,$$

which gives

$$\varepsilon_{rr} = \frac{\partial u_1}{\partial r} + \frac{1}{2}\left[\left(\frac{\partial u_1}{\partial r}\right)^2 + \left(\frac{\partial u_2}{\partial r}\right)^2 + \left(\frac{\partial u_3}{\partial r}\right)^2\right]. \tag{1.166}$$

Similarly,

$$\left(\frac{\partial a_1}{\partial \theta}\right)^2 + \left(\frac{\partial a_2}{\partial \theta}\right)^2 + \left(\frac{\partial a_3}{\partial \theta}\right)^2 = r^2 + 2r\left(\frac{\partial u_2}{\partial \theta} + u_1\right) + \left(\frac{\partial u_2}{\partial \theta} + u_1\right)^2 + \left(\frac{\partial u_1}{\partial \theta} - u_2\right)^2 + \left(\frac{\partial u_3}{\partial \theta}\right)^2,$$

which, after some manipulations, gives

$$\varepsilon_{\theta\theta} = \frac{1}{r}\left(\frac{\partial u_2}{\partial \theta} + u_1\right) + \frac{1}{2r^2}\left[\left(\frac{\partial u_2}{\partial \theta} + u_1\right)^2 + \left(\frac{\partial u_1}{\partial \theta} - u_2\right)^2 + \left(\frac{\partial u_3}{\partial \theta}\right)^2\right]. \tag{1.167}$$

Finally,

$$\frac{\partial a_1}{\partial r}\frac{\partial a_1}{\partial \theta} + \frac{\partial a_2}{\partial r}\frac{\partial a_2}{\partial \theta} + \frac{\partial a_3}{\partial r}\frac{\partial a_3}{\partial \theta} = \frac{\partial u_1}{\partial \theta} + r\frac{\partial u_2}{\partial r} - u_2 + \left[\frac{\partial u_1}{\partial r}\left(\frac{\partial u_1}{\partial \theta} - u_2\right) + \frac{\partial u_2}{\partial r}\left(\frac{\partial u_2}{\partial \theta} + u_1\right) + \frac{\partial u_3}{\partial r}\frac{\partial u_3}{\partial \theta}\right],$$

which, after some manipulations, gives

$$\gamma_{r\theta} = \left(\frac{\partial u_1}{r\,\partial \theta} + \frac{\partial u_2}{\partial r} - \frac{u_2}{r}\right) + \frac{1}{r}\left[\frac{\partial u_1}{\partial r}\left(\frac{\partial u_1}{\partial \theta} - u_2\right) + \frac{\partial u_2}{\partial r}\left(\frac{\partial u_2}{\partial \theta} + u_1\right) + \left(\frac{\partial u_3}{\partial r}\frac{\partial u_3}{\partial \theta}\right)\right], \tag{1.168}$$

where the traditional symbol $\gamma_{r\theta}$ is used instead of $\varepsilon_{r\theta}$. Equations (1.166–1.168) give the Green's strains in the coordinates system (r, θ, z).

1.4.2 Strains for Finite Deflection of Circular Plates: Von Kármán Theory

The von Kármán hypotheses, previously described, are assumed. By using hypothesis (H5) in the von Kármán nonlinear theory of plates,

$$u_1 = u(r, \theta) - z\frac{\partial w}{\partial r}, \tag{1.169}$$

$$u_2 = v(r, \theta) - z\frac{\partial w}{r\,\partial \theta}, \tag{1.170}$$

$$u_3 = w(r, \theta). \tag{1.171}$$

Equations (1.169–1.171) are linear expressions that are obtained exactly in the same way as for rectangular plates. These equations are inserted into Green's strain tensor, equations (1.166–1.168), in order to obtain the strain-displacement relations; moreover, $\varepsilon_{zz} \simeq 0$, $\gamma_{rz} \simeq 0$, $\gamma_{\theta z} \simeq 0$. The strain components ε_{rr}, $\varepsilon_{\theta\theta}$ and $\gamma_{r\theta}$ at an arbitrary point of the circular plate are related to the middle surface strains $\varepsilon_{r,0}$, $\varepsilon_{\theta,0}$ and $\gamma_{r\theta,0}$ and to the changes in the curvature and torsion of the middle surface k_r, k_θ and $k_{r\theta}$ by the following three relations:

$$\varepsilon_{rr} = \varepsilon_{r,0} + z\,k_r, \tag{1.172}$$

$$\varepsilon_{\theta\theta} = \varepsilon_{\theta,0} + z\,k_\theta, \tag{1.173}$$

$$\gamma_{r\theta} = \gamma_{r\theta,0} + z\,k_{r\theta}, \tag{1.174}$$

where z is, as usual, the distance of the arbitrary point of the plate from the middle surface. If the von Kármán hypothesis (H6) is used, the following expressions for the middle surface strains and the changes in the curvature and torsion of the middle surface are obtained:

$$\varepsilon_{r,0} = \frac{\partial u}{\partial r} + \frac{1}{2}\left(\frac{\partial w}{\partial r}\right)^2, \tag{1.175}$$

$$\varepsilon_{\theta,0} = \frac{\partial v}{r\,\partial\theta} + \frac{u}{r} + \frac{1}{2}\left(\frac{\partial w}{r\,\partial\theta}\right)^2, \tag{1.176}$$

$$\gamma_{r\theta,0} = \frac{\partial u}{r\,\partial\theta} + \frac{\partial v}{\partial r} - \frac{v}{r} + \frac{\partial w}{\partial r}\frac{\partial w}{r\,\partial\theta}, \tag{1.177}$$

$$k_r = -\frac{\partial^2 w}{\partial r^2}, \tag{1.178}$$

$$k_\theta = -\frac{\partial w}{r\,\partial r} - \frac{\partial^2 w}{r^2 \partial\theta^2}, \tag{1.179}$$

$$k_{r\theta} = -2\left(\frac{\partial^2 w}{r\,\partial r\,\partial\theta} - \frac{\partial w}{r^2 \partial\theta}\right). \tag{1.180}$$

1.4.3 Von Kármán Equation of Motion for Circular Plates

The full derivation is omitted in this case. It can be deduced from the derivations for rectangular plates and circular cylindrical shells, neglecting in-plane inertia. By introducing the in-plane stress function F, the von Kármán equation of motion for circular plates is

$$D\nabla^4 w + ch\dot{w} + \rho h\ddot{w} = f + p + \left[\frac{\partial^2 w}{\partial r^2}\left(\frac{\partial F}{r\partial r} + \frac{\partial^2 F}{r^2 \partial\theta^2}\right) + \left(\frac{\partial w}{r\,\partial r} + \frac{\partial^2 w}{r^2 \partial\theta^2}\right)\frac{\partial^2 F}{\partial r^2} - 2\left(\frac{\partial^2 w}{r\,\partial r\,\partial\theta} - \frac{\partial w}{r^2 \partial\theta}\right)\left(\frac{\partial^2 F}{r\,\partial r\,\partial\theta} - \frac{\partial F}{r^2 \partial\theta}\right)\right], \tag{1.181}$$

which makes a pair with the compatibility equation

$$\frac{1}{Eh}\nabla^4 F = \left[\left(\frac{\partial^2 w}{r\,\partial r \partial\theta} - \frac{\partial w}{r^2 \partial\theta}\right)^2 - \frac{\partial^2 w}{\partial r^2}\left(\frac{\partial w}{r\,\partial r} + \frac{\partial^2 w}{r^2 \partial\theta^2}\right)\right], \tag{1.182}$$

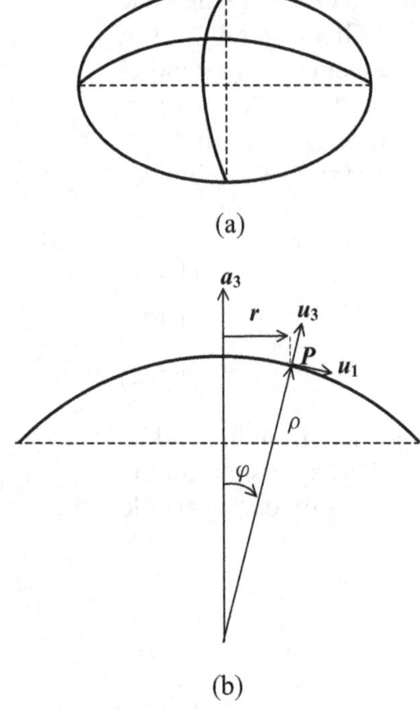

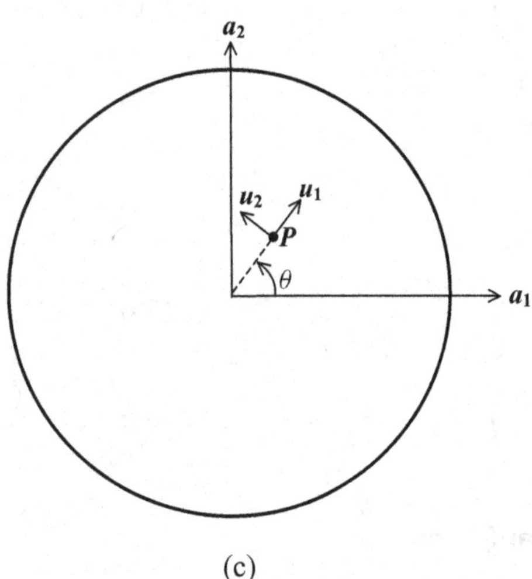

Figure 1.15. Spherical cap. (a) Three-dimensional drawing. (b) Cross-section. (c) View from the top.

where the biharmonic operator in polar coordinates is in this case defined as $\nabla^4 = [\partial^2/\partial r^2 + (1/r)(\partial/\partial r) + \partial^2/(r^2\partial\theta^2)]^2$, F is the in-plane stress function, f is the external pressure load, p is the fluid pressure in the case of fluid-structure interaction, $\rho h\ddot{w}$ is the inertia force per unit area and $ch\dot{w}$ is the viscous damping force per unit area of the plate.

1.5 Large Deflection of Spherical Caps

Spherical caps are segments of spherical shells with a circular base, as shown in Figure 1.15. Spherical shells are a special case of the doubly curved shell addressed in Chapter 2; however, in Chapter 2, spherical shells are not described in spherical coordinates.

1.5.1 Green's Strain Tensor in Spherical Coordinates

Three orthogonal displacements u_1, u_2, u_3, in φ, θ and ρ direction, respectively, are given to a point P of coordinates (φ, θ, ρ) in a spherical coordinate system, so that a new point P' is obtained (see Figure 1.15). The coordinates of P' in the rectangular Cartesian system (a_1, a_2, a_3) are

$$a_1 = \rho \sin\varphi \cos\theta + u_1 \cos\varphi \cos\theta - u_2 \sin\theta + u_3 \sin\varphi \cos\theta, \tag{1.183a}$$

$$a_2 = \rho \sin\varphi \sin\theta + u_1 \cos\varphi \sin\theta + u_2 \cos\theta + u_3 \sin\varphi \sin\theta, \tag{1.183b}$$

$$a_3 = \rho \cos\varphi - u_1 \sin\varphi + u_3 \cos\varphi. \tag{1.183c}$$

The components of the Euclidean metric tensor are

$$g_{\varphi\varphi} = \rho^2, \quad g_{\theta\theta} = \rho^2 \sin^2\varphi, \quad g_{\rho\rho} = 1. \tag{1.184}$$

Green's strain tensor in the spherical coordinates (φ, θ, ρ) is obtained by using equations (1.92), in which (1.183) and (1.184) must be substituted. In particular,

$$\begin{aligned}\varepsilon_{\varphi\varphi} &= \frac{1}{2\rho^2}\left[\left(\frac{\partial a_1}{\partial\varphi}\right)^2 + \left(\frac{\partial a_2}{\partial\varphi}\right)^2 + \left(\frac{\partial a_3}{\partial\varphi}\right)^2 - \rho^2\right] \\ &= \frac{\partial u_1}{\rho\,\partial\varphi} + \frac{u_3}{\rho} + \frac{1}{2}\left[\left(\frac{\partial u_3}{\rho\,\partial\varphi} - \frac{u_1}{\rho}\right)^2 + \left(\frac{\partial u_1}{\rho\,\partial\varphi} + \frac{u_3}{\rho}\right)^2 + \left(\frac{\partial u_2}{\rho\,\partial\varphi}\right)^2\right];\end{aligned} \tag{1.185}$$

similarly,

$$\begin{aligned}\varepsilon_{\theta\theta} &= \frac{1}{2\rho^2\sin^2\varphi}\left[\left(\frac{\partial a_1}{\partial\theta}\right)^2 + \left(\frac{\partial a_2}{\partial\theta}\right)^2 + \left(\frac{\partial a_3}{\partial\theta}\right)^2 - \rho^2\sin^2\varphi\right] \\ &= \frac{1}{\sin\varphi}\frac{\partial u_2}{\rho\,\partial\theta} + \cot\varphi\frac{u_1}{\rho} + \frac{u_3}{\rho} + \frac{1}{2\rho^2\sin^2\varphi}\left[\left(\frac{\partial u_2}{\partial\theta} + u_1\cos\varphi + u_3\sin\varphi\right)^2\right. \\ &\quad \left. + \left(\frac{\partial u_1}{\partial\theta} - u_2\cos\varphi\right)^2 + \left(\frac{\partial u_3}{\partial\theta} - u_2\sin\varphi\right)^2\right];\end{aligned} \tag{1.186}$$

finally,

$$\begin{aligned}\gamma_{\varphi\theta} &= \frac{1}{\rho^2\sin\varphi}\left[\frac{\partial a_1}{\partial\varphi}\frac{\partial a_1}{\partial\theta} + \frac{\partial a_2}{\partial\varphi}\frac{\partial a_2}{\partial\theta} + \frac{\partial a_3}{\partial\varphi}\frac{\partial a_3}{\partial\theta} - \rho^2\sin\varphi\right] \\ &= \frac{\partial u_2}{\rho\,\partial\varphi} - \cot\varphi\frac{u_2}{\rho} + \frac{1}{\sin\varphi}\frac{\partial u_1}{\rho\,\partial\theta} + \frac{1}{\rho^2\sin\varphi}\left[\left(\frac{\partial u_1}{\partial\varphi} + u_3\right)\left(\frac{\partial u_1}{\partial\theta} - u_2\cos\varphi\right)\right. \\ &\quad \left. + \frac{\partial u_2}{\partial\varphi}\left(\frac{\partial u_2}{\partial\theta} + u_1\cos\varphi + u_3\sin\varphi\right) + \left(\frac{\partial u_3}{\partial\varphi} - u_1\right)\left(\frac{\partial u_3}{\partial\theta} - u_2\sin\varphi\right)\right].\end{aligned} \tag{1.187}$$

1.5.2 Strains for Finite Deflection of Spherical Caps: Donnell's Nonlinear Theory

Donnell's hypotheses, previously described, are assumed. By using hypothesis (H5),

$$u_1 = u(\varphi, \theta) - z\frac{\partial w}{R\,\partial\varphi}, \tag{1.188}$$

$$u_2 = v(\varphi, \theta) - z\frac{\partial w}{R \sin\varphi\, \partial\theta}, \tag{1.189}$$

$$u_3 = w(r, \theta). \tag{1.190}$$

Equations (1.188–1.190) are linear expressions that are obtained exactly in the same way as for plates and circular cylindrical shells; these equations are inserted into Green's strain tensor, equations (1.185–1.187) with $\rho = R$ (R being the radius of the middle surface of the spherical shell), in order to obtain the strain-displacement relations; moreover, $\varepsilon_{zz} \simeq 0$, $\gamma_{\varphi z} \simeq 0$, $\gamma_{\theta z} \simeq 0$. The strain components $\varepsilon_{\varphi\varphi}$, $\varepsilon_{\theta\theta}$ and $\gamma_{\varphi\theta}$ at an arbitrary point of the spherical shell are related to the middle surface strains $\varepsilon_{\varphi,0}$, $\varepsilon_{\theta,0}$ and $\gamma_{\varphi\theta,0}$ and to the changes in the curvature and torsion of the middle surface k_φ, k_θ and $k_{\varphi\theta}$ by the following three relations:

$$\varepsilon_{\varphi\varphi} = \varepsilon_{\varphi,0} + z\,k_\varphi, \tag{1.191}$$

$$\varepsilon_{\theta\theta} = \varepsilon_{\theta,0} + z\,k_\theta, \tag{1.192}$$

$$\gamma_{\varphi\theta} = \gamma_{\varphi\theta,0} + z\,k_{\varphi\theta}. \tag{1.193}$$

If Donnell's hypothesis (H6) is used, and z $(-h/2 \le z \le h/2)$ is neglected with respect to R and $R\sin\varphi$, the following expressions for the middle surface strains and the changes in the curvature and torsion of the middle surface are obtained:

$$\varepsilon_{\varphi,0} = \frac{\partial u}{R\partial\varphi} + \frac{w}{R} + \frac{1}{2}\left(\frac{\partial w}{R\partial\varphi}\right)^2, \tag{1.194}$$

$$\varepsilon_{\theta,0} = \frac{1}{\sin\varphi}\frac{\partial v}{R\partial\theta} + \cot\varphi\frac{u}{R} + \frac{w}{R} + \frac{1}{2}\left(\frac{\partial w}{R\sin\varphi\,\partial\theta}\right)^2, \tag{1.195}$$

$$\gamma_{\varphi\theta,0} = \frac{\partial v}{R\partial\varphi} - \cot\varphi\frac{v}{R} + \frac{1}{\sin\varphi}\frac{\partial u}{R\partial\theta} + \frac{\partial w}{R\partial\varphi}\frac{\partial w}{R\sin\varphi\,\partial\theta}, \tag{1.196}$$

$$k_\varphi = -\frac{\partial^2 w}{R^2\,\partial\varphi^2}, \tag{1.197}$$

$$k_\theta = -\frac{\cot\varphi}{R}\frac{\partial w}{R\partial\varphi} - \frac{\partial^2 w}{R^2\sin^2\varphi\,\partial\theta^2}, \tag{1.198}$$

$$k_{\varphi\theta} = -2\left(\frac{1}{R^2\sin\varphi}\frac{\partial^2 w}{\partial\varphi\,\partial\theta} - \frac{\cot\varphi}{R}\frac{\partial w}{R\sin\varphi\,\partial\theta}\right). \tag{1.199}$$

In equations (1.189) and (1.195–1.199), the quantity $R\sin\varphi$ is the radius of the parallel circle (parallel and meridian circles are the coordinates used to identify positions on the heart globe).

1.5.3 Donnell's Equation of Motion for Shallow Spherical Caps

Shallow-shells are curved panels with small rise compared to the smallest radius of curvature. In the case of shallow spherical caps, the quantities $\sin\varphi \cong \varphi$, $\cos\varphi \cong 1$ and the radius of the parallel circle is $r = R\sin\varphi \cong R\varphi$.

Donnell's nonlinear shallow-shell equation of motion for spherical shallow-shells with circular base is discussed here. It is obtained similarly to equations for rectangular plates and circular cylindrical shells; also in this case, in-plane inertia is neglected.

With the assumption that the shell is shallow, any point of the middle surface of the shell can be identified by the polar coordinates (r, θ) of the projection on its base. The equation of motion is

$$D\nabla^4 w + ch\dot{w} + \rho h\ddot{w} = f - p - \frac{1}{R}\nabla^2 F + \left[\frac{\partial^2 w}{\partial r^2}\left(\frac{\partial F}{r\partial r} + \frac{\partial^2 F}{r^2\,\partial\theta^2}\right)\right.$$
$$\left. + \left(\frac{\partial w}{r\,\partial r} + \frac{\partial^2 w}{r^2\,\partial\theta^2}\right)\frac{\partial^2 F}{\partial r^2} - 2\left(\frac{\partial^2 w}{r\,\partial r\,\partial\theta} - \frac{\partial w}{r^2\,\partial\theta}\right)\left(\frac{\partial^2 F}{r\,\partial r\,\partial\theta} - \frac{\partial F}{r^2\,\partial\theta}\right)\right], \tag{1.200}$$

which makes a pair with the compatibility equation

$$\frac{1}{Eh}\nabla^4 F = \frac{1}{R}\nabla^2 w + \left[\left(\frac{\partial^2 w}{r\,\partial r\,\partial\theta} - \frac{\partial w}{r^2\,\partial\theta}\right)^2 - \frac{\partial^2 w}{\partial r^2}\left(\frac{\partial w}{r\,\partial r} + \frac{\partial^2 w}{r^2\,\partial\theta^2}\right)\right], \tag{1.201}$$

where the Laplacian operator in polar coordinates is defined as $\nabla^2 = [\partial^2/\partial r^2 + (1/r)(\partial/\partial r) + \partial^2/(r^2\,\partial\theta^2)]$, R is the radius of curvature, F is the in-plane stress function, f is the external pressure loading, p is the internal pressure in the case of fluid-structure interaction, $\rho h\ddot{w}$ is the inertia force per unit area and $ch\dot{w}$ is the viscous damping force per unit area of the shell.

1.5.4 The Flügge-Lur'e-Byrne Nonlinear Shell Theory

In order to give more accurate strain-displacement relations, the Flügge-Lur'e-Byrne nonlinear shell theory is presented. According with this theory, equations (1.188–1.190) are replaced by

$$u_1 = \left(1 + \frac{z}{R}\right)u(\varphi, \theta) - z\frac{\partial w}{R\,\partial\varphi}, \tag{1.202}$$

$$u_2 = \left(1 + \frac{z}{R\sin\varphi}\right)v(\varphi, \theta) - z\frac{\partial w}{R\sin\varphi\,\partial\theta}, \tag{1.203}$$

$$u_3 = w(r, \theta). \tag{1.204}$$

The strain-displacement relationships for a generic point of the shell are given by equations (1.185–1.187), where $\rho = R + z$. By introducing equations (1.202–1.204) into equations (1.185–1.187) and using approximations (1.116) and (1.117), the middle surface strain-displacement relationships, changes in the curvature and torsion are obtained for the Flügge-Lur'e-Byrne nonlinear shell theory for spherical caps:

$$\varepsilon_{\varphi,0} = \frac{\partial u}{R\partial\varphi} + \frac{w}{R} + \frac{1}{2}\left[\left(\frac{\partial w}{R\partial\varphi} - \frac{u}{R}\right)^2 + \left(\frac{\partial u}{R\partial\varphi} + \frac{w}{R}\right)^2 + \left(\frac{\partial v}{R\partial\varphi}\right)^2\right], \tag{1.205}$$

$$\varepsilon_{\theta,0} = \frac{1}{\sin\varphi}\frac{\partial v}{R\partial\theta} + \cot\varphi\frac{u}{R} + \frac{w}{R} + \frac{1}{2R^2\sin^2\varphi}\left[\left(\frac{\partial v}{\partial\theta} + u\cos\varphi + w\sin\varphi\right)^2\right.$$
$$\left. + \left(\frac{\partial u}{\partial\theta} - v\cos\varphi\right)^2 + \left(\frac{\partial w}{\partial\theta} - v\sin\varphi\right)^2\right], \tag{1.206}$$

$$\gamma_{\varphi\theta,0} = \frac{\partial v}{R\partial\varphi} - \cot\varphi\frac{v}{R} + \frac{1}{\sin\varphi}\frac{\partial u}{R\partial\theta} + \frac{1}{R^2\sin\varphi}\left[\left(\frac{\partial u}{\partial\varphi} + w\right)\left(\frac{\partial u}{\partial\theta} - v\cos\varphi\right)\right.$$
$$\left. + \frac{\partial v}{\partial\varphi}\left(\frac{\partial v}{\partial\theta} + u\cos\varphi + w\sin\varphi\right) + \left(\frac{\partial w}{\partial\varphi} - u\right)\left(\frac{\partial w}{\partial\theta} - v\sin\varphi\right)\right], \tag{1.207}$$

$$k_\varphi = -\frac{\partial^2 w}{R^2\,\partial \varphi^2} - \frac{w}{R^2}, \tag{1.208}$$

$$k_\theta = -\frac{\cot\varphi}{R}\frac{\partial w}{R\,\partial\varphi} - \frac{\partial^2 w}{R^2 \sin^2\varphi\,\partial\theta^2} - \frac{w}{R^2} + \frac{1-\sin\varphi}{R^2\sin^2\varphi}\frac{\partial v}{\partial\theta}, \tag{1.209}$$

$$k_{\varphi\theta} = -2\left(\frac{1}{R^2\sin\varphi}\frac{\partial^2 w}{\partial\varphi\,\partial\theta} - \frac{\cot\varphi}{R}\frac{\partial w}{R\sin\varphi\,\partial\theta}\right) - \frac{\cot\varphi(2-\sin\varphi)}{R^2\sin\varphi}v + \frac{1-\sin\varphi}{R^2\sin\varphi}\frac{\partial v}{\partial\varphi}. \tag{1.210}$$

In the equation above, nonlinear terms in changes of curvature and torsion have been neglected.

REFERENCES

M. Amabili 2003 *Journal of Sound and Vibration* **264**, 1091–1125. Comparison of shell theories for large-amplitude vibrations of circular cylindrical shells: Lagrangian approach.

M. Amabili and M. P. Païdoussis 2003 *Applied Mechanics Reviews* **56**, 349–381. Review of studies on geometrically nonlinear vibrations and dynamics of circular cylindrical shells and panels, with and without fluid-structure interaction.

B. Budiansky 1968 *Journal of Applied Mechanics* **35**, 393–401. Notes on nonlinear shell theory.

A. L. Cauchy 1828 *Exercises de Mathematique* **3**, 328–355. Sur l'équilibre et le mouvement d'une plaque solid.

H-N. Chu and G. Herrmann 1956 *Journal of Applied Mechanics* **23**, 532–540. Influence of large amplitude on free flexural vibrations of rectangular elastic plates.

S. T. Dennis and A. N. Palazotto 1990 *International Journal of Non-Linear Mechanics* **25**, 67–85. Large displacement and rotation formulation for laminated shells including parabolic transverse shear.

L. H. Donnell 1934 *Transactions of the ASME* **56**, 795–806. A new theory for the buckling of thin cylinders under axial compression and bending.

Y. C. Fung 1965 *Foundations of Solid Mechanics*. Prentice-Hall, Englewood Cliffs, NJ, USA.

J. H. Ginsberg 1973 *ASME Journal of Applied Mechanics* **40**, 471–477. Large-amplitude forced vibrations of simply supported thin cylindrical shells.

L. N. B. Gummadi and A. N. Palazotto 1999 *AIAA Journal* **37**, 1489–1494. Nonlinear dynamic finite element analysis of composite cylindrical shells considering large rotations.

T. von Kármán 1910 Festigkeitsprobleme im Maschinenbau. Encyklopadie der Mathematischen Wissenschaften. Vol. 4, Heft 4, 311–385.

T. von Kármán and H.-S. Tsien 1941 *Journal of the Aeronautical Sciences* **8**, 303–312. The buckling of thin cylindrical shells under axial compression.

G. Kirchhoff 1850 *Journal für die Reine und Angewandte Mathematik* (Crelle's) **40**, 51–88. Uber das gleichgewicht und die bewegung einer elastischen scheibe.

Y. Kobayashi and A. W. Leissa 1995 *International Journal of Non-Linear Mechanics* **30**, 57–66. Large-amplitude free vibration of thick shallow shells supported by shear diaphragms.

W. T. Koiter 1966 *Proceedings Koninklijke Nederlandse Akademie van Wetenschappen* **B 69**, 1–54. On the nonlinear theory of thin elastic shells. I, II, III.

A. Libai and J. G. Simmonds 1988 *The Nonlinear Theory of Elastic Shells*, 2nd edition 1998. Academic Press, London, UK.

L. Librescu 1987 *Quarterly of Applied Mathematics* **45**, 1–22. Refined geometrically nonlinear theories of anisotropic laminated shells.

Kh. M. Mushtari and K. Z. Galimov 1957 *Non-Linear Theory of Thin Elastic Shells*. Academy of Sciences (Nauka), Kazan'; English version, NASA-TT-F62 in 1961.

P. M. Naghdi and R. P. Nordgren 1963 *Quarterly of Applied Mathematics* **21**, 49–59. On the nonlinear theory of elastic shells under the Kirchhoff hypothesis.

V. V. Novozhilov 1953 *Foundations of the Nonlinear Theory of Elasticity*. Graylock Press, Rochester, NY, USA (now available from Dover, NY, USA).

P. F. Pai and A. H. Nayfeh 1994 *Nonlinear Dynamics* **6**, 459–500. A unified nonlinear formulation for plate and shell theories.

S. D. Poisson 1829 *Mémoires de l'Académie Royale des Sciences de l'Institut* **8**, 357–570. Mémoire sur l'eguilibre et le mouvement des corp élastique.

J. N. Reddy 1984 *Journal of Engineering Mechanics* **110**, 794–809. Exact solutions of moderately thick laminated shells.

J. N. Reddy 1990 *International Journal of Non-Linear Mechanics* **25**, 677–686. A general non-linear third-order theory of plates with moderate thickness.

J. N. Reddy 2003 *Mechanics of Laminated Composite Plates and Shells: Theory and Analysis*, 2nd edition. CRC Press, Boca Raton, FL, USA.

J. N. Reddy and K. Chandrashekhara 1985 *International Journal of Non-Linear Mechanics* **20**, 79–90. Geometrically non-linear transient analysis of laminated, doubly curved shells.

J. N. Reddy and C. F. Liu 1985 *International Journal of Engineering Science* **23**, 319–330. A higher-order shear deformation theory of laminated elastic shells.

J. L. Sanders Jr. 1963 *Quarterly of Applied Mathematics* **21**, 21–36. Nonlinear theories for thin shells.

C. Sansour, P. Wriggers and J. Sansour 1997 *Nonlinear Dynamics* **13**, 279–305. Nonlinear dynamics of shells: theory, finite element formulation, and integration schemes.

K. P. Soldatos 1992 *Journal of Pressure Vessel Technology* **114**, 105–109. Nonlinear analysis of transverse shear deformable laminated composite cylindrical shells. Part I: derivation of governing equations.

J. J. Stoker 1968 *Nonlinear Elasticity*. Gordon and Breach, New York, USA.

C. T. Tsai and A. N. Palazotto 1991 *International Journal of Non-Linear Mechanics* **26**, 379–388. On the finite element analysis of non-linear vibration for cylindrical shells with high-order shear deformation theory.

I. I. Vorovich 1999 *Nonlinear Theory of Shallow Shells*. Springer-Verlag, New York, USA.

N. Yamaki 1984 *Elastic Stability of Circular Cylindrical Shells*. North-Holland, Amsterdam.

Y.-H. Zhou and H. S. Tzou 2000 *International Journal of Solids and Structures* **37**, 1663–1677. Active control of nonlinear piezoelectric circular shallow spherical shells.

2 Nonlinear Theories of Doubly Curved Shells for Conventional and Advanced Materials

2.1 Introduction

In this chapter, more advanced problems of finite deformation (geometric nonlinearity) of shells and plates are considered. Initially, Donnell's and Novozhilov's nonlinear theories for doubly curved shells with constant curvature are presented. Then, the classical theory for thin shells of arbitrary shape is presented, which makes use of the theory of surfaces. Composite, sandwich and innovative functionally graded materials are introduced in the next section. In order to deal with these special materials and with moderately thick shells, nonlinear shear deformation theories are introduced. These theories, formulated for shells, can easily be modified to be applied to laminated, sandwich and functionally graded plates by setting the surface curvature equal to zero. Finally, the effect of thermal stresses is addressed.

2.2 Doubly Curved Shells of Constant Curvature

A doubly curved shell with rectangular base is considered, as shown in Figure 2.1. A curvilinear coordinate system (O; x, y, z) having the origin O at one edge of the panel is assumed; the curvilinear coordinates are defined as $x = \psi R_x$ and $y = \theta R_y$, where ψ and θ are the angular coordinates and R_x and R_y are principal radii of curvature (constant); a and b are the curvilinear lengths of the edges and h is the shell thickness. The smallest radius of curvature at every point of the shell is larger than the greatest lengths measured along the middle surface of the shell. The displacements of an arbitrary point of coordinates (x, y) on the middle surface of the shell are denoted by u, v and w, in the x, y and z directions, respectively; w is taken positive outward from the center of the smallest radius of curvature. Initial imperfections of the shell associated with zero initial tension are denoted by displacement w_0 in normal direction, also positive outward.

Special cases of doubly curved shells are $R_x/R_y = 1$, spherical shell; $R_x/R_y = 0$, circular cylindrical shell; and $R_x/R_y = -1$, hyperbolic paraboloid shell.

Two different strain-displacement relationships for thin shells are presented. They are based on the Kirchhoff-Love first-approximation assumptions. These theories are (i) Donnell's and (ii) Novozhilov's nonlinear shell theories. According to these two theories, the strain components ε_x, ε_y and γ_{xy} at an arbitrary point of the panel are related to the middle surface strains $\varepsilon_{x,0}$, $\varepsilon_{y,0}$, $\gamma_{xy,0}$ and to the changes in

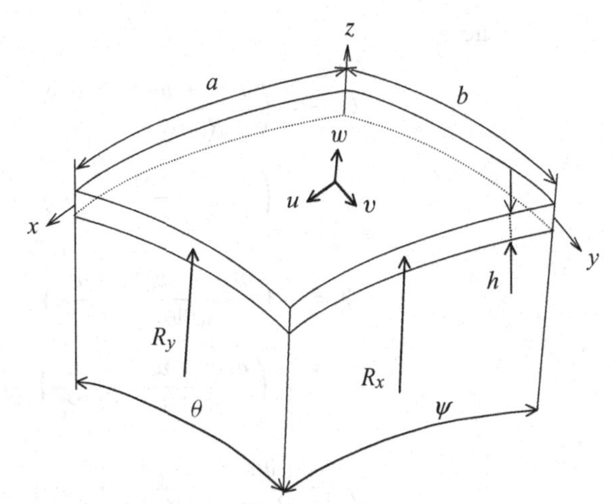

Figure 2.1. Doubly curved shell with rectangular base.

the curvature and torsion of the middle surface k_x, k_y and k_{xy} by the following three relationships

$$\varepsilon_x = \varepsilon_{x,0} + zk_x, \quad \varepsilon_y = \varepsilon_{y,0} + zk_y, \quad \gamma_{xy} = \gamma_{xy,0} + zk_{xy}, \tag{2.1}$$

where z is the distance of the arbitrary point of the panel from the middle surface. The middle surface strain-displacement relationships and changes in the curvature and torsion have different expressions for Donnell's and Novozhilov's theories.

According to Donnell's nonlinear shell theory, the middle surface strain-displacement relationships and changes in the curvature and torsion for a doubly curved shallow-shell are

$$\varepsilon_{x,0} = \frac{\partial u}{R_x \, \partial \psi} + \frac{w}{R_x} + \frac{1}{2}\left(\frac{\partial w}{R_x \, \partial \psi}\right)^2 + \frac{\partial w}{R_x \, \partial \psi}\frac{\partial w_0}{R_x \, \partial \psi}, \tag{2.2a}$$

$$\varepsilon_{y,0} = \frac{\partial v}{R_y \, \partial \theta} + \frac{w}{R_y} + \frac{1}{2}\left(\frac{\partial w}{R_y \, \partial \theta}\right)^2 + \frac{\partial w}{R_y \, \partial \theta}\frac{\partial w_0}{R_y \, \partial \theta}, \tag{2.2b}$$

$$\gamma_{xy,0} = \frac{\partial u}{R_y \, \partial \theta} + \frac{\partial v}{R_x \, \partial \psi} + \frac{\partial w}{R_x \, \partial \psi}\frac{\partial w}{R_y \, \partial \theta} + \frac{\partial w}{R_x \, \partial \psi}\frac{\partial w_0}{R_y \, \partial \theta} + \frac{\partial w_0}{R_x \, \partial \psi}\frac{\partial w}{R_y \, \partial \theta}, \tag{2.2c}$$

$$k_x = -\frac{\partial^2 w}{R_x^2 \, \partial \psi^2}, \tag{2.2d}$$

$$k_y = -\frac{\partial^2 w}{R_y^2 \, \partial \theta^2}, \tag{2.2e}$$

$$k_{xy} = -2\frac{\partial^2 w}{R_x R_y \, \partial \psi \, \partial \theta}. \tag{2.2f}$$

According to Novozhilov's nonlinear theory, already presented in Section 1.3.7 for circular cylindrical shells, displacements u_1, u_2 and u_3 of points at distance z from the middle surface are introduced; they are related to displacements on the middle surface by the relationships (Novozhilov 1953):

$$u_1 = u + z\Theta, \tag{2.3a}$$

$$u_2 = v + z\Psi, \tag{2.3b}$$

$$u_3 = w + w_0 + z\chi, \tag{2.3c}$$

where

$$\Theta = -\left(\frac{\partial(w+w_0)}{R_x\,\partial\psi} - \frac{u}{R_x}\right)\left(1 + \frac{\partial v}{R_y\,\partial\theta} + \frac{w}{R_y}\right)$$
$$+\left(\frac{\partial(w+w_0)}{R_y\,\partial\theta} - \frac{v}{R_y}\right)\frac{\partial u}{R_y\,\partial\theta} - \left(\frac{\partial w}{R_x\,\partial\psi} - \frac{u}{R_x}\right)\frac{w_0}{R_y}, \tag{2.3d}$$

$$\Psi = -\left(\frac{\partial(w+w_0)}{R_y\,\partial\theta} - \frac{v}{R_y}\right)\left(1 + \frac{\partial u}{R_x\,\partial\psi} + \frac{w}{R_x}\right)$$
$$+\left(\frac{\partial(w+w_0)}{R_x\,\partial\psi} - \frac{u}{R_x}\right)\frac{\partial v}{R_x\,\partial\psi} - \left(\frac{\partial w}{R_y\,\partial\theta} - \frac{v}{R_y}\right)\frac{w_0}{R_x}, \tag{2.3e}$$

$$\chi \cong \frac{\partial u}{R_x\,\partial\psi} + \frac{\partial v}{R_y\,\partial\theta} + (w+w_0)\left(\frac{1}{R_x} + \frac{1}{R_y}\right)$$
$$+\left(\frac{\partial u}{R_x\,\partial\psi} + \frac{w+w_0}{R_x}\right)\left(\frac{\partial v}{R_y\,\partial\theta} + \frac{w}{R_y}\right) \tag{2.3f}$$
$$+\left(\frac{\partial u}{R_x\,\partial\psi} + \frac{w}{R_x}\right)\frac{w_0}{R_y} - \frac{\partial u}{R_y\,\partial\theta}\frac{\partial v}{R_x\,\partial\psi}.$$

The strain-displacement relationships for a generic point of the shell are (Novozhilov 1953)

$$\varepsilon_x = \frac{1}{R_x+z}\left(\frac{\partial u_1}{\partial\psi} + u_3\right) + \frac{1}{2\,(R_x+z)^2}\left[\left(\frac{\partial u_1}{\partial\psi} + u_3\right)^2 + \left(\frac{\partial u_2}{\partial\psi}\right)^2 + \left(\frac{\partial u_3}{\partial\psi} - u_1\right)^2\right], \tag{2.4a}$$

$$\varepsilon_y = \frac{1}{R_y+z}\left(\frac{\partial u_2}{\partial\theta} + u_3\right) + \frac{1}{2\,(R_y+z)^2}\left[\left(\frac{\partial u_1}{\partial\theta}\right)^2 + \left(\frac{\partial u_2}{\partial\theta} + u_3\right)^2 + \left(\frac{\partial u_3}{\partial\theta} - u_2\right)^2\right], \tag{2.4b}$$

$$\gamma_{xy} = \frac{1}{R_x+z}\frac{\partial u_2}{\partial\psi} + \frac{1}{R_y+z}\frac{\partial u_1}{\partial\theta} + \frac{1}{R_x+z}\frac{1}{R_y+z}$$
$$\times\left[\frac{\partial u_1}{\partial\theta}\left(\frac{\partial u_1}{\partial\psi} + u_3\right) + \frac{\partial u_2}{\partial\psi}\left(\frac{\partial u_2}{\partial\theta} + u_3\right) + \left(\frac{\partial u_3}{\partial\psi} - u_1\right)\left(\frac{\partial u_3}{\partial\theta} - u_2\right)\right]. \tag{2.4c}$$

The following thinness approximations are now introduced:

$$1/(R+z) = (1/R)\{1 - z/R + O(z/R)^2\}, \tag{2.5a}$$
$$1/(R+z)^2 = (1/R^2)\{1 - 2z/R + O(z/R)^2\}, \tag{2.5b}$$
$$1/[(1+z/R_1)(1+z/R_2)] = \{1 - z/R_1 - z/R_2 + O(z/R)^2\}, \tag{2.5c}$$

where $O\,(z/R)^2$ is a small quantity of order $(z/R)^2$, which is neglected. By introducing equations (2.3) into equations (2.4) and using approximations (2.5a–c), the middle surface strain-displacement relationships, changes in the curvature and torsion are obtained for the Novozhilov theory of shells, as follows:

$$\varepsilon_{x,0} = \frac{\partial u}{R_x\,\partial\psi} + \frac{w}{R_x} + \frac{1}{2R_x^2}\left[\left(\frac{\partial u}{\partial\psi} + w\right)^2 + \left(\frac{\partial v}{\partial\psi}\right)^2 + \left(\frac{\partial w}{\partial\psi} - u\right)^2\right] + \frac{1}{R_x^2}\left[\frac{\partial w_0}{\partial\psi}\left(\frac{\partial w}{\partial\psi} - u\right) + w_0\left(\frac{\partial u}{\partial\psi} + w\right)\right], \tag{2.6a}$$

$$\varepsilon_{y,0} = \frac{\partial v}{R_y\,\partial\theta} + \frac{w}{R_y} + \frac{1}{2R_y^2}\left[\left(\frac{\partial u}{\partial\theta}\right)^2 + \left(\frac{\partial v}{\partial\theta} + w\right)^2 + \left(\frac{\partial w}{\partial\theta} - v\right)^2\right] + \frac{1}{R_y^2}\left[\frac{\partial w_0}{\partial\theta}\left(\frac{\partial w}{\partial\theta} - v\right) + w_0\left(w + \frac{\partial v}{\partial\theta}\right)\right], \tag{2.6b}$$

$$\gamma_{xy,0} = \frac{\partial u}{R_y\,\partial\theta} + \frac{\partial v}{R_x\,\partial\psi} + \frac{1}{R_x R_y}\left[\frac{\partial u}{\partial\theta}\left(\frac{\partial u}{\partial\psi} + w + w_0\right) + \frac{\partial v}{\partial\psi}\left(\frac{\partial v}{\partial\theta} + w + w_0\right) + \left(\frac{\partial w}{\partial\psi} - u\right)\left(\frac{\partial w}{\partial\theta} - v\right) + \frac{\partial w_0}{\partial\psi}\left(\frac{\partial w}{\partial\theta} - v\right) + \frac{\partial w_0}{\partial\theta}\left(\frac{\partial w}{\partial\psi} - u\right)\right], \tag{2.6c}$$

$$k_x = -\frac{\partial^2 w}{R_x^2\,\partial\psi^2} + \frac{w}{R_x R_y} + \frac{\partial v}{R_x R_y\,\partial\theta} + \frac{\partial u}{R_x^2\,\partial\psi}, \tag{2.6d}$$

$$k_y = -\frac{\partial^2 w}{R_y^2\,\partial\theta^2} + \frac{w}{R_x R_y} + \frac{\partial u}{R_x R_y\,\partial\psi} + \frac{\partial v}{R_y^2\,\partial\theta}, \tag{2.6e}$$

$$k_{xy} = -2\frac{\partial^2 w}{R_x R_y\,\partial\psi\,\partial\theta} + \frac{\partial v}{R_x\,\partial\psi}\left(-\frac{1}{R_x} + \frac{1}{R_y}\right) + \frac{\partial u}{R_y\,\partial\theta}\left(\frac{1}{R_x} - \frac{1}{R_y}\right). \tag{2.6f}$$

Equations (2.6a–f) are an improved version of the Novozhilov nonlinear shell theory obtained by Amabili (2005); in fact, approximations (2.5) have been used instead of neglecting terms in $(z/R)^2$ in the derivation, as originally done by Novozhilov (1953). Nonlinearities in changes of curvature and torsion have been neglected in this case for simplicity. It is interesting to observe that the derived equations are valid both for regular and shallow doubly curved shells (shallow-shells are curved panels with small rise, compared to the smallest radius of curvature).

2.2.1 Elastic Strain Energy

The elastic strain energy U_S of the shell, neglecting σ_z, τ_{zx}, τ_{zy}, as stated by the Kirchhoff-Love first-approximation assumptions, is given by

$$U_S = \frac{1}{2}\int_0^a\int_0^b\int_{-h/2}^{h/2} (\sigma_x\varepsilon_x + \sigma_y\varepsilon_y + \tau_{xy}\gamma_{xy})(1 + z/R_x)(1 + z/R_y)\,\mathrm{d}x\,\mathrm{d}y\,\mathrm{d}z, \tag{2.7}$$

where $x = R_x\psi$ and $y = R_y\theta$; the stresses σ_x, σ_y and τ_{xy} are related to the strains for homogeneous and isotropic material ($\sigma_z = 0$, case of plane stress) by

$$\sigma_x = \frac{E}{1-\nu^2}(\varepsilon_x + \nu\varepsilon_y), \quad \sigma_y = \frac{E}{1-\nu^2}(\varepsilon_y + \nu\varepsilon_x), \quad \tau_{xy} = \frac{E}{2(1+\nu)}\gamma_{xy}, \tag{2.8}$$

where E is Young's modulus and ν is Poisson's ratio. By using equations (2.1, 2.7, 2.8), the following expression is obtained for shells of uniform thickness:

$$U_S = \frac{1}{2}\frac{Eh}{1-\nu^2}\int_0^a\int_0^b\left(\varepsilon_{x,0}^2+\varepsilon_{y,0}^2+2\nu\varepsilon_{x,0}\varepsilon_{y,0}+\frac{1-\nu}{2}\gamma_{xy,0}^2\right)\mathrm{d}x\,\mathrm{d}y+\frac{1}{2}\frac{Eh^3}{12(1-\nu^2)}$$
$$\times\int_0^a\int_0^b\left(k_x^2+k_y^2+2\nu k_x k_y+\frac{1-\nu}{2}k_{xy}^2\right)\mathrm{d}x\,\mathrm{d}y+\frac{1}{2}\left(\frac{1}{R_x}+\frac{1}{R_y}\right)\frac{Eh^3}{6(1-\nu^2)}$$
$$\times\int_0^a\int_0^b\left(\varepsilon_{x,0}\,k_x+\varepsilon_{y,0}\,k_y+\nu\varepsilon_{x,0}\,k_y+\nu\varepsilon_{y,0}\,k_x+\frac{1-\nu}{2}\gamma_{xy,0}\,k_{xy}\right)\mathrm{d}x\,\mathrm{d}y+O(h^4),\tag{2.9}$$

where $O(h^4)$ is a higher-order term in h, and the last term in h^3 disappears if z/R is neglected with respect to unity in equation (2.7), as must be done for Donnell's theory. If this term is neglected, the right-hand side of equation (2.9) can be easily interpreted: the first term is the membrane (also referred to as stretching) energy and the second one is the bending energy. If the last term is retained, membrane and bending energies are coupled.

2.3 General Theory of Doubly Curved Shells

In this section, the classical results of the theory of surfaces are summarized initially, following Novozhilov (1964). Then, the strain-displacement relationships in local curvilinear coordinates are obtained for thin shells of arbitrary curvature.

2.3.1 Theory of Surfaces

Every smooth surface may be determined in a rectangular Cartesian coordinate system by three equations:

$$x = f_1(\alpha_1,\alpha_2),\quad y = f_2(\alpha_1,\alpha_2),\quad z = f_3(\alpha_1,\alpha_2),\tag{2.10a–c}$$

where f_i are definite, continuous, single-valued functions of the two parameters α_1, α_2. The values of these parameters are assumed to be restricted, in order to have a one-to-one correspondence between the surface points and α_1, α_2. These two parameters are the curvilinear coordinates of the surface.

The coordinate lines α_1 are obtained by setting a constant value for α_2 and varying α_1; similarly, the coordinate lines α_2 are obtained by setting a constant value for α_1 and varying α_2. Hence, every point of the surface may be regarded as the intersection of the two coordinate lines α_1 and α_2 (see Figure 2.2).

For each surface, there exist an infinite number of possible choices for the functions f_1, f_2, f_3, and, depending on their choice, the form of the coordinate lines on the surface will change. However, it is convenient to choose the functions that give the simplest form of the curvilinear coordinates. This simplest choice is the one that gives the two families of coordinate lines, which are simultaneously lines of principal curvature of the surface. Such a system is orthogonal, because the directions of

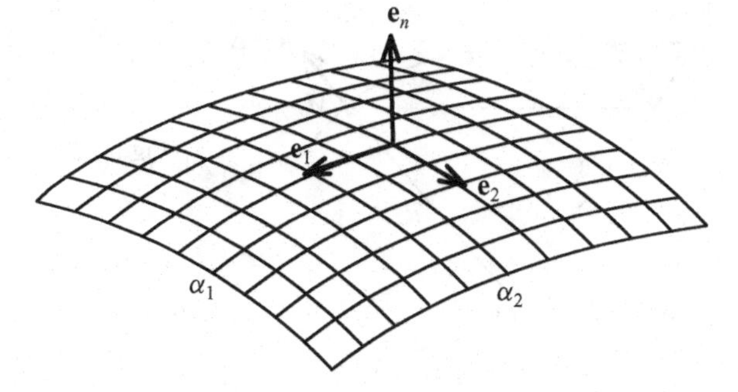

Figure 2.2. Coordinate lines on a surface.

principal curvature on a surface are mutually perpendicular. Therefore, this system of curvilinear coordinates will be always used.

The three scalar equations (2.10a–c) can be rewritten in vectorial form

$$\mathbf{r} = \mathbf{r}(\alpha_1, \alpha_2), \tag{2.11}$$

which after projection on the rectangular Cartesian coordinate system gives equations (2.10a–c).

The derivatives of $\mathbf{r}$ with respect to α_1 and α_2 are indicated as

$$\mathbf{r}_{\alpha_1} = \frac{\partial \mathbf{r}}{\partial \alpha_1}, \quad \mathbf{r}_{\alpha_2} = \frac{\partial \mathbf{r}}{\partial \alpha_2}, \tag{2.12}$$

where the vectors $\mathbf{r}_{\alpha_1}$ and $\mathbf{r}_{\alpha_2}$ are at each point of the surface tangent to the α_1 and α_2 coordinate lines, respectively. Therefore, an increase of one of the curvilinear coordinates corresponds to a shift of the vector $\mathbf{r}$ on the surface in the direction corresponding to this coordinate line. The vector $\mathbf{ds}$ joining the surface points of coordinates (α_1, α_2) and $(\alpha_1 + \mathrm{d}\alpha_1, \alpha_2 + \mathrm{d}\alpha_2)$ is given by (see Figure 2.3)

$$\mathbf{ds} = \frac{\partial \mathbf{r}}{\partial \alpha_1}\mathrm{d}\alpha_1 + \frac{\partial \mathbf{r}}{\partial \alpha_2}\mathrm{d}\alpha_2. \tag{2.13}$$

The square of the length of this vector is given by

$$|\mathbf{ds}|^2 = |\mathbf{r}_{\alpha_1}|^2 \mathrm{d}\alpha_1^2 + 2\,\mathbf{r}_{\alpha_1} \cdot \mathbf{r}_{\alpha_2}\,\mathrm{d}\alpha_1\,\mathrm{d}\alpha_2 + |\mathbf{r}_{\alpha_2}|^2 \mathrm{d}\alpha_2^2, \tag{2.14}$$

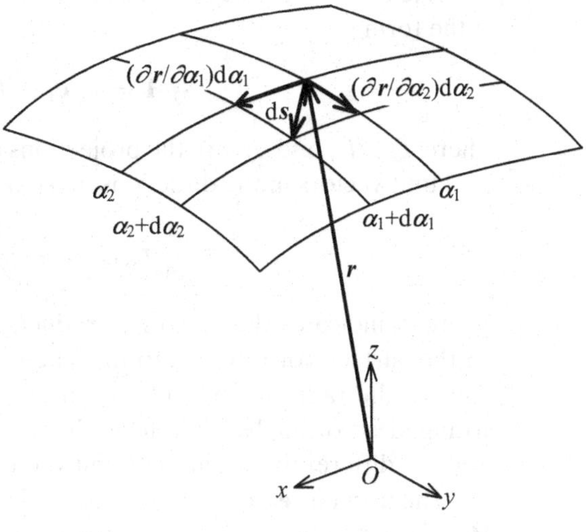

Figure 2.3. Vector $\mathbf{ds}$ joining the surface points of coordinates (α_1, α_2) and $(\alpha_1 + \mathrm{d}\alpha_1, \alpha_2 + \mathrm{d}\alpha_2)$.

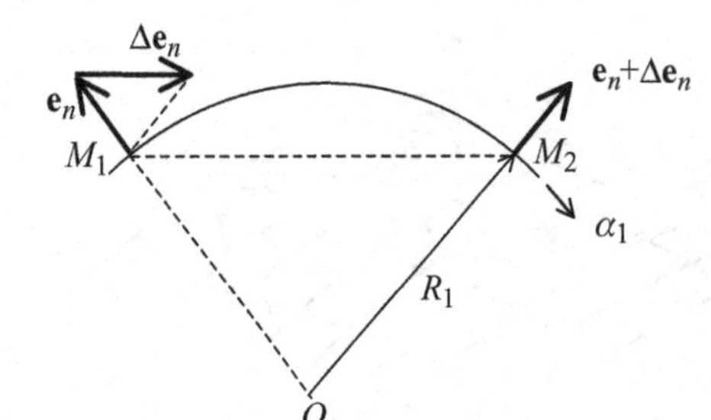

Figure 2.4. Variation of the vector $\mathbf{e}_n$ with α_1.

where the dot indicates the scalar product. By using the introduced orthogonality of the curvilinear coordinates, $\mathbf{r}_{\alpha_1} \cdot \mathbf{r}_{\alpha_2} = 0$. Introducing the following notation:

$$|\mathbf{r}_{\alpha_1}|^2 = A_1^2 = \left(\frac{\partial x}{\partial \alpha_1}\right)^2 + \left(\frac{\partial y}{\partial \alpha_1}\right)^2 + \left(\frac{\partial z}{\partial \alpha_1}\right)^2, \tag{2.15a}$$

$$|\mathbf{r}_{\alpha_2}|^2 = A_2^2 = \left(\frac{\partial x}{\partial \alpha_2}\right)^2 + \left(\frac{\partial y}{\partial \alpha_2}\right)^2 + \left(\frac{\partial z}{\partial \alpha_2}\right)^2, \tag{2.15b}$$

equation (2.14) can be rewritten as

$$\mathrm{d}s^2 = A_1^2\,\mathrm{d}\alpha_1^2 + A_2^2\,\mathrm{d}\alpha_2^2. \tag{2.16}$$

In particular, when only one coordinate at the time is varied, it yields

$$\mathrm{d}s_1 = A_1\,\mathrm{d}\alpha_1, \quad \mathrm{d}s_2 = A_2\,\mathrm{d}\alpha_2. \tag{2.17a,b}$$

The quantities A_1 and A_2 are called the Lamé parameters and are directly related to the more general Euclidean metric tensor introduced in Section 1.3.1. In fact, $A_1 = \sqrt{g_{11}}$ and $A_2 = \sqrt{g_{22}}$ in the case of a two-dimensional orthogonal coordinate system.

It is convenient to describe all vectors that are functions of surface points in terms of their components along the tangents to the coordinate lines α_1, α_2, and on the normal to the surface at the point considered. Because these reference axes change their direction for a passage from one point of the surface to another, it is necessary to investigate the differentiation of vectors with respect to α_1 and α_2.

Each vector $\mathbf{T}$, referred to a given orthogonal coordinate system, can be written in the form

$$\mathbf{T} = T_{\alpha_1}\mathbf{e}_1 + T_{\alpha_2}\mathbf{e}_2 + T_{\alpha_n}\mathbf{e}_n, \tag{2.18}$$

where T_{α_1}, T_{α_2} and T_{α_n} are the projections and $\mathbf{e}_1$, $\mathbf{e}_2$ and $\mathbf{e}_n$ are right-handed orthogonal unit vectors in the directions of α_1, α_2 and orthogonal to the surface

$$\mathbf{e}_1 = \frac{1}{A_1}\mathbf{r}_{\alpha_1}, \quad \mathbf{e}_2 = \frac{1}{A_2}\mathbf{r}_{\alpha_2}, \quad \mathbf{e}_n = \mathbf{e}_1 \wedge \mathbf{e}_2, \tag{2.19a–c}$$

where $\wedge$ indicates the vectorial product. It is assumed that $\mathbf{e}_n$ is directed outward on the surface (or directed to the side of the centers of negative curvature, if the signs of the radii of curvature are different). The positive directions of $\mathbf{e}_1$, $\mathbf{e}_2$ are arranged accordingly. Therefore, the study of differentiation of $\mathbf{T}$ reduces to the study of differentiation rules of unit vectors (2.19).

The normals $\mathbf{e}_n$ are constructed in Figure 2.4 for two neighboring points M_1 and M_2 on the coordinate line α_1. These normals intersect at point O; the corresponding principal radius of curvature of the surface is denoted by R_1. Figure 2.4 shows that in the passage from point M_1 to the infinitely close point M_2, the increase of the unit vector $\mathbf{e}_n$ is in good approximation parallel to the chord M_1M_2. At the limit it is parallel to vector $\mathbf{e}_1$ (the normals remain coplanar to small terms of second order and rotate about the corresponding center of curvature, as a result of the property of the lines of principal curvature known as the Rodrigues Theorem). By using similar triangles, the magnitude of the increase of the unit vector $\mathbf{e}_n$ is

$$\frac{|\Delta \mathbf{e}_n|}{|\mathbf{e}_n|} = \frac{\overline{M_1 M_2}}{R_1}. \tag{2.20}$$

By observing that $|\mathbf{e}_n| = 1$ and $\overline{M_1 M_2} = A_1\, \mathrm{d}\alpha_1$, equation (2.20) leads to the first of the differentiation rules:

$$\frac{\partial \mathbf{e}_n}{\partial \alpha_1} = \frac{A_1}{R_1}\mathbf{e}_1. \tag{2.21}$$

Similarly,

$$\frac{\partial \mathbf{e}_n}{\partial \alpha_2} = \frac{A_2}{R_2}\mathbf{e}_2 \tag{2.22}$$

is obtained, where R_2 is the second principal radius of curvature at the point considered.

Next, the relationship

$$\frac{\partial \mathbf{r}_{\alpha_1}}{\partial \alpha_2} = \frac{\partial \mathbf{r}_{\alpha_2}}{\partial \alpha_1} \tag{2.23}$$

is introduced, which is equivalent to the identity

$$\frac{\partial^2 \mathbf{r}}{\partial \alpha_2 \partial \alpha_1} = \frac{\partial^2 \mathbf{r}}{\partial \alpha_1 \partial \alpha_2}. \tag{2.24}$$

The following identity is obtained by introducing (2.19a,b) into equation (2.23):

$$\frac{\partial A_1 \mathbf{e}_1}{\partial \alpha_2} = \frac{\partial A_2 \mathbf{e}_2}{\partial \alpha_1}. \tag{2.25}$$

The derivative $\partial \mathbf{e}_1/\partial \alpha_1$ is studied by considering its components along $\mathbf{e}_1$, $\mathbf{e}_2$ and $\mathbf{e}_n$. In the direction $\mathbf{e}_1$, this derivative has zero projection, because the unit vector has constant length and its change is perpendicular to $\mathbf{e}_1$. The projection of $\partial \mathbf{e}_1/\partial \alpha_1$ on $\mathbf{e}_2$ is given by

$$\mathbf{e}_2 \cdot \frac{\partial \mathbf{e}_1}{\partial \alpha_1} = \frac{\partial (\mathbf{e}_2 \cdot \mathbf{e}_1)}{\partial \alpha_1} - \mathbf{e}_1 \cdot \frac{\partial \mathbf{e}_2}{\partial \alpha_1} = -\mathbf{e}_1 \cdot \frac{\partial \mathbf{e}_2}{\partial \alpha_1}, \tag{2.26}$$

because $(\mathbf{e}_2 \cdot \mathbf{e}_1) = 0$ due to the orthogonality of the coordinate lines α_1 and α_2. By using equation (2.25), one obtains

$$\frac{\partial \mathbf{e}_2}{\partial \alpha_1} = \frac{1}{A_2}\frac{\partial A_1 \mathbf{e}_1}{\partial \alpha_2} - \frac{1}{A_2}\frac{\partial A_2}{\partial \alpha_1}\mathbf{e}_2; \tag{2.27}$$

substituting (2.27) into the right-hand side of equation (2.26), the following expression is found:

$$\mathbf{e}_2 \cdot \frac{\partial \mathbf{e}_1}{\partial \alpha_1} = -\frac{1}{A_2}\mathbf{e}_1 \cdot \frac{\partial A_1 \mathbf{e}_1}{\partial \alpha_2} + \frac{1}{A_2}\frac{\partial A_2}{\partial \alpha_1}\mathbf{e}_1 \cdot \mathbf{e}_2 = -\frac{A_1}{A_2}\mathbf{e}_1 \cdot \frac{\partial \mathbf{e}_1}{\partial \alpha_2} - \frac{1}{A_2}\frac{\partial A_1}{\partial \alpha_2}\mathbf{e}_1 \cdot \mathbf{e}_1. \tag{2.28}$$

Obviously $\mathbf{e}_1 \cdot \mathbf{e}_1 = 1$ and $\mathbf{e}_1 \cdot \partial \mathbf{e}_1/\partial \alpha_2 = 0$, as discussed above, and it yields

$$\mathbf{e}_2 \cdot \frac{\partial \mathbf{e}_1}{\partial \alpha_1} = -\frac{1}{A_2}\frac{\partial A_1}{\partial \alpha_2}. \tag{2.29}$$

The remaining projection of $\partial \mathbf{e}_1/\partial \alpha_1$ on $\mathbf{e}_n$ is obtained as follows:

$$\mathbf{e}_n \cdot \frac{\partial \mathbf{e}_1}{\partial \alpha_1} = \frac{\partial(\mathbf{e}_n \cdot \mathbf{e}_1)}{\partial \alpha_1} - \mathbf{e}_1 \cdot \frac{\partial \mathbf{e}_n}{\partial \alpha_1} = -\mathbf{e}_1 \cdot \frac{\partial \mathbf{e}_n}{\partial \alpha_1}, \tag{2.30}$$

because $(\mathbf{e}_n \cdot \mathbf{e}_1) = 0$. By using (2.21) the final expression,

$$\mathbf{e}_1 \cdot \frac{\partial \mathbf{e}_n}{\partial \alpha_1} = \frac{A_1}{R_1}, \tag{2.31}$$

is found, which gives the projection of $\partial \mathbf{e}_n/\partial \alpha_1$ on $\mathbf{e}_1$. As previously discussed (Theorem of Rodrigues), $\partial \mathbf{e}_n/\partial \alpha_1$ has zero projections on both $\mathbf{e}_2$ and $\mathbf{e}_n$.

Similarly, it is possible to determinate the other derivatives of the vector $\mathbf{e}_1$. Starting with $\partial \mathbf{e}_1/\partial \alpha_2$, it is observed that its projection on $\mathbf{e}_1$ will be equal to zero. By using (2.25), one obtains

$$\mathbf{e}_2 \cdot \frac{\partial \mathbf{e}_1}{\partial \alpha_2} = \frac{1}{A_1}\frac{\partial A_1}{\partial \alpha_2}\mathbf{e}_1 \cdot \mathbf{e}_2 + \frac{1}{A_1}\mathbf{e}_2 \cdot \frac{\partial A_2 \mathbf{e}_2}{\partial \alpha_1} = \frac{1}{A_1}\frac{\partial A_2}{\partial \alpha_1}. \tag{2.32}$$

The remaining projection of $\partial \mathbf{e}_1/\partial \alpha_2$ on $\mathbf{e}_n$ is calculated as follows:

$$\mathbf{e}_n \cdot \frac{\partial \mathbf{e}_1}{\partial \alpha_2} = \frac{\partial(\mathbf{e}_n \cdot \mathbf{e}_1)}{\partial \alpha_2} - \mathbf{e}_1 \cdot \frac{\partial \mathbf{e}_n}{\partial \alpha_2} = -\mathbf{e}_1 \cdot \frac{\partial \mathbf{e}_n}{\partial \alpha_2} = -\mathbf{e}_1 \cdot \mathbf{e}_2 \frac{A_2}{R_2} = 0, \tag{2.33}$$

where (2.22) has been used. Equation (2.33) clarifies that $\partial \mathbf{e}_1/\partial \alpha_2$ has zero component on $\mathbf{e}_n$.

Similarly, other projections of derivatives $\partial \mathbf{e}_2/\partial \alpha_1$, $\partial \mathbf{e}_2/\partial \alpha_2$ and $\partial \mathbf{e}_n/\partial \alpha_2$ can be obtained. The final rules of differentiation are summarized as follows:

$$\frac{\partial \mathbf{e}_1}{\partial \alpha_1} = -\frac{1}{A_2}\frac{\partial A_1}{\partial \alpha_2}\mathbf{e}_2 - \frac{A_1}{R_1}\mathbf{e}_n, \tag{2.34a}$$

$$\frac{\partial \mathbf{e}_1}{\partial \alpha_2} = \frac{1}{A_1}\frac{\partial A_2}{\partial \alpha_1}\mathbf{e}_2, \tag{2.34b}$$

$$\frac{\partial \mathbf{e}_2}{\partial \alpha_1} = \frac{1}{A_2}\frac{\partial A_1}{\partial \alpha_2}\mathbf{e}_1, \tag{2.34c}$$

$$\frac{\partial \mathbf{e}_2}{\partial \alpha_2} = -\frac{1}{A_1}\frac{\partial A_2}{\partial \alpha_1}\mathbf{e}_1 - \frac{A_2}{R_2}\mathbf{e}_n, \tag{2.34d}$$

$$\frac{\partial \mathbf{e}_n}{\partial \alpha_1} = \frac{A_1}{R_1}\mathbf{e}_1, \tag{2.34e}$$

$$\frac{\partial \mathbf{e}_n}{\partial \alpha_2} = \frac{A_2}{R_2}\mathbf{e}_2. \tag{2.34f}$$

These formulae permit differentiation of any vector with given projections on the local axes $\mathbf{e}_1$, $\mathbf{e}_2$, $\mathbf{e}_n$ of a curvilinear coordinate system. From these equations, it is possible to formulate fundamental relations between the Lamé parameters and the principal radii of curvature. From the identity

$$\frac{\partial^2 \mathbf{e}_n}{\partial \alpha_1 \partial \alpha_2} = \frac{\partial^2 \mathbf{e}_n}{\partial \alpha_2 \partial \alpha_1}, \tag{2.35}$$

it follows that

$$\frac{\partial}{\partial \alpha_1}\left(\frac{A_2}{R_2}\mathbf{e}_2\right) = \frac{\partial}{\partial \alpha_2}\left(\frac{A_1}{R_1}\mathbf{e}_1\right), \tag{2.36}$$

or

$$\frac{\partial}{\partial \alpha_1}\left(\frac{A_2}{R_2}\right)\mathbf{e}_2 + \frac{A_2}{R_2}\frac{\partial \mathbf{e}_2}{\partial \alpha_1} = \frac{\partial}{\partial \alpha_2}\left(\frac{A_1}{R_1}\right)\mathbf{e}_1 + \frac{A_1}{R_1}\frac{\partial \mathbf{e}_1}{\partial \alpha_2}. \tag{2.37}$$

Replacing $\partial \mathbf{e}_1/\partial \alpha_2$ and $\partial \mathbf{e}_2/\partial \alpha_1$ by using equations (2.34), it yields

$$\left[\frac{\partial}{\partial \alpha_2}\left(\frac{A_1}{R_1}\right) - \frac{1}{R_2}\frac{\partial A_1}{\partial \alpha_2}\right]\mathbf{e}_1 - \left[\frac{\partial}{\partial \alpha_1}\left(\frac{A_2}{R_2}\right) - \frac{1}{R_1}\frac{\partial A_2}{\partial \alpha_1}\right]\mathbf{e}_2 = 0. \tag{2.38}$$

Because the vector equation (2.38) must be satisfied identically, the two following relations, known as the conditions of Codazzi, are obtained:

$$\frac{\partial}{\partial \alpha_1}\left(\frac{A_2}{R_2}\right) = \frac{1}{R_1}\frac{\partial A_2}{\partial \alpha_1}, \tag{2.39a}$$

$$\frac{\partial}{\partial \alpha_2}\left(\frac{A_1}{R_1}\right) = \frac{1}{R_2}\frac{\partial A_1}{\partial \alpha_2}. \tag{2.39b}$$

Next, from the identity

$$\frac{\partial^2 \mathbf{e}_1}{\partial \alpha_1 \partial \alpha_2} = \frac{\partial^2 \mathbf{e}_1}{\partial \alpha_2 \partial \alpha_1}, \tag{2.40}$$

and by using (2.34), it yields

$$\left[\frac{\partial}{\partial \alpha_1}\left(\frac{1}{A_1}\frac{\partial A_2}{\partial \alpha_1}\right) + \frac{\partial}{\partial \alpha_2}\left(\frac{1}{A_2}\frac{\partial A_1}{\partial \alpha_2}\right) + \frac{A_1}{R_1}\frac{A_2}{R_2}\right]\mathbf{e}_2 + \left[\frac{\partial}{\partial \alpha_2}\left(\frac{A_1}{R_1}\right) - \frac{1}{R_2}\frac{\partial A_1}{\partial \alpha_2}\right]\mathbf{e}_n = 0. \tag{2.41}$$

In order to satisfy identically this equation, each of the expressions in square brackets must vanish separately. This gives two additional differential equations, only one of which is new; the other coincides with one of equations (2.39). From the first term in square brackets in equation (2.41), the following Gauss condition is obtained:

$$\frac{\partial}{\partial \alpha_1}\left(\frac{1}{A_1}\frac{\partial A_2}{\partial \alpha_1}\right) + \frac{\partial}{\partial \alpha_2}\left(\frac{1}{A_2}\frac{\partial A_1}{\partial \alpha_2}\right) = -\frac{A_1}{R_1}\frac{A_2}{R_2}. \tag{2.42}$$

Equations (2.39) and (2.42) give three differential equations linking the Lamé parameters and the principal radii of curvature. Therefore, the four functions A_1, A_2, R_1 and R_2 of the curvilinear coordinates α_1, α_2 are not arbitrary but must satisfy the Codazzi and Gauss conditions in order to define a surface.

Finally, the Gaussian curvature K of a surface is introduced as

$$K = \frac{1}{R_1 R_2}. \tag{2.43}$$

Points of the surface can be classified into elliptic ($K > 0$), parabolic ($K = 0$) and hyperbolic ($K < 0$). If all points of the surface are elliptic, the surface has positive Gaussian curvature. Similarly, if all points are parabolic, the surface has zero Gaussian curvature; if all points are hyperbolic, it has negative Gaussian curvature.

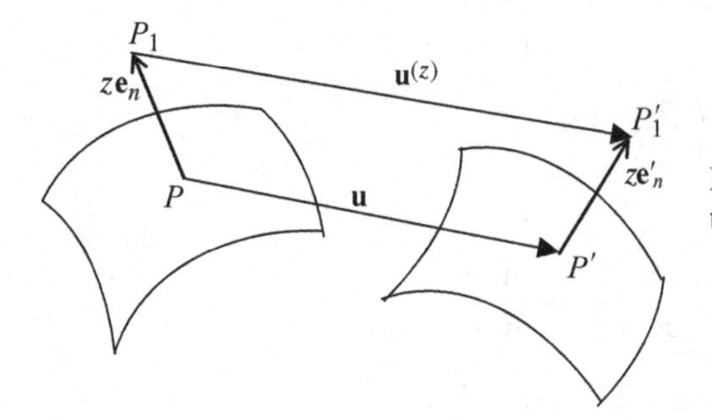

Figure 2.5. Shell in the original and displaced positions.

2.3.2 Green's Strain Tensor for a Shell in Curvilinear Coordinates

A point P on the middle surface of the shell is identified by the curvilinear coordinates α_1 and α_2, which are assumed to be lines of principal curvature. After deformation of the surface, the point P will undergo a displacement $\mathbf{u}$; this vector $\mathbf{u}$ is a function of α_1 and α_2. As usual, the projections of this vector on the directions of the local axes $\mathbf{e}_1$, $\mathbf{e}_2$, $\mathbf{e}_n$ are denoted by u, v, w.

The point P_1 is a point of the shell at distance z along the normal $\mathbf{e}_n$ from the point P, which is on the middle surface. P_1 will undergo a displacement $\mathbf{u}^{(z)}$, which is a function of α_1, α_2 and z. As usual, the projections of this vector on the directions of the local axes $\mathbf{e}_1$, $\mathbf{e}_2$, $\mathbf{e}_n$ are denoted by u_1, u_2 and u_3. Observing Figure 2.5, it is clear that the following relationship can be written between the middle surface displacement $\mathbf{u}$ and the displacement $\mathbf{u}^{(z)}$ of a point at distance z from the middle surface:

$$z\mathbf{e}_n + \mathbf{u}^{(z)} = z\mathbf{e}'_n + \mathbf{u}, \tag{2.44}$$

where $\mathbf{e}'_n$ is the unit vector normal to the deformed middle surface. From equation (2.44), it is possible to obtain the displacement $\mathbf{u}^{(z)}$ of a generic point of the shell as a function of the middle surface displacement $\mathbf{u}$

$$\mathbf{u}^{(z)} = \mathbf{u} + z(\mathbf{e}'_n - \mathbf{e}_n), \tag{2.45}$$

where $\mathbf{e}'_n = \mathbf{e}'_1 \wedge \mathbf{e}'_2$, $\mathbf{e}'_1$ and $\mathbf{e}'_2$ being the unit vectors tangent to the lines α_1, α_2 on the deformed middle surface.

The vectorial equation of the deformed shell, considering points at distance z from the middle surface, is

$$\mathbf{R}^{(z)} = \mathbf{r} + z\mathbf{e}_n + \mathbf{u}^{(z)} = \mathbf{r}^{(z)} + u_1\mathbf{e}_1 + u_2\mathbf{e}_2 + u_3\mathbf{e}_n, \tag{2.46}$$

where $\mathbf{r}$ is the equation of the undeformed middle surface and $\mathbf{r}^{(z)} = \mathbf{r} + z\mathbf{e}_n$ is the equation of the undeformed shell layer at distance z from the middle surface.

The principal radii of curvature of points on the shell layer at distance z from the middle surface are

$$R_1^{(z)} = R_1 + z, \quad R_2^{(z)} = R_2 + z. \tag{2.47a,b}$$

Equations (2.17a,b) define the Lamé parameters referred to the middle surface. Because the principal radii for points at distance z from the middle surface are given by (2.47a,b), these equations are modified into

$$\mathrm{d}s_1^{(z)} = A_1\left(1+\frac{z}{R_1}\right)\mathrm{d}\alpha_1, \quad \mathrm{d}s_2^{(z)} = A_2\left(1+\frac{z}{R_2}\right)\mathrm{d}\alpha_2, \tag{2.48a,b}$$

which give the following expressions for the Lamé parameters of the shell layer at distance z from the middle surface:

$$A_1^{(z)} = A_1\left(1+\frac{z}{R_1}\right), \quad A_2^{(z)} = A_2\left(1+\frac{z}{R_2}\right). \tag{2.49a,b}$$

By using equations (2.34a–f), where the new expressions (2.48a,b) and (2.49a,b) for the radii of curvature and the Lamé parameters have been inserted, it is possible to evaluate the following derivative of the equation of the deformed shell:

$$\begin{aligned}\frac{1}{A_1(1+z/R_1)}\frac{\partial \mathbf{R}^{(z)}}{\partial \alpha_1} &= \frac{1}{A_1(1+z/R_1)}\frac{\partial \mathbf{r}^{(z)}}{\partial \alpha_1} + \frac{1}{A_1(1+z/R_1)}\left(\frac{\partial u_1}{\partial \alpha_1}\mathbf{e}_1 + u_1\frac{\partial \mathbf{e}_1}{\partial \alpha_1}\right.\\ &\quad \left. + \frac{\partial u_2}{\partial \alpha_1}\mathbf{e}_2 + u_2\frac{\partial \mathbf{e}_2}{\partial \alpha_1} + \frac{\partial u_3}{\partial \alpha_1}\mathbf{e}_n + u_3\frac{\partial \mathbf{e}_n}{\partial \alpha_1}\right)\\ &= \left(1 + \frac{1}{A_1(1+z/R_1)}\frac{\partial u_1}{\partial \alpha_1} + \frac{1}{A_1(1+z/R_1)}\frac{1}{A_2(1+z/R_2)}\right.\\ &\quad \left.\times \frac{\partial A_1(1+z/R_1)}{\partial \alpha_2}u_2 + \frac{u_3}{R_1+z}\right)\mathbf{e}_1 + \left(\frac{1}{A_1(1+z/R_1)}\frac{\partial u_2}{\partial \alpha_1}\right.\\ &\quad \left. - \frac{1}{A_1(1+z/R_1)}\frac{1}{A_2(1+z/R_2)}\frac{\partial A_1(1+z/R_1)}{\partial \alpha_2}u_1\right)\mathbf{e}_2\\ &\quad + \left(\frac{1}{A_1(1+z/R_1)}\frac{\partial u_3}{\partial \alpha_1} - \frac{u_1}{R_1+z}\right)\mathbf{e}_n,\end{aligned} \tag{2.50}$$

where equation (2.19a) has been used to evaluate the first term after equality. It is possible to simplify equation (2.50) by using the relationship

$$\frac{\partial A_1(1+z/R_1)}{\partial \alpha_2} = \frac{\partial A_1}{\partial \alpha_2} + z\frac{\partial}{\partial \alpha_2}\left(\frac{A_1}{R_1}\right) = \frac{\partial A_1}{\partial \alpha_2}(1+z/R_2), \tag{2.51}$$

where the second of the Codazzi conditions (2.39b) has been used to obtain the expression on the right-hand side. Equation (2.50) can be rewritten in the following form:

$$\frac{1}{A_1(1+z/R_1)}\frac{\partial \mathbf{R}^{(z)}}{\partial \alpha_1} = (1+\bar{\varepsilon}_1)\mathbf{e}_1 + \omega_1\mathbf{e}_2 - \bar{\Theta}\mathbf{e}_n, \tag{2.52}$$

where

$$\bar{\varepsilon}_1 = \frac{1}{1+z/R_1}\left(\frac{1}{A_1}\frac{\partial u_1}{\partial \alpha_1} + \frac{1}{A_1 A_2}\frac{\partial A_1}{\partial \alpha_2}u_2 + \frac{u_3}{R_1}\right), \tag{2.53a}$$

$$\omega_1 = \frac{1}{1+z/R_1}\left(\frac{1}{A_1}\frac{\partial u_2}{\partial \alpha_1} - \frac{1}{A_1 A_2}\frac{\partial A_1}{\partial \alpha_2}u_1\right), \tag{2.53b}$$

$$\bar{\Theta} = \frac{1}{1+z/R_1}\left(-\frac{1}{A_1}\frac{\partial u_3}{\partial \alpha_1} + \frac{u_1}{R_1}\right). \tag{2.53c}$$

The other derivative of the equation of the deformed shell is similarly obtained:

$$\frac{1}{A_2(1+z/R_2)}\frac{\partial \mathbf{R}^{(z)}}{\partial \alpha_2} = \omega_2 \mathbf{e}_1 + (1+\bar{\varepsilon}_2)\mathbf{e}_2 - \bar{\Psi}\mathbf{e}_n, \tag{2.54}$$

where

$$\bar{\varepsilon}_2 = \frac{1}{1+z/R_2}\left(\frac{1}{A_2}\frac{\partial u_2}{\partial \alpha_2} + \frac{1}{A_1 A_2}\frac{\partial A_2}{\partial \alpha_1}u_1 + \frac{u_3}{R_2}\right), \tag{2.55a}$$

$$\omega_2 = \frac{1}{1+z/R_2}\left(\frac{1}{A_2}\frac{\partial u_1}{\partial \alpha_2} - \frac{1}{A_1 A_2}\frac{\partial A_2}{\partial \alpha_1}u_2\right), \tag{2.55b}$$

$$\bar{\Psi} = \frac{1}{1+z/R_2}\left(-\frac{1}{A_2}\frac{\partial u_3}{\partial \alpha_2} + \frac{u_2}{R_2}\right). \tag{2.55c}$$

The vector $\mathrm{d}\mathbf{R}^{(z)}$, joining the points of coordinates (α_1, α_2, z) to the points of coordinates $(\alpha_1 + \mathrm{d}\alpha_1, \alpha_2 + \mathrm{d}\alpha_2, z)$ of the deformed shell layer at distance z from the middle surface, is given by

$$\mathrm{d}\mathbf{R}^{(z)} = \frac{1}{A_1(1+z/R_1)}\frac{\partial \mathbf{R}^{(z)}}{\partial \alpha_1}A_1(1+z/R_1)\,\mathrm{d}\alpha_1 + \frac{1}{A_2(1+z/R_2)}\frac{\partial \mathbf{R}^{(z)}}{\partial \alpha_2}A_2(1+z/R_2)\,\mathrm{d}\alpha_2, \tag{2.56}$$

where equations (2.52–2.55) can be substituted into (2.56) in order to have the explicit expression.

The vector $\mathrm{d}\mathbf{r}^{(z)}$ joins the points of coordinates (α_1, α_2, z) to the points of coordinates $(\alpha_1 + \mathrm{d}\alpha_1, \alpha_2 + \mathrm{d}\alpha_2, z)$ on the undeformed shell layer at distance z from the middle surface; then, the following expression can be obtained as generalization of equation (2.16) to points outside the middle surface

$$\mathrm{d}\mathbf{r}^{(z)} \cdot \mathrm{d}\mathbf{r}^{(z)} = A_1^2(1+z/R_1)^2\,\mathrm{d}\alpha_1^2 + A_2^2(1+z/R_2)^2\,\mathrm{d}\alpha_2^2. \tag{2.57}$$

By using equations (2.56) and (2.57), it is possible to write the generalization of equation (1.7), as follows:

$$\mathrm{d}\mathbf{R}^{(z)} \cdot \mathrm{d}\mathbf{R}^{(z)} - \mathrm{d}\mathbf{r}^{(z)} \cdot \mathrm{d}\mathbf{r}^{(z)} = 2\sum_{i=1}^{2}\sum_{j=1}^{2}\varepsilon_{ij}A_i(1+z/R_i)A_j(1+z/R_j)\,\mathrm{d}\alpha_i\,\mathrm{d}\alpha_j. \tag{2.58}$$

By using equation (2.58), it is possible to write the expressions of the Green's strain tensor ε_{ij} in the curvilinear coordinate system, namely

$$\varepsilon_{11} = \bar{\varepsilon}_1 + \frac{1}{2}\left(\bar{\varepsilon}_1^2 + \omega_1^2 + \bar{\Theta}^2\right), \tag{2.59a}$$

$$\varepsilon_{22} = \bar{\varepsilon}_2 + \frac{1}{2}\left(\omega_2^2 + \bar{\varepsilon}_2^2 + \bar{\Psi}^2\right), \tag{2.59b}$$

$$\gamma_{12} = \omega_1 + \omega_2 + (\bar{\varepsilon}_1\omega_2 + \bar{\varepsilon}_2\omega_1 + \bar{\Theta}\bar{\Psi}), \tag{2.59c}$$

where in equation (2.59c) the symbol γ has been used instead of ε, as usual. Equations (2.59a–c) show that $\bar{\varepsilon}_1$, $\bar{\varepsilon}_2$ and $\omega_1 + \omega_2$ are the linear components of the strains ε_{11}, ε_{22} and γ_{12}, respectively.

2.3.3 Strain-Displacement Relationships for Novozhilov's Nonlinear Shell Theory

Assuming for the moment that the shell is a fully three-dimensional solid, it is possible to obtain the expressions of other Green's strains. In particular, the derivative of the equation of the deformed shell in the z direction is given by

$$\frac{\partial \mathbf{R}^{(z)}}{\partial z} = \frac{\partial \mathbf{r}^{(z)}}{\partial z} + \frac{\partial}{\partial z}(u_1\mathbf{e}_1 + u_2\mathbf{e}_2 + u_3\mathbf{e}_n) = \frac{\partial u_1}{\partial z}\mathbf{e}_1 + \frac{\partial u_2}{\partial z}\mathbf{e}_2 + \left(1 + \frac{\partial u_3}{\partial z}\right)\mathbf{e}_n. \tag{2.60a}$$

By extending equations (2.56–2.58) to three dimensions and using (2.60a), the following expressions of the additional Green's strains are obtained:

$$\begin{aligned}\varepsilon_{1n} &= \frac{1}{A_1(1+z/R_1)}\frac{\partial \mathbf{R}^{(z)}}{\partial \alpha_1} \cdot \frac{\partial \mathbf{R}^{(z)}}{\partial z} \\ &= \frac{\partial u_1}{\partial z} + \frac{1}{1+z/R_1}\left[\left(\frac{1}{A_1}\frac{\partial u_3}{\partial \alpha_1} - \frac{u_1}{R_1}\right) + \left(\frac{1}{A_1}\frac{\partial u_1}{\partial \alpha_1} + \frac{1}{A_1A_2}\frac{\partial A_1}{\partial \alpha_2}u_2 + \frac{u_3}{R_1}\right)\frac{\partial u_1}{\partial z}\right. \\ &\quad \left. + \left(\frac{1}{A_1}\frac{\partial u_2}{\partial \alpha_1} - \frac{1}{A_1A_2}\frac{\partial A_1}{\partial \alpha_2}u_1\right)\frac{\partial u_2}{\partial z} + \left(\frac{1}{A_1}\frac{\partial u_3}{\partial \alpha_1} - \frac{u_1}{R_1}\right)\frac{\partial u_3}{\partial z}\right],\end{aligned} \tag{2.60b}$$

$$\begin{aligned}\varepsilon_{2n} &= \frac{1}{A_2(1+z/R_2)}\frac{\partial \mathbf{R}^{(z)}}{\partial \alpha_2} \cdot \frac{\partial \mathbf{R}^{(z)}}{\partial z} \\ &= \frac{\partial u_2}{\partial z} + \frac{1}{1+z/R_2}\left[\left(\frac{1}{A_2}\frac{\partial u_3}{\partial \alpha_2} - \frac{u_2}{R_2}\right) + \left(\frac{1}{A_2}\frac{\partial u_2}{\partial \alpha_2} + \frac{1}{A_1A_2}\frac{\partial A_2}{\partial \alpha_1}u_1 + \frac{u_3}{R_2}\right)\frac{\partial u_2}{\partial z}\right. \\ &\quad \left. + \left(\frac{1}{A_2}\frac{\partial u_1}{\partial \alpha_2} - \frac{1}{A_1A_2}\frac{\partial A_2}{\partial \alpha_1}u_2\right)\frac{\partial u_1}{\partial z} + \left(\frac{1}{A_2}\frac{\partial u_3}{\partial \alpha_2} - \frac{u_2}{R_2}\right)\frac{\partial u_3}{\partial z}\right],\end{aligned} \tag{2.60c}$$

$$\varepsilon_{nn} = \frac{1}{2}\left(\frac{\partial \mathbf{R}^{(z)}}{\partial z} \cdot \frac{\partial \mathbf{R}^{(z)}}{\partial z} - 1\right) = \frac{\partial u_3}{\partial z} + \frac{1}{2}\left[\left(\frac{\partial u_1}{\partial z}\right)^2 + \left(\frac{\partial u_2}{\partial z}\right)^2 + \left(\frac{\partial u_3}{\partial z}\right)^2\right]. \tag{2.60d}$$

The assumption that the straight fibers, which are normal to the middle surface of the shell before the deformation, remain straight and normal to the middle surface after deformation can be formulated as

$$\varepsilon_{1n} = \varepsilon_{2n} = 0. \tag{2.61}$$

The assumption that these fibers are not elongated is expressed by

$$\varepsilon_{nn} = 0. \tag{2.62}$$

The displacement of a point P_1 of the shell at distance z along the normal from the middle surface is assumed to be related to the middle surface displacement u, v, w by the following linear relationships in z:

$$u_1 = u + z\Theta, \tag{2.63}$$

$$u_2 = v + z\Psi, \tag{2.64}$$

$$u_3 = w + z\chi. \tag{2.65}$$

After substituting (2.63–2.65) into equations (2.60b–d) and satisfying equations (2.61) and (2.62), the following relations are obtained for Θ, ψ and χ:

$$\Theta = \hat{\Theta}(1 + \hat{\varepsilon}_2) - \hat{\Psi}\hat{\omega}_2, \tag{2.66a}$$

$$\Psi = \hat{\Psi}(1 + \hat{\varepsilon}_1) - \hat{\Theta}\hat{\omega}_1, \tag{2.66b}$$

$$\chi \cong \hat{\varepsilon}_1 + \hat{\varepsilon}_2 + \hat{\varepsilon}_1\hat{\varepsilon}_2 - \hat{\omega}_1\hat{\omega}_2, \tag{2.66c}$$

where the expressions below, obtained evaluating equations (2.53a–c) and (2.55a–c) for $z = 0$, have been used:

$$\hat{\varepsilon}_1 = \frac{1}{A_1}\frac{\partial u}{\partial \alpha_1} + \frac{1}{A_1 A_2}\frac{\partial A_1}{\partial \alpha_2} v + \frac{w}{R_1}, \tag{2.67a}$$

$$\hat{\omega}_1 = \frac{1}{A_1}\frac{\partial v}{\partial \alpha_1} - \frac{1}{A_1 A_2}\frac{\partial A_1}{\partial \alpha_2} u, \tag{2.67b}$$

$$\hat{\Theta} = -\frac{1}{A_1}\frac{\partial w}{\partial \alpha_1} + \frac{u}{R_1}, \tag{2.67c}$$

$$\hat{\varepsilon}_2 = \frac{1}{A_2}\frac{\partial v}{\partial \alpha_2} + \frac{1}{A_1 A_2}\frac{\partial A_2}{\partial \alpha_1} u + \frac{w}{R_2}, \tag{2.67d}$$

$$\hat{\omega}_2 = \frac{1}{A_2}\frac{\partial u}{\partial \alpha_2} - \frac{1}{A_1 A_2}\frac{\partial A_2}{\partial \alpha_1} v, \tag{2.67e}$$

$$\hat{\Psi} = -\frac{1}{A_2}\frac{\partial w}{\partial \alpha_2} + \frac{v}{R_2}. \tag{2.67f}$$

Equations (2.66a–c) show that $\hat{\Theta}$ and $\hat{\Psi}$ are the linear components (in terms of displacements) of the rotations Θ and Ψ, respectively.

The strain components $\varepsilon_{11}, \varepsilon_{22}$ and γ_{12} at an arbitrary point of the shell are related to the middle surface strains $\varepsilon_{1,0}, \varepsilon_{2,0}$ and $\gamma_{12,0}$ and to the changes in the curvature and torsion of the middle surface k_1, k_2 and k_{12} by the following three relationships:

$$\varepsilon_{11} = \varepsilon_{1,0} + zk_1, \quad \varepsilon_{22} = \varepsilon_{2,0} + zk_2, \quad \gamma_{12} = \gamma_{12,0} + zk_{12}. \tag{2.68a–c}$$

By inserting equations (2.63–2.67) into (2.59a–c), using the approximations $1 + z/R_1 \cong 1 + z/R_2 \cong 1$ and writing the results in the form given by equations (2.68a–c) neglecting higher-order terms in z, the middle surface strains and the changes in the curvature and torsion are obtained, as follows:

$$\begin{aligned}\varepsilon_{1,0} &= \hat{\varepsilon}_1 + \frac{1}{2}\left(\hat{\varepsilon}_1^2 + \hat{\omega}_1^2 + \hat{\Theta}^2\right)\\ &= \frac{1}{A_1}\frac{\partial u}{\partial \alpha_1} + \frac{1}{A_1 A_2}\frac{\partial A_1}{\partial \alpha_2} v + \frac{w}{R_1} + \frac{1}{2}\left[\left(\frac{1}{A_1}\frac{\partial u}{\partial \alpha_1} + \frac{1}{A_1 A_2}\frac{\partial A_1}{\partial \alpha_2} v + \frac{w}{R_1}\right)^2\right.\\ &\quad \left. + \left(\frac{1}{A_1}\frac{\partial v}{\partial \alpha_1} - \frac{1}{A_1 A_2}\frac{\partial A_1}{\partial \alpha_2} u\right)^2 + \left(\frac{1}{A_1}\frac{\partial w}{\partial \alpha_1} - \frac{u}{R_1}\right)^2\right],\end{aligned} \tag{2.69a}$$

$$\varepsilon_{2,0} = \hat{\varepsilon}_2 + \frac{1}{2}\left(\hat{\omega}_2^2 + \hat{\varepsilon}_2^2 + \hat{\Psi}^2\right)$$
$$= \frac{1}{A_2}\frac{\partial v}{\partial \alpha_2} + \frac{1}{A_1 A_2}\frac{\partial A_2}{\partial \alpha_1}u + \frac{w}{R_2} + \frac{1}{2}\left[\left(\frac{1}{A_2}\frac{\partial u}{\partial \alpha_2} - \frac{1}{A_1 A_2}\frac{\partial A_2}{\partial \alpha_1}v\right)^2\right.$$
$$\left. + \left(\frac{1}{A_2}\frac{\partial v}{\partial \alpha_2} + \frac{1}{A_1 A_2}\frac{\partial A_2}{\partial \alpha_1}u + \frac{w}{R_2}\right)^2 + \left(\frac{1}{A_2}\frac{\partial w}{\partial \alpha_2} - \frac{v}{R_2}\right)^2\right], \tag{2.69b}$$

$$\gamma_{12,0} = \hat{\omega}_1 + \hat{\omega}_2 + \left(\hat{\varepsilon}_1\hat{\omega}_2 + \hat{\varepsilon}_2\hat{\omega}_1 + \hat{\Theta}\hat{\Psi}\right)$$
$$= \frac{1}{A_1}\frac{\partial v}{\partial \alpha_1} + \frac{1}{A_2}\frac{\partial u}{\partial \alpha_2} - \frac{1}{A_1 A_2}\frac{\partial A_1}{\partial \alpha_2}u - \frac{1}{A_1 A_2}\frac{\partial A_2}{\partial \alpha_1}v$$
$$+ \left(\frac{1}{A_1}\frac{\partial u}{\partial \alpha_1} + \frac{1}{A_1 A_2}\frac{\partial A_1}{\partial \alpha_2}v + \frac{w}{R_1}\right)\left(\frac{1}{A_2}\frac{\partial u}{\partial \alpha_2} - \frac{1}{A_1 A_2}\frac{\partial A_2}{\partial \alpha_1}v\right)$$
$$+ \left(\frac{1}{A_2}\frac{\partial v}{\partial \alpha_2} + \frac{1}{A_1 A_2}\frac{\partial A_2}{\partial \alpha_1}u + \frac{w}{R_2}\right)\left(\frac{1}{A_1}\frac{\partial v}{\partial \alpha_1} - \frac{1}{A_1 A_2}\frac{\partial A_1}{\partial \alpha_2}u\right)$$
$$+ \left(\frac{1}{A_1}\frac{\partial w}{\partial \alpha_1} - \frac{u}{R_1}\right)\left(\frac{1}{A_2}\frac{\partial w}{\partial \alpha_2} - \frac{v}{R_2}\right), \tag{2.69c}$$

$$k_1 = (1 + \hat{\varepsilon}_1)\left(\frac{1}{A_1}\frac{\partial \Theta}{\partial \alpha_1} + \frac{1}{A_1 A_2}\frac{\partial A_1}{\partial \alpha_2}\Psi + \frac{\chi}{R_1}\right) + \hat{\omega}_1\left(\frac{1}{A_1}\frac{\partial \Psi}{\partial \alpha_1} - \frac{1}{A_1 A_2}\frac{\partial A_1}{\partial \alpha_2}\Theta\right)$$
$$+ \hat{\Theta}\left(\frac{1}{A_1}\frac{\partial \chi}{\partial \alpha_1} - \frac{\Theta}{R_1}\right), \tag{2.69d}$$

$$k_2 = (1 + \hat{\varepsilon}_2)\left(\frac{1}{A_2}\frac{\partial \Psi}{\partial \alpha_2} + \frac{1}{A_1 A_2}\frac{\partial A_2}{\partial \alpha_1}\Theta + \frac{\chi}{R_2}\right) + \hat{\omega}_2\left(\frac{1}{A_2}\frac{\partial \Theta}{\partial \alpha_2} - \frac{1}{A_1 A_2}\frac{\partial A_2}{\partial \alpha_1}\Psi\right)$$
$$+ \hat{\Psi}\left(\frac{1}{A_2}\frac{\partial \chi}{\partial \alpha_2} - \frac{\Psi}{R_2}\right), \tag{2.69e}$$

$$k_{12} = (1 + \hat{\varepsilon}_1)\left(\frac{1}{A_2}\frac{\partial \Theta}{\partial \alpha_2} - \frac{1}{A_1 A_2}\frac{\partial A_2}{\partial \alpha_1}\Psi\right) + (1 + \hat{\varepsilon}_2)\left(\frac{1}{A_1}\frac{\partial \Psi}{\partial \alpha_1} - \frac{1}{A_1 A_2}\frac{\partial A_1}{\partial \alpha_2}\Theta\right)$$
$$+ \hat{\omega}_2\left(\frac{1}{A_1}\frac{\partial \Theta}{\partial \alpha_1} + \frac{1}{A_1 A_2}\frac{\partial A_1}{\partial \alpha_2}\Psi + \frac{\chi}{R_1}\right) + \hat{\omega}_1\left(\frac{1}{A_2}\frac{\partial \Psi}{\partial \alpha_2} + \frac{1}{A_1 A_2}\frac{\partial A_2}{\partial \alpha_1}\Theta + \frac{\chi}{R_2}\right)$$
$$+ \hat{\Psi}\left(\frac{1}{A_1}\frac{\partial \chi}{\partial \alpha_1} - \frac{\Theta}{R_1}\right) + \hat{\Theta}\left(\frac{1}{A_2}\frac{\partial \chi}{\partial \alpha_2} - \frac{\Psi}{R_2}\right). \tag{2.69f}$$

Geometric imperfections are not introduced in this case, because this theory is developed for a shell of arbitrary shape, so that the actual shell geometry can be used in the numerical calculations.

2.3.4 Strain-Displacement Relationships for an Improved Version of the Novozhilov Shell Theory

The approximation $1 + z/R_1 \cong 1 + z/R_2 \cong 1$ is not very accurate. In fact, it is conveniently replaced by the approximations (2.5a–c), which give consistent terms of order z. By inserting equations (2.63–2.67) into (2.59a–c), using approximations (2.5a–c) and writing the results in the form given by equations (2.68a–c) neglecting higher-order terms in z, the following results are obtained for the changes in the curvature and torsion:

$$k_1 = \frac{w}{R_1 R_2} + \frac{1}{A_1 A_2 R_1}\frac{\partial A_2}{\partial \alpha_1}u + \frac{1}{A_1 A_2 R_2}\frac{\partial A_1}{\partial \alpha_2}v + \frac{1}{A_1 R_1}\frac{\partial u}{\partial \alpha_1} + \frac{1}{A_2 R_1}\frac{\partial v}{\partial \alpha_2}$$
$$- \frac{1}{A_1 A_2^2}\frac{\partial A_1}{\partial \alpha_2}\frac{\partial w}{\partial \alpha_2} + \frac{1}{A_1^3}\frac{\partial A_1}{\partial \alpha_1}\frac{\partial w}{\partial \alpha_1} - \frac{1}{A_1^2}\frac{\partial^2 w}{\partial \alpha_1^2}, \tag{2.70a}$$

$$k_2 = \frac{w}{R_1 R_2} + \frac{1}{A_1 A_2 R_1}\frac{\partial A_2}{\partial \alpha_1}u + \frac{1}{A_1 A_2 R_2}\frac{\partial A_1}{\partial \alpha_2}v + \frac{1}{A_1 R_2}\frac{\partial u}{\partial \alpha_1} + \frac{1}{A_2 R_2}\frac{\partial v}{\partial \alpha_2}$$
$$- \frac{1}{A_1^2 A_2}\frac{\partial A_2}{\partial \alpha_1}\frac{\partial w}{\partial \alpha_1} + \frac{1}{A_2^3}\frac{\partial A_2}{\partial \alpha_2}\frac{\partial w}{\partial \alpha_2} - \frac{1}{A_2^2}\frac{\partial^2 w}{\partial \alpha_2^2}, \tag{2.70b}$$

$$k_{12} = \frac{1}{A_2}\frac{\partial u}{\partial \alpha_2}\left(\frac{1}{R_1} - \frac{1}{R_2}\right) + \frac{1}{A_1}\frac{\partial v}{\partial \alpha_1}\left(\frac{1}{R_2} - \frac{1}{R_1}\right)$$
$$+ \frac{2}{A_1 A_2^2}\frac{\partial A_2}{\partial \alpha_1}\frac{\partial w}{\partial \alpha_2} + \frac{2}{A_1^2 A_2}\frac{\partial A_1}{\partial \alpha_2}\frac{\partial w}{\partial \alpha_1} + \frac{2}{A_1 A_2}\frac{\partial^2 w}{\partial \alpha_1 \partial \alpha_2}, \tag{2.70c}$$

where only linear terms in the changes in the curvature and torsion have been reported for the sake of brevity. The middle surface strains are still given by equations (2.69a–c). It is interesting to observe that the use of the approximations (2.5a–c) results in different linear and nonlinear terms in k_1, k_2, k_{12}.

2.3.5 Simplified Strain-Displacement Relationships

Equations obtained with the improved version of the Novozhilov nonlinear shell theory can be simplified by retaining only the main nonlinear terms; this gives rise to a version analogous to Donnell's nonlinear shell theory for doubly curved shells, reported in equations (2.2), but valid for arbitrary shells and with the advantage of having an extremely accurate linear part. These equations are the following:

$$\varepsilon_{1,0} = \frac{1}{A_1}\frac{\partial u}{\partial \alpha_1} + \frac{1}{A_1 A_2}\frac{\partial A_1}{\partial \alpha_2}v + \frac{w}{R_1} + \frac{1}{2}\left(\frac{1}{A_1}\frac{\partial w}{\partial \alpha_1}\right)^2, \tag{2.71a}$$

$$\varepsilon_{2,0} = \frac{1}{A_2}\frac{\partial v}{\partial \alpha_2} + \frac{1}{A_1 A_2}\frac{\partial A_2}{\partial \alpha_1}u + \frac{w}{R_2} + \frac{1}{2}\left(\frac{1}{A_2}\frac{\partial w}{\partial \alpha_2}\right)^2, \tag{2.71b}$$

$$\gamma_{12,0} = \frac{1}{A_1}\frac{\partial v}{\partial \alpha_1} + \frac{1}{A_2}\frac{\partial u}{\partial \alpha_2} - \frac{1}{A_1 A_2}\frac{\partial A_1}{\partial \alpha_2}u - \frac{1}{A_1 A_2}\frac{\partial A_2}{\partial \alpha_1}v + \frac{1}{A_1 A_2}\frac{\partial w}{\partial \alpha_1}\frac{\partial w}{\partial \alpha_2}, \tag{2.71c}$$

and the changes in the curvature and torsion are still given by the linear equations (2.70a–c).

2.3.6 Elastic Strain Energy

The elastic strain energy U_S of the shell, neglecting σ_n, τ_{n1}, τ_{n2} as stated by Kirchhoff-Love first-approximation assumptions, is given by

$$U_S = \frac{1}{2}\int_0^{\alpha}\int_0^{\beta}\int_{-h/2}^{h/2} (\sigma_{11}\varepsilon_{11} + \sigma_{22}\,\varepsilon_{22} + \tau_{12}\,\gamma_{12})(1 + z/R_1)(1 + z/R_2)A_1A_2\,\mathrm{d}\alpha_1\,\mathrm{d}\alpha_2\,\mathrm{d}z, \tag{2.72}$$

where it is assumed that the shell is rectangular in the curvilinear coordinates and is described by the coordinate intervals (0,α) and (0,β), for α_1 and α_1, respectively; the stresses σ_{11}, σ_{22} and τ_{12} are related to the strains for homogeneous and isotropic material by

$$\sigma_{11} = \frac{E}{1-\nu^2}\left(\varepsilon_{11} + \nu\varepsilon_{22}\right), \quad \sigma_{22} = \frac{E}{1-\nu^2}\left(\varepsilon_{22} + \nu\varepsilon_{11}\right), \quad \tau_{12} = \frac{E}{2\,(1+\nu)}\gamma_{12}, \tag{2.73}$$

where E is the Young's modulus and ν is the Poisson's ratio. By using equations (2.68a–c), (2.72) and (2.73), the following expression is obtained in the case of uniform thickness:

$$\begin{aligned} U_S = {} & \frac{1}{2}\frac{Eh}{1-\nu^2}\int_0^{\alpha}\int_0^{\beta}\left(\varepsilon_{1,0}^2 + \varepsilon_{2,0}^2 + 2\nu\varepsilon_{1,0}\varepsilon_{2,0} + \frac{1-\nu}{2}\gamma_{12,0}^2\right) A_1A_2\,\mathrm{d}\alpha_1\,\mathrm{d}\alpha_2 \\ & + \frac{1}{2}\frac{Eh^3}{12\,(1-\nu^2)}\int_0^{\alpha}\int_0^{\beta}\left(k_1^2 + k_2^2 + 2\nu k_1k_2 + \frac{1-\nu}{2}k_{12}^2\right) A_1A_2\,\mathrm{d}\alpha_1\,\mathrm{d}\alpha_2 \\ & + \frac{1}{2}\left(\frac{1}{R_1} + \frac{1}{R_2}\right)\frac{Eh^3}{6\,(1-\nu^2)} \\ & \times\int_0^{\alpha}\int_0^{\beta}\left(\varepsilon_{1,0}k_1 + \varepsilon_{2,0}k_2 + \nu\varepsilon_{1,0}k_2 + \nu\varepsilon_{2,0}k_1 + \frac{1-\nu}{2}\gamma_{12,0}k_{12}\right) \\ & \times A_1A_2\,\mathrm{d}\alpha_1\,\mathrm{d}\alpha_2 + O(h^4), \end{aligned} \tag{2.74}$$

where $O(h^4)$ is a higher-order term in h. If the last term is retained, membrane and bending energies are coupled.

2.3.7 Kinetic Energy

The kinetic energy T_S of the shell, by neglecting rotary inertia, is simply given by

$$T_S = \frac{1}{2}\rho_S h\int_0^{\alpha}\int_0^{\beta}(\dot{u}^2 + \dot{v}^2 + \dot{w}^2)A_1\,A_2\,\mathrm{d}\alpha_1\,\mathrm{d}\alpha_2, \tag{2.75}$$

where ρ_S is the mass density of the shell and the overdot denotes time derivative.

(a)

(b)

Figure 2.6. Military aircraft with large use of composite materials. (a) B-2 stealth bomber made by Northrop Grumman. (b) F-117A stealth fighter made by Lockheed Martin. Courtesy of NASA.

2.4 Composite and Functionally Graded Materials

Composite materials present high stiffness-to-weight and high strength-to-weight ratios. For this reason, they are largely used in aircraft and spacecraft. The use of these materials introduced innovative design. Typical examples are the B-2 stealth bomber made by Northrop Grumman (Figure 2.6[a]) and the F-117A stealth fighter made by Lockheed Martin (Figure 2.6[b]), where virtually all the external parts are made of various composite materials that have radar-absorption characteristics and make possible a design reducing the radar visibility of the plane.

A composite material consists of two or more phases on a microscopic scale and is designed to have mechanical properties and performance superior to those of the constituent materials acting independently. One of the phases is usually discontinuous, stiffer and stronger, and is called reinforcement, whereas the less stiff

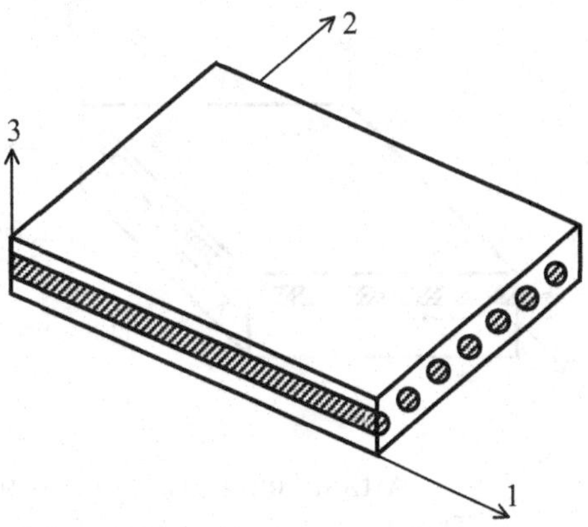

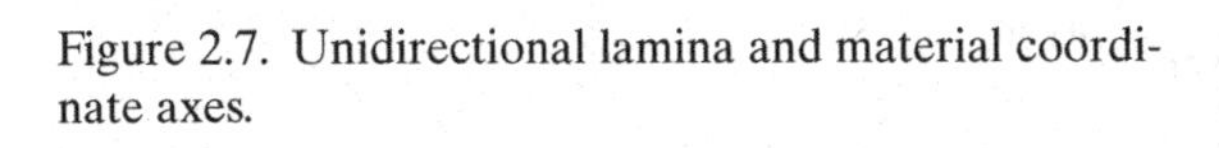
Figure 2.7. Unidirectional lamina and material coordinate axes.

and weaker phase is continuous and is called the matrix. Sometimes, an interphase exists between reinforcement and matrix. One of the most important parameters characterizing the mechanical properties of the composite is the volume fraction of reinforcement. The distribution of the reinforcement determines the homogeneity or the strong anisotropy of the material. In high-performance structural components, the usually continuous fiber reinforcement determines the stiffness and strength in the direction of the fibers. The matrix phase provides protection and support to the fibers and local stress transfer from one fiber to another.

Composite materials can be classified according to type, geometry and orientation of the reinforcement fibers. A particulate filler gives a particulate composite, which is quasi-homogeneous and quasi-isotropic. Discontinuous (short) fibers give (i) unidirectional discontinuous fiber composites and (ii) randomly oriented discontinuous fiber composites. Continuous fibers give (i) unidirectional continuous fiber composites (ii) cross-ply or fabric continuous fiber composites and (iii) multidirectional continuous fiber composites. Continuous fiber composites are the most efficient for stiffness and strength.

Fiber-reinforced composites can be classified according to the matrix (Daniel and Ishai 1994): polymer, metal, ceramic and carbon. Polymer matrix composites include thermoset (epoxy, polyimide, polyester) or thermoplastic resins reinforced with glass, graphite, aramid (Kevlar) or boron fibers and are used mainly in low-temperature applications. For higher temperatures, metal matrix composites reinforced with boron, graphite or ceramic fibers can be applied. For very high temperatures, ceramic matrix composites with ceramic fibers should be used. Carbon matrix composites have a carbon or graphite matrix reinforced by graphite fabric, presenting the advantage of high strength at high temperatures with low thermal expansion and low weight.

Composites can consist of thin layers of different materials (or the same material with different orientation of fibers) bonded together; these are laminated composites.

2.4.1 Stress-Strain Relations for a Thin Lamina

A lamina, or ply, is a plane or curved layer of unidirectional fibers (see Figure 2.7) or woven fabric in a matrix. A laminate is made of two or more laminae (see Figure 2.8) stacked together at various orientations. The laminae can be of various thicknesses and different materials.

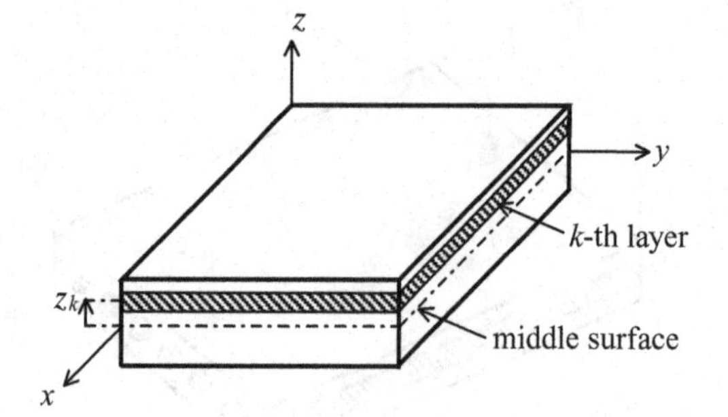

Figure 2.8. Laminate with definition of the k-th layer.

A thin lamina is assumed to be under plane stress, that is, $\sigma_3 = \tau_{13} = \tau_{23} = 0$. The stress-strain relations for a thin lamina in plane stress, in the material principal coordinates, are given by (Daniel and Ishai 1994)

$$\begin{Bmatrix} \sigma_1 \\ \sigma_2 \\ \tau_{12} \end{Bmatrix} = \begin{bmatrix} c_{11} & c_{12} & 0 \\ c_{21} & c_{22} & 0 \\ 0 & 0 & G_{12} \end{bmatrix} \begin{Bmatrix} \varepsilon_1 \\ \varepsilon_2 \\ \gamma_{12} \end{Bmatrix}, \tag{2.76}$$

where G_{12} is the shear modulus in 1–2 directions and

$$c_{11} = \frac{E_1}{1 - \nu_{12}\nu_{21}}, \quad c_{12} = c_{21} = \frac{E_2 \nu_{12}}{1 - \nu_{12}\nu_{21}}, \quad c_{22} = \frac{E_2}{1 - \nu_{12}\nu_{21}}, \quad \nu_{ij} E_j = \nu_{ji} E_i. \tag{2.77}$$

Usually, the lamina material axes (1,2) do not coincide with the plate (or shell) reference axes (x,y), as shown in Figure 2.9. Then, the strains and stresses on material axes can be related to the reference axes by using the following invertible expressions:

$$\begin{Bmatrix} \sigma_1 \\ \sigma_2 \\ \tau_{12} \end{Bmatrix} = \mathbf{T} \begin{Bmatrix} \sigma_x \\ \sigma_y \\ \tau_{xy} \end{Bmatrix}, \quad \begin{Bmatrix} \varepsilon_1 \\ \varepsilon_2 \\ \frac{1}{2}\gamma_{12} \end{Bmatrix} = \mathbf{T} \begin{Bmatrix} \varepsilon_x \\ \varepsilon_y \\ \frac{1}{2}\gamma_{xy} \end{Bmatrix}, \tag{2.78a,b}$$

where

$$\mathbf{T} = \begin{bmatrix} \cos^2\theta & \sin^2\theta & 2\sin\theta\cos\theta \\ \sin^2\theta & \cos^2\theta & -2\sin\theta\cos\theta \\ -\sin\theta\cos\theta & \sin\theta\cos\theta & \cos^2\theta - \sin^2\theta \end{bmatrix}. \tag{2.79}$$

Therefore, equation (2.76) is transformed into the reference axes

$$\begin{Bmatrix} \sigma_x \\ \sigma_y \\ \tau_{xy} \end{Bmatrix} = [Q] \begin{Bmatrix} \varepsilon_x \\ \varepsilon_y \\ \gamma_{xy} \end{Bmatrix}, \tag{2.80}$$

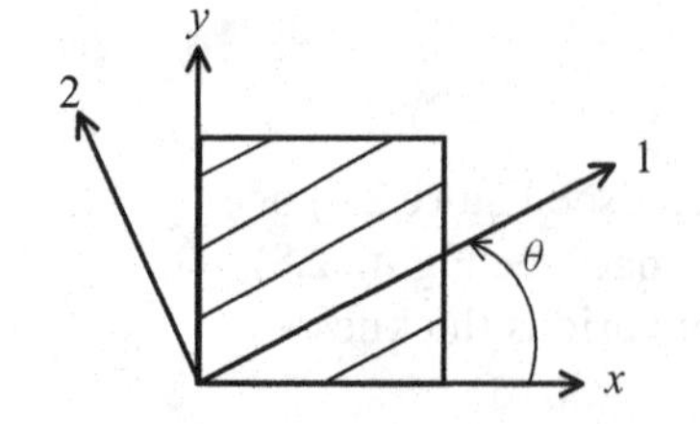

Figure 2.9. Material and reference axes for a unidirectional lamina.

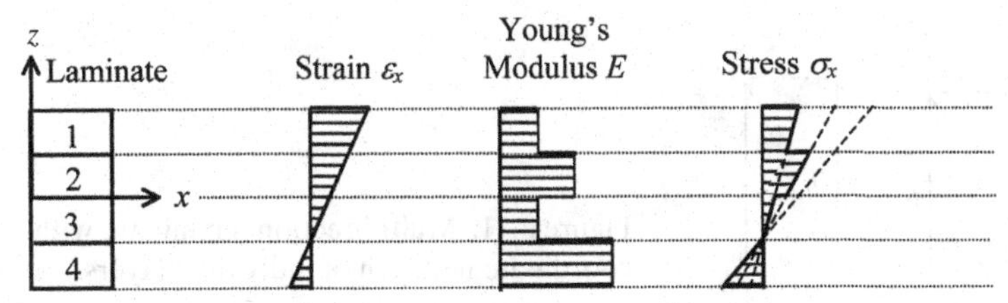

Figure 2.10. Linear strain distribution giving discontinuous stress in multidirectional laminate.

where

$$[Q] = \begin{bmatrix} Q_{11} & Q_{12} & Q_{13} \\ Q_{21} & Q_{22} & Q_{23} \\ Q_{31} & Q_{32} & Q_{33} \end{bmatrix}, \tag{2.81}$$

and

$$\begin{bmatrix} Q_{11} & Q_{12} & 2Q_{13} \\ Q_{21} & Q_{22} & 2Q_{23} \\ Q_{31} & Q_{32} & 2Q_{33} \end{bmatrix} = \mathbf{T}^{-1} \begin{bmatrix} c_{11} & c_{12} & 0 \\ c_{21} & c_{22} & 0 \\ 0 & 0 & 2G_{12} \end{bmatrix} \mathbf{T}. \tag{2.82}$$

2.4.2 Stress-Strain Relations for a Layer within a Laminate

Equation (2.80) can be rewritten in the following vectorial notation for the k-th layer (see Figure 2.8):

$$\{\sigma^{(k)}\} = [Q^{(k)}]\{\varepsilon\}, \tag{2.83}$$

where $Q_{ij}^{(k)}$ are the material properties of the k-th layer in the plate or shell principal coordinates.

As a consequence of the discontinuous variation of the stiffness matrix $Q_{ij}^{(k)}$ from layer to layer, the stresses may be discontinuous layer to layer. This is shown for a four-layer laminate in Figure 2.10 under a linear strain variation.

Equation (2.83) can be rewritten by using equation (2.1) in the following form:

$$\{\sigma^{(k)}\} = [Q^{(k)}] \begin{Bmatrix} \varepsilon_{x,0} \\ \varepsilon_{y,0} \\ \gamma_{xy,0} \end{Bmatrix} + z[Q^{(k)}] \begin{Bmatrix} k_x \\ k_y \\ k_{xy} \end{Bmatrix}. \tag{2.84}$$

2.4.3 Elastic Strain Energy for Laminated Shells

The elastic strain energy U_S of a doubly curved shell, by assuming the Kirchhoff-Love hypothesis, is given by

$$U_S = \frac{1}{2}\sum_{k=1}^{K} \int_0^{\alpha}\int_0^{\beta}\int_{h^{(k-1)}}^{h^{(k)}} \left(\sigma_x^{(k)}\varepsilon_x + \sigma_y^{(k)}\varepsilon_y + \tau_{xy}^{(k)}\gamma_{xy}\right)(1 + z/R_x)(1 + z/R_y)A_x A_y \,\mathrm{d}\alpha_x \,\mathrm{d}\alpha_y \,\mathrm{d}z, \tag{2.85}$$

where equation (2.85) is the extension of (2.72) with the axes (x,y) used instead of $(1,2)$; K is the total number of layers in the laminated shell and $(h^{(k-1)}, h^{(k)})$ are the z coordinates of the k-th layer, as shown in Figure 2.11.

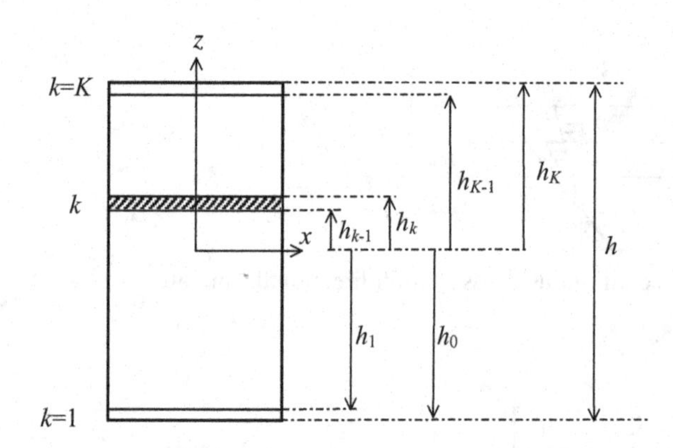

Figure 2.11. Multidirectional laminate with coordinate notation of individual layers.

2.4.4 Elastic Strain Energy for Orthotropic and Cross-Ply Shells

By using a classical shell theory, such as Donnell's or Novozhilov's, the elastic strain energy of an orthotropic and laminated shell of uniform thickness must be rewritten. By using equations (2.7), (2.76) and (2.77), the following expression for the strain energy of orthotropic shells is obtained:

$$U_S = \frac{1}{2}\frac{E_x h}{1-\nu_{yx}^2 E_x/E_y}\int_0^a\int_0^b\left[\varepsilon_{x,0}^2 + \frac{E_y}{E_x}\varepsilon_{y,0}^2 + 2\nu_{yx}\varepsilon_{x,0}\varepsilon_{y,0} + \frac{G_{xy}}{E_x}\left(1-\nu_{yx}^2 E_x/E_y\right)\gamma_{xy,0}^2\right]\mathrm{d}x\,\mathrm{d}y$$
$$+\frac{1}{2}\frac{E_x h^3}{12\left(1-\nu_{yx}^2 E_x/E_y\right)}\int_0^a\int_0^b\left[k_x^2 + \frac{E_y}{E_x}k_y^2 + 2\nu_{yx}k_x k_y + \frac{G_{xy}}{E_x}\left(1-\nu_{yx}^2 E_x/E_y\right)k_{xy}^2\right]\mathrm{d}x\,\mathrm{d}y$$
$$+\frac{1}{2}\left(\frac{1}{R_x}+\frac{1}{R_y}\right)\frac{E_x h^3}{6\left(1-\nu_{yx}^2 E_x/E_y\right)}\int_0^a\int_0^b\left[\varepsilon_{x,0}\,k_x + \frac{E_y}{E_x}\varepsilon_{y,0}\,k_y + \nu_{yx}\varepsilon_{x,0}\,k_y + \nu_{yx}\varepsilon_{y,0}\,k_x + \frac{G_{xy}}{E_x}\left(1-\nu_{yx}^2 E_x/E_y\right)\gamma_{xy,0}k_{xy}\right]\mathrm{d}x\,\mathrm{d}y. \quad (2.86)$$

Now attention is focused on symmetric, laminated composite shells. A laminate is called symmetric when, for each layer, on one side of the middle surface there is a corresponding identical layer on the other side. It is assumed that each lamina is orthotropic and that each lamina's orthotropic principal directions coincide with the shell coordinates. This kind of laminate is referred to as a symmetric cross-ply laminate. It is convenient to introduce the distance h_j of the upper surface of the j-th lamina from the middle surface of the shell (see Figure 2.12), where $j = 1, \ldots H$ and H is the total number of layers of the half-laminate (the central half-lamina in a laminate with an odd number of layers counts for one). For a laminate with an even number of laminae, the first surface of the half-laminate coincides with the middle

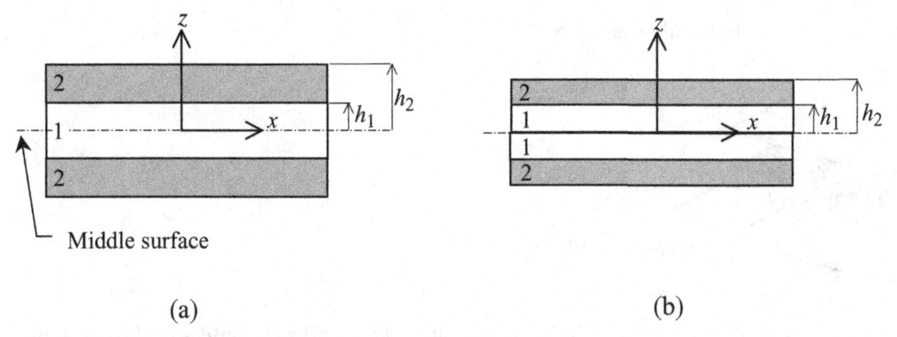

Figure 2.12. Symmetric laminated composite shell: distance h_j of the upper surface of the j-th lamina from the middle surface of the shell and numeration of layers of half-laminate. (a) Shell with odd number of laminae. (b) Shell with even number of laminae.

surface of the shell and is not considered. The expression of the strain energy for this laminated composite shell is obtained as a generalization of equation (2.86):

$$
\begin{aligned}
U_S &= \sum_{j=1}^{H} \frac{E_{x_j}(h_j - h_{j-1})}{1 - \nu_{yx_j}^2 E_{x_j}/E_{y_j}} \\
&\times \int_0^a \int_0^b \left[\varepsilon_{x,0}^2 + \frac{E_{y_j}}{E_{x_j}} \varepsilon_{y,0}^2 + 2\nu_{yx_j} \varepsilon_{x,0}\, \varepsilon_{y,0} + \frac{G_{xy_j}}{E_{x_j}} \left(1 - \nu_{yx_j}^2 \frac{E_{x_j}}{E_{y_j}}\right) \gamma_{xy,0}^2 \right] \mathrm{d}x\, \mathrm{d}y \\
&+ \sum_{j=1}^{H} \frac{E_{x_j}\left(h_j^3 - h_{j-1}^3\right)}{3\left(1 - \nu_{yx_j}^2 E_{x_j}/E_{y_j}\right)} \\
&\times \int_0^a \int_0^b \left[k_x^2 + \frac{E_{y_j}}{E_{x_j}} k_y^2 + 2\nu_{yx_j} k_x k_y + \frac{G_{xy_j}}{E_{x_j}} \left(1 - \nu_{yx_j}^2 \frac{E_{x_j}}{E_{y_j}}\right) k_{xy}^2 \right] \mathrm{d}x\, \mathrm{d}y \\
&+ \sum_{j=1}^{H} \frac{2E_{x_j}\left(h_j^3 - h_{j-1}^3\right)}{3\left(1 - \nu_{yx_j}^2 E_{x_j}/E_{y_j}\right)} \left(\frac{1}{R_x} + \frac{1}{R_y}\right) \\
&\times \int_0^a \int_0^b \left[\varepsilon_{x,0}\, k_x + \frac{E_{y_j}}{E_{x_j}} \varepsilon_{y,0}\, k_y + \nu_{yx_j} \varepsilon_{x,0}\, k_y + \nu_{yx_j} \varepsilon_{y,0}\, k_x \right. \\
&\left. + \frac{G_{xy_j}}{E_{x_j}} \left(1 - \nu_{yx_j}^2 E_{x_j}/E_{y_j}\right) \gamma_{xy,0}\, k_{xy} \right] \mathrm{d}x\, \mathrm{d}y, \qquad (2.87)
\end{aligned}
$$

where $h_0 = 0$ and the subscript j refers to the material properties of the j-th lamina.

2.4.5 Sandwich Plates and Shells

In order to obtain structural elements with high stiffness and low weight, sandwich plates and shells have been developed. A typical sandwich is composed by external layers made of high-strength materials (aluminium, steel or composite), bonded to a thick core made of low-weight material (honeycomb metal, balsa wood, porous rubber, corrugated metal sheet or plastic), as shown in Figure 2.13. As a consequence

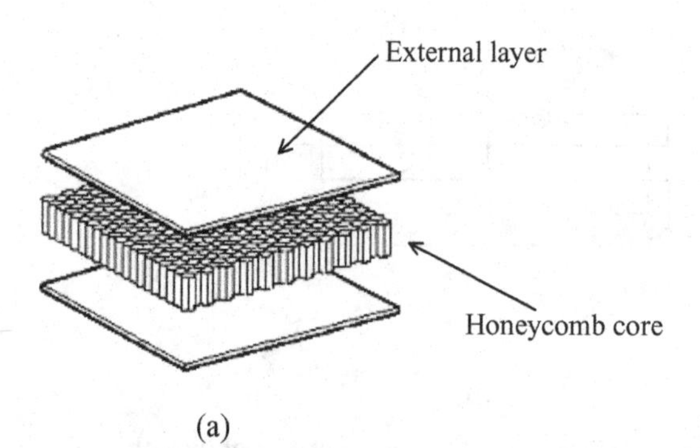

(a)

(b)

Figure 2.13. Sandwich plate with honeycomb core. (a) Schematic drawing. (b) Photo of a symmetric sandwich plate with external layers made of graphite/epoxy plain-weave fabric composite and aluminium honeycomb core.

of their structure, sandwiches offer increased bending stiffness with low weight and excellent thermal and sound insulation. Sandwich plates and shells can be classified in two main classes: (i) weak-core sandwich, characterized by low weight, and (ii) strong-core sandwich.

In a weak-core sandwich, the in-plane stresses are carried by the external layers and are negligible in the core, which carries only transverse shear and transverse normal stresses. Although the external layers can be treated with classical or shear deformation shell theories, special formulations can be developed for the core (Librescu 1975).

For a strong-core sandwich, the in-plane stresses cannot be neglected in the core, and theories for moderately thick shells or plates with shear deformation must be used to model the core.

2.4.6 Functionally Graded Materials and Thermal Effects

Functionally graded materials are those in which the volume fractions of two or more phases are varied continuously as a function of the position along certain directions. Thermal barrier plates for high-temperature applications may be formed from a mixture of ceramic and metal. The composition is varied from a ceramic-rich surface to a metal-rich surface, with a desired variation of the volume fractions of the two materials through the thickness. The gradual change of the material properties avoids discontinuities of stresses that are associated with composite laminates and

Table 2.1. *Temperature-dependent coefficients of material properties*

Property	Material	P_{-1}	P_0	P_1	P_2	P_3
E	Si_3N_4	0	385×10^9 Pa	-3×10^{-4}	2.2×10^{-7}	-8.9×10^{-11}
	AISI 304	0	192×10^9 Pa	3×10^{-4}	-6.5×10^{-7}	0
ρ	Si_3N_4	0	2370 kg/m^3	0	0	0
	AISI 304	0	7850 kg/m^3	0	0	0
ν	Si_3N_4	0	0.24	0	0	0
	AISI 304	0	0.3	0	0	0

are the origin of delamination, cracks and interface stresses due to different thermal expansion of fiber and matrix. In functionally graded materials, these problems are avoided or reduced by gradually varying the volume fraction of the constituents.

A material with property graduation through the panel thickness h is considered (Reddy 2000)

$$P(z) = (P_t - P_b)\,V + P_b, \tag{2.88}$$

where the profile of volume fraction variation with z is expressed by

$$V = \left(\frac{z}{h} + \frac{1}{2}\right)^n. \tag{2.89}$$

P is the generic material property (e.g. elastic moduli E and G, density ρ and thermal coefficient of expansion α); P_t and P_b are the property at the top and bottom surface of the panel, respectively; and n is a parameter that dictates the material variation profile through the thickness. For $n = 1$, the material property varies linearly through the thickness. For very small n, the property varies very quickly close to the bottom surface; the opposite is obtained for a very large n value.

The generic material properties P_t, P_b and P are temperature dependent, and this dependence may be formulated as

$$P = P_0(P_{-1}T^{-1} + 1 + P_1 T + P_2 T^2 + P_3 T^3),$$

in which P_0, P_{-1}, P_1, P_2 and P_3 are the coefficients of the dependence form the temperature T(K) and are specific for each material. Typical coefficients for silicon nitride ceramic (Si_3N_4) and AISI 304 stainless steel are reported in Table 2.1.

The stress-strain relations for a functionally graded material in plane stress are given by (Reddy 2000)

$$\begin{Bmatrix} \sigma_x \\ \sigma_y \\ \tau_{xy} \end{Bmatrix} = \begin{bmatrix} \frac{E}{1-\nu^2} & \frac{\nu E}{1-\nu^2} & 0 \\ \frac{\nu E}{1-\nu^2} & \frac{E}{1-\nu^2} & 0 \\ 0 & 0 & \frac{E}{2(1+\nu)} \end{bmatrix} \left(\begin{Bmatrix} \varepsilon_x \\ \varepsilon_y \\ \gamma_{xy} \end{Bmatrix} - \begin{Bmatrix} 1 \\ 1 \\ 0 \end{Bmatrix} \alpha \Delta T \right), \tag{2.90}$$

where α is the thermal coefficient of expansion and ΔT is the temperature change from a stress-free state. The Young's modulus E and the thermal coefficient of expansion α vary through the panel thickness according to equation (2.88), also as a function of the nonuniform temperature distribution, but ν can be assumed to be constant in many applications. The temperature change ΔT usually is a function of z.

2.5 Nonlinear Shear Deformation Theories for Moderately Thick, Laminated and Functionally Graded, Doubly Curved Shells

In the case of moderately thick laminated shells, the classical shell theories presented in Chapter 1 and Sections 2.2 and 2.3 can become inaccurate. In fact, the hypotheses of negligible shear deformation and rotary inertia for thick laminated shells can be a rough approximation. For laminated composite shells, because of the anisotropic material, there is a coupling between bending and stretching. The use of the Kirchhoff-Love hypothesis, which assumes that the normals to the middle surface after deformation remain straight and undergo no thickness stretching, gives rise to an underprediction of the potential stress energy in laminated shells; this is due to the neglect of the shear strains. For this reason, the nonlinear first-order shear deformation theory and the nonlinear higher-order shear deformation theory are introduced here. Analogous theories exist for plates; however, these can be obtained as a special case of doubly curved shells setting to infinity the two principal radii of curvature.

2.5.1 Nonlinear First-Order Shear Deformation Theory for Doubly Curved Shells of Constant Curvature

The nonlinear first-order shear deformation theory of shells, introduced by Reddy and Chandrashekhara (1985), is presented. Five independent variables, three displacements u, v, w and two rotations ϕ_1 and ϕ_2, are used to describe the shell deformation. This theory may be regarded as the thick-shell version of the Sanders shell theory for linear terms and of the Donnell nonlinear shell theory for nonlinear terms.

A doubly curved (with constant curvature) laminated shell made of a finite number of orthotropic layers, oriented arbitrarily with respect to the shell curvilinear coordinate system (x, y, z), is considered. The coordinate system is chosen such that x and y are principal lines of curvature of the middle surface, which is obtained for $z = 0$, and the coordinate z is taken always perpendicular to the middle surface.

The hypotheses are (i) the thickness of the shell is small compared with the principal radii of curvature, so that only moderately thick shells can be considered; (ii) the transverse normal stress σ_z is negligible; in general, it is verified that σ_z is small compared to τ_{xz} and τ_{yz}, except near the shell edges, so that the hypothesis is a good approximation of the actual behavior of moderately thick shells; and (iii) the normal to the middle surface of the shell before deformation remains straight, but not necessarily normal, after deformation; this is a relaxed version of the Kirchhoff-Love hypothesis (H5) in Section 1.3.4.

The displacements of a generic point are related to the middle surface displacements by

$$u_1 = (1 + z/R_x)\,u + z\phi_1, \quad u_2 = (1 + z/R_y)\,v + z\phi_2, \quad u_3 = w + w_0, \qquad (2.91\text{a–c})$$

where ϕ_1 and ϕ_2 are the rotations of the transverse normals about the y and x axes, respectively, and w_0 is the geometrical imperfection in z direction, as usual. A higher-order (in z) displacement field can be assumed in equations (2.91a–c); however, a linear field in z is assumed for the first-order shear deformation theory; ψ and θ are the angular coordinates and R_x and R_y are principal radii of curvature (assumed to be constant). In equation (2.91c), it is assumed that the normal displacement is constant through the thickness, which means that $\varepsilon_z = 0$ is assumed.

Keeping only the linear-displacement terms in equations (2.60b,c), the following expressions are obtained for the transverse shear strains:

$$\gamma_{xz} = \frac{\partial u_1}{\partial z} + \frac{1}{1 + z/R_x}\left(\frac{\partial u_3}{R_x\,\partial\psi} - \frac{u_1}{R_x}\right), \tag{2.92a}$$

$$\gamma_{yz} = \frac{\partial u_2}{\partial z} + \frac{1}{1 + z/R_y}\left(\frac{\partial u_3}{R_y\,\partial\theta} - \frac{u_2}{R_y}\right). \tag{2.92b}$$

The strain-displacement equations for the first-order shear deformation theory, obtained neglecting terms z/R_x and z/R_y, are given by

$$\varepsilon_x = \varepsilon_{x,0} + z k_x, \tag{2.93a}$$

$$\varepsilon_y = \varepsilon_{y,0} + z k_y, \tag{2.93b}$$

$$\gamma_{xy} = \gamma_{xy,0} + z k_{xy}, \tag{2.93c}$$

$$\gamma_{xz} = \gamma_{xz,0}, \tag{2.93d}$$

$$\gamma_{yz} = \gamma_{yz,0}, \tag{2.93e}$$

where

$$\varepsilon_{x,0} = \frac{\partial u}{R_x\,\partial\psi} + \frac{w}{R_x} + \frac{1}{2}\left(\frac{\partial w}{R_x\,\partial\psi}\right)^2 + \frac{\partial w}{R_x\,\partial\psi}\frac{\partial w_0}{R_x\,\partial\psi}, \tag{2.94a}$$

$$\varepsilon_{y,0} = \frac{\partial v}{R_y\,\partial\theta} + \frac{w}{R_y} + \frac{1}{2}\left(\frac{\partial w}{R_y\,\partial\theta}\right)^2 + \frac{\partial w}{R_y\,\partial\theta}\frac{\partial w_0}{R_y\,\partial\theta}, \tag{2.94b}$$

$$\gamma_{xy,0} = \frac{\partial u}{R_y\,\partial\theta} + \frac{\partial v}{R_x\,\partial\psi} + \frac{\partial w}{R_x\,\partial\psi}\frac{\partial w}{R_y\,\partial\theta} + \frac{\partial w}{R_x\,\partial\psi}\frac{\partial w_0}{R_y\,\partial\theta} + \frac{\partial w_0}{R_x\,\partial\psi}\frac{\partial w}{R_y\,\partial\theta}, \tag{2.94c}$$

$$\gamma_{xz,0} = \phi_1 + \frac{\partial w}{R_x\,\partial\psi} - \frac{u}{R_x}, \tag{2.94d}$$

$$\gamma_{yz,0} = \phi_2 + \frac{\partial w}{R_y\,\partial\theta} - \frac{v}{R_y}, \tag{2.94e}$$

$$k_x = \frac{\partial\phi_1}{R_x\,\partial\psi}, \tag{2.94f}$$

$$k_y = \frac{\partial\phi_2}{R_y\,\partial\theta}, \tag{2.94g}$$

$$k_{xy} = \frac{\partial\phi_1}{R_y\,\partial\theta} + \frac{\partial\phi_2}{R_x\,\partial\psi} + \frac{1}{2}\left(\frac{1}{R_y} - \frac{1}{R_x}\right)\left(\frac{\partial v}{R_x\,\partial\psi} - \frac{\partial u}{R_y\,\partial\theta}\right). \tag{2.94h}$$

Equations (2.93d,e) show a uniform distribution of shear strains through the shell thickness, which gives uniform shear stresses. The top and bottom surfaces of the shell can clearly not support shear stresses; therefore, the result is only a first approximation. The actual distribution of shear stresses is close to a parabolic distribution through the thickness, taking zero value at the top and bottom surfaces. For this reason, for equilibrium considerations, it is necessary to introduce a shear correction factor with the first-order shear deformation theory in order not to overestimate the shear forces.

2.5.2 Elastic Strain Energy for Laminated Shells

The stress-strain relations for the k-th orthotropic lamina of the shell, in the material principal coordinates, are obtained by modifying equation (2.76) in order to take into account shear strains and stresses, under the hypothesis $\sigma_3 = 0$,

$$\begin{Bmatrix} \sigma_1 \\ \sigma_2 \\ \tau_{12} \\ \tau_{13} \\ \tau_{23} \end{Bmatrix}^{(k)} = \begin{bmatrix} c_{11} & c_{12} & 0 & 0 & 0 \\ c_{21} & c_{22} & 0 & 0 & 0 \\ 0 & 0 & G_{12} & 0 & 0 \\ 0 & 0 & 0 & G_{13} & 0 \\ 0 & 0 & 0 & 0 & G_{23} \end{bmatrix}^{(k)} \begin{Bmatrix} \varepsilon_1 \\ \varepsilon_2 \\ \gamma_{12} \\ \gamma_{13} \\ \gamma_{23} \end{Bmatrix}, \tag{2.95}$$

where G_{12}, G_{13} and G_{23} are the shear moduli in 1–2, 1–3 and 2–3 directions and the coefficients c_{ij} are given in equation (2.77); τ_{13} and τ_{23} are the shear stresses. Equation (2.95) is obtained (i) under the transverse isotropy assumption with respect to planes parallel to the 2-3 plane, that is, assuming fibers in the direction parallel to axis 1, so that $E_2 = E_3$, $G_{12} = G_{13}$ and $\nu_{12} = \nu_{13}$; and (ii) solving the constitutive equations for ε_3 as function of ε_1 and ε_2 and then eliminating it, thus approximately removing the inconsistency of $\sigma_3 = \varepsilon_3 = 0$ introduced in the shell theory.

Equation (2.95) can be transformed to the shell coordinates by using equation (2.82) for the 3×3 submatrix obtained by taking the first three lines and rows and leaving unchanged the other terms, and rewritten in the following vectorial notation:

$$\left\{\sigma^{(k)}\right\} = \left[Q^{(k)}\right]\{\varepsilon\}, \tag{2.96}$$

where $Q_{ij}^{(k)}$ are the material properties of the k-th layer in the shell principal coordinates.

The elastic strain energy U_S of the shell, introducing the variables $x = R_x\psi$ and $y = R_y\theta$, is given by

$$U_S = \frac{1}{2}\sum_{k=1}^{K}\int_0^a\int_0^b\int_{h^{(k-1)}}^{h^{(k)}} \left(\sigma_x^{(k)}\varepsilon_x + \sigma_y^{(k)}\varepsilon_y + \tau_{xy}^{(k)}\gamma_{xy} + K_x^2\tau_{xz}^{(k)}\gamma_{xz} + K_y^2\tau_{yz}^{(k)}\gamma_{yz}\right)$$
$$\times (1 + z/R_x)(1 + z/R_y)\,\mathrm{d}x\,\mathrm{d}y\,\mathrm{d}z, \tag{2.97}$$

where K is the total number of layers in the laminated shell, $(h^{(k\text{-}1)}, h^{(k)})$ are the z coordinates of the k-th layer and K_x and K_y are the shear correction factors. A typical value of the shear correction factor is $K_x^2 = K_y^2 = 3/4$ or $5/6$. For simplicity, a shell of rectangular base has been considered.

If required, the shear force resultants per unit length Q_i are given by

$$Q_i = K_i^2\sum_{k=1}^{K}\int_{h^{(k-1)}}^{h^{(k)}} \tau_{iz}^{(k)}\mathrm{d}z, \quad i = x, y. \tag{2.98}$$

2.5.3 Kinetic Energy with Rotary Inertia for Laminated Shells

The kinetic energy T_S of the shell, including rotary inertia, is given by

$$T_S = \frac{1}{2}\sum_{k=1}^{K} \rho_S^{(k)} \int_0^a \int_0^b \int_{h^{(k-1)}}^{h^{(k)}} (\dot{u}_1^2 + \dot{u}_2^2 + \dot{u}_3^2)(1 + z/R_x)(1 + z/R_y)\mathrm{d}x\,\mathrm{d}y\,\mathrm{d}z$$

$$= \frac{1}{2}\sum_{k=1}^{K} \rho_S^{(k)} \int_0^a \int_0^b \int_{h^{(k-1)}}^{h^{(k)}} \{[(1 + z/R_x)\dot{u} + z\dot{\phi}_1]^2 + [(1 + z/R_y)\dot{v} + z\dot{\phi}_2]^2 + \dot{w}^2\}$$
$$\times (1 + z/R_x)(1 + z/R_y)\,\mathrm{d}x\,\mathrm{d}y\,\mathrm{d}z, \tag{2.99}$$

where $\rho_S^{(k)}$ is the mass density of the k-th layer of the shell and the overdot denotes time derivative. After simple calculation, equation (2.99) becomes

$$T_S = \frac{1}{2}\sum_{k=1}^{K} \rho_S^{(k)} \int_0^a \int_0^b \int_{h^{(k-1)}}^{h^{(k)}} \Bigg\{\dot{u}^2 + \dot{v}^2 + \dot{w}^2 + z$$
$$\times \left[2\dot{\phi}_1\dot{u} + 2\dot{\phi}_2\dot{v} + \dot{u}^2\left(\frac{3}{R_x} + \frac{1}{R_y}\right) + \dot{v}^2\left(\frac{1}{R_x} + \frac{3}{R_y}\right) + \dot{w}^2\left(\frac{1}{R_x} + \frac{1}{R_y}\right)\right]$$
$$+ z^2\left[\dot{\phi}_1^2 + \dot{\phi}_2^2 + 2\dot{\phi}_1\dot{u}\left(\frac{2}{R_x} + \frac{1}{R_y}\right) + 2\dot{\phi}_2\dot{v}\left(\frac{1}{R_x} + \frac{2}{R_y}\right)\right.$$
$$\left. + 3\left(\frac{\dot{u}^2}{R_x} + \frac{\dot{v}^2}{R_y}\right)\left(\frac{1}{R_x} + \frac{1}{R_y}\right) + \frac{\dot{w}^2}{R_x R_y}\right] + O(z^2)\Bigg\}\,\mathrm{d}x\,\mathrm{d}y\,\mathrm{d}z, \tag{2.100}$$

where $O(z^2)$ are small terms compared with z^2. The z term vanishes after integration on z in the case of laminate with symmetric density with respect to the z axis. In particular, for a laminate with the same density for all the layers and uniform thickness, the following simplified expression is obtained:

$$T_S = \frac{1}{2}\rho_S h \int_0^a \int_0^b \Bigg\{\dot{u}^2 + \dot{v}^2 + \dot{w}^2 + \frac{h^2}{12}\left[\dot{\phi}_1^2 + \dot{\phi}_2^2 + 2\dot{\phi}_1\dot{u}\left(\frac{2}{R_x} + \frac{1}{R_y}\right)\right.$$
$$\left. + 2\dot{\phi}_2\dot{v}\left(\frac{1}{R_x} + \frac{2}{R_y}\right) + 3\left(\frac{\dot{u}^2}{R_x} + \frac{\dot{v}^2}{R_y}\right)\left(\frac{1}{R_x} + \frac{1}{R_y}\right) + \frac{\dot{w}^2}{R_x R_y}\right] + O(h^4)\Bigg\}\,\mathrm{d}x\,\mathrm{d}y. \tag{2.101}$$

2.5.4 Nonlinear Higher-Order Shear Deformation Theory for Laminated, Doubly Curved Shells

A linear higher-order shear deformation theory of shells has been introduced by Reddy (1984) and Reddy and Liu (1985). Dennis and Palazotto (1990) and Palazotto and Dennis (1992) have extended this theory to nonlinear deformation. The reason for introducing this theory is to overcome the limit of the uniform shear strain and stress distribution through the thickness, obtained with the first-order shear deformation theory.

A laminated shell of arbitrary shape, made of a finite number of orthotropic layers, oriented arbitrarily with respect to the shell curvilinear coordinate system (α_1, α_2, z), is considered; however, the theory is unchanged for isotropic, orthotropic and functionally graded materials. The displacements of a generic point are related to the middle surface displacements by

$$u_1 = (1 + z/R_1)\, u + z\phi_1 + z^2\psi_1 + z^3\gamma_1 + z^4\theta_1, \tag{2.102a}$$

$$u_2 = (1 + z/R_2)\, v + z\phi_2 + z^2\psi_2 + z^3\gamma_2 + z^4\theta_2, \tag{2.102b}$$

$$u_3 = w + w_0, \tag{2.102c}$$

where ϕ_1 and ϕ_2 are the rotations of the transverse normals at $z = 0$ about the $\mathbf{e}_2$ and $\mathbf{e}_1$ axes, respectively. Keeping only the linear displacement terms in equations (2.60b,c), the following expressions are obtained for the transverse shear strains:

$$\gamma_{13} = \frac{\partial u_1}{\partial z} + \frac{1}{1 + z/R_1}\left(\frac{\partial u_3}{A_1 \partial \alpha_1} - \frac{u_1}{R_1}\right), \tag{2.103a}$$

$$\gamma_{23} = \frac{\partial u_2}{\partial z} + \frac{1}{1 + z/R_2}\left(\frac{\partial u_3}{A_2\, \partial \alpha_2} - \frac{u_2}{R_2}\right), \tag{2.103b}$$

where A_1 and A_2 are the Lamé parameters of the middle surface and R_1 and R_2 are the principal radii of curvature.

The expressions for the transverse shear strains are obtained by substituting equations (2.102a–c) into (2.103a,b); the vanishing of the shear strains at the top and the bottom surfaces of the shell requires

$$\gamma_{13}|_{z=\pm h/2} = 0, \qquad \gamma_{23}|_{z=\pm h/2} = 0. \tag{2.104a,b}$$

Equations (2.104a,b) give

$$\psi_1 = \psi_2 = 0, \tag{2.105a,b}$$

$$\gamma_1 = \frac{-4}{3h^2}\left(\phi_1 + \frac{\partial w}{A_1\, \partial \alpha_1}\right), \quad \gamma_2 = \frac{-4}{3h^2}\left(\phi_2 + \frac{\partial w}{A_2\, \partial \alpha_2}\right), \tag{2.106a,b}$$

$$\theta_1 \cong \frac{\gamma_1}{2R_1}, \quad \theta_2 \cong \frac{\gamma_2}{2R_2}. \tag{2.107a,b}$$

In particular, in equations (2.107a,b), h/R_2 has been neglected with respect to unity. Also negligible are the fourth-order terms in θ_1 and θ_2. Finally,

$$u_1 = (1 + z/R_1)\, u + z\phi_1 - \frac{4}{3h^2} z^3 \left(\phi_1 + \frac{\partial w}{A_1\, \partial \alpha_1}\right), \tag{2.108a}$$

$$u_2 = (1 + z/R_2)\, v + z\phi_2 - \frac{4}{3h^2} z^3 \left(\phi_2 + \frac{\partial w}{A_2\, \partial \alpha_2}\right), \tag{2.108b}$$

$$u_3 = w + w_0, \tag{2.108c}$$

where the geometric imperfection w_0 in the normal direction has been introduced. Equations (2.108a,b) represent the parabolic distribution of shear effects through the thickness and satisfy the zero shear boundary condition at both the top and bottom surfaces of the shell. This is the justification for the use of a third-order shear deformation theory.

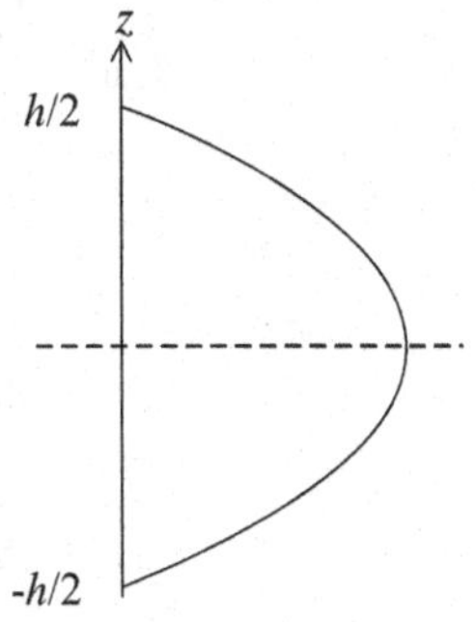

Figure 2.14. Shear strain and stress distribution through the shell thickness.

By substituting equations (2.108a–c) into (2.103a,b), and using approximation (2.5a), the following strain-displacement relationships are obtained for the shear strains:

$$\gamma_{13} = \left(\frac{\partial w}{A_1 \partial \alpha_1} + \phi_1\right)\left(1 - \frac{z}{R_1} - z^2\frac{4}{h^2} + z^3\frac{4}{3R_1h^2} - z^4\frac{4}{3R_1^2h^2}\right) + \left(\frac{z}{R_1}\right)^2 \phi_1$$
$$\cong \left(\frac{\partial w}{A_1 \partial \alpha_1} + \phi_1\right)\left(1 - \frac{z}{R_1} - z^2\frac{4}{h^2} + z^3\frac{4}{3R_1h^2}\right), \qquad (2.109a)$$

$$\gamma_{23} = \left(\frac{\partial w}{A_2 \partial \alpha_2} + \phi_2\right)\left(1 - \frac{z}{R_2} - z^2\frac{4}{h^2} + z^3\frac{4}{3R_2h^2} - z^4\frac{4}{3R_2^2h^2}\right) + \left(\frac{z}{R_2}\right)^2 \phi_2$$
$$\cong \left(\frac{\partial w}{A_2 \partial \alpha_2} + \phi_2\right)\left(1 - \frac{z}{R_2} - z^2\frac{4}{h^2} + z^3\frac{4}{3R_2h^2}\right). \qquad (2.109b)$$

Among the terms kept in equations (2.109a,b), those in z and z^3 are at least one order of magnitude smaller than the others (i.e. of the order h/R_2 with respect to them). These terms have been neglected by Reddy and Liu (1985), Dennis and Palazotto (1990) and Palazotto and Dennis (1992). The strain and stress distribution through the thickness is almost a parabolic distribution, as shown in Figure 2.14, and the shear correction factor is no longer required.

The strain-displacement equations for the higher-order shear deformation theory, keeping terms up to z^3 and using approximation (2.5), can be written as

$$\varepsilon_1 = \varepsilon_{1,0} + z\left(k_1^{(0)} + zk_1^{(1)} + z^2k_1^{(2)}\right), \qquad (2.110a)$$

$$\varepsilon_2 = \varepsilon_{2,0} + z\left(k_2^{(0)} + zk_2^{(1)} + z^2k_2^{(2)}\right), \qquad (2.110b)$$

$$\gamma_{12} = \gamma_{12,0} + z\left(k_{12}^{(0)} + zk_{12}^{(1)} + z^2k_{12}^{(2)}\right), \qquad (2.110c)$$

$$\gamma_{13} = \gamma_{13,0} + z\left(k_{13}^{(0)} + zk_{13}^{(1)} + z^2k_{13}^{(2)}\right), \qquad (2.110d)$$

$$\gamma_{23} = \gamma_{23,0} + z\left(k_{23}^{(0)} + zk_{23}^{(1)} + z^2k_{23}^{(2)}\right), \qquad (2.110e)$$

where

$$\varepsilon_{1,0} = \frac{1}{A_1}\frac{\partial u}{\partial \alpha_1} + \frac{1}{A_1A_2}\frac{\partial A_1}{\partial \alpha_2}v + \frac{w}{R_1} + \frac{1}{2}\left(\frac{\partial w}{A_1\,\partial \alpha_1}\right)^2 + \frac{\partial w}{A_1\,\partial \alpha_1}\frac{\partial w_0}{A_1\,\partial \alpha_1}, \qquad (2.111a)$$

$$\varepsilon_{2,0} = \frac{1}{A_2}\frac{\partial v}{\partial \alpha_2} + \frac{1}{A_1A_2}\frac{\partial A_2}{\partial \alpha_1}u + \frac{w}{R_2} + \frac{1}{2}\left(\frac{\partial w}{A_2\,\partial \alpha_2}\right)^2 + \frac{\partial w}{A_2\,\partial \alpha_2}\frac{\partial w_0}{A_2\,\partial \alpha_2}, \qquad (2.111b)$$

$$\gamma_{12,0} = \frac{1}{A_1}\frac{\partial v}{\partial \alpha_1} + \frac{1}{A_2}\frac{\partial u}{\partial \alpha_2} - \frac{1}{A_1 A_2}\frac{\partial A_1}{\partial \alpha_2}u - \frac{1}{A_1 A_2}\frac{\partial A_2}{\partial \alpha_1}v + \frac{\partial w}{A_1\,\partial \alpha_1}\frac{\partial w}{A_2\,\partial \alpha_2}$$
$$+ \frac{\partial w}{A_1\,\partial \alpha_1}\frac{\partial w_0}{A_2\,\partial \alpha_2} + \frac{\partial w_0}{A_1\,\partial \alpha_1}\frac{\partial w}{A_2\,\partial \alpha_2}, \tag{2.111c}$$

$$\gamma_{13,0} = \phi_1 + \frac{\partial w}{A_1\,\partial \alpha_1}, \tag{2.111d}$$

$$\gamma_{23,0} = \phi_2 + \frac{\partial w}{A_2\,\partial \alpha_2}, \tag{2.111e}$$

$$k_1^{(0)} = \frac{\partial \phi_1}{A_1\,\partial \alpha_1} - \frac{w}{R_1^2} + \frac{v}{A_1 A_2}\frac{\partial A_1}{\partial \alpha_2}\left(-\frac{1}{R_1} + \frac{1}{R_2}\right) + \frac{\phi_2}{A_1 A_2}\frac{\partial A_1}{\partial \alpha_2}, \tag{2.112a}$$

$$k_1^{(1)} = -\frac{1}{R_1 A_1}\left(\frac{\partial \phi_1}{\partial \alpha_1} + \frac{\partial u}{R_1\,\partial \alpha_1}\right) - \frac{\partial A_1}{R_1 A_1 A_2\,\partial \alpha_2}\left(\phi_2 + \frac{v}{R_2}\right), \tag{2.112b}$$

$$k_1^{(2)} = -\frac{4}{3h^2}\left(\frac{\partial \phi_1}{A_1\,\partial \alpha_1} + \frac{\partial^2 w}{A_1^2\,\partial \alpha_1^2} + \frac{\phi_2}{A_1 A_2}\frac{\partial A_1}{\partial \alpha_2} - \frac{\partial w}{A_1^3\,\partial \alpha_1}\frac{\partial A_1}{\partial \alpha_1} + \frac{\partial w}{A_1 A_2^2\,\partial \alpha_2}\frac{\partial A_1}{\partial \alpha_2}\right), \tag{2.112c}$$

$$k_2^{(0)} = \frac{\partial \phi_2}{A_2\,\partial \alpha_2} - \frac{w}{R_2^2} + \frac{u}{A_1 A_2}\frac{\partial A_2}{\partial \alpha_1}\left(\frac{1}{R_1} - \frac{1}{R_2}\right) + \frac{\phi_1}{A_1 A_2}\frac{\partial A_2}{\partial \alpha_1}, \tag{2.112d}$$

$$k_2^{(1)} = -\frac{1}{R_2 A_2}\left(\frac{\partial \phi_2}{\partial \alpha_2} + \frac{\partial v}{R_2\,\partial \alpha_2}\right) - \frac{\partial A_2}{R_2 A_1 A_2\,\partial \alpha_1}\left(\phi_1 + \frac{u}{R_1}\right), \tag{2.112e}$$

$$k_2^{(2)} = -\frac{4}{3h^2}\left(\frac{\partial \phi_2}{A_2\,\partial \alpha_2} + \frac{\partial^2 w}{A_2^2\,\partial \alpha_2^2} + \frac{\phi_1}{A_1 A_2}\frac{\partial A_2}{\partial \alpha_1} - \frac{\partial w}{A_2^3\,\partial \alpha_2}\frac{\partial A_2}{\partial \alpha_2} + \frac{\partial w}{A_1^2 A_2\,\partial \alpha_1}\frac{\partial A_2}{\partial \alpha_1}\right), \tag{2.112f}$$

$$k_{12}^{(0)} = \frac{\partial \phi_1}{A_2\,\partial \alpha_2} + \frac{\partial \phi_2}{A_1\,\partial \alpha_1} - \frac{\phi_1}{A_1 A_2}\frac{\partial A_1}{\partial \alpha_2} - \frac{\phi_2}{A_1 A_2}\frac{\partial A_2}{\partial \alpha_1} + \frac{\partial u}{A_2\,\partial \alpha_2}\left(\frac{1}{R_1} - \frac{1}{R_2}\right)$$
$$+ \frac{\partial v}{A_1\,\partial \alpha_1}\left(-\frac{1}{R_1} + \frac{1}{R_2}\right), \tag{2.112g}$$

$$k_{12}^{(1)} = -\frac{\partial \phi_1}{R_2 A_2\,\partial \alpha_2} - \frac{\partial \phi_2}{R_1 A_1\,\partial \alpha_1} - \frac{\partial u}{R_1 R_2 A_2\,\partial \alpha_2} - \frac{\partial v}{R_1 R_2 A_1\,\partial \alpha_1}$$
$$+ \frac{1}{A_1 A_2}\left(\frac{\phi_1}{R_1}\frac{\partial A_1}{\partial \alpha_2} + \frac{\phi_2}{R_2}\frac{\partial A_2}{\partial \alpha_1} + \frac{u}{R_1^2}\frac{\partial A_1}{\partial \alpha_2} + \frac{v}{R_2^2}\frac{\partial A_2}{\partial \alpha_1}\right), \tag{2.112h}$$

$$k_{12}^{(2)} = -\frac{4}{3h^2}\left(\frac{\partial \phi_1}{A_2\,\partial \alpha_2} + \frac{\partial \phi_2}{A_1\,\partial \alpha_1} + 2\frac{\partial^2 w}{A_1 A_2\,\partial \alpha_1\,\partial \alpha_2} - 2\frac{\partial w}{A_1^2 A_2\,\partial \alpha_1}\frac{\partial A_1}{\partial \alpha_2}\right.$$
$$\left. - 2\frac{\partial w}{A_1 A_2^2\,\partial \alpha_2}\frac{\partial A_2}{\partial \alpha_1} - \frac{\phi_1}{A_1 A_2}\frac{\partial A_1}{\partial \alpha_2} - \frac{\phi_2}{A_1 A_2}\frac{\partial A_2}{\partial \alpha_1}\right), \tag{2.112i}$$

$$k_{13}^{(0)} = -\frac{1}{R_1}\gamma_{13,0}, \quad k_{13}^{(1)} = -\frac{4}{h^2}\gamma_{13,0}, \quad k_{13}^{(2)} = \frac{4}{3R_1 h^2}\gamma_{13,0}, \tag{2.113a–c}$$

$$k_{23}^{(0)} = -\frac{1}{R_2}\gamma_{23,0}, \quad k_{23}^{(1)} = -\frac{4}{h^2}\gamma_{23,0}, \quad k_{23}^{(2)} = \frac{4}{3R_2 h^2}\gamma_{23,0}. \tag{2.113d–f}$$

Equations (2.111a–c) and (2.112a–i) have been obtained by substituting equations (2.108a–c) into (2.59a–c) and retaining only nonlinear terms in $\partial w/\partial\alpha_i$. It can be observed that $k_{13}^{(0)}, k_{13}^{(2)}, k_{23}^{(0)}, k_{23}^{(2)}, k_1^{(1)}, k_2^{(1)}, k_{12}^{(1)}$ are negligible for shells that are not very thick.

If all the nonlinear terms are retained in the membrane strains, equations (2.111a–c), in the case of zero geometric imperfections, are replaced by (2.69a–c) obtained with the Novozhilov theory, neglecting shear deformation.

2.5.5 Elastic Strain and Kinetic Energies, Including Rotary Inertia, for Laminated Shells According with Higher-Order Shear Deformation Theory

The elastic strain energy of the shell is still given by equation (2.97) with the shear correction factors equal to one, but the stresses are now given by

$$\{\sigma^{(k)}\} = [Q^{(k)}]\begin{Bmatrix}\varepsilon_{1,0}\\ \varepsilon_{2,0}\\ \gamma_{12,0}\\ \gamma_{13,0}\\ \gamma_{23,0}\end{Bmatrix} + z[Q^{(k)}]\begin{Bmatrix}k_1^{(0)}\\ k_2^{(0)}\\ k_{12}^{(0)}\\ 0\\ 0\end{Bmatrix} + z^2[Q^{(k)}]\begin{Bmatrix}0\\ 0\\ 0\\ k_{13}^{(1)}\\ k_{23}^{(1)}\end{Bmatrix} + z^3[Q^{(k)}]\begin{Bmatrix}k_1^{(2)}\\ k_2^{(2)}\\ k_{12}^{(2)}\\ 0\\ 0\end{Bmatrix}; \tag{2.114}$$

for the first-order shear deformation theory, the terms in z^2 and z^3 do not appear. In equation (2.114), $k_{13}^{(0)}$, $k_{13}^{(2)}$, $k_{23}^{(0)}$, $k_{23}^{(2)}$, $k_1^{(1)}$, $k_2^{(1)}$, $k_{12}^{(1)}$ have been neglected, supposing a moderately thick shell.

The kinetic energy T_S of the shell, including rotary inertia, is given by

$$\begin{aligned} T_S &= \frac{1}{2}\sum_{k=1}^{K}\rho_S^{(k)}\int_0^a\int_0^b\int_{h^{(k-1)}}^{h^{(k)}}(\dot{u}_1^2+\dot{u}_2^2+\dot{u}_3^2)(1+z/R_x)(1+z/R_y)\,\mathrm{d}x\,\mathrm{d}y\,\mathrm{d}z \\ &= \frac{1}{2}\sum_{k=1}^{K}\rho_S^{(k)}\int_0^a\int_0^b\int_{h^{(k-1)}}^{h^{(k)}}\left\{\left[(1+z/R_1)\,\dot{u}+z\dot{\phi}_1-\frac{4}{3h^2}z^3\left(\dot{\phi}_1+\frac{\partial\dot{w}}{A_1\partial\alpha_1}\right)\right]^2\right. \\ &\quad \left.+\left[(1+z/R_2)\,\dot{v}+z\dot{\phi}_2-\frac{4}{3h^2}z^3\left(\dot{\phi}_2+\frac{\partial\dot{w}}{A_2\,\partial\alpha_2}\right)\right]^2+\dot{w}^2\right\} \\ &\quad \times(1+z/R_x)(1+z/R_y)\,\mathrm{d}x\,\mathrm{d}y\,\mathrm{d}z. \end{aligned} \tag{2.115}$$

After simple manipulation, equation (2.115) becomes

$$\begin{aligned} T_S = \frac{1}{2}\sum_{k=1}^{K}\rho_S^{(k)}\int_0^a\int_0^b\int_{h^{(k-1)}}^{h^{(k)}}&\left\{\dot{u}^2+\dot{v}^2+\dot{w}^2\right. \\ &+z\left[2\dot{\phi}_1\dot{u}+2\dot{\phi}_2\dot{v}+\dot{u}^2\left(\frac{3}{R_x}+\frac{1}{R_y}\right)+\dot{v}^2\left(\frac{1}{R_x}+\frac{3}{R_y}\right)+\dot{w}^2\left(\frac{1}{R_x}+\frac{1}{R_y}\right)\right] \\ &+z^2\left[\dot{\phi}_1^2+\dot{\phi}_2^2+2\dot{\phi}_1\dot{u}\left(\frac{2}{R_x}+\frac{1}{R_y}\right)+2\dot{\phi}_2\dot{v}\left(\frac{1}{R_x}+\frac{2}{R_y}\right)\right. \end{aligned}$$

$$+3\left(\frac{\dot{u}^2}{R_x}+\frac{\dot{v}^2}{R_y}\right)\left(\frac{1}{R_x}+\frac{1}{R_y}\right)+\frac{\dot{w}^2}{R_x R_y}\Bigg]$$
$$+z^3\left[(\dot{\phi}_1^2+\dot{\phi}_2^2)\left(\frac{1}{R_x}+\frac{1}{R_y}\right)+\frac{2\dot{\phi}_1\dot{u}}{R_x}\left(\frac{1}{R_x}+\frac{2}{R_y}\right)\right.$$
$$+\frac{2\dot{\phi}_2\dot{v}}{R_y}\left(\frac{2}{R_x}+\frac{1}{R_y}\right)+\frac{\dot{u}^2}{R_x^2}\left(\frac{1}{R_x}+\frac{3}{R_y}\right)+\frac{\dot{v}^2}{R_y^2}\left(\frac{3}{R_x}+\frac{1}{R_y}\right)$$
$$\left.-2\frac{4}{3h^2}\dot{u}\left(\dot{\phi}_1+\frac{\partial\dot{w}}{A_1\,\partial\alpha_1}\right)-2\frac{4}{3h^2}\dot{v}\left(\dot{\phi}_2+\frac{\partial\dot{w}}{A_2\,\partial\alpha_2}\right)\right]+o(z^3)\Bigg\}\,\mathrm{d}x\,\mathrm{d}y\,\mathrm{d}z, \tag{2.116}$$

where $o(z^3)$ are small terms compared with z^3. The z and z^3 terms vanish after integration on z in the case of a laminate with symmetric density with respect to the z axis. In particular, for a laminate with the same density for all the layers and uniform thickness, the following simplified expression is obtained:

$$T_S=\frac{1}{2}\rho sh\int_0^a\int_0^b\left\{\dot{u}^2+\dot{v}^2+\dot{w}^2+\frac{h^2}{12}\left[\dot{\phi}_1^2+\dot{\phi}_2^2+2\dot{\phi}_1\dot{u}\left(\frac{2}{R_x}+\frac{1}{R_y}\right)\right.\right.$$
$$\left.\left.+2\dot{\phi}_2\dot{v}\left(\frac{1}{R_x}+\frac{2}{R_y}\right)+3\left(\frac{\dot{u}^2}{R_x}+\frac{\dot{v}^2}{R_y}\right)\left(\frac{1}{R_x}+\frac{1}{R_y}\right)+\frac{\dot{w}^2}{R_x R_y}\right]\right\}\mathrm{d}x\,\mathrm{d}y+O(h^4), \tag{2.117}$$

where $O(h^4)$ are small terms of the order of h^4.

2.5.6 Elastic Strain Energy for Heated, Functionally Graded Shells

The stress-strain relations for a functionally graded material, generalizing equation (2.90) for shear strains and stresses, are given by

$$\begin{Bmatrix}\sigma_1\\ \sigma_2\\ \tau_{12}\\ \tau_{13}\\ \tau_{23}\end{Bmatrix}=E(z)\begin{bmatrix}\frac{1}{1-\nu^2} & \frac{\nu}{1-\nu^2} & 0 & 0 & 0\\ \frac{\nu}{1-\nu^2} & \frac{1}{1-\nu^2} & 0 & 0 & 0\\ 0 & 0 & \frac{1}{2(1+\nu)} & 0 & 0\\ 0 & 0 & 0 & \frac{1}{2(1+\nu)} & 0\\ 0 & 0 & 0 & 0 & \frac{1}{2(1+\nu)}\end{bmatrix}\left(\begin{Bmatrix}\varepsilon_1\\ \varepsilon_2\\ \gamma_{12}\\ \gamma_{13}\\ \gamma_{23}\end{Bmatrix}-\begin{Bmatrix}1\\ 1\\ 0\\ 0\\ 0\end{Bmatrix}\alpha(z)\Delta T(z)\right), \tag{2.118}$$

where thermal effects are taken into account by the thermal expansion coefficient $\alpha(z)$, and a nonuniform change of temperature ΔT is assumed through the thickness; the Young's modulus of the functionally graded shell can be expressed as

$$E(z)=(E_t-E_b)\left(\frac{z}{h}+\frac{1}{2}\right)^n+E_b, \tag{2.119}$$

where E_t and E_b are the Young's moduli at the top and bottom surface of the panel, respectively, and n is a parameter that dictates the material variation profile through the thickness. Similarly,

$$\alpha(z)=(\alpha_t-\alpha_b)\left(\frac{z}{h}+\frac{1}{2}\right)^n+\alpha_b. \tag{2.120}$$

Both E and α are also a function of the nonuniform temperature T, which varies as function of z; for the sake of brevity, the full expression is not derived here.

The strains of a generic point at distance z from the middle surface of the shell are given by

$$\begin{Bmatrix} \varepsilon_1 \\ \varepsilon_2 \\ \gamma_{12} \\ \gamma_{13} \\ \gamma_{23} \end{Bmatrix} = \begin{Bmatrix} \varepsilon_{1,0} \\ \varepsilon_{2,0} \\ \gamma_{12,0} \\ \gamma_{13,0} \\ \gamma_{23,0} \end{Bmatrix} + z \begin{Bmatrix} k_1^{(0)} \\ k_2^{(0)} \\ k_{12}^{(0)} \\ 0 \\ 0 \end{Bmatrix} + z^2 \begin{Bmatrix} 0 \\ 0 \\ 0 \\ k_{13}^{(1)} \\ k_{23}^{(1)} \end{Bmatrix} + z^3 \begin{Bmatrix} k_1^{(2)} \\ k_2^{(2)} \\ k_{12}^{(2)} \\ 0 \\ 0 \end{Bmatrix}, \quad (2.121)$$

where $k_{13}^{(0)}, k_{13}^{(2)}, k_{23}^{(0)}, k_{23}^{(2)}, k_1^{(1)}, k_2^{(1)}, k_{12}^{(1)}$ have been neglected. Equation (2.121) can be rewritten in vectorial notation as

$$\{\varepsilon\} = \{\varepsilon^{(0)}\} + z\{k^{(0)}\} + z^2\{k^{(1)}\} + z^3\{k^{(2)}\}. \quad (2.122)$$

By using (2.122), equation (2.118) can be rewritten in vectorial notation as

$$\{\sigma\} = E(z)\,[G]\left(\{\varepsilon^{(0)}\} + z\{k^{(0)}\} + z^2\{k^{(1)}\} + z^3\{k^{(2)}\} - \{\lambda\}\,\alpha(z)\Delta T(z)\right), \quad (2.123a)$$

where

$$\{\lambda\}^T = \{1, 1, 0, 0, 0\}^T, \quad (2.123b)$$

and

$$[G] = \begin{bmatrix} \frac{1}{1-\nu^2} & \frac{\nu}{1-\nu^2} & 0 & 0 & 0 \\ \frac{\nu}{1-\nu^2} & \frac{1}{1-\nu^2} & 0 & 0 & 0 \\ 0 & 0 & \frac{1}{2(1+\nu)} & 0 & 0 \\ 0 & 0 & 0 & \frac{1}{2(1+\nu)} & 0 \\ 0 & 0 & 0 & 0 & \frac{1}{2(1+\nu)} \end{bmatrix}. \quad (2.123c)$$

The elastic strain energy U_S of the functionally graded shell is given by

$$U_S = \frac{1}{2}\int_0^a \int_0^b \int_{-h/2}^{h/2} [\sigma_1(\varepsilon_1 - \alpha\Delta T) + \sigma_2(\varepsilon_2 - \alpha\Delta T) + \tau_{12}\gamma_{12} + \tau_{13}\gamma_{13} + \tau_{23}\gamma_{23}]$$
$$\times (1 + z/R_1)(1 + z/R_2)\,\mathrm{d}x\,\mathrm{d}y\,\mathrm{d}z, \quad (2.124)$$

where strains and stresses are given by equations (2.122) and (2.123a), respectively.

2.5.7 Kinetic Energy with Rotary Inertia for Functionally Graded Shells

The density of the functionally graded shell can be expressed by

$$\rho_S(z) = (\rho_t - \rho_b)\left(\frac{z}{h} + \frac{1}{2}\right)^n + \rho_b, \quad (2.125)$$

where ρ_t and ρ_b are the density at the top and bottom surface of the panel, respectively, and n is a parameter that dictates the material variation profile through the thickness.

The kinetic energy T_S of the shell, including rotary inertia, is given by

$$T_S = \frac{1}{2}\int_0^a\int_0^b\int_{-h/2}^{h/2} \rho_S\,(\dot{u}_1^2 + \dot{u}_2^2 + \dot{u}_3^2)(1 + z/R_x)(1 + z/R_y)\,\mathrm{d}x\,\mathrm{d}y\,\mathrm{d}z$$
$$= \frac{1}{2}\int_0^a\int_0^b\int_{-h/2}^{h/2} \rho_S\left\{\left[(1 + z/R_1)\,\dot{u} + z\dot{\phi}_1 - \frac{4}{3h^2}z^3\left(\dot{\phi}_1 + \frac{\partial \dot{w}}{A_1\,\partial\alpha_1}\right)\right]^2\right.$$
$$\left. + \left[(1 + z/R_2)\,\dot{v} + z\dot{\phi}_2 - \frac{4}{3h^2}z^3\left(\dot{\phi}_2 + \frac{\partial \dot{w}}{A_2\,\partial\alpha_2}\right)\right]^2 + \dot{w}^2\right\}$$
$$\times (1 + z/R_x)(1 + z/R_y)\,\mathrm{d}x\,\mathrm{d}y\,\mathrm{d}z. \qquad (2.126)$$

After simple manipulations, equation (2.126) becomes

$$T_S = \frac{1}{2}\int_0^a\int_0^b\int_{-h/2}^{h/2}\left[(\rho_t - \rho_b)\left(\frac{z}{h} + \frac{1}{2}\right)^n + \rho_b\right]\left\{\dot{u}^2 + \dot{v}^2 + \dot{w}^2 + z\left[2\dot{\phi}_1\dot{u} + 2\dot{\phi}_2\dot{v}\right.\right.$$
$$\left. + \dot{u}^2\left(\frac{3}{R_x} + \frac{1}{R_y}\right) + \dot{v}^2\left(\frac{1}{R_x} + \frac{3}{R_y}\right) + \dot{w}^2\left(\frac{1}{R_x} + \frac{1}{R_y}\right)\right]$$
$$+ z^2\left[\dot{\phi}_1^2 + \dot{\phi}_2^2 + 2\dot{\phi}_1\dot{u}\left(\frac{2}{R_x} + \frac{1}{R_y}\right) + 2\dot{\phi}_2\dot{v}\left(\frac{1}{R_x} + \frac{2}{R_y}\right)\right.$$
$$\left. + 3\left(\frac{\dot{u}^2}{R_x} + \frac{\dot{v}^2}{R_y}\right)\left(\frac{1}{R_x} + \frac{1}{R_y}\right) + \frac{\dot{w}^2}{R_x R_y}\right]$$
$$+ z^3\left[(\dot{\phi}_1^2 + \dot{\phi}_2^2)\left(\frac{1}{R_x} + \frac{1}{R_y}\right) + \frac{2\dot{\phi}_1\dot{u}}{R_x}\left(\frac{1}{R_x} + \frac{2}{R_y}\right)\right.$$
$$+ \frac{2\dot{\phi}_2\dot{v}}{R_y}\left(\frac{2}{R_x} + \frac{1}{R_y}\right) + \frac{\dot{u}^2}{R_x^2}\left(\frac{1}{R_x} + \frac{3}{R_y}\right) + \frac{\dot{v}^2}{R_y^2}\left(\frac{3}{R_x} + \frac{1}{R_y}\right) - 2\frac{4}{3h^2}\dot{u}$$
$$\left.\left.\times\left(\dot{\phi}_1 + \frac{\partial \dot{w}}{A_1\,\partial\alpha_1}\right) - 2\frac{4}{3h^2}\dot{v}\left(\dot{\phi}_2 + \frac{\partial \dot{w}}{A_2\,\partial\alpha_2}\right)\right] + O(z^3)\right\}\mathrm{d}x\,\mathrm{d}y\,\mathrm{d}z. \qquad (2.127)$$

2.6 Thermal Effects on Plates and Shells

In Sections 2.4 and 2.5, the effect of nonuniform thermal expansion has been taken into account for functionally graded materials, as they are used for thermal barriers. However, shells and plates of traditional or composite materials can be subjected to significant changes of temperature. Thermal expansion produces a strain growth that gives rise to thermal stresses. Assuming a nonuniform change of temperature ΔT from a stress-free state, the strain-stress relations for a plate or shell made of isotropic material are given by

$$\varepsilon_{11} = \frac{1}{E}(\sigma_{11} - \nu\sigma_{22}) + \alpha\Delta T, \qquad (2.128a)$$

$$\varepsilon_{22} = \frac{1}{E}(\sigma_{22} - \nu\sigma_{11}) + \alpha\Delta T, \tag{2.128b}$$

$$\gamma_{12} = \frac{1}{G}\tau_{12}, \quad \gamma_{13} = \frac{1}{G}\tau_{13}, \quad \gamma_{23} = \frac{1}{G}\tau_{23}, \tag{2.128c–e}$$

where α is the thermal expansion coefficient. Usually ΔT varies on the surface and through the thickness, that is, it is a function of α_1, α_2 and z. Solving equations (2.128) for the stresses, the following equation is obtained:

$$\begin{Bmatrix} \sigma_1 \\ \sigma_2 \\ \tau_{12} \\ \tau_{13} \\ \tau_{23} \end{Bmatrix} = \begin{bmatrix} \frac{E}{1-\nu^2} & \frac{\nu E}{1-\nu^2} & 0 & 0 & 0 \\ \frac{\nu E}{1-\nu^2} & \frac{E}{1-\nu^2} & 0 & 0 & 0 \\ 0 & 0 & \frac{E}{2(1+\nu)} & 0 & 0 \\ 0 & 0 & 0 & \frac{E}{2(1+\nu)} & 0 \\ 0 & 0 & 0 & 0 & \frac{E}{2(1+\nu)} \end{bmatrix} \left(\begin{Bmatrix} \varepsilon_1 \\ \varepsilon_2 \\ \gamma_{12} \\ \gamma_{13} \\ \gamma_{23} \end{Bmatrix} - \begin{Bmatrix} 1 \\ 1 \\ 0 \\ 0 \\ 0 \end{Bmatrix} \alpha\Delta T \right), \tag{2.129}$$

which is analogous to (2.118), and (2.90) in the case of plane stress, obtained for functionally graded materials. In equation (2.129), E and α can be assumed to be a function of the nonuniform temperature T. Extension to orthotropic and laminated plates and shells is straightforward.

REFERENCES

M. Amabili 2005 *International Journal of Non-Linear Mechanics* **40**, 683–710. Non-linear vibrations of doubly curved shallow shells.

I. M. Daniel and O. Ishai 1994 *Engineering Mechanics of Composite Materials*. Oxford University Press, New York, USA.

S. T. Dennis and A. N. Palazotto 1990 *International Journal of Non-Linear Mechanics* **25**, 67–85. Large displacement and rotation formulation for laminated shells including parabolic transverse shear.

L. Librescu 1975 *Elastostatics and Kinetics of Anisotropic and Heterogeneous Shell-type Structures*. Noordhoff, Leyden, The Netherlands.

V. V. Novozhilov 1953 *Foundations of the Nonlinear Theory of Elasticity*. Graylock Press, Rochester, NY, USA (now available from Dover, NY, USA).

V. V. Novozhilov 1964 *Thin Shell Theory*, 2nd edition. Noordhoff, Groningen, The Netherlands.

A. N. Palazotto and S. T. Dennis 1992 *Nonlinear Analysis of Shell Structures*. AIAA Educational Series, Washington, DC, USA.

J. N. Reddy 1984 *Journal of Engineering Mechanics* **110**, 794–809. Exact solutions of moderately thick laminated shells.

J. N. Reddy 2000 *International Journal for Numerical Methods in Engineering* **47**, 663–684. Analysis of functionally graded plates.

J. N. Reddy and K. Chandrashekhara 1985 *International Journal of Non-Linear Mechanics* **20**, 79–90. Geometrically non-linear transient analysis of laminated, doubly curved shells.

J. N. Reddy and C. F. Liu 1985 *International Journal of Engineering Science* **23**, 319–330. A higher-order shear deformation theory of laminated elastic shells.

3 Introduction to Nonlinear Dynamics

3.1 Introduction

Different from linear systems, where the superposition theorem holds true, vibrations of nonlinear systems can give rise to multiple coexisting solutions, bifurcations and very complex dynamics. This chapter introduces nonlinear dynamics, focusing on the concepts and tools used in the following parts of the book to study nonlinear vibrations of plates and shells.

The resonance frequency of a nonlinear system changes with the vibration amplitude. For very small amplitudes, it coincides with the natural frequency of the linear approximation. For larger amplitudes, the resonance frequency decreases with amplitude for softening systems and increases with amplitude for hardening systems.

Static solutions and their bifurcations can be analyzed by using a local geometric theory. For periodic solutions, the Floquet theory may be used to study the stability and to classify the bifurcations. Numerical techniques are used to find solutions and to follow solution branches, such as the arclength and pseudo-arclength methods.

Hints are given on internal resonances, quasi-periodic and chaotic vibrations, Poincaré maps, Lyapunov exponents and the Lyapunov dimension, which are used to identify complex dynamics.

At the end of this chapter, two different methods are introduced to discretize partial differential equations, which govern vibrations of continuous systems, such as plates and shells. These are the Galerkin method and the energy approach that leads to the Lagrange equations of motion; they allow us to obtain a finite set of ordinary differential equations from the original partial differential equations.

3.2 Periodic Nonlinear Vibrations: Softening and Hardening Systems

A classical example of nonlinear system, originally studied by Duffing (1918), is the forced mass-spring system with viscous damping, where the restoring force of the spring is nonlinear:

$$m\ddot{x} + c\dot{x} + k_1 x + k_2 x^2 + k_3 x^3 = f(t), \tag{3.1}$$

where m is the mass, c is the viscous damping coefficient, k_1 is the linear spring stiffness, k_2 is the quadratic stiffness, k_3 is the cubic stiffness, x is the vibration amplitude, f is the force excitation and t is time. In this case, there is a symmetric nonlinearity, that is, with equal effects in tension and compression. Equation (3.1), which represents a nonautonomous system for the explicit dependence on time, can be rewritten in the form

$$\ddot{x} + 2\zeta\omega_n\dot{x} + \omega_n^2 x + (k_2/m)x^2 + (k_3/m)x^3 = f(t)/m, \tag{3.2}$$

where ζ is the damping ratio and ω_n is the natural circular frequency of the linearized system, that is, the resonance circular frequency for infinitesimal vibration amplitude x. If the excitation is harmonic,

$$f(t) = f_0 \cos(\omega t); \tag{3.3}$$

the first expectation is that the response x is harmonic, by analogy to linear systems. Harmonic vibration with the same frequency of the excitation is given by

$$x = x_0 \cos(\omega t + \varphi), \tag{3.4}$$

where x_0 is the vibration amplitude, ω_n is the excitation circular frequency and φ is the phase. In this case, the frequency-response curve around the linear resonance ω_n is given in Figures 3.1(a,b). In particular, in Figure 3.1(a) the peak of the vibration amplitude moves to the left of ω_n, indicating a softening behavior of the system; this behavior is given by (i) the quadratic nonlinearity in the stiffness, not depending on the sign of k_2, (ii) the cubic nonlinearity with negative sign of k_3 or (iii) a combination of quadratic and cubic nonlinearities. Figure 3.1(b) shows the opposite behavior, with prevalence of the cubic nonlinearity with positive sign of k_3 on the quadratic one. This gives hardening behavior of the system and the peak of the response moves to the right with respect to ω_n. In Figure 3.1(a,b), the continuous line indicates stable periodic response (showing the maximum of the oscillation amplitude in a period), and the dashed line gives unstable response. It is interesting to note that the nonlinear response in Figure 3.1 is not a single-valued function, and a hysteretic effect arises for increasing and decreasing excitation frequency. This gives rise to a jump phenomenon, indicated by arrows in Figures 3.1(a,b).

Other periodic solutions x can be found, such as subharmonic and superharmonic vibrations. The subharmonic vibration has the form

$$x = x_0 \cos(\omega t/n + \varphi), \tag{3.5}$$

where n is an integer. In addition, there are higher harmonics. Similarly, it is possible to define a superharmonic vibration.

If the external excitation is removed in equations (3.1) and (3.2), an autonomous system is obtained. The chain-dotted curve in Figure 3.2 represents the free vibration behavior of the nonlinear system, showing the dependence of the resonance on the vibration amplitude; this curve is called the backbone curve. The forced response of Figure 3.1(b) is superimposed on this curve to show that the backbone curve lies "in the middle" of the forced response curve; that is, it is equidistant from the forced response curve, where the distance is measured, as a first approximation, orthogonal to the backbone curve. Therefore, an experimental technique to obtain the backbone

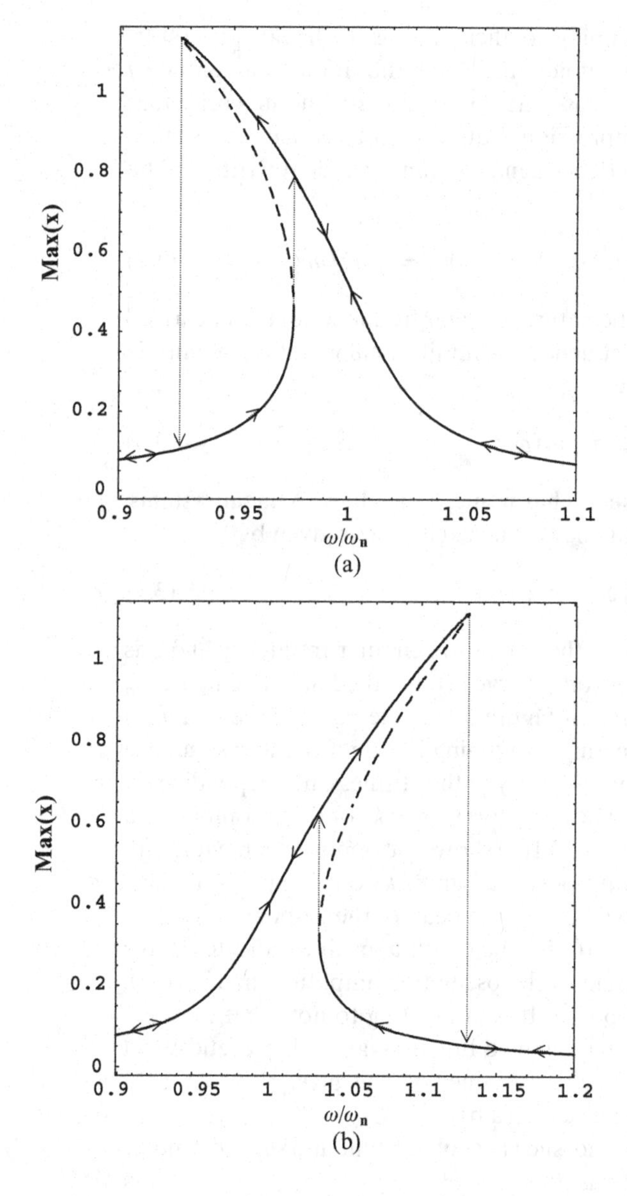

Figure 3.1. Frequency-response curve of the forced, damped Duffing equation; —, stable solution; – –, unstable solution; ↑ or ↓, jump. (a) Softening nonlinear response. (b) Hardening nonlinear response.

curve is to measure the system response to different excitation levels, and then to draw the backbone curve. In practice, the backbone curve can be obtained by joining the peaks of the measured responses at different excitation levels. The backbone curve does not depend on damping. In contrast, damping changes the shape of the forced response curve, which becomes wider and more rounded for increasing damping, up to the point where jumps can disappear altogether.

Many other types of nonlinearities, different from those introduced in equation (3.1), must be used to describe different vibration problems. For example, stiffness terms kx^{α}, with α different from two and three, piecewise stiffness and viscosity, nonlinear viscosity and mass, nonlinear friction forces. However, equation (3.1) is particularly important for shells and plates as a consequence that the nonlinear models introduced in Chapters 1 and 2 give rise to nonlinear equations of motion with linear, quadratic and cubic stiffness terms.

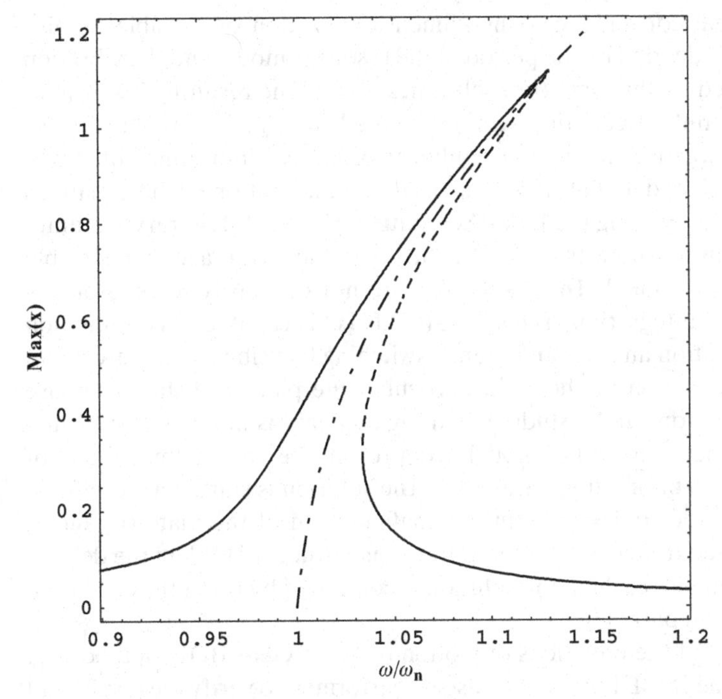

Figure 3.2. Backbone curve and forced response curve; – · –, backbone curve; —, stable forced response; – –, unstable forced response.

3.3 Numerical Integration of the Equations of Motion

It is assumed that a set of N second-order ordinary differential equations of motion describe the system:

$$\ddot{x}_j + 2\zeta_j\,\omega_j\,\dot{x}_j + \sum_{i=1}^{N} z_{j,i}x_i + \sum_{i=1}^{N}\sum_{k=1}^{N} z_{j,i,k}x_i x_k + \sum_{i=1}^{N}\sum_{k=1}^{N}\sum_{l=1}^{N} z_{j,i,k,l}x_i x_k x_l = f_j\cos(\omega\,t), \quad \text{for } j = 1, \ldots, N \tag{3.6}$$

where N is an integer representing the number of degrees of freedom used to discretize the original partial differential equations, $z_{j,i}$ are the coefficients associated with the linear stiffness terms (transformed in the elimination of the mass form the equations), $z_{j,i,k}$ are the coefficients associated with quadratic stiffness terms and $z_{j,i,k,l}$ are the coefficients associated with cubic stiffness terms. In inertial coupling, a linear transformation, described in Section 4.6, converts the equations of motion to the form given in equation (3.6).

The generic j-th equation of motion can be recast into the following two first-order equations by using the dummy variable y_j:

$$\begin{cases} \dot{x}_j = y_j \\ \dot{y}_j = -2\zeta_j\,\omega_j\,y_j - \sum\limits_{i=1}^{N} z_{j,i}x_i - \sum\limits_{i=1}^{N}\sum\limits_{k=1}^{N} z_{j,i,k}x_i x_k \quad \text{for } j = 1, \ldots, N, \\ \qquad - \sum\limits_{i=1}^{N}\sum\limits_{k=1}^{N}\sum\limits_{l=1}^{N} z_{j,i,k,l}x_i x_k x_l + f_j\cos(\omega t) \end{cases} \tag{3.7}$$

For computational convenience a nondimensionalization of variables is also performed: the time is divided by the period of the resonant mode, and the vibration amplitudes are divided by the plate (or shell) thickness h. The resulting $2 \times N$ first-order nonlinear differential equations can be studied by using (i) a software for continuation and bifurcation analysis of nonlinear ordinary differential equations, for example, AUTO (Doedel et al. 1998), or (ii) direct integration of the equations of motion, for example, by using the DIVPAG routine of the IMSL library. Continuation methods allow us to follow the solution path, with the advantage that unstable solutions can also be obtained. These solutions are not ordinarily worked out by using direct numerical integration. The software AUTO is capable of continuation of the solution, bifurcation analysis and branch switching by using pseudo-arclength continuation and collocation methods. In particular, the plate and shell response under harmonic excitation can be studied by using an analysis in two steps: (i) first the excitation frequency is fixed far enough from resonance and the magnitude of the excitation is used as bifurcation parameter. The solution is started at zero force where the solution is the trivial undisturbed configuration of the plate (or shell), and then is incremented to reach the desired force magnitude. (ii) When the desired magnitude of excitation is reached, the solution is continued by using the excitation frequency as bifurcation parameter.

Direct integration of the equations of motion by using Gear BDF method (e.g. routine DIVPAG of the IMSL library) can also be performed to verify the results and obtain the time behavior. The Adams-Gear algorithm can be used in the case of a relatively high dimension of the dynamical system. Indeed, when a high-dimensional phase space is analyzed, the problem can exhibit stiff characteristics due to the presence of different timescales in the response. In simulations with adaptive step-size Runge-Kutta methods, spurious nonstationary and divergent motions can be obtained. Therefore, the Adams-Gear method, designed for stiff equations, should be preferred in this case.

3.4 Local Geometric Theory

Interpretation of nonlinear dynamics is often presented in geometric terms or pictures (Moon 1992). For example, the undamped free vibrations of a single-degree-of-freedom linear system are described by the following equation of motion:

$$\ddot{x} + \omega_n^2 x = 0, \tag{3.8}$$

which gives in the phase plane $(x, \dot{x})$ an ellipse. In this picture, time is implicit, and the time history runs clockwise around the ellipse; the size of the ellipse is determined by the initial conditions for $(x, \dot{x})$. If the time axis is added to the plot, a helix, with elliptic projection on the phase plane, is obtained.

A more general procedure, applicable to nonlinear problems, is to find the equilibrium points of the system and then to study the motion around each equilibrium point. The local motion around equilibrium points is characterized by the nature of the eigenvalues of the linearized system. It is assumed that the system is described by the following set of first-order differential equations (for plate and shell vibrations, the second-order differential equations can be recast in first-order form with the technique shown in equation [3.7]):

$$\dot{\mathbf{x}} = \mathbf{F}(\mathbf{x}), \tag{3.9}$$

where the vectorial notation has been used and $\mathbf{x}$ is a vector whose $2N$ components are the state variables x_j and $\dot{x}_j$. The equilibrium points are those for $\dot{\mathbf{x}} = 0$ and are given by

$$\mathbf{F}(\mathbf{x}_e) = 0. \tag{3.10}$$

In the previous example of equation (3.8), there is only one equilibrium point: $\mathbf{x}_e = (x_e = 0,\ \dot{x}_e = 0)$. In order to determine the dynamics around the equilibrium point, the vectorial function $\mathbf{F}$ is expanded in a Taylor series about $\mathbf{x}_e$ in order to study the linearized problem. A point $\mathbf{x}$ adjacent to $\mathbf{x}_e$ is obtained giving a small perturbation $\bar{\mathbf{x}}$ to the equilibrium point $\mathbf{x}_e$:

$$\mathbf{x} = \mathbf{x}_e + \bar{\mathbf{x}}. \tag{3.11}$$

Substituting (3.11) into equation (3.10) and expanding $\mathbf{F}$ in a Taylor series about $\mathbf{x}_e$, the following expression is obtained:

$$\dot{\mathbf{x}} = \dot{\mathbf{x}}_e + \dot{\bar{\mathbf{x}}} = \mathbf{F}(\mathbf{x}_e) + \left.\frac{\partial \mathbf{F}}{\partial \mathbf{x}}\right|_{\mathbf{x}_e} \bar{\mathbf{x}} + O(\bar{\mathbf{x}}^2), \tag{3.12}$$

where the Jacobian matrix is given by

$$\left.\frac{\partial \mathbf{F}}{\partial \mathbf{x}}\right|_{\mathbf{x}_e} = \begin{bmatrix} \dfrac{\partial F_1}{\partial x_1} & \cdots & \dfrac{\partial F_1}{\partial x_{2N}} \\ \vdots & \cdots & \vdots \\ \dfrac{\partial F_{2N}}{\partial x_1} & \cdots & \dfrac{\partial F_{2N}}{\partial x_{2N}} \end{bmatrix}_{\mathbf{x}=\mathbf{x}_e}. \tag{3.13}$$

The nature of the motion about the equilibrium point is characterized by the eigenvalues λ_i of the Jacobian matrix. The sign of the real part of the eigenvalues determines the stability of the solution. An equilibrium point for which the eigenvalues of the corresponding Jacobian matrix do not possess vanishing real parts is called a hyperbolic or nondegenerate fixed point. If there is at least one eigenvalue with zero real part (i.e. with pure imaginary eigenvalues $\pm\, i\omega$), the equilibrium point is called a nonhyperbolic or degenerate fixed point and stability is determined by the nonlinear terms. If the real part of one of the eigenvalues is positive, the motion about the equilibrium point is unstable.

Figure 3.3 shows the classification of fixed points and corresponding phase plane plots for $N=1$ (i.e. in the case of only two eigenvalues λ_1 and λ_2). If all the eigenvalues of the Jacobian matrix have negative real part, the fixed point is asymptotically stable and is called a sink, which is an attractor. A sink can be classified either as a stable focus if the eigenvalues are complex conjugate, or as a stable node if all the eigenvalues are real (this is the case in supercritical damping). If one or more eigenvalues have positive real part, the fixed point is asymptotically unstable and is called source, which is a repellor. A source can be classified as an unstable focus if the eigenvalues are complex conjugate, or as an unstable node if all the eigenvalues are real. If some of the eigenvalues have positive real part, but the others have negative real part, the fixed point is called a saddle point, which is unstable. A nonhyperbolic fixed point is unstable if one or more eigenvalues have positive real part. If some of the eigenvalues have negative real part, but the others have zero real part, the point is neutrally or marginally stable. Only if all the eigenvalues of the Jacobian matrix are purely imaginary and different from zero is the nonhyperbolic fixed point called center.

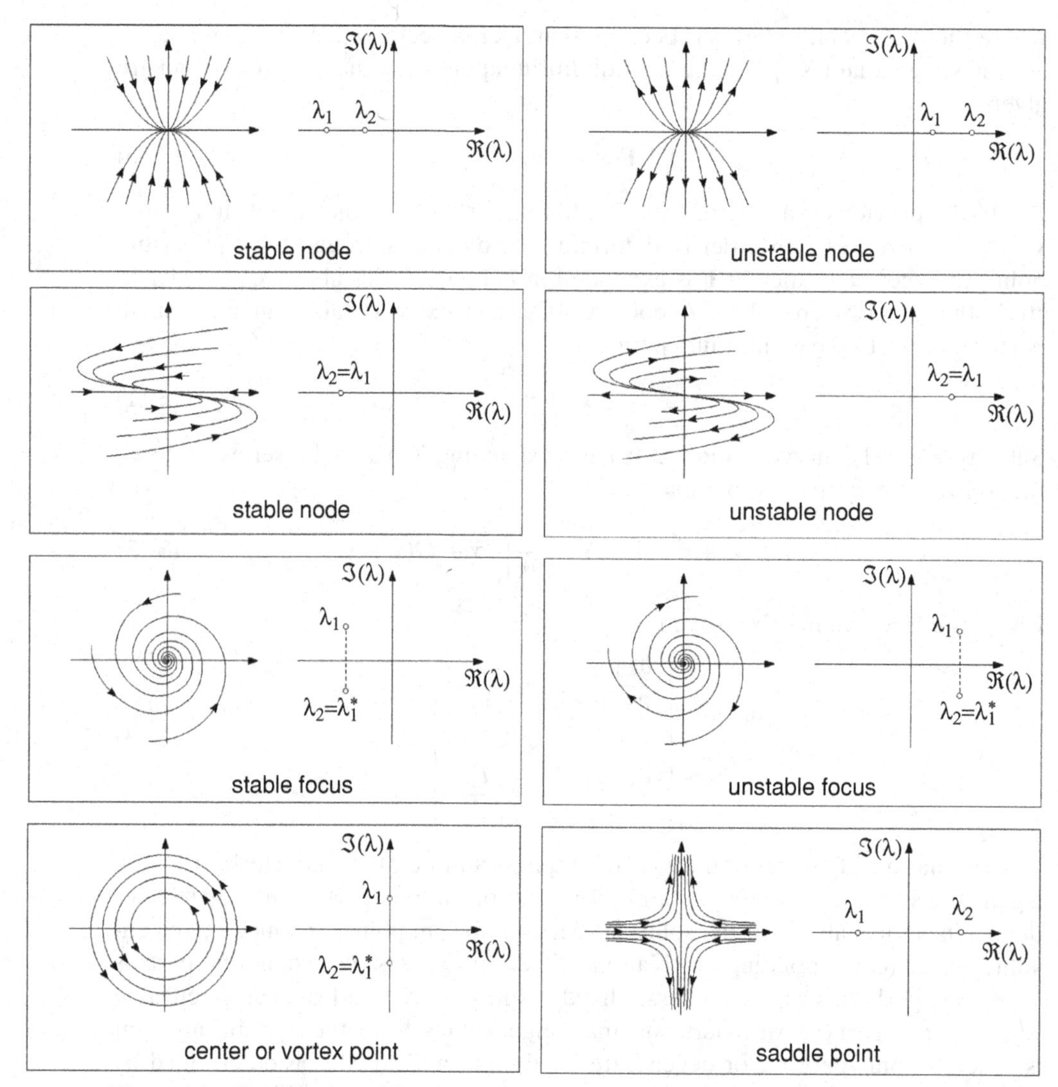

Figure 3.3. Classification of fixed points in a two-dimensional phase space.

3.5 Bifurcations of Equilibrium

The variation of one or more parameters in a nonlinear system can affect the stability, as well as the number of equilibrium points. The study of these changes in nonlinear systems, when system parameters are varied, is referred to as bifurcation theory. The value of the parameters giving a qualitative or topological change in the nature of motion or equilibrium is called the critical or bifurcation value. The local bifurcations arise as a result of the variation of one or more control parameters. By applying a linear stability analysis, it is possible to determine the bifurcation value μ_{cr} for which the fixed points become nonhyperbolic (Argyris et al. 1994). As a consequence of the variation of system parameters, global bifurcations can arise in nonlinear systems. Global bifurcations give global qualitative changes in the dynamic characteristics that cannot be deducted from local information. For example, a global bifurcation arises in the nonforced Duffing equation when small damping is added.

Bifurcations can also be classified into continuous and discontinuous or catastrophic, depending on whether the states of the system vary continuously or discontinuously when the control parameter is varied gradually through its critical value. Bifurcations of autonomous systems can be static or dynamic. Two different branches of solution meet at bifurcation points. In the case of saddle-node, pitchfork and transcritical bifurcations, these solution branches are constituted by fixed points or static solutions. In fact, saddle-node, pitchfork and transcritical bifurcations are static bifurcations. In contrast, a branch of fixed points meets a branch of periodic solutions at a Hopf bifurcation point, which is a dynamic bifurcation (Nayfeh and Balachandran 1995).

Discontinuous bifurcations can be divided into dangerous and explosive, depending on whether the system response jumps to a remote disconnected attractor or explodes into a larger attractor, where the new attractor includes the old one. Upon reversal of the control parameter, the new strange attractor implodes into the old one at the same critical bifurcation value without hysteresis. The new large attractor may be chaotic or not (Nayfeh and Balachandran 1995), where chaos is introduced in Sections 3.6 and 3.10 and indicates the loss of deterministic solutions. A dangerous bifurcation, where the periodic response jumps to a remote chaotic attractor, which may be bounded or unbounded, is also known as "blue sky catastrophe." Usually, reversing the control parameter, a bounded response remains on the path of the new attractor, resulting in hysteresis.

In the next sections, local continuous bifurcations of fixed points in autonomous systems are presented.

3.5.1 Saddle-Node Bifurcation

A classical example of saddle-node bifurcating system is considered:

$$\dot{x} = F(x, \mu) = \mu - x^2, \tag{3.14}$$

where μ is the control parameter. The equilibrium points are

$$x_e = \pm\sqrt{\mu}. \tag{3.15}$$

The equilibrium points are plotted in Figure 3.4, which is a typical bifurcation diagram. No equilibrium exists for $\mu < 0$. The only eigenvalue λ of the Jacobian matrix is $\lambda = -2x|_{x=x_e}$; it shows that the branch $x_e = \sqrt{\mu}$ is stable, whereas the branch $x_e = -\sqrt{\mu}$ is unstable. For μ approaching zero, the stable and unstable branches approach each other until they merge at $\mu = 0$. For $x = 0$ and $\mu = 0$, there is a particular situation: $F(x, \mu) = 0$ and $\lambda = 0$. Therefore, there is a nonhyperbolic fixed point at $\mu = 0$. Passing through $\mu = 0$ there is a change in the number of fixed points. The point at $x = 0$ and $\mu_{cr} = 0$ is a static bifurcation point, referred to as a saddle-node bifurcation; the solution displays a fold at the saddle-node bifurcation.

3.5.2 Pitchfork Bifurcation

The pitchfork bifurcation can be illustrated by the following system:

$$\dot{x} = F(x, \mu) = \mu x - x^3. \tag{3.16}$$

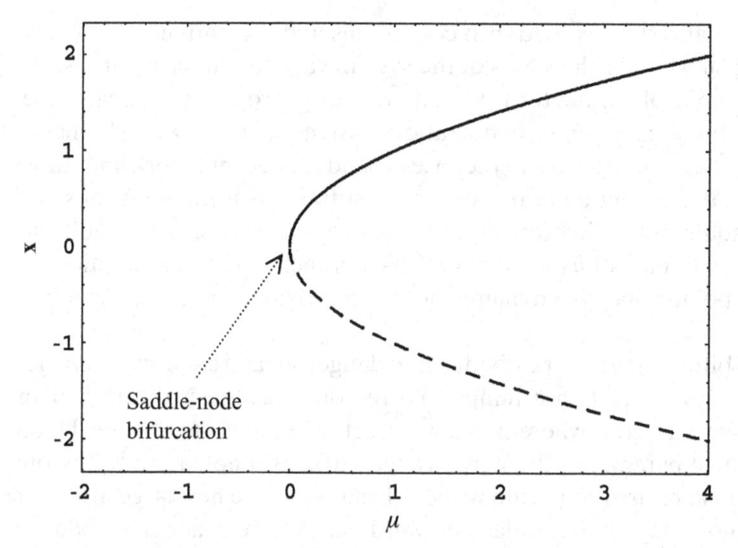

Figure 3.4. Saddle-node (fold) bifurcation. —, stable solution; – –, unstable solution.

Three equilibrium solutions are obtained:

$$x_{e,0} = 0, \quad x_{e,1,2} = \pm\sqrt{\mu}. \tag{3.17}$$

The three solutions are shown in the bifurcation diagram in Figure 3.5. For positive values of μ, at $\mu = 0$ two new solutions $x_{e,1,2}$ branch off from the solution $x_{e,0} = 0$. The only eigenvalue λ of the Jacobian matrix is $\lambda = (-3x^2 + \mu)|_{x=x_e}$, indicating that the solution $x_{e,0} = 0$ is stable for $\mu < 0$ and unstable for $\mu > 0$. The solution pair $x_{e,1,2} = \pm\sqrt{\mu}$ is stable for $\mu > 0$ (i.e. in the whole range where they exist). The bifurcation value is $\mu_{cr} = 0$, and for the typical shape this bifurcation is named a pitchfork bifurcation. This type of bifurcation can be supercritical, as the one in Figure 3.5, or subcritical, as shown in Figure 3.6. Moreover, at a pitchfork bifurcation point, for an axisymmetric shell, instead of a bifurcating branch, a bifurcating surface

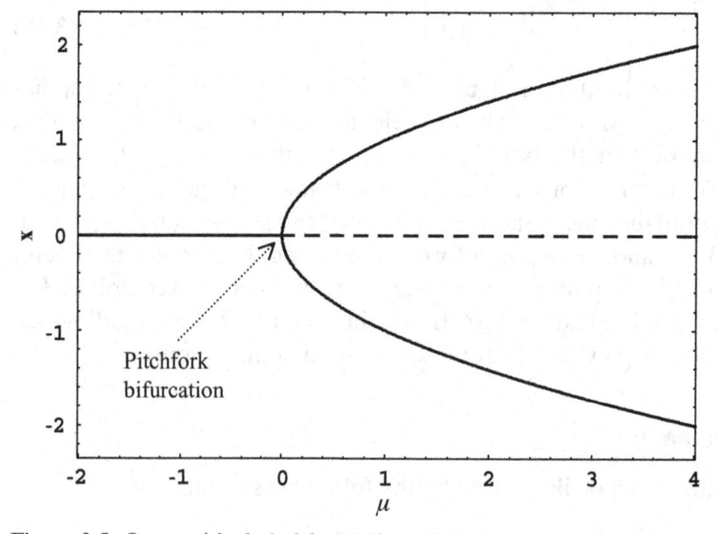

Figure 3.5. Supercritical pitchfork bifurcation.

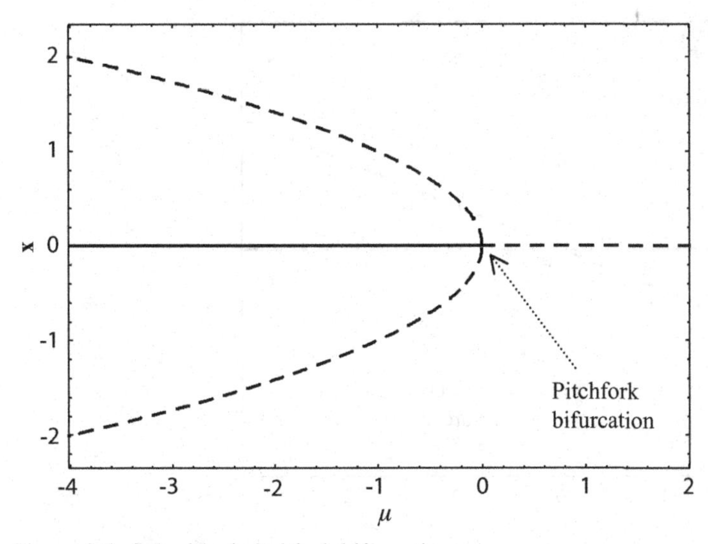

Figure 3.6. Subcritical pitchfork bifurcation.

can come out. This is the case for divergence of circular cylindrical shells conveying liquid.

3.5.3 Transcritical Bifurcation

The following system is considered:

$$\dot{x} = F(x, \mu) = \mu x - x^2. \tag{3.18}$$

The equilibrium solutions are the two straight lines:

$$x_{e,0} = 0, \quad x_{e,1} = \mu. \tag{3.19}$$

The only eigenvalue λ of the Jacobian matrix is $\lambda = (-2x + \mu)|_{x=x_e}$ and indicates that the solution $x_{e,0} = 0$ is stable for $\mu < 0$ and unstable for $\mu > 0$. The solution $x_{e,1} = \mu$ is stable for $\mu > 0$; therefore, the equilibrium solutions interchange stability at $\mu_{cr} = 0$, where a so-called transcritical bifurcation occurs. Figure 3.7 shows the corresponding bifurcation diagram.

3.5.4 Hopf Bifurcation

This classical bifurcation is named after the mathematician Eberhard Hopf. The following two-dimensional system is considered in order to illustrate this dynamic bifurcation:

$$\dot{x} = -y + x[\mu - (x^2 + y^2)], \tag{3.20a}$$

$$\dot{y} = x + y[\mu - (x^2 + y^2)]. \tag{3.20b}$$

The only equilibrium solution is $x_e = y_e = 0$. The Jacobian matrix is

$$\left.\frac{\partial \mathbf{F}}{\partial \mathbf{x}}\right|_{\mathbf{x}_e=(0,0)} = \begin{bmatrix} \mu & -1 \\ 1 & \mu \end{bmatrix}_{\mathbf{x}=(0,0)}, \tag{3.21}$$

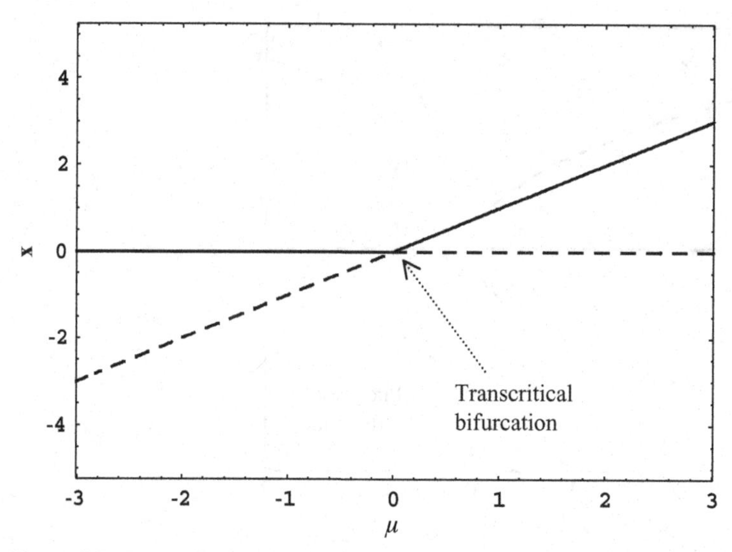

Figure 3.7. Transcritical bifurcation.

which has the complex conjugate eigenvalues $\lambda_{1,2} = \mu \pm i$. For $\mu < 0$, the equilibrium solution is a stable focus as a consequence of the fact that $\mathrm{Re}(\lambda_{1,2}) < 0$, which becomes an unstable focus for $\mu > 0$; therefore, for $\mu = 0$, there is a change in stability.

In order to show that the solution escapes from the focus and is attracted by a limit cycle, which is a dynamic attractor similar to a simple oscillation with specific frequency, equations (3.20) are transformed into polar coordinates. By using the transformations

$$x = r\cos\varphi, \quad y = r\sin\varphi, \tag{3.22}$$

system (3.20) is transformed into

$$\dot{r}\cos\varphi - r\dot{\varphi}\sin\varphi = -r\sin\varphi + r(\mu - r^2)\cos\varphi, \tag{3.23a}$$

$$\dot{r}\sin\varphi + r\dot{\varphi}\cos\varphi = r\cos\varphi + r(\mu - r^2)\sin\varphi. \tag{3.23b}$$

Multiplying the first equation by $\cos\varphi$ (or $\sin\varphi$) and the second by $\sin\varphi$ (respectively, $-\cos\varphi$) and adding the results, the following uncoupled system is obtained in polar coordinates:

$$\dot{r} = -r^3 + \mu r, \quad \dot{\varphi} = 1. \tag{3.24a,b}$$

In particular, equation (3.24a) resembles (3.16). The states of equilibrium are

$$r_{e,0} = 0, \quad r_{e,1} = \sqrt{\mu}, \quad \text{for any } \varphi. \tag{3.25}$$

The negative radius solution has been ignored because it is meaningless in polar coordinates. The solution $r_{e,1}$ gives a closed circle, a limit cycle, that is, a periodic orbit with amplitude given by the radius $\sqrt{\mu}$. By means of stability analysis for periodic solutions, it is possible to observe that for $r < \sqrt{\mu}$, inside the limit cycle, $\dot{r} > 0$; for $r > \sqrt{\mu}$, outside the limit cycle, $\dot{r} < 0$; therefore, the periodic orbit is stable. The Hopf bifurcation is therefore the bifurcation of a state of equilibrium, a focus, into a dynamic state, a limit cycle, and is represented in Figure 3.8 for a supercritical case. The frequency of oscillation in Hz $1/(2\pi)$ is obtained by using equation (3.24b);

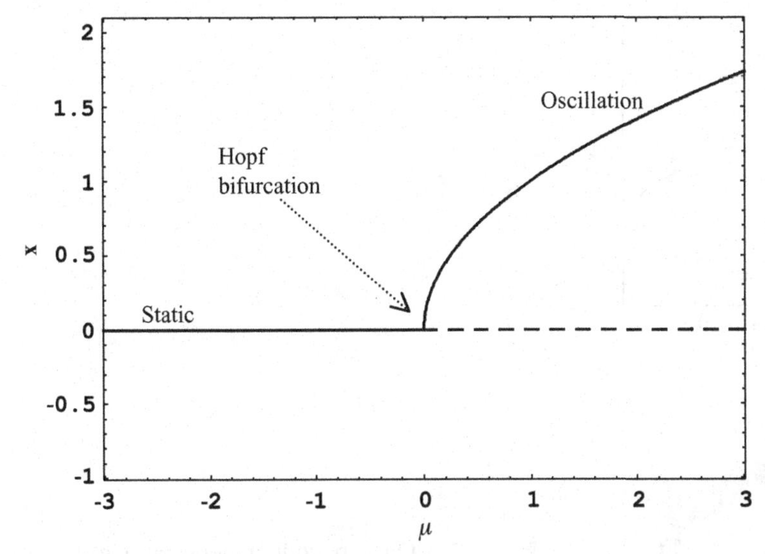

Figure 3.8. Supercritical Hopf bifurcation.

in general, the oscillation frequency is given by $\mathrm{Im}(\lambda)/(2\pi)$. A classical example of Hopf bifurcation is given by plates and shells in axial airflow. By increasing the flow velocity, a Hopf bifurcation can arise, giving rise to the phenomenon known as flutter.

3.6 Poincaré Maps

A map for dynamical systems refers to a time-sampled sequence of data $\{x(t_1), x(t_2), \ldots, x(t_N)\}$; usually the notation $x_n = x(t_n)$ is used. Instead of looking at the motion of a particle in the phase plane $(x(t), \dot{x}(t))$ continuously, the dynamics is observed only at discrete times. The motion will appear as a sequence of dots in the phase plane. Setting $x_n = x(t_n)$ and $y_n = \dot{x}(t_n)$, this sequence of points in the phase plane represents a two-dimensional map:

$$x_{n+1} = f(x_n, y_n), \qquad y_{n+1} = g(x_n, y_n). \tag{3.26}$$

Equations (3.26) are difference equations. When there is a driving motion of period T and the sampling rule is taken as $t_n = nT + t_0$, with n an integer, the two-dimensional map is called a Poincaré map (Moon 1992). If a harmonic motion with the same frequency of excitation is plotted in a Poincaré map, a single point is obtained. If a subharmonic motion, with circular frequency $\omega t/n$, ω being the excitation circular frequency, is plotted, n points are obtained in the Poincaré map.

If the motion is obtained as superposition of two oscillations with incommensurate frequencies,

$$x(t) = C_1 \sin(\omega_1 t + \varphi_1) + C_2 \sin(\omega_2 t + \varphi_2), \tag{3.27}$$

where ω_1/ω_2 is an irrational number, sampling at either frequency, the Poincaré map will show a closed orbit (see Figure 3.9[a]). This motion is called almost-periodic or quasi-periodic, or motion on a torus. The last name can easily be explained by looking at Figure 3.9(b). Defining $\theta_1 = \omega_1 t + \varphi_1$ and $\theta_2 = \omega_2 t + \varphi_2$, the quasi-periodic motion

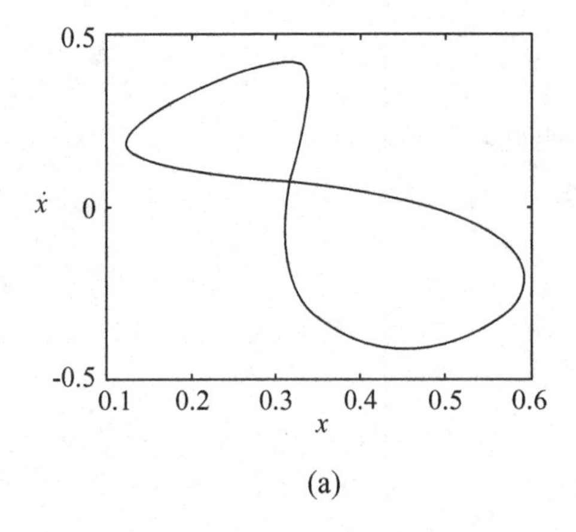

(a)

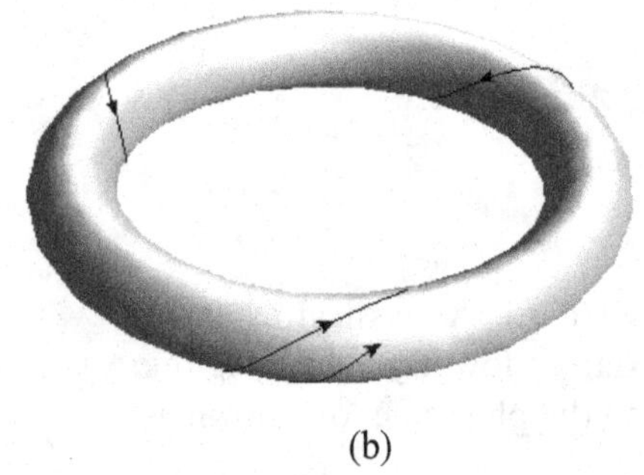

(b)

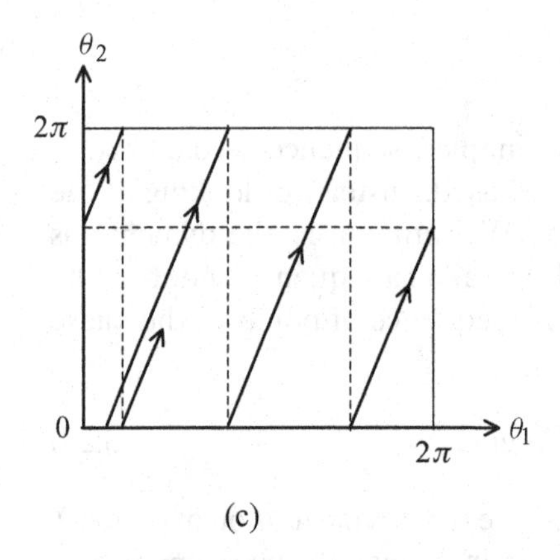

(c)

Figure 3.9. Quasi-periodic motion on a torus for a tub-dimensional system. (a) Poincaré map. (b) Motion on a torus. (c) Plot in the (θ_1, θ_2) plane.

is described in Figure 3.9(c) in the (θ_1, θ_2) plane; with continuing time, the curve will eventually completely fill the plane. In order to avoid discontinuities in Figure 3.9(c) at 0 and 2π for both θ_1 and θ_2, the plane is folded in both the θ_1 and θ_2 directions and opposite sides are joined. This operation gives the torus shown in Figure 3.9(b).

If the Poincaré map does not display a finite number of points or a closed orbit, the motion can be chaotic (see Figure 3.10), which means the loss of deterministic solutions. In undamped or lightly damped systems, the points of chaotic motion appear in the Poincaré map as disorganized points; such motion is called stochastic. In damped systems, the points in the Poincaré map will appear as highly organized. In particular, it is possible to observe particular substructures composed by parallel curved lines. Zooming in on the map, if it is possible to observe a structured set of points, the map is said to be fractal-like; the motion behaves as a strange attractor.

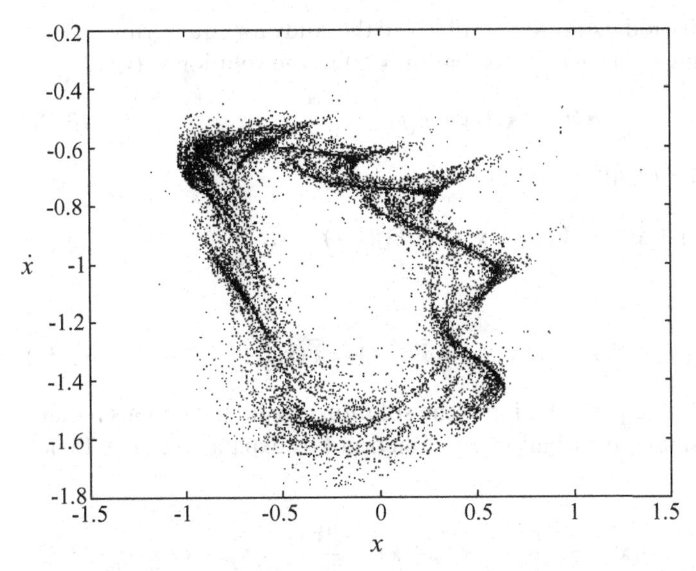

Figure 3.10. Poincaré map for damped forced vibrations of a circular cylindrical shell conveying water flow.

The appearance of fractal-like patterns in Poincaré maps is a strong indicator of chaotic motion (Moon 1992).

3.7 Bifurcations of Periodic Solutions

Unlike equilibrium solutions studied in Sections 3.4 and 3.5, periodic solutions are characterized by time-varying states. A periodic solution is characterized by a fundamental frequency, a zero frequency component and superharmonics. The stability of periodic solutions can be studied by using Floquet theory; in particular, the Floquet multipliers replace the eigenvalues of the Jacobian matrix (3.13) obtained for equilibrium solution. Bifurcations of the periodic solutions of dissipative systems can be classified similarly to bifurcations of fixed points of autonomous systems. For local bifurcations, saddle-node, transcritical and pitchfork bifurcations are analogous to those obtained in Section 3.5 for the static case, but now they are related to periodic solutions. No Hopf bifurcation arises from a periodic solution, but two additional bifurcations appear: the period doubling (or flip) and the Neimark-Sacker (or secondary Hopf or torus) bifurcations.

3.7.1 Floquet Theory

The Floquet theory represents a linear stability analysis of periodic processes (Argyris et al. 1994). Assuming a solution $\mathbf{x}_p(t)$ with periodic motion of period T, it is observed that

$$\mathbf{x}_p(t) = \mathbf{x}_p(t + T), \tag{3.28}$$

where $\mathbf{x}_p(t)$ is the solution of a periodic, externally excited system, described by an inhomogeneous set of nonautonomous, nonlinear differential equations

$$\dot{\mathbf{x}} = \mathbf{F}(\mathbf{x}, t) = \mathbf{F}(\mathbf{x}, t + T), \tag{3.29}$$

where $\mathbf{F}(\mathbf{0}, t) \neq \mathbf{0}$.

The stability of the periodic solution, similarly to the study for equilibrium points, can be investigated by giving a small perturbation $\bar{\mathbf{x}}\,(t)$ to the solution $\mathbf{x}_p(t)$:

$$\mathbf{x}(t) = \mathbf{x}_p(t) + \bar{\mathbf{x}}(t). \tag{3.30}$$

Substituting equation (3.30) into (3.29) yields

$$\dot{\mathbf{x}}_p(t) + \dot{\bar{\mathbf{x}}}(t) = \mathbf{F}(\mathbf{x}_p + \bar{\mathbf{x}}, t) = \mathbf{F}(\mathbf{x}_p, t) + \dot{\bar{\mathbf{x}}}(t). \tag{3.31}$$

From (3.31)

$$\dot{\bar{\mathbf{x}}}(t) = \mathbf{F}(\mathbf{x}_p + \bar{\mathbf{x}}, t) - \mathbf{F}(\mathbf{x}_p, t) = \bar{\mathbf{F}}(\bar{\mathbf{x}}, t) = \bar{\mathbf{F}}(\bar{\mathbf{x}}, t + T), \tag{3.32}$$

where $\bar{\mathbf{F}}(\mathbf{0}, t) = \mathbf{0}$; that is, the perturbation is the solution of a homogeneous nonautonomous system. For small perturbation, the function $\bar{\mathbf{F}}$ is expanded into a Taylor series:

$$\dot{\bar{\mathbf{x}}}(t) = \left.\frac{\partial \bar{\mathbf{F}}(\bar{\mathbf{x}}, t)}{\partial \bar{\mathbf{x}}}\right|_{\bar{\mathbf{x}}=0} \bar{\mathbf{x}} + O(\bar{\mathbf{x}}^2) = \left.\frac{\partial \mathbf{F}}{\partial \mathbf{x}}\right|_{\mathbf{x}=\mathbf{x}_p} (\mathbf{x}_p + \bar{\mathbf{x}}) - \left.\frac{\partial \mathbf{F}}{\partial \mathbf{x}}\right|_{\mathbf{x}=\mathbf{x}_p} \mathbf{x}_p + O(\bar{\mathbf{x}}^2), \tag{3.33}$$

where the time-dependent coefficient matrix $(\partial \bar{\mathbf{F}}(\bar{\mathbf{x}}, t)/\partial \bar{\mathbf{x}})|_{\bar{\mathbf{x}}=0}$ is defined analogously to equation (3.13). By using equation (3.33), it is possible to write

$$\left.\frac{\partial \bar{\mathbf{F}}(\bar{\mathbf{x}}, t)}{\partial \bar{\mathbf{x}}}\right|_{\bar{\mathbf{x}}=0} = \left.\frac{\partial \mathbf{F}(\mathbf{x}, t)}{\partial \mathbf{x}}\right|_{\mathbf{x}=\mathbf{x}_p(t)}. \tag{3.34}$$

Moreover, this time-dependent coefficient matrix is periodic:

$$\left.\frac{\partial \mathbf{F}(\mathbf{x}, t)}{\partial \mathbf{x}}\right|_{\mathbf{x}=\mathbf{x}_p(t)} = \left.\frac{\partial \mathbf{F}(\mathbf{x}, t + T)}{\partial \mathbf{x}}\right|_{\mathbf{x}=\mathbf{x}_p(t)}. \tag{3.35}$$

The periodicity of the coefficient matrix (3.35) allows the reduction of the system (3.33), neglecting the term $O(\bar{\mathbf{x}}^2)$, to a system of differential equations with constant coefficients in order to observe the behavior of $\bar{\mathbf{x}}\,(t)$ only at discrete times $t = 0$, T, $2T, \ldots$. The system of n differential equations (3.33) has n independent vector solutions $\mathbf{\Phi}_i(t)$, which can be used to describe the space of the solutions. These solutions can be assembled in the $n \times n$ matrix $\mathbf{\Phi}(t)$, defined as

$$\mathbf{\Phi}(t) = \{\mathbf{\Phi}_1(t), \mathbf{\Phi}_2(t), \ldots, \mathbf{\Phi}_n(t)\}. \tag{3.36}$$

For convenience, among the possible $\mathbf{\Phi}_i(t)$, those having the following initial conditions at time $t = 0$ are chosen:

$$\mathbf{\Phi}(t = 0) = \mathbf{I}. \tag{3.37}$$

Then, the perturbation $\bar{\mathbf{x}}(t)$ can be rewritten as a linear combination of the base vectors

$$\bar{\mathbf{x}}(t) = \mathbf{\Phi}(t)\,\mathbf{c}, \tag{3.38}$$

where $\mathbf{c}$ is a constant vector, $\mathbf{c}^{\mathrm{T}} = \{c_1, c_2, \ldots, c_n\}$. Also $\bar{\mathbf{x}}(t + T)$, which is different from $\bar{\mathbf{x}}(t)$, can be represented as a linear combination of the base vectors. Moreover, it is possible to find a constant $n \times n$ matrix $\mathbf{C}$ mapping $\mathbf{\Phi}(t)$ onto $\mathbf{\Phi}(t+T)$, which allows one to write

$$\mathbf{\Phi}(t+T) = \mathbf{\Phi}(t)\,\mathbf{C}. \tag{3.39}$$

By using (3.39) and (3.37), it is obvious that

$$\mathbf{\Phi}(T) = \mathbf{\Phi}(t=0)\,\mathbf{C} = \mathbf{C}. \tag{3.40}$$

Equation (3.40) allows to determine the matrix $\mathbf{C}$, which is called the monodromy matrix. This matrix can be built up numerically.

The matrix $\mathbf{C}$ can be used to transform a generic vector $\mathbf{x}$,

$$\mathbf{x}' = \mathbf{C}\,\mathbf{x}, \tag{3.41}$$

which has the inverse transformation

$$\mathbf{x} = \mathbf{C}^{-1}\mathbf{x}'. \tag{3.42}$$

In particular, it is interesting to study the transformation of the hypersphere S of unit radius, where the initial conditions of the n independent vector solutions $\mathbf{\Phi}_i(t=0)$ lie, after a period T. The transformation of S gives a hyperellipsoid with the independent vector solutions $\mathbf{\Phi}_i(t=T)$ as semiaxes. The equation of the hypersphere S is

$$\mathbf{x}^{\mathrm{T}}\mathbf{x} = 1, \tag{3.43}$$

which, after substituting equation (3.42) into (3.43), is transformed into

$$\mathbf{x}'^{\mathrm{T}}\mathbf{A}\mathbf{x}' = 1, \tag{3.44}$$

where

$$\mathbf{A} = (\mathbf{C}^{-1})^{\mathrm{T}}\mathbf{C}^{-1}. \tag{3.45}$$

The eigenvalues of $\mathbf{A}$ coincide with the reciprocal values of the square of the semiaxes of the hyperellipsoid. Therefore, for equation (3.45), the semiaxes of the hyperellipsoid agree with the eigenvalues λ_j, called Floquet multipliers, of the monodromy matrix $\mathbf{C}$. Therefore, the Floquet multipliers provide all the necessary information on the stability of the periodic solution $\mathbf{x}_p(t)$. In fact, if

$$|\lambda_j| < 1 \quad \text{for } j = 1, \ldots, n, \tag{3.46}$$

the hyperellipsoid is contained within the sphere. Because the same transformation $\mathbf{C}$ operates at every period T, as established by equation (3.39), the perturbation tends asymptotically to zero. Therefore, if equation (3.46) holds true, the periodic solution $\mathbf{x}_p(t)$ is stable. If for one of the eigenvalues it is verified that $|\lambda_j| > 1$, the periodic solution $\mathbf{x}_p(t)$ is unstable, as shown in Figure 3.11(a). A periodic solution can lose stability in three different ways:

$$\text{Case (a): } \lambda_j = 1 \text{ with } \operatorname{Im}(\lambda_j) = 0, \tag{3.47a}$$

$$\text{Case (b): } \lambda_j = -1 \text{ with } \operatorname{Im}(\lambda_j) = 0, \tag{3.47b}$$

$$\text{Case (c): } |\lambda_j| = 1 \text{ with } \operatorname{Im}(\lambda_j) \neq 0; \tag{3.47c}$$

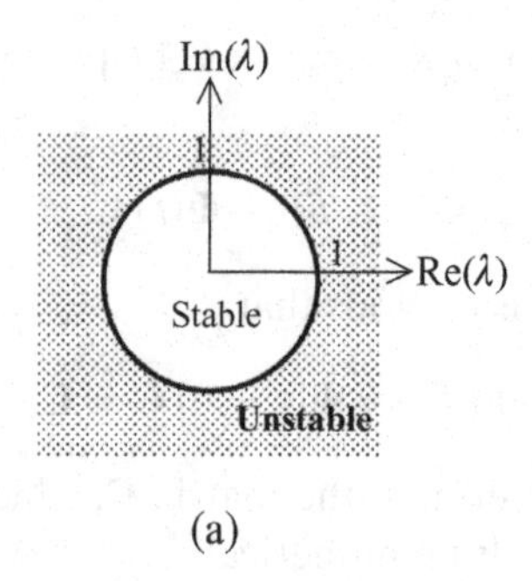

(a)

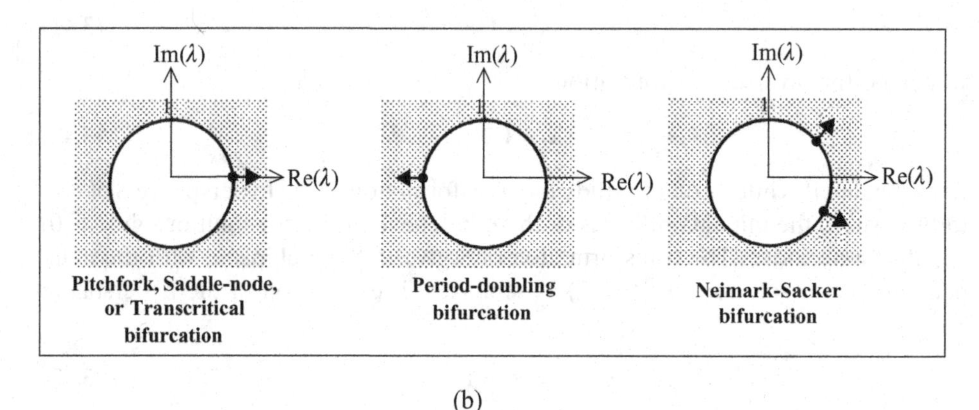

(b)

Figure 3.11. (a) Unit circle representing the stability limit for Floquet multipliers. (b) Classification of the bifurcation points of periodic solutions based on the Floquet multipliers.

in case (c) there is a pair of complex conjugate eigenvalues. In particular, case (a) gives saddle-node or fold, transcritical and pitchfork bifurcations; case (b) gives a period-doubling bifurcation; finally, case (c) gives the Neimark-Sacker bifurcation. This classification is shown in Figure 3.11(b).

3.7.2 Period-Doubling Bifurcation

When a Floquet multiplier becomes equal to –1, case (b) above, as in equation (3.47b), a period-doubling bifurcation takes place. The periodic solution with oscillation frequency ω, which exists before the bifurcation value μ_{cr}, continues as an unstable solution after μ_{cr}. At the bifurcation, a new stable branch is created in the case of supercritical bifurcation, with oscillation frequency $\omega/2$ (double period). In this case, the bifurcation is continuous. In the case of a subcritical bifurcation, a branch of unstable period-doubled solution is destroyed at the bifurcation point. In this second case, the bifurcation is catastrophic and the behavior of the system for $\mu > \mu_{cr}$ can be dangerous or explosive (Nayfeh and Balachandran 1995). An example of supercritical period-doubling bifurcation is given by

$$\ddot{x} + 0.4\dot{x} + x - x^3 = F\cos(\omega t), \tag{3.48}$$

which is a forced and damped Duffing equation with cubic nonlinearity; the excitation frequency is kept constant, $\omega = 0.4$. Equation (3.48) can be recast into two first-order differential equations, as usual, by using the dummy variable introduced in equation (3.7). The bifurcation parameter is the excitation amplitude F. At around $F = 0.38$, a period-doubling bifurcation appears. Before the bifurcation, equation (3.48) presents a stable periodic solution with frequency ω, giving one dot in the Poincaré map and a closed loop in the phase plane. After the bifurcation, the only stable solution is the bifurcated branch with frequency $\omega/2$, giving two dots in the Poincaré map and a single close curved with two loops in the phase plane (Nayfeh and Balachandran 1995).

Dynamical systems can present an infinite sequence of period-doubling bifurcations under the variation of a single parameter, culminating in a chaotic motion.

3.7.3 Neimark-Sacker Bifurcation

A Hopf bifurcation of a periodic solution is called a secondary Hopf or Neimark-Sacker bifurcation. This bifurcation happens when two complex conjugate Floquet multipliers reach unit modulus. The bifurcating solution may be periodic or quasi-periodic, depending on the ratio between the two new frequencies introduced after the bifurcation and the frequency of the periodic oscillation existing before the bifurcation. There are supercritical and subcritical Neimark-Sacker bifurcations. In both cases, the stable solution existing before the bifurcation becomes unstable after. A new branch of stable quasi-periodic solution appears in the case of a supercritical bifurcation, giving a continuous bifurcation. A branch of unstable quasi-periodic solution is destroyed in the case of a subcritical bifurcation, giving a catastrophic bifurcation. As a consequence of a quasi-periodic solution being a motion on a torus, as discussed in Section 3.6, the Neimark-Sacker bifurcation can be also called a torus bifurcation.

An example of supercritical Neimark-Sacker bifurcation arises in large-amplitude vibrations of circular cylindrical shells, which are studied in Chapter 5 (Section 5.6.1).

3.8 Numerical Continuation Methods

In the case of an autonomous system (e.g. representing nonlinear stability of plates and shells under a static load), the pseudo-arclength continuation used in the software AUTO (Doedel et al. 1998) can be successfully used for continuation of equilibrium solutions (fixed points). Once a fixed point is found, for example, the point corresponding to the trivial undeformed configuration of the unloaded system, a parameter can be varied (e.g. the load on the plate or shell) until a bifurcation occurs, representing the instability point. After the bifurcation, the postcritical (e.g. post-buckling) behavior can also be investigated by using the same technique; however, a specific algorithm for branch switching must be used at the bifurcation point, which requires the calculation of the second derivatives of $\mathbf{F}(\mathbf{x}, \mu)$.

A successful algorithm for continuation of periodic solutions is also presented.

3.8.1 Arclength Continuation of Fixed Points

In the arclength continuation, the arclength s along a branch of solutions is used as the continuation parameter (Nayfeh and Balachandran 1995). The system of equations

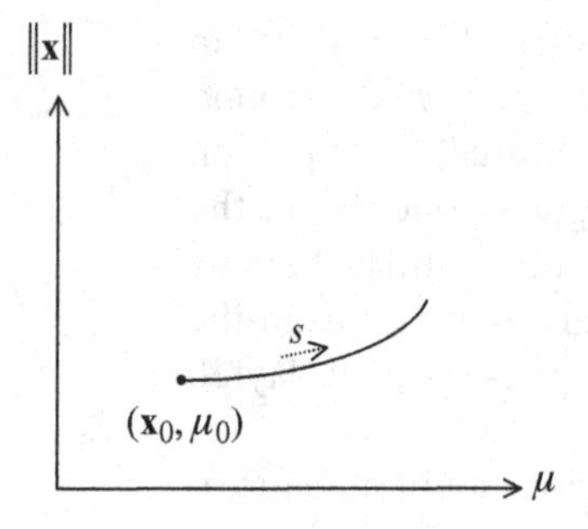

Figure 3.12. Arclength continuation of static solution.

of motion is assumed to be written in the form $\dot{\mathbf{x}} = \mathbf{F}(\mathbf{x}, \mu)$. Both the n-dimensional solution $\mathbf{x}$ and the parameter μ are considered as a function of s: $\mathbf{x} = \mathbf{x}(s)$ and $\mu = \mu(s)$, as shown in Figure 3.12. The equilibrium positions are given by

$$\mathbf{F}(\mathbf{x}(s), \mu(s)) = \mathbf{0}. \tag{3.49}$$

The following expression is obtained by differentiation of (3.49) with respect to s:

$$\frac{\partial \mathbf{F}}{\partial \mathbf{x}}(\mathbf{x}, \mu)\frac{\mathrm{d}\mathbf{x}}{\mathrm{d}s} + \frac{\partial \mathbf{F}}{\partial \mu}(\mathbf{x}, \mu)\frac{\mathrm{d}\mu}{\mathrm{d}s} = \mathbf{0}, \tag{3.50}$$

which is a system of n equations in the $n+1$ unknown $\mathbf{x}$ and μ. By introducing the symbols $\mathbf{x}' = \mathrm{d}\mathbf{x}/\mathrm{d}s$, $\mu' = \mathrm{d}\mu/\mathrm{d}s$, the Jacobian matrix $\mathbf{F}_{\mathbf{x}} = \partial\mathbf{F}/\partial\mathbf{x}$ and $\mathbf{F}_{\mu} = \partial\mathbf{F}/\partial\mu$, equation (3.50) can be rewritten as

$$\{\mathbf{F}_{\mathbf{x}}, \mathbf{F}_{\mu}\}\begin{Bmatrix} \mathbf{x}' \\ \mu' \end{Bmatrix} = \{\mathbf{F}_{\mathbf{x}}, \mathbf{F}_{\mu}\}\mathbf{t} = \mathbf{0}, \tag{3.51}$$

where $\mathbf{t}$ is the $(n+1)$-dimensional vector tangent to the $(\mathbf{x}, \mu)$ curve. In order to solve equation (3.50), or its equivalent, equation (3.51), it is necessary to add an equation. The following non-homogeneous equation is chosen, giving the Euclidean arclength normalization, that is, giving unit length to the tangent vector $\mathbf{t}$:

$$\mathbf{x}'^{\mathrm{T}}\mathbf{x}' + \mu'^2 = x_1'^2 + x_2'^2 + \ldots + x_n'^2 + \mu'^2 = 1. \tag{3.52}$$

It is assumed that the initial equilibrium solution to be continued is

$$\mathbf{x} = \mathbf{x}_0, \quad \mu = \mu_0 \quad \text{at } s = 0. \tag{3.53}$$

If the Jacobian matrix $\mathbf{F}_{\mathbf{x}}$ is nonsingular and $\mathbf{F}_{\mu}$ is a vector different from zero, it is possible to solve the system given by (3.50) and (3.52) and to calculate the tangent vector $\mathbf{t}$. The system (3.50) gives

$$\mathbf{x}' = -\mu'\,\mathbf{F}_{\mathbf{x}}^{-1}(\mathbf{x}, \mu)\mathbf{F}_{\mu}(\mathbf{x}, \mu) = -\mu'\mathbf{z}. \tag{3.54}$$

Substituting (3.54) into (3.52) yields

$$\mu' = \pm\frac{1}{\sqrt{1 + \mathbf{z}^{\mathrm{T}}\mathbf{z}}}, \tag{3.55}$$

where the plus or minus sign determines the continuation direction. Equations (3.54) and (3.55) identify the tangent vector $\mathbf{t}$, which is used to predict the value of $\mathbf{x}$ and μ at the next step $s + \Delta s$,

$$\mathbf{x} = \mathbf{x}_0 + \mathbf{x}'\Delta s, \quad \mu = \mu_0 + \mu'\Delta s. \tag{3.56}$$

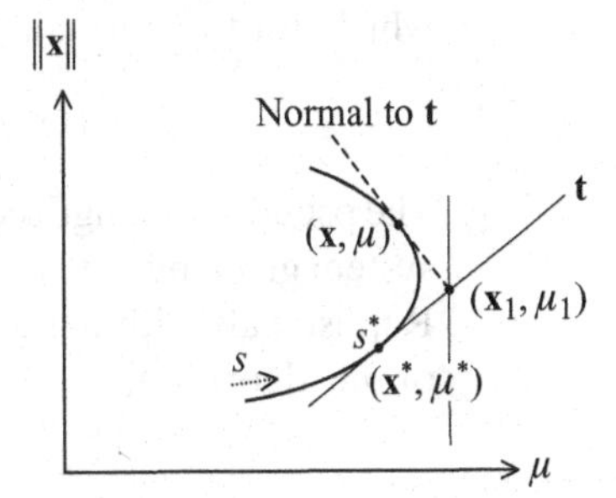

Figure 3.13. Pseudo-arclength continuation of static solution.

Equation (3.56) is called the tangent predictor, which is a first-order predictor. The predicted values are corrected through a Newton-Raphson algorithm. Crucial is the choice of the step-size Δs, which depends on the convergence of the predictor and the curvature of the equilibrium solution curve. It is convenient that the step-size is adaptively varied during the continuation; this may be done by algorithms for step-size control.

In the arclength continuation method it is possible to overpass folding points and to end up at a value of the parameter μ where there are no equilibrium solutions. In order to overcome this problem, the pseudo-arclength continuation has been introduced.

3.8.2 Pseudo-Arclength Continuation of Fixed Points

The pseudo-arclength continuation, which is the continuation technique implemented in the software AUTO (Doedel et al. 1998), also uses the arclength s as continuation parameter. The solution $\mathbf{x}^*$ and μ^* is obtained for $s = s^*$ near the folding with the arclength continuation (3.54–3.56), as shown in Figure 3.13. By using the tangent predictor (3.56) for prediction at $s = s^* + \Delta s$, yields (Nayfeh and Balachandran 1995)

$$\mu_1 = \mu^* + \mu^{*\prime}\Delta s, \quad \mathbf{x}_1 = \mathbf{x}^* + \mathbf{x}^{*\prime}\Delta s. \tag{3.57}$$

According to Figure 3.13, the point $(\mathbf{x}_1, \mu_1)$ corresponds to a parameter value μ_1 for which there are no equilibrium solutions. The correction of the predictor is left to the Newton-Raphson algorithm, which would seek for the solution along the vertical line $\mu = \mu_1$ through the point $(\mathbf{x}_1, \mu_1)$. As a consequence of no solution existing for $\mu = \mu_1$, the correction would fail. To overcome this problem, the correction is introduced on the normal to $\mathbf{t}$ at the point $(\mathbf{x}_1, \mu_1)$, which is indicated by a dashed line in Figure 3.13.

For a generic solution point $(\mathbf{x}, \mu)$ close to folding, which satisfies $\mathbf{F}(\mathbf{x}, \mu) = \mathbf{0}$, the vector

$$\mathbf{X} = \begin{Bmatrix} \mathbf{x} - \mathbf{x}_1 \\ \mu - \mu_1 \end{Bmatrix}, \tag{3.58}$$

is normal to the tangent $\mathbf{t} = \{^{\mathbf{x}^{*\prime}}_{\mu^{*\prime}}\}$; that is,

$$\mathbf{X} \cdot \mathbf{t} = 0. \tag{3.59}$$

Substituting equation (3.57) into (3.59)

$$(\mathbf{x} - \mathbf{x}^*)^{\mathrm{T}}\mathbf{x}^{*\prime} + (\mu - \mu^*)\mu^{*\prime} - [\mu^{*\prime 2} + (\mathbf{x}^{*\prime})^{\mathrm{T}}\mathbf{x}^{*\prime}]\Delta s = 0, \tag{3.60}$$

which, for the normalization (3.52), is equivalent to

$$f(\mathbf{x}, \mu) = (\mathbf{x} - \mathbf{x}^*)^{\mathrm{T}}\mathbf{x}^{*\prime} + (\mu - \mu^*)\mu^{*\prime} - \Delta s = 0. \tag{3.61}$$

The pseudo-arclength continuation consists in the solution of the $(n+1)$-dimensional system given by $\mathbf{F}(\mathbf{x}, \mu) = \mathbf{0}$ and by the additional scalar equation (3.61). A Newton-Raphson algorithm is applied to find a solution to this system. At each iteration $k+1$ of the algorithm,

$$\mathbf{x}^{k+1} = \mathbf{x}^k + r\Delta\mathbf{x}^{k+1}, \quad \mu^{k+1} = \mu^k + r\Delta\mu^{k+1}, \tag{3.62}$$

where r is the relaxation parameter and $\Delta\mathbf{x}^{k+1}$ and $\Delta\mu^{k+1}$ are determined by

$$\mathbf{F}_{\mathbf{x}}(\mathbf{x}^k, \mu^k)\Delta\mathbf{x}^{k+1} + \mathbf{F}_{\mu}(\mathbf{x}^k, \mu^k)\Delta\mu^{k+1} = -\mathbf{F}(\mathbf{x}^k, \mu^k), \tag{3.63}$$

$$(\mathbf{x}^{*\prime})^{\mathrm{T}}\Delta\mathbf{x}^{k+1} + \mu^{*\prime}\Delta\mu^{k+1} = -f(\mathbf{x}^k, \mu^k). \tag{3.64}$$

If the Jacobian matrix $\mathbf{F}_{\mathbf{x}}$ is nonsingular, equation (3.63) can be solved by using the following bordering algorithm, which is based on the superposition method. Two vectors $\mathbf{z}_1$ and $\mathbf{z}_2$ are found in order to satisfy the following equations:

$$\mathbf{F}_{\mathbf{x}}(\mathbf{x}^k, \mu^k)\mathbf{z}_2 = -\mathbf{F}_{\mu}(\mathbf{x}^k, \mu^k), \tag{3.65}$$

and

$$\mathbf{F}_{\mathbf{x}}(\mathbf{x}^k, \mu^k)\mathbf{z}_1 = -\mathbf{F}(\mathbf{x}^k, \mu^k). \tag{3.66}$$

Then, substituting (3.65) and (3.66) into (3.63) yields

$$\Delta\mathbf{x}^{k+1} = \mathbf{z}_1 + \mathbf{z}_2\Delta\mu^{k+1}. \tag{3.67}$$

Substituting (3.67) into (3.64) finally gives

$$\Delta\mu^{k+1} = -\frac{f(\mathbf{x}^k, \mu^k) + \mathbf{z}_1^{\mathrm{T}}\mathbf{x}^{*\prime}}{\mu^{*\prime} + \mathbf{z}_2^{\mathrm{T}}\mathbf{x}^{*\prime}}. \tag{3.68}$$

Once $\Delta\mu^{k+1}$ is obtained from (3.68), the value of $\Delta\mathbf{x}^{k+1}$ is computed by using equation (3.67).

3.8.3 Pseudo-Arclength Continuation of Periodic Solutions

The simplest numerical approach for constructing periodic solutions is the numerical integration starting from an initial condition. Integrating over a long time, the solution will converge to an attractor. However, even if the method is very general, there is no guarantee of converging to the desired attractor. Moreover, (i) the convergence can be very slow in lightly damped systems, (ii) only some unstable solutions can be obtained by reversing the direction of integration, and (iii) achievement of steady-state solutions may be difficult (Nayfeh and Balachandran 1995). Therefore, specific methods have been introduced to determine periodic solutions of systems of n first-order differential equations:

$$\dot{\mathbf{x}} = \mathbf{F}(\mathbf{x}, \mu). \tag{3.69}$$

Frequency domain, as the harmonic balance, and time domain, as the finite difference, shooting, Poincaré map and collocation methods can be used.

A very effective continuation method of periodic solutions is the pseudo-arclength continuation, which is implemented in the software program AUTO. Time-domain methods allow finding an initial condition $\mathbf{x}(t = 0, \mu) = \eta$ and a periodic solution $\mathbf{x}(t, \eta, \mu)$ with a minimal period T that satisfies

$$\mathbf{x}(T, \eta, \mu) = \eta, \tag{3.70}$$

where μ is a system parameter, as usual. In the pseudo-arclength continuation, the arclength s is the continuation parameter. Therefore, equation (3.70) can be rewritten as (Nayfeh and Balachandran 1995)

$$\mathbf{x}[T(s), \eta(s), \mu(s)] = \eta(s). \tag{3.71}$$

The n-dimensional equation (3.71) gives the n boundary conditions for equation (3.69); however, there are $n+2$ unknowns, which are η, T and μ. Therefore, it is necessary to add two equations. The first of these two additional equations gives a phase condition in the form of an integral equation. In particular, two consecutive points $\mathbf{x}_0 = \mathbf{x}(T_0, \eta_0, \mu_0)$ and $\hat{\mathbf{x}}$ are considered on the same branch of solution. For any time shift σ, $\hat{\mathbf{x}}(t + \sigma)$ is also a solution. The phase condition consists in the minimization of the distance $\|\hat{\mathbf{x}} - \mathbf{x}_0\|$ with respect to σ, that is, in the minimization of

$$D(\sigma) = \int_0^T \|\hat{\mathbf{x}}(t + \sigma) - \mathbf{x}_0(t)\|^2 \mathrm{d}t. \tag{3.72}$$

Setting $\mathrm{d}D/\mathrm{d}\sigma = 0$ yields

$$\int_0^T \frac{\mathrm{d}}{\mathrm{d}\sigma}[\|\hat{\mathbf{x}}(t + \sigma) - \mathbf{x}_0(t)\|^2]\mathrm{d}t = 0, \tag{3.73}$$

which is satisfied for $\sigma = \sigma^*$. Equation (3.73) can be rewritten as

$$\int_0^T [\mathbf{x}(t) - \mathbf{x}_0(t)]^{\mathrm{T}}\dot{\mathbf{x}}(t)\mathrm{d}t = \frac{1}{2}\mathbf{x}^{\mathrm{T}}(t)\mathbf{x}(t)\,\Big|_0^T - \int_0^T \mathbf{x}_0^{\mathrm{T}}(t)\,\dot{\mathbf{x}}(t)\mathrm{d}t = -\int_0^T \mathbf{x}_0^{\mathrm{T}}(t)\,\dot{\mathbf{x}}(t)\mathrm{d}t, \tag{3.74}$$

where $\mathbf{x}(t) = \hat{\mathbf{x}}(t + \sigma^*)$ and integration by parts has been used. By substituting (3.69) into (3.74), the following integral equation is finally obtained:

$$\int_0^T \mathbf{x}^{\mathrm{T}}\dot{\mathbf{x}}_0\,\mathrm{d}t = \int_0^T \mathbf{x}^{\mathrm{T}}\mathbf{F}[\mathbf{x}(T_0, \eta_0, \mu_0),\ \mu_0]\,\mathrm{d}t = 0. \tag{3.75}$$

The second additional equation is the pseudo-arclength constraint, which is an extension of equation (3.61) to periodic solutions:

$$\int_0^T (\mathbf{x} - \mathbf{x}_0)^{\mathrm{T}}\mathbf{x}_0'\mathrm{d}t + (T - T_0)T_0' + (\mu - \mu_0)\mu_0' - \Delta s = 0, \tag{3.76}$$

where the prime denotes a derivative with respect to s and Δs is the step. In the program AUTO, the integrals (3.75) and (3.76) are approximated by back-forward difference formulas and the $(n+2)$-dimensional system of equations given by (3.69),

(3.75) and (3.76) with the boundary conditions (3.71) is discretized by using a collocation algorithm.

3.9 Nonlinear and Internal Resonances

Nonlinear systems under harmonic excitation can have superharmonic and subharmonic responses, as previously discussed. Moreover, more resonance peaks can be expected with respect to those obtained from the linearized systems. In particular, if ω is the circular frequency of harmonic excitation, and Ω_j, $j = 1, \ldots, N$, are the natural circular frequencies of the system (in the case of a continuous system, it must be discretized, so that it presents a finite number N of natural frequencies), the resonant solutions for linearized system are

$$\omega = \Omega_j, \quad j = 1, \ldots, N. \tag{3.77}$$

In the case of a nonlinear system, the resonant solutions can be expected near the excitation circular frequencies ω satisfying the following relations (Schmidt and Tondl 1986):

$$k\omega = \sum_{j=1}^{N} n_j \Omega_j, \quad j = 1, \ldots, N, \quad k = 1, 2, \ldots, \quad n_j = 0, \pm 1, \pm 2, \ldots. \tag{3.78}$$

Particular cases of equation (3.78) are (i) subharmonic resonance $\omega = n_j \Omega_j$ with $n_j = 2, 3, \ldots$; (ii) superharmonic resonance $\omega = \Omega_j / k$ with $k = 2, 3, \ldots$; (iii) sub-superharmonic resonance $\omega = n_j \Omega_j / k$ with $n_j / k \neq 1, 2, \ldots$ and $k / n_j \neq 1, 2, \ldots$; (iv) combination resonances $\frac{1}{k} \sum_{j=1}^{N} n_j \Omega_j$ with $j = 1, \ldots, N, k = 1, 2, \ldots, n_j = 0, \pm 1, \pm 2, \ldots$. Nonlinear resonances can similarly be obtained for periodic, but nonharmonic, excitations.

Note that most of the systems that are studied in this book can have additional resonance peaks only in the presence of particular conditions. Complex responses with additional peaks are often observed for nonlinear systems in the presence of internal resonances, which are obtained when the ratio of two or several natural frequencies is close to a ratio of small integers:

$$\Omega_i / \Omega_j = n, \quad n = 1, 2, \ldots, \quad i, j = 1, \ldots, N. \tag{3.79}$$

Special cases are (i) one-to-one (1:1) internal resonance, which is characteristic of doubly symmetric systems, such as circular cylindrical shells, circular and square plates, which have pairs of modes with the same natural frequency; (ii) one-to-two (1:2) internal resonance when $\Omega_i \cong 2\Omega_j$; (iii) one-to-three (1:3) internal resonance when $\Omega_i \simeq 3\Omega_j$; and (iv) multiple internal resonances involving more than two modes. An example of 1:1:1:2 internal resonances for a water-filled circular cylindrical shell is shown in Figure 3.14. Because of the high modal density (i.e. to the high number of natural modes in a given frequency range), plates and shells are structural systems that can easily exhibit internal resonances.

Internal resonances not only can give additional peaks in the vibration response, but can also give additional branches in the solution through bifurcations when the excitation circular frequency ω approaches Ω_i. Classical examples are given by circular cylindrical shells and circular plates (see Figures 5.11 and 12.3, respectively).

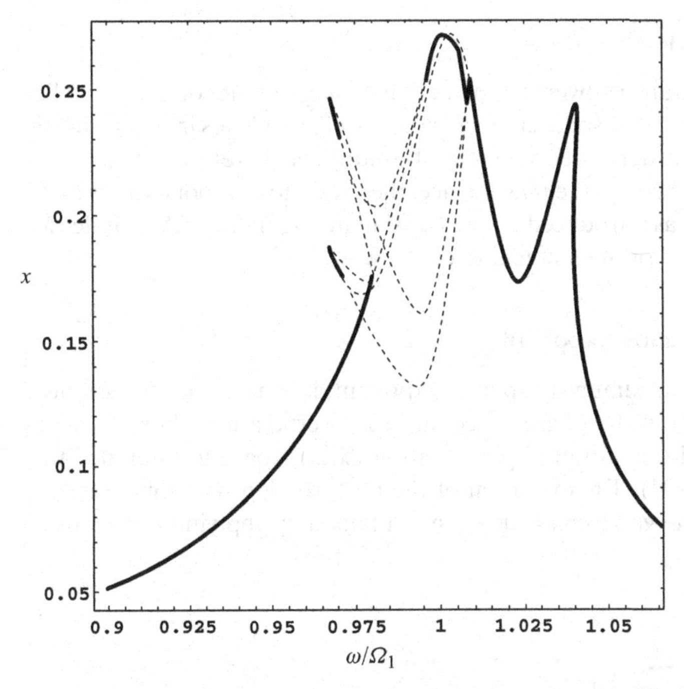

Figure 3.14. Forced response x of a simply supported, water-filled circular cylindrical shell in the case of 1:1:1:2 internal resonance; see Amabili et al. (2000).

In these cases, because of the axial symmetry of the systems, there are two orthogonal modes with the same frequency (1:1 internal resonance), angularly described by $\sin(n\theta)$ and $\cos(n\theta)$, respectively, n being the circumferential wavenumber. Even if an external excitation drives only one of these two modes, close to the resonance a pitchfork bifurcation of the periodic solution arises, and a traveling wave periodic solution appears, given by a combination of both the two modes with a specific phase shift. In correspondence to this bifurcated branch, energy is transferred from the driven mode to the so-called companion mode, which is not directly excited, because of the internal resonance.

3.10 Chaotic Vibrations

In Section 3.6 on the Poincaré maps, strange attractors and chaotic oscillation have been introduced. Only nonlinear systems can exhibit chaos. Chaotic vibrations have some peculiar characteristics, which are synthesized as follows (Moon 1992): (i) high sensitivity to small changes in the initial conditions; (ii) broad frequency spectrum of vibration, even if produced by a harmonic excitation; (iii) fractal properties of the oscillation in the phase space, denoting a strange attractor, which can be observed in the Poincaré maps; (iv) increasing complexity of regular motion by varying one or more parameters, following a particular route to chaos; for example, period-doubling bifurcations; (v) and transient or intermittent chaotic motion, in particular, nonperiodic bursts of irregular motion can appear. All the indicated points are good symptoms of chaos. However, in order to have a reliable identification of chaotic vibrations, it is necessary to introduce sophisticated diagnostic tests, such as the calculation of the maximum Lyapunov exponent.

3.11 Lyapunov Exponents

The Lyapunov exponents are a powerful tool for diagnosing whether or not an oscillation is chaotic and to determine some characteristics of chaos. Chaos in deterministic systems implies a sensitive dependence on initial conditions; therefore, if two trajectories start close to each other in the phase space, they will move exponentially away (Moon 1992). The idea was introduced by the Russian mathematician Alexander M. Lyapunov around the turn of the nineteenth century.

3.11.1 Maximum Lyapunov Exponent

In order to evaluate the maximum Lyapunov exponent, it is necessary to assume a reference trajectory $\mathbf{x}_r(t)$ in the phase space and to observe a neighboring trajectory originating at infinitesimal initial perturbation $\delta\mathbf{x}(t_0)$ from the reference trajectory (Argyris et al. 1994). The evolution of the perturbation with time, $\delta\mathbf{x}(t)$, is governed by the following variational equations, obtained by applying variations to equations (3.7):

$$\begin{cases} \dfrac{\mathrm{d}}{\mathrm{d}t}\delta x_j = \delta y_j \\ \dfrac{\mathrm{d}}{\mathrm{d}t}\delta y_j = -2\zeta_j\,\omega_j\,\delta y_j - \displaystyle\sum_{i=1}^{N} z_{j,i}\,\delta x_i \\ \qquad - \displaystyle\sum_{n=1}^{N}\sum_{i=1}^{N}\sum_{k=1}^{N} z_{j,i,k}\,\delta x_n\,(\delta_{k,n}x_i + \delta_{i,n}x_k) \qquad \text{for } j = 1, \ldots, N, \\ \qquad - \displaystyle\sum_{n=1}^{N}\sum_{i=1}^{N}\sum_{k=1}^{N}\sum_{l=1}^{N} z_{j,i,k,l}\,\delta x_n\,(\delta_{i,n}x_k x_l + \delta_{k,n}x_i x_l + \delta_{l,n}x_k x_i) \end{cases} \tag{3.80}$$

where $\delta_{k,n}$ is the Kronecker delta. Taking δx_j and δy_j as new variables, the simultaneous integration of the $4 \times N$ first-order differential equations (3.7) plus (3.80) (note that equations [3.7] are nonlinear and can be integrated by using the DIVPAG IMSL routine; and equations [3.80] are linear, but with time-varying coefficients, and can be integrated by using the adaptive step-size fourth- to fifth-order Runge-Kutta method) gives the solution of the equations of motion and the variational equations.

As usual, equation (3.7) can be rewritten in the compact form (3.69). Analogously, equation (3.80) can be rewritten as

$$\frac{\mathrm{d}}{\mathrm{d}t}\delta\mathbf{x} = \frac{\partial\mathbf{F}(\mathbf{x},\mu)}{\partial\mathbf{x}}\delta\mathbf{x}, \tag{3.81}$$

where in (3.81) $\delta\mathbf{x} = \{\delta x_1, \ldots, \delta x_N, \delta y_1, \ldots, \delta y_N\}^{\mathrm{T}}$ and $\mathbf{x} = \{x_1, \ldots, x_N, y_1, \ldots, y_N\}^{\mathrm{T}}$.

A specific software has been developed by Amabili (2005) for calculation of Lyapunov exponents of shells and plates. In a typical application, the excitation period has been divided in 10,000 integration steps Δt in order to have accurate evaluation of the time-varying coefficients in equations (3.80) that are obtained at each step by integration of equations (3.7). To find a reference trajectory, 6×10^6 steps are skipped in order to eliminate the transient and 1×10^6 steps are skipped to eliminate the transitory on the variational equations (3.80). Then 1×10^6 or 1×10^7 steps are used for evaluation of the maximum Lyapunov exponent σ_1 for the reference trajectory $\mathbf{x}(t)$.

For a suitably chosen initial perturbation $\delta\mathbf{x}(t_0)$, the rate of exponential expansion or contraction in the direction of $\delta\mathbf{x}(t_0)$ on the trajectory is measured by the maximum Lyapunov exponent σ_1, which is defined as

$$\delta\mathbf{x}(t) = \delta\mathbf{x}(t_0)\, e^{\sigma_1 t}, \tag{3.82}$$

or

$$\sigma_1 = \lim_{t\to\infty} \sup \frac{1}{t} \ln \frac{|\delta\mathbf{x}(t)|}{|\delta\mathbf{x}(t_0)|}. \tag{3.83}$$

Assuming the initial perturbation of unitary amplitude, equation (3.83) is simplified into

$$\sigma_1 = \lim_{t\to\infty} \sup \frac{1}{t} \ln|\delta\mathbf{x}(t)|. \tag{3.84}$$

Then, by restoring at each integration time step k the amplitude of $\delta\mathbf{x}(t)$ to its original unitary measure by the following re-normalization:

$$\delta\bar{\mathbf{x}}(t)_k = \frac{\delta\mathbf{x}(t)_k}{d_k}, \tag{3.85}$$

where $|\delta\mathbf{x}(t)|_k = d_k$, the following formula for the maximum Lyapunov exponent, evaluated at step k, is obtained:

$$\sigma_{1,k} = \frac{1}{k\,\Delta t} \sum_{i=1}^{k} \ln d_i. \tag{3.86}$$

In the numerical calculation of the maximum Lyapunov exponent, the nondimensional time is obtained by dividing by the period of the resonant mode. A perturbation vector $\boldsymbol{f}_1$ is determined with the maximum Lyapunov exponent σ_1.

It can be observed that:

(i) for periodic forced vibrations, $\sigma_1 < 0$,
(ii) for quasi-periodic response, $\sigma_1 = 0$,
(iii) for chaotic response, $\sigma_1 > 0$.

Therefore, σ_1 can conveniently be used for identification of system dynamics. The physical reason for $\sigma_1 < 0$ for periodic response of nonautonomous forced systems is that any perturbation in displacement or velocity is damped; in fact, a perturbation in the direction of the trajectory in the phase space will be canceled because it changes the phase relationship between excitation and response. This is the difference with respect to an autonomous system, for which a perturbation in the direction of the periodic trajectory will remain, giving $\sigma_1 = 0$.

3.11.2 Lyapunov Spectrum

The $2N$ numbers designating the spectrum of the Lyapunov exponents (Argyris et al. 1994) can be ordered according to their magnitude

$$\sigma_1 \geq \sigma_2 \geq \ldots \geq \sigma_{2N}. \tag{3.87}$$

It is clear that, for any perturbation, the largest Lyapunov exponent σ_1 will emerge and the perturbation vector will orient as $\mathbf{f}_1$. In order to calculate all the Lyapunov

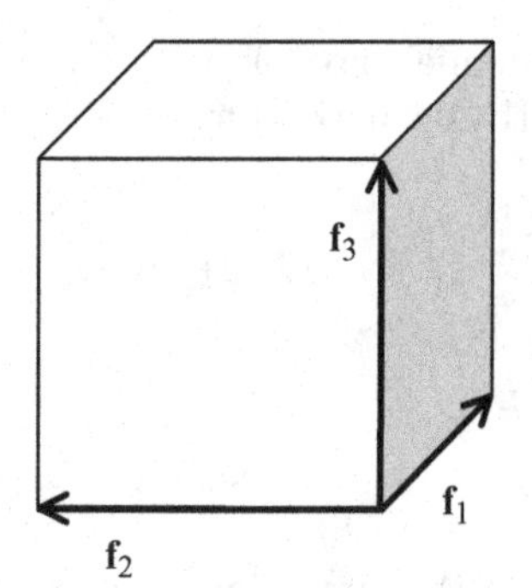

Figure 3.15. The p-dimensional parallelepiped ($p = 3$ in this case) included in the $2N$-dimensional space.

exponents, giving the so-called Lyapunov spectrum, it is necessary to introduce the Lyapunov exponent of p-th order, indicated as $\sigma^{(p)}$, which is the mean exponential growth rate of the volume $V^{(p)}$ of a p-dimensional parallelepiped ($p = 2, \ldots, 2N$) included in the $2N$-dimensional space (see Figure 3.15):

$$\sigma^{(p)} = \lim_{t\to\infty} \sup \frac{1}{t} \ln \frac{V^{(p)}(t)}{V^{(p)}(t_0)}. \tag{3.88}$$

This p-dimensional parallelepiped is spanned by the p orthogonal perturbation vectors $\mathbf{f}_1, \ldots, \mathbf{f}_p$ in which the generic perturbation can be decomposed. As the volume $V^{(p)}$ is obtained as the product of the lengths of these orthogonal perturbation vectors, which are directly related to the Lyapunov exponents, the Lyapunov exponent of p-th order is equal to the sum of the corresponding Lyapunov exponents; that is,

$$\sigma^{(p)} = \sigma_1 + \ldots + \sigma_p. \tag{3.89}$$

Therefore, once $\sigma^{(p)}$ is obtained,

$$\begin{aligned} \sigma_2 &= \sigma^{(2)} - \sigma_1, \\ &\vdots \\ \sigma_p &= \sigma^{(p)} - \sigma_1 - \ldots - \sigma_{p-1} \end{aligned}, \tag{3.90}$$

Consequently, in order to obtain all the Lyapunov exponents, the Lyapunov exponent of p-th order must be computed up to $p = 2N$. The main computational problem is to keep all the perturbation vectors orthogonal each other. The set of p orthogonal perturbations is indicated with $\{\mathbf{f}_1, \ldots, \mathbf{f}_p\}$. After choosing an initial set of unitary and orthogonal perturbation vectors, the following Gram-Schmidt orthonormalization procedure is applied at each integration time step k in order to restore the amplitude of each vector to its original unitary measure and keeping all vectors orthogonal to one another (Argyris et al. 1994):

$$(\bar{\mathbf{f}}_1)_k = (\mathbf{f}_1)_k / |\mathbf{f}_1|_k, \tag{3.91a}$$

$$(\bar{\mathbf{f}}_i)_k = \left(\mathbf{f}_i + \sum_{m=1}^{i-1} c_{i,m}\bar{\mathbf{f}}_m\right)_k \Bigg/ \left|\left(\mathbf{f}_i + \sum_{m=1}^{i-1} c_{i,m}\bar{\mathbf{f}}_m\right)\right|_k \tag{3.91b}$$

for $i = 2, \ldots, p$, with $(c_{i,m})_k = (-\mathbf{f}_i \cdot \bar{\mathbf{f}}_m)_k$.

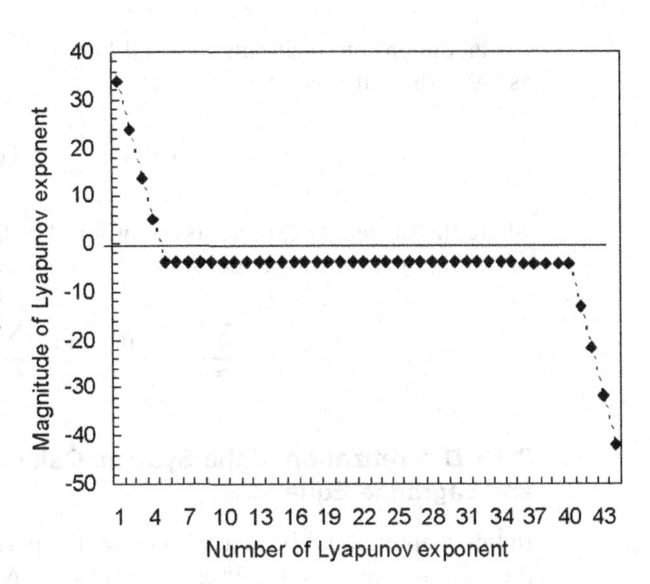

Figure 3.16. Lyapunov spectrum for a forced, damped spherical panel with rectangular base. Hyperchaos with four positive Lyapunov exponents is detected.

The volume $V_k^{(p)}$ at step k is given by

$$V_k^{(p)} = |\mathbf{f}_1|_k \wedge \ldots \wedge |\mathbf{f}_p|_k. \tag{3.92}$$

By using this computational procedure, the Lyapunov exponent of p-th order is given by

$$\sigma_k^{(p)} = \frac{1}{k\,\Delta t} \sum_{i=1}^{k} \ln V_i^{(p)}. \tag{3.93}$$

By using equations (3.92) and (3.90), all the Lyapunov exponents can be calculated.

The Lyapunov spectrum allows us to identify characteristics of chaotic solutions. In particular, an attractor with two or more positive Lyapunov exponents is associated with so-called hyperchaos. The phase space of a dynamic system is stretched in two or more directions in the case of hyperchaos. The Lyapunov spectrum of chaotic undamped systems is antisymmetric. This antisymmetry is almost conserved for slightly damped systems, as shown in Figure 3.16 for large harmonic excitation of a shell, where it is possible to observe a small translation of the antisymmetry axis in the ordinate direction due to damping.

3.12 Lyapunov Dimension

The long-term behavior of dissipative systems is characterized by attractors of the most varied types if the trajectories are not drawn toward infinity. After a transient state, in which some modes of motion finally vanish due to damping, the state of the system approaches an attractor where the number of independent variables, which determine the dimension of the phase space, is, in general, reduced considerably (Argyris et al. 1994). The fractal dimension is a measure of the strangeness of an attractor (Moon 1992) and indicates the number of effective independent variables determining the long-term behavior of a motion. There exist several measures of the fractal dimension, including the well-known Lyapunov dimension, which is defined as (Argyris et al. 1994; Moon 1992)

$$d_L = s + \sum_{r=1}^{s} \sigma_r / |\sigma_{s+1}|, \tag{3.94}$$

where the Lyapunov exponents are ordered by their magnitude, and s is obtained by satisfying the following conditions:

$$\sum_{r=1}^{s} \sigma_r > 0 \quad \text{and} \quad \sum_{r=1}^{s+1} \sigma_r < 0. \tag{3.95}$$

3.13 Discretization of the System: Galerkin Method and Lagrange Equations

In this chapter, it has been assumed that the partial differential equations governing the system dynamics are discretized into the N second-order ordinary differential equations of motion in the form (3.6). Two different methods are used in the following chapters in order to obtain the discretized equations (3.6) of continuous systems: the Galerkin method and an energy approach leading to the Lagrange equations of motion. In both methods, the displacements (and rotations for shear deformation theories) are expanded by using a sum of trial functions that satisfy the boundary conditions. For example, in the case of Donnell's nonlinear shallow-shell theory only the displacement w is used and can be expanded with the following expression:

$$w = \sum_{j=1}^{N} q_j(t)\, \varphi_j(\alpha_1, \alpha_2), \tag{3.96}$$

where $\varphi_j(\alpha_1, \alpha_2)$ are the N trial functions of the curvilinear coodinates α_1, α_2, and $q_j(t)$ are the corresponding generalized coordinates, that is, the only unknowns, which are functions of time only. The choice of the number and shape of the trial functions is crucial in order to obtain accurate solution.

The Galerkin method (see also Section 5.5.2) successively weighs the partial differential equation of motion, for example, equation (1.162), with the trial functions $\varphi_j(\alpha_1, \alpha_2)$. The Galerkin projection $\langle \ldots, \varphi_j \rangle$ of equation (1.162) is defined as

$$\langle D\nabla^4 w + ch\dot{w} + \rho h\ddot{w} = \ldots,\ \varphi_j \rangle = \int_0^{\alpha} \int_0^{\beta} (D\nabla^4 w + ch\dot{w} + \rho h\ddot{w} = \ldots)\varphi_j(\alpha_1, \alpha_2)\, A_1 A_2 \,\mathrm{d}\alpha_1 \,\mathrm{d}\alpha_2, \quad \text{for } j = 1, \ldots, N, \tag{3.97}$$

where A_1 and A_2 are the Lamé parameters defined in equations (2.17a,b). The Galerkin method gives N second-order ordinary differential equations of motion in the form of equation (3.6), that is, a problem with N degrees of freedom, which are the generalized coordinates.

A more versatile method to obtain the discretized equations is an energy approach, called here the Lagrange equations. Instead of discretizing the partial differential equations, the discretized equations of motion (see also Section 4.3.3) are directly obtained by minimizing the energy of the system. The Lagrange equations of motion are given by

$$\frac{\mathrm{d}}{\mathrm{d}t}\left(\frac{\partial T}{\partial \dot{q}_j}\right) - \frac{\partial T}{\partial q_j} + \frac{\partial U}{\partial q_j} = Q_j, \quad j = 1, \ldots, N, \tag{3.98}$$

where T is the kinetic energy of the system, $\partial T/\partial q_j$ is zero in most of the applications addressed in the present book, U is the potential energy of the system (e.g. the elastic strain energy) and Q are the generalized forces obtained by differentiation of Rayleigh's dissipation function F and of the virtual work W done by external forces:

$$Q_j = -\frac{\partial F}{\partial \dot{q}_j} + \frac{\partial W}{\partial q_j}. \tag{3.99}$$

Rayleigh's dissipation function takes into account nonconservative damping forces proportional to generalized velocities. For a doubly curved shell (neglecting shear deformation and rotary inertia), if viscous damping is assumed, F is given by

$$F = \frac{1}{2}c \int_0^{\alpha}\int_0^{\beta} (\dot{u}^2 + \dot{v}^2 + \dot{w}^2) A_1 A_2 \, \mathrm{d}\alpha_1 \, \mathrm{d}\alpha_2, \tag{3.100}$$

where c is the viscous damping coefficient. Therefore, all the terms in equation (3.98) are computed after the choice of the appropriate expansions of all the displacements (and eventually rotations) of the plate (or shell) middle surface.

REFERENCES

M. Amabili 2005 *International Journal of Non-Linear Mechanics* **40**, 683–710. Non-linear vibrations of doubly curved shallow shells.

M. Amabili, F. Pellicano and A. F. Vakakis 2000 *ASME Journal of Vibration and Acoustics* **122**, 346–354. Nonlinear vibrations and multiple resonances of fluid-filled, circular shells. Part 1: equations of motion and numerical results.

J. Argyris, G. Faust and M. Haase 1994 *An Exploration of Chaos*. North-Holland, Amsterdam, The Netherlands.

E. J. Doedel, A. R. Champneys, T. F. Fairgrieve, Y. A. Kuznetsov, B. Sandstede and X. Wang 1998 *AUTO 97: Continuation and Bifurcation Software for Ordinary Differential Equations (with HomCont)*. Concordia University, Montreal, Canada.

G. Duffing 1918 *Erzwungene Schwingungen bei veranderlicher Eigenfrequenz ihre technische bedeutung*. Vieweg, Braunschweig, Germany.

F. C. Moon 1992 *Chaotic and Fractal Dynamics*. Wiley, New York, USA.

A. H. Nayfeh and B. Balachandran 1995 *Applied Nonlinear Dynamics*. Wiley, New York, USA.

G. Schmidt and A. Tondl 1986 *Non-linear Vibrations*. Cambridge University Press, Cambridge, UK.

4 Vibrations of Rectangular Plates

4.1 Introduction

Flat rectangular plates with restrained normal displacement at the four edges exhibit a strong hardening-type nonlinearity for vibration amplitude of the order of the plate thickness. In order to have a behavior correctly described by a linear theory, the vibration amplitude of thin plates must be of the order of 1/10 of the thickness, or smaller. In-plane constraints largely enhance the nonlinear behavior; consequently, in-plane stretching is produced for large-amplitude deflection, differently from what is stated for linear theory.

Rectangular plates with normal displacement that is not restrained at all the edges can present a linear behavior for larger vibration amplitude. This is the case of the cantilever plate (clamped at one edge and free on the other three edges). In fact, quite large displacement w can be associated with very small rotations $\partial w/\partial x$ and $\partial w/\partial y$ for these boundary conditions, so that nonlinear terms in the strains can be neglected.

Geometric imperfections play an important role; they transform the flat plate in a curved panel (even if very shallow), which exhibits an initial weak softening behavior, turning to strong hardening nonlinearity for larger vibration amplitude.

In this chapter, the linear vibrations of simply supported rectangular plates are first addressed; numerical and experimental results are presented. Then, nonlinear forced vibrations of plates with different boundary conditions are studied by using the Lagrange equations of motion and the von Kármán theory. The effect of geometric imperfections is investigated. Numerical and experimental results are presented and satisfactorily compared. Finally, the problem of inertia coupling among the equations of motion, arising in the case of added lumped or distributed masses, is addressed.

4.1.1 Literature Review

Linear vibration of plates of different forms and subjected to several sets of boundary conditions are studied in the book by Leissa (1969).

A literature review of work on the nonlinear vibrations of plates is given by Chia (1980; 1988) and Sathyamoorthy (1987). The fundamental study in the analysis of large-amplitude vibrations of rectangular plates is by Chu and Herrmann (1956), who were the pioneers in the field. They studied simply supported rectangular plates with immovable edges and obtained the backbone curve for the fundamental mode. The solution was obtained by using a perturbation procedure and shows strong hardening-type nonlinearity.

A series of interesting papers (Ganapathi et al. 1991; Rao et al. 1993; Leung and Mao 1995; Shi and Mei 1996) compared different results for the backbone curve of the fundamental mode of isotropic plates with those of Chu and Herrmann (1956). All of them are in good agreement with the original results of Chu and Herrmann. In particular, Leung and Mao (1995) also studied simply supported rectangular plates with movable edges, which present reduced hardening-type nonlinearity with respect to simply supported plates with immovable edges. Different boundary conditions have been studied by Amabili (2004, 2006).

The effect of geometric imperfection was investigated by Hui (1984) and Amabili (2006). Recent studies by Han and Petyt (1997a,b) and Ribeiro and Petyt (1999a,b; 2000) used the hierarchical finite-element method to deeply investigate the nonlinear response of clamped rectangular plates. A similar approach was used by Ribeiro (2001) to investigate the forced response of simply supported plates with immovable edges. El Kadiri and Benamar (2003) developed a simplified analytical approach for the case studied by Chu and Herrmann (1956).

Nonlinear vibrations of rectangular laminated plates were thoroughly investigated, for example, by Noor et al. (1993), Abe et al. (1998) and Harras et al. (2002). Internal resonances in nonlinear vibration of rectangular plates were studied by Chang et al. (1993).

Even if a large number of theoretical studies on large-amplitude vibrations of plates are available in the scientific literature, experimental results are very scarce. Complete experimental results are given by Amabili (2004; 2006), and only experimental mode shapes for vibration amplitude 0.5 and 0.8 h are given by Harras et al. (2002) for the fundamental mode of a clamped laminated rectangular plate.

4.2 Linear Vibrations with Classical Plate Theory

In the case of small-amplitude vibrations, the von Kármán equations of motion of rectangular plates, given by equations (1.72) and (1.73), can be simplified by deleting all the nonlinear terms. No geometric imperfections are considered here, so that the plate is perfectly flat. In particular, equation (1.73) becomes $\nabla^4 F = 0$, which is independent from the normal deflection w. This states that no in-plane stretching is produced for small-amplitude deflections and that statics and dynamics are only governed by equation (1.72) where F is set equal to zero. Therefore, the equation of motion of rectangular plates becomes

$$D\nabla^4 w + ch\dot{w} + \rho h\ddot{w} = f + p, \tag{4.1}$$

where $\nabla^4 = [\partial^2/\partial x^2 + \partial^2/(\partial y^2)]^2$. Equation (4.1) is usually referred to as the Kirchhoff plate theory or the classical plate theory. In linear vibrations, the main interest is the calculation of natural frequencies and modes. In order to perform this calculation, viscous damping c, external excitation f and pressure p due to fluid-structure interaction are all set to zero in equation (4.1), giving

$$D\nabla^4 w + \rho h\ddot{w} = 0. \tag{4.2}$$

In order to solve equation (4.2), it is necessary to assign boundary conditions that represent the constraints at the four edges of the plate; only boundary conditions on the deflection w must be assigned. This shows that in-plane boundary conditions do not change natural frequencies and mode shapes of a plate, according with the classical plate theory.

The simplest case is a simply supported rectangular plate, having the following boundary conditions:

$$w = 0, \quad M_x = \frac{Eh^3}{12(1-\nu^2)}(k_x + \nu k_y) = 0, \quad \text{at } x = 0, a, \tag{4.3a,b}$$

$$w = 0 \quad M_y = \frac{Eh^3}{12(1-\nu^2)}(k_y + \nu k_x) = 0, \quad \text{at } y = 0, b, \tag{4.4a,b}$$

where M is the bending moment per unit length. Substituting the expressions for k_x and k_y given by equations (1.28) and (1.29), the boundary conditions (4.3b) and (4.4b) can be rewritten as

$$\left(\frac{\partial^2 w}{\partial x^2} + \nu \frac{\partial^2 w}{\partial y^2}\right) = 0, \quad \text{at } x = 0, a, \tag{4.5a}$$

$$\left(\nu \frac{\partial^2 w}{\partial x^2} + \frac{\partial^2 w}{\partial y^2}\right) = 0, \quad \text{at } y = 0, b. \tag{4.5b}$$

The expression of w that exactly satisfies the boundary conditions is

$$w(x, y, t) = w_{m,n} \sin(m\pi x/a) \sin(n\pi y/b) f(t), \tag{4.6}$$

where m and n are integers representing the number of half-waves of the mode shape in the x and y directions, respectively, and $w_{m,n}$ is the vibration amplitude. The time function is, as usual in the study of natural vibrations,

$$f(t) = \cos(\omega_{m,n} t), \tag{4.7}$$

where $\omega_{m,n}$ are the natural circular frequencies.

Substituting equations (4.6) and (4.7) into (4.2), the following result is obtained

$$\omega_{m,n} = \pi^2 \left[\left(\frac{m}{a}\right)^2 + \left(\frac{n}{b}\right)^2\right] \sqrt{\frac{D}{\rho h}}. \tag{4.8}$$

Equation (4.8) gives the natural circular frequencies and (4.6) gives the mode shapes of natural vibrations. There is an infinite number of natural frequencies, representing plate resonances, which are obtained setting $m = 1, 2, \ldots$ and $n = 1, 2, \ldots$ in equation (4.8).

The case of perfectly flat, simply supported rectangular plates is particularly simple. For more complex boundary conditions and geometric imperfections, the solution is more complex; for example, the technique presented in Section 4.3 for nonlinear vibrations can be used, canceling all the nonlinear terms.

4.2.1 Theoretical and Experimental Results

Calculations and experiments have been performed for a simply supported, rectangular aluminum plate, shown in Figure 4.1, with the following dimensions and material properties: $a = 0.515$ m, $b = 0.184$ m, $h = 0.0003$ m, $E = 69$ GPa, $\rho = 2700$ kg/m^3 and $\nu = 0.33$. The plate was subjected to a burst-random excitation to identify the natural frequencies and perform a modal analysis by measuring the

Figure 4.1. Photograph of the experimental setup.

plate response on a grid of points. The excitation was provided by an electrodynamical exciter (shaker). A piezoelectric miniature force transducer placed on the plate and connected to the shaker with a stinger measured the force transmitted. The plate response was measured by using a subminiature accelerometer.

The sum of the measured frequency-response functions (FRFs) is shown in Figure 4.2 with identification of natural modes in correspondence to peaks in the response. Comparison of experimental and theoretical natural frequencies, computed by using formula (4.8), is given in Figure 4.3. The agreement of results is satisfactory and assures that the experimental boundary conditions approximate simply supported edges. The theoretical mode shapes for $n = 1, 2$, $m = 1, 2$ are shown in Figure 4.4.

4.3 Nonlinear Vibrations with Von Kármán Plate Theory

Geometrically nonlinear vibrations, that is, large-amplitude vibrations, are studied in this section by using an energy approach. The elastic strain energy given in equation (1.52) is used. It is important to remark that the solution can alternatively be obtained by directly approaching the von Kármán equation of motion (1.72) and (1.73). The energy approach used here involves more degrees of freedom than the von Kármán equation of motion; in fact, the two in-plane displacements u and v are used instead of the single in-plane stress function F and in-plane inertia is retained. However, the energy approach allows to take into account complicating effects more easily, resulting in a more flexible technique.

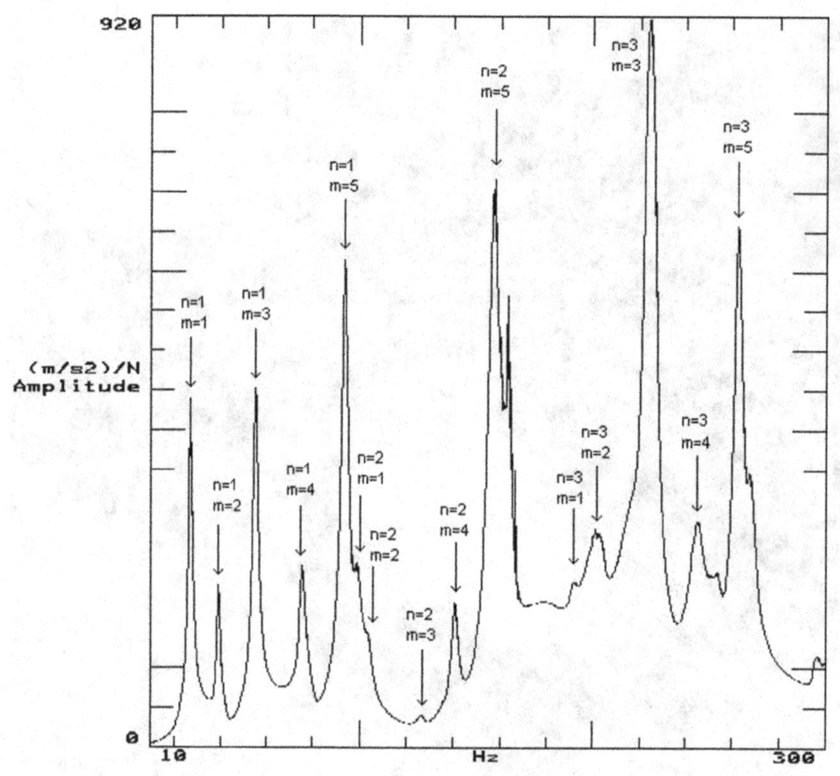

Figure 4.2. Sum of the measured FRFs with identification of natural modes.

4.3.1 Boundary Conditions, Kinetic Energy, External Loads and Mode Expansion

The kinetic energy T_P of a rectangular plate, by neglecting rotary inertia, is given by

$$T_P = \frac{1}{2}\rho h \int_0^a \int_0^b (\dot{u}^2 + \dot{v}^2 + \dot{w}^2)\,\mathrm{d}x\,\mathrm{d}y, \tag{4.9}$$

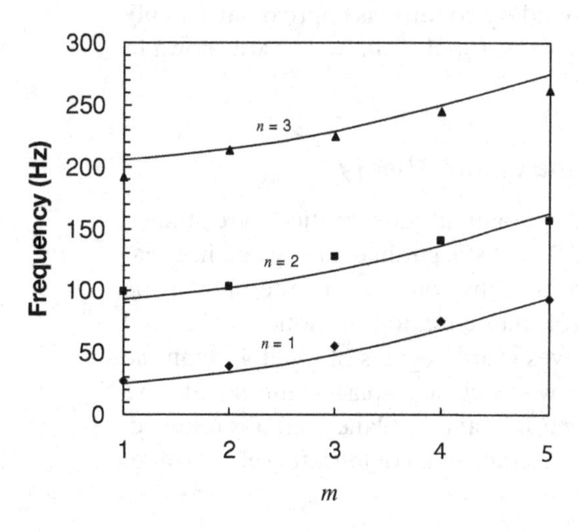

Figure 4.3. Theoretical and experimental natural frequencies of the plate. Theoretical results: —. Experimental results: ♦, $n = 1$; ■, $n = 2$; ▲, $n = 3$.

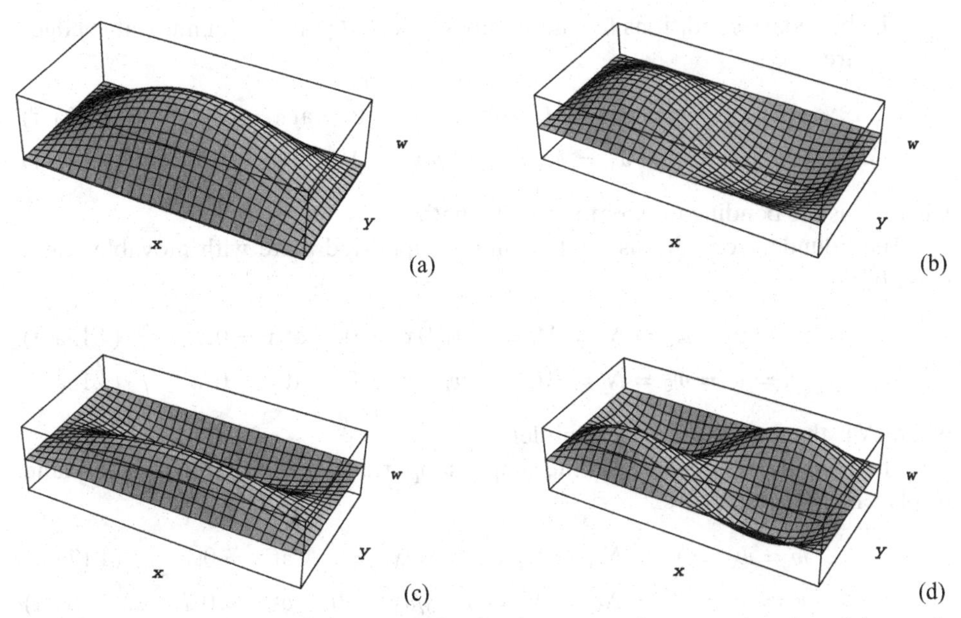

Figure 4.4. Theoretical mode shapes of the simply supported rectangular plate. (a) $n = 1$, $m = 1$. (b) $n = 1, m = 2$. (c) $n = 2, m = 1$. (d) $n = 2, m = 2$.

where ρ is the mass density of the plate. In equation (4.9), the overdot denotes time derivative.

The virtual work W done by the external forces is written as

$$W = \int_0^a \int_0^b (q_x u + q_y v + q_z w) \, \mathrm{d}x \, \mathrm{d}y, \tag{4.10}$$

where q_x, q_y and q_z are the distributed forces per unit area acting in x, y and z directions, respectively. Initially, only a single harmonic force orthogonal to the plate is considered; therefore, $q_x = q_y = 0$. The external distributed load q_z applied to the plate, due to the out-of-plane concentrated force $\tilde{f}$, is given by

$$q_z = \tilde{f} \, \delta(y - \tilde{y}) \, \delta(x - \tilde{x}) \cos(\omega t), \tag{4.11}$$

where ω is the excitation frequency, t is the time, δ is the Dirac delta function, $\tilde{f}$ gives the force magnitude positive in z direction and $\tilde{x}$ and $\tilde{y}$ give the position of the point of application of the force. Here, the point excitation is located at the center of plate, that is, $\tilde{x} = a/2$ and $\tilde{y} = b/2$. Equation (4.10) can be rewritten in the following form:

$$W = \tilde{f} \cos(\omega t) \, (w)_{x=a/2, \; y=b/2} \, . \tag{4.12}$$

In order to reduce the system to finite dimensions, the middle surface displacements u, v and w are expanded by using approximate functions.

Four different boundary conditions are analyzed in the present study: case (a), simply supported plate with immovable edges; case (b), simply supported plate with movable edges; case (c), simply supported plate with fully free in-plane edges; case (d), clamped plate.

The boundary conditions for the simply supported plate with immovable edges (case a) are

$$u = v = w = w_0 = M_x = \partial^2 w_0/\partial x^2 = 0, \quad \text{at } x = 0, a, \tag{4.13a–f}$$

$$u = v = w = w_0 = M_y = \partial^2 w_0/\partial y^2 = 0, \quad \text{at } y = 0, b, \tag{4.14a–f}$$

where M is the bending moment per unit length.

The boundary conditions for the simply supported plate with movable edges (case b) are

$$v = w = w_0 = N_x = M_x = \partial^2 w_0/\partial x^2 = 0, \quad \text{at } x = 0, a, \tag{4.15a–f}$$

$$u = w = w_0 = N_y = M_y = \partial^2 w_0/\partial y^2 = 0, \quad \text{at } y = 0, b, \tag{4.16a–f}$$

where N is the normal force per unit length.

The boundary conditions for the simply supported plate with fully free in-plane displacements (case c) are

$$w = w_0 = N_x = N_{xy} = M_x = \partial^2 w_0/\partial x^2 = 0, \quad \text{at } x = 0, a, \tag{4.17a–f}$$

$$w = w_0 = N_y = N_{yx} = M_y = \partial^2 w_0/\partial y^2 = 0, \quad \text{at } y = 0, b. \tag{4.18a–f}$$

The boundary conditions for the clamped plate (case d) are

$$u = v = w = w_0 = \partial w/\partial x = \partial w_0/\partial x = 0, \quad \text{at } x = 0, a, \tag{4.19a–f}$$

$$u = v = w = w_0 = \partial w/\partial y = \partial w_0/\partial y = 0, \quad \text{at } y = 0, b. \tag{4.20a–f}$$

Three bases of panel displacements are used to discretize the system for different boundary conditions. For cases (a) and (d), the displacements u, v and w are expanded by using the following expressions, which satisfy identically the geometric boundary conditions (4.13a–c) and (4.14a–c):

$$u(x, y, t) = \sum_{m=1}^{M}\sum_{n=1}^{N} u_{2m,n}(t) \sin(2m\pi x/a) \sin(n\pi y/b), \tag{4.21a}$$

$$v(x, y, t) = \sum_{m=1}^{M}\sum_{n=1}^{N} v_{m,2n}(t) \sin(m\pi x/a) \sin(2n\pi y/b), \tag{4.21b}$$

$$w(x, y, t) = \sum_{m=1}^{\hat{M}}\sum_{n=1}^{\hat{N}} w_{m,n}(t) \sin(m\pi x/a) \sin(n\pi y/b), \tag{4.21c}$$

where m and n are the numbers of half-waves in x and y directions, respectively, and t is the time; $u_{m,n}(t)$, $v_{m,n}(t)$ and $w_{m,n}(t)$ are the generalized coordinates, which are unknown functions of t. M and N indicate the terms necessary in the expansion of the in-plane displacements and, in general, are larger than $\hat{M}$ and $\hat{N}$, respectively, which indicate the terms in the expansion of w.

For case (b), the displacements u, v and w are expanded by using the following expressions, which satisfy identically the geometric boundary conditions (4.15a,b) and (4.16a,b):

$$u(x, y, t) = \sum_{m=1}^{M}\sum_{n=1}^{N} u_{m,n}(t) \cos(m\pi x/a) \sin(n\pi y/b), \tag{4.22a}$$

$$v(x, y, t) = \sum_{m=1}^{M} \sum_{n=1}^{N} v_{m,n}(t) \sin(m\pi x/a) \cos(n\pi y/b), \tag{4.22b}$$

$$w(x, y, t) = \sum_{m=1}^{\hat{M}} \sum_{n=1}^{\hat{N}} w_{m,n}(t) \sin(m\pi x/a) \sin(n\pi y/b). \tag{4.22c}$$

Case (b) represents the classical case, and the expansions of the displacements take a simple and intuitive form.

For case (c), the expansions of the displacements u, v and w, which satisfy identically the geometric boundary conditions (4.17a) and (4.18a) are given by:

$$u(x, y, t) = \sum_{m=1}^{M} \sum_{n=1}^{N} u_{m,n}(t) \cos(m\pi x/a) \cos(n\pi y/b), \tag{4.23a}$$

$$v(x, y, t) = \sum_{m=1}^{M} \sum_{n=1}^{N} v_{m,n}(t) \cos(m\pi x/a) \cos(n\pi y/b), \tag{4.23b}$$

$$w(x, y, t) = \sum_{m=1}^{\hat{M}} \sum_{n=1}^{\hat{N}} w_{m,n}(t) \sin(m\pi x/a) \sin(n\pi y/b). \tag{4.23c}$$

By using a different number of terms in the expansions, it is possible to study the convergence and the accuracy of the solution. It will be shown in the following that a sufficiently accurate model for the fundamental mode of the plate (case b) has 9 degrees of freedom. In particular, it is necessary to use the following terms: $m, n = 1, 3$ in equations (4.22a,b) and $m, n = 1$ in equation (4.22c). For the boundary conditions of other cases, more terms are necessary in the expansion to achieve the same accuracy.

Initial geometric imperfections of the rectangular plate are considered only in z direction. They are assumed to be associated with zero initial stress. The imperfection w_0 is expanded in the same form of w, that is, in a double Fourier sine series satisfying the boundary conditions (4.13d,f) and (4.14d,f) at the plate edges,

$$w_0(x, y) = \sum_{m=1}^{\tilde{M}} \sum_{n=1}^{\tilde{N}} A_{m,n} \sin(m\pi x/a) \sin(n\pi y/b), \tag{4.24}$$

where $A_{m,n}$ are the modal amplitudes of imperfections; $\tilde{N}$ and $\tilde{M}$ are integers indicating the number of terms in the expansion.

4.3.2 Satisfaction of Boundary Conditions

4.3.2.1 Case (a)

The geometric boundary conditions, equations (4.13a–d,f) and (4.14a–d,f), are exactly satisfied by the expansions of u, v, w and w_0. On the other hand, equations (4.13e) and (4.14e) can be rewritten in the form (4.3b) and (4.4b), which are identically satisfied for the expressions of k_x and k_y given in equations (1.28) and (1.29). Therefore, all the boundary conditions are exactly satisfied in this case.

4.3.2.2 Case (b)

In addition to the geometric boundary conditions and constraints on bending moment, the following constraints must be satisfied:

$$N_x = \frac{Eh}{1-\nu^2}(\varepsilon_{x,0} + \nu\varepsilon_{y,0}) = 0, \quad \text{at } x = 0, a, \tag{4.25}$$

$$N_y = \frac{Eh}{1-\nu^2}(\varepsilon_{y,0} + \nu\,\varepsilon_{x,0}) = 0, \quad \text{at } y = 0, b. \tag{4.26}$$

Equations (4.25) and (4.26) are not identically satisfied. Eliminating null terms at the panel edges, equations (4.25) and (4.26) can be rewritten as

$$\left[\frac{\partial \hat{u}}{\partial x} + \frac{1}{2}\left(\frac{\partial w}{\partial x}\right)^2 + \frac{\partial w}{\partial x}\frac{\partial w_0}{\partial x} + \nu\frac{\partial \hat{v}}{\partial y}\right]_{x=0,a} = 0, \tag{4.27}$$

$$\left[\frac{\partial \hat{v}}{\partial y} + \frac{1}{2}\left(\frac{\partial w}{\partial y}\right)^2 + \frac{\partial w}{\partial y}\frac{\partial w_0}{\partial y} + \nu\frac{\partial \hat{u}}{\partial x}\right]_{y=0,b} = 0, \tag{4.28}$$

where $\hat{u}$ and $\hat{v}$ are terms added to the expansion of u and v, given in equation (4.22a–c), in order to satisfy exactly the boundary conditions $N_x = 0$ and $N_y = 0$. Because $\hat{u}$ and $\hat{v}$ are second-order terms in the panel displacement, they have not been inserted in the second-order terms that involve u and v.

Nontrivial calculations (see Chapter 10, Section 10.2.1, for details on the mathematics in the case of doubly curved shallow-shells) give

$$\hat{u}(t) = -\sum_{n=1}^{\hat{N}}\sum_{m=1}^{\hat{M}}(m\pi/a)\left\{\frac{1}{2}w_{m,n}(t)\sin(n\pi y/b)\sum_{k=1}^{\hat{N}}\sum_{s=1}^{\hat{M}}\frac{s}{m+s}\,w_{s,k}(t)\sin(k\pi y/b)\right.$$
$$\times\sin[(m+s)\pi x/a] + w_{m,n}(t)\sin(n\pi y/b)\sum_{j=1}^{\hat{N}}\sum_{i=1}^{\hat{M}}\frac{i}{m+i}$$
$$\left.\times A_{i,j}\sin(j\pi y/b)\,\sin[(m+i)\pi x/a]\right\}, \tag{4.29}$$

$$\hat{v}(t) = -\sum_{n=1}^{\hat{N}}\sum_{m=1}^{\hat{M}}(n\pi/b)\left\{\frac{1}{2}w_{m,n}(t)\sin(m\pi x/a)\sum_{k=1}^{\hat{N}}\sum_{s=1}^{\hat{M}}\frac{k}{n+k}\,w_{s,k}(t)\,\sin(s\pi x/a)\right.$$
$$\times\sin[(n+k)\pi y/b] + w_{m,n}(t)\sin(m\pi x/a)\sum_{j=1}^{\hat{N}}\sum_{i=1}^{\hat{M}}\frac{j}{n+j}$$
$$\left.\times A_{i,j}\sin(i\pi x/a)\sin[(n+j)\pi y/b]\right\}. \tag{4.30}$$

4.3.2.3 Case (c)

In addition to boundary conditions (4.25) and (4.26), which give the terms (4.29) and (4.30) to be added to the expansions, the boundary conditions (4.17d) and (4.18d) must be also satisfied. They give

$$N_{xy} = N_{yx} = \frac{Eh}{2(1+\nu)}\gamma_{xy} = 0, \quad \text{at } x = 0, a, \quad \text{and at } y = 0, b. \tag{4.31}$$

Eliminating null terms at the plate edges, equation (4.31) can be rewritten as

$$\left[\frac{\partial u}{\partial y} + \frac{\partial v}{\partial x}\right]_{\substack{x=0,a \\ \text{or} \\ y=0,b}} = 0. \tag{4.32}$$

Equation (4.32) is a linear condition, which is satisfied by using minimization of energy in the Lagrange equations of motion. Therefore, no additional terms in the expansion are introduced. Actually, equations (4.25) and (4.26) can also be satisfied by energy minimization by avoiding to introduce equations (4.29) and (4.30). But, in this case, the choice of the expansions of u and v becomes very tricky, that is, all the terms involved in equations (4.29) and (4.30) must be inserted in the expansion in order to predict the system behavior with accuracy.

4.3.2.4 Case (d)

As discussed in Section 4.3.1, equations (4.19a–d,f) and (4.20a–f) are identically satisfied by the expansions of u, v, w and w_0. On the other hand, equations (4.19e) and (4.20e) can be rewritten in the following form:

$$M_x = \frac{Eh^3}{12(1-\nu^2)}(k_x + \nu k_y) = k\,\partial w/\partial x \quad \text{at } x = 0, a, \tag{4.33}$$

$$M_y = \frac{Eh^3}{12(1-\nu^2)}(k_y + \nu k_x) = k\,\partial w/\partial y \quad \text{at } y = 0, b, \tag{4.34}$$

where k is the stiffness per unit length of the elastic, distributed rotational springs placed at the four edges, $x = 0, a$ and $y = 0, b$. Equations (4.33) and (4.34) represent the case of an elastic rotational constraint at the shell edges. They give any rotational constraint from zero bending moment ($M_x = 0$ and $M_y = 0$, unconstrained rotation, obtained for $k = 0$) to perfectly clamped plate ($\partial w/\partial x = 0$ and $\partial w/\partial y = 0$, obtained as limit for $k \to \infty$), according to the value of k. If k is different from zero, an additional potential energy stored by the elastic rotational springs at the plate edges must be added. This potential energy U_R is given by

$$U_R = \frac{1}{2}\int_0^b k\left\{\left[\left(\frac{\partial w}{\partial x}\right)_{x=0}\right]^2 + \left[\left(\frac{\partial w}{\partial x}\right)_{x=a}\right]^2\right\}\mathrm{d}y + \frac{1}{2}\int_0^a k\left\{\left[\left(\frac{\partial w}{\partial y}\right)_{y=0}\right]^2 + \left[\left(\frac{\partial w}{\partial y}\right)_{y=b}\right]^2\right\}\mathrm{d}x. \tag{4.35}$$

In equation (4.35) a nonuniform stiffness k (function of x or y, simulating a nonuniform constraint) can be assumed. In order to simulate clamped edges in numerical calculations, a very high value of the stiffness k must be assumed. This approach is usually referred to as the artificial spring method (Yuan and Dickinson 1992; Amabili and Garziera 1999), which can be regarded as a variant of the classical penalty method. The values of the spring stiffness simulating a clamped plate can be obtained by studying the convergence of the natural frequencies of the linearized solution by increasing the value of k. In fact, it was found that the natural frequencies of the system converge asymptotically with those of a clamped plate when k becomes very large.

4.3.3 Lagrange Equations of Motion

The nonconservative damping forces are assumed to be of viscous type and are taken into account by using the Rayleigh dissipation function

$$F = \frac{1}{2}c \int_0^a \int_0^b (\dot{u}^2 + \dot{v}^2 + \dot{w}^2)\,\mathrm{d}x\,\mathrm{d}y, \tag{4.36}$$

where c has a different value for each term of the mode expansion. Simple calculations give

$$F = \frac{1}{2}(ab/4)\left[\sum_{n=1}^{N}\sum_{m=1}^{M} c_{m,n}\left(\dot{u}_{m,n}^2 + \dot{v}_{m,n}^2\right) + \sum_{n=1}^{\hat{N}}\sum_{m=1}^{\hat{M}} c_{m,n}\dot{w}_{m,n}^2\right]. \tag{4.37}$$

The damping coefficient $c_{m,n}$ is related to the modal damping ratio that can be evaluated from experiments by $\zeta_{m,n} = c_{m,n}/(2\,\mu_{m,n}\,\omega_{m,n})$, where $\omega_{m,n}$ is the natural circular frequency of mode (m, n) and $\mu_{m,n}$ is the modal mass of this mode, given by $\mu_{m,n} = \rho h\,(ab/4)$.

The following notation is introduced for brevity:

$$\mathbf{q} = \{u_{m,n}, v_{m,n}, w_{m,n}\}^{\mathrm{T}}, \quad m = 1, \ldots, M \text{ or } \hat{M} \quad \text{and} \quad n = 1, \ldots, N \text{ or } \hat{N}. \tag{4.38}$$

The generic element of the time-dependent vector $\mathbf{q}$ is referred to as q_j; the dimension of $\mathbf{q}$ is $\overline{N}$, which is the number of degrees of freedom (dofs) used in the mode expansion.

The generalized forces Q_j are obtained by differentiation of the Rayleigh dissipation function and of the virtual work done by external forces:

$$Q_j = -\frac{\partial F}{\partial \dot{q}_j} + \frac{\partial W}{\partial q_j} = -(ab/4)\,c_j\,\dot{q}_j \tag{4.39}$$

$$+ \begin{cases} 0 & \text{if } q_j = u_{m,n},\ v_{m,n};\ \text{or } w_{m,n} \quad \text{with } m \text{ or } n = 4, 8, \ldots, \\ \alpha_{mn}\,\tilde{f}\,\cos(\omega t) & \text{if } q_j = w_{m,n} \quad \text{with } m \text{ or } n \neq 4, 8, \ldots, \end{cases}$$

where α_{mn} is a coefficient taking the opportune numerical value (± 1, $\pm\sqrt{2}/2$, $\pm 1/2$) according with the value of (m, n).

The Lagrange equations of motion are

$$\frac{\mathrm{d}}{\mathrm{d}t}\left(\frac{\partial T_P}{\partial \dot{q}_j}\right) - \frac{\partial T_P}{\partial q_j} + \frac{\partial U}{\partial q_j} = Q_j, \quad j = 1, \ldots, \overline{N}, \tag{4.40}$$

where $\partial T_P/\partial q_j = 0$ and the potential energy is given by $U = U_P + U_R$, where U_P is given by equation (1.52) and U_R by (4.35); U_R is different from zero only for boundary conditions expressed in case (d). These second-order equations have very long expressions containing quadratic and cubic nonlinear terms. In particular,

$$\frac{\mathrm{d}}{\mathrm{d}t}\left(\frac{\partial T_P}{\partial \dot{q}_j}\right) = \rho h\,(ab/4)\,\ddot{q}_j, \tag{4.41}$$

which shows that no inertial coupling among the Lagrange equations exists for the plate with the mode expansion used.

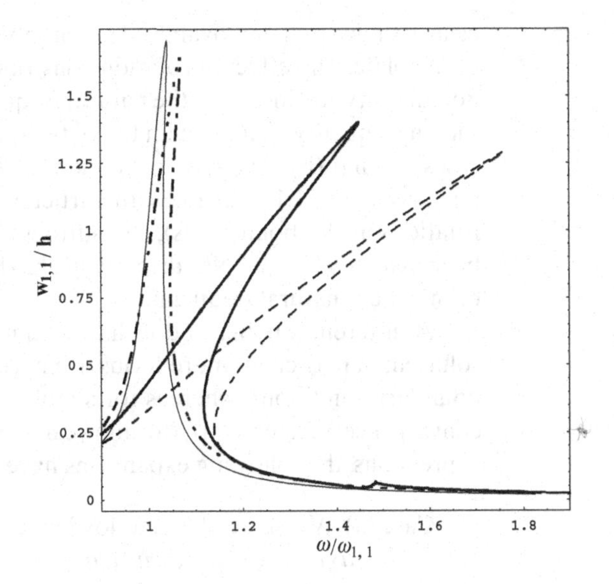

Figure 4.5. Response of the plate with different boundary conditions; mode (1, 1). --, case (a), simply supported plate with immovable edges, $\tilde{f} = 0.007\,\text{N}$, $\zeta_{1,1} = 0.0147$, 16 dofs; ---, case (b), simply supported plate with movable edges, $\tilde{f} = 0.007\,\text{N}$, $\zeta_{1,1} = 0.0117$, 12 dofs. —, case (c), simply supported plate with fully free in-plane displacements, $\tilde{f} = 0.007\,\text{N}$, $\zeta_{1,1} = 0.0117$, 22 dofs. **—**, case (d), clamped plate, $\tilde{f} = 0.0467\,\text{N}$, $\zeta_{1,1} = 0.0147$, 39 dofs.

The complicated term, derived from the maximum potential energy of the system, giving quadratic and cubic nonlinearities can be written in the form

$$\frac{\partial U}{\partial q_j} = \sum_{i=1}^{\overline{N}} f_{j,i}\, q_i + \sum_{i,k=1}^{\overline{N}} f_{j,i,k}\, q_i\, q_k + \sum_{i,k,l=1}^{\overline{N}} f_{j,i,k,l}\, q_i\, q_k\, q_l, \tag{4.42}$$

where the coefficients f have long expressions that include also geometric imperfections. It is interesting to observe that in equation (4.42) there are quadratic and cubic terms. In particular, quadratic terms appear in all the equations of motion as a consequence of including in-plane generalized coordinates. If the simpler von Kármán equation of motion (1.72, 1.73) is used, which neglects in-plane inertia, only cubic nonlinearities are obtained for perfectly flat plates. The presence of quadratic nonlinearities in the discretized equations of motion lead to the appearance of second-order harmonic components in the response to harmonic excitation in the neighborhood of a plate resonance. Third-order harmonics are due to cubic nonlinearities and are obtained both retaining and neglecting in-plane inertia. In the presence of geometric imperfections, plates become shallow shells; due to the curvature of the middle surface, stronger second-order harmonics appear.

4.4 Numerical Results for Nonlinear Vibrations

Calculations have been initially performed for the same rectangular aluminum plate without imperfections studied in Section 4.2.1, having the following dimensions and material properties: $a = 0.515\,\text{m}$, $b = 0.184\,\text{m}$, $h = 0.0003\,\text{m}$, $E = 69$ GPa, $\rho = 2700$ kg/m^3 and $\nu = 0.33$. This plate has fundamental mode ($n = 1, m = 1$) with radian frequency $\omega_{1,1} = 24.26 \times 2\pi$ rad/s and is subjected to harmonic excitation around the fundamental resonance, at the plate center. Calculations have been performed by using the computer program AUTO 97 (Doedel et al. 1998) based on the pseudo-arclength continuation method.

The maximum oscillation at the center of the plate (almost coincident with $w_{1,1}$ in this case) is presented in Figure 4.5 for the four boundary conditions studied. It is clearly shown that immovable edges largely enhance the hardening-type nonlinearity of the plate. The effect of rotation constraints, case (d), reduces the hardening-type nonlinearity and increases the natural frequency. It must be observed that the natural radian frequency changes with the out-of-plane constraints, but is left unchanged for cases (a) and (b). The radian frequency of the fundamental mode of the clamped plate is $\omega_{1,1} = 51.1 \times 2\pi$ rad/s. In particular, to simulate the clamped plate (i.e. zero rotations at the boundaries), the stiffness of the rotational distributed springs has been assumed $k = 10^9$ N/rad; this value has been determined by convergence analysis of the linear natural frequency.

A different number of dofs has been used in order to reach convergence of the solution; in particular, more terms are necessary to reach convergence for clamped boundary conditions, whereas for simply supported plate with movable edges, the convergence is faster. By performing calculations in order to choose the most suitable expressions, the following expansions have been obtained:

Case (a), $\overline{N} = 16$, with the following terms in equations (4.21a–c): $w_{1,1}$, $w_{3,1}$, $w_{1,3}$, $w_{3,3}$, $u_{2,1}$, $u_{4,1}$, $u_{6,1}$, $u_{8,1}$, $u_{2,3}$, $u_{4,3}$, $v_{1,2}$, $v_{1,4}$, $v_{1,6}$, $v_{1,8}$, $v_{3,2}$, $v_{3,4}$;

Case (b), $\overline{N} = 12$, with the following terms in equations (4.22a–c): $w_{1,1}$, $u_{1,1}$, $v_{1,1}$, $u_{3,1}$, $u_{1,3}$, $u_{3,3}$, $v_{1,3}$, $v_{3,1}$, $v_{3,3}$, $w_{1,3}$, $w_{3,1}$, $w_{3,3}$;

Case (c), $\overline{N} = 22$, with the following terms in equations (4.23a–c): $w_{1,1}$, $w_{1,3}$, $w_{3,1}$, $w_{3,3}$, $u_{1,0}$, $u_{1,2}$, $u_{1,4}$, $u_{3,0}$, $u_{3,2}$, $u_{3,4}$, $u_{5,0}$, $u_{5,2}$, $u_{5,4}$, $v_{0,1}$, $v_{2,1}$, $v_{4,1}$, $v_{0,3}$, $v_{2,3}$, $v_{4,3}$, $v_{0,5}$, $v_{2,5}$, $v_{4,5}$;

Case (d), $\overline{N} = 39$, with the following terms in equations (4.21a–c): $w_{i,j}$, $i, j = 1, 3, 5$; $u_{i,j}$, $v_{j,i}$, $i = 2, 4, 6$ and $j = 1, 3, 5$; with in addition: $w_{7,1}$, $w_{7,3}$, $w_{1,7}$, $w_{3,7}$, $u_{8,1}$, $u_{8,3}$, $u_{2,7}$, $u_{4,7}$, $v_{1,8}$, $v_{3,8}$, $v_{7,2}$, $v_{7,4}$.

Symmetry considerations have been applied in the choice of the generalized coordinates, and geometric imperfections have been assumed to be symmetric. Case (c) shows that the expansions of the in-plane displacements present terms $u_{m,n}$ and $v_{m,n}$ with exchanged indexes with respect to the expansions of cases (a, d). Moreover, the terms with $n = 0$ appear for u, and those with $m = 0$ for v, as a consequence of the free in-plane condition.

4.5 Comparison of Numerical and Experimental Results

Experimental tests have been conducted on an almost squared, stainless steel plate with the following dimensions and material properties: $a = 0.2085$ m, $b = 0.21$ m, $h = 0.0003$ m, $E = 198$ GPa, $\rho = 7850$ kg/m^3 and $\nu = 0.3$. The plate was inserted into a heavy rectangular steel frame (see Figure 4.6), with grooves designed with a V shape to hold the plate and avoid out-of-plane displacements at the edges. Silicon was placed into the grooves to better fill any possible gap between the grooves and the plate. Almost all the in-plane displacements at the edges were allowed because the constraint given by silicon on in-plane displacements is small. Therefore, the experimental boundary conditions are close to case (c).

The plate has been subjected to (i) burst-random excitation to identify the natural frequencies and perform a modal analysis by measuring the plate response on a grid of points, (ii) harmonic excitation, increasing or decreasing by very small

Figure 4.6. Photograph of the experimental plate, connected to the shaker by the stinger and the load cell.

steps the excitation frequency in the spectral neighborhood of the lowest natural frequencies to characterize nonlinear responses in the presence of large-amplitude vibrations (step-sine excitation). The excitation has been provided by an electrodynamical exciter (shaker), model *B&K* 4810, as shown in Figure 4.6. A piezoelectric miniature force transducer *B&K* 8203 of the weight of 3.2 g, placed on the plate at $x = a/4$ and $y = b/4$ and connected to the shaker with a stinger, measured the force transmitted. The plate response has been measured by using a very accurate laser Doppler vibrometer Polytec (sensor head OFV-505 and controller OFV-5000) in order to have noncontact measurement with no introduction of inertia. The time responses have been measured by using the Difa Scadas II front-end connected to a workstation and the software CADA-X of LMS for signal processing, data analysis and excitation control. The same front-end has been used to generate the excitation signal. The CADA-X closed-loop control has been used to keep constant the value of the excitation force for any excitation frequency, during the measurement of the nonlinear response.

Geometric imperfections of the plate have been measured by using a three-dimensional laser scanning system. The contour plot indicating the deviation from the ideal flat surface is reported in Figure 4.7. This figure shows that the actual shape of the plate is closer to a very shallow spherical shell; this is probably due to residual stresses of lamination in the steel foil and laser cut. Geometric imperfections are always present in actual plates.

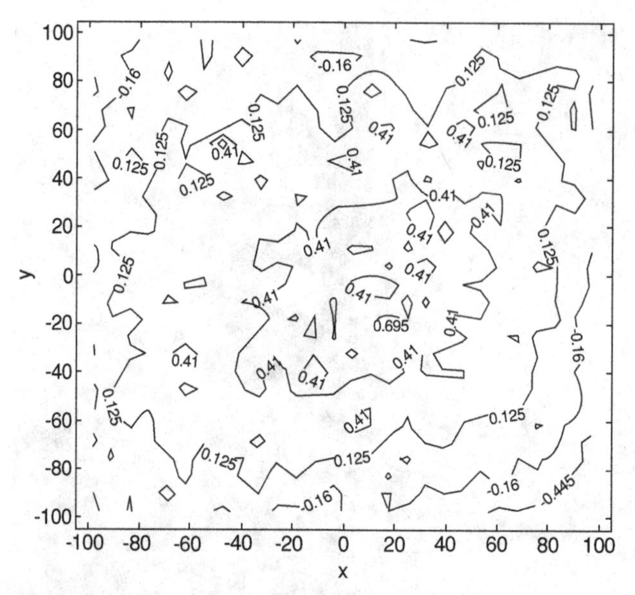

Figure 4.7. Contour plot indicating measured geometric imperfections as deviations from the flat surface. Deviations are in millimeters.

A comparison of theoretical (22 dofs model for case [c]) and experimental results is shown in Figure 4.8(a) for two force levels (damping $\zeta_{1,1} = 0.0105$, forces 0.01 and 0.04 N) and in Figure 4.8(b) for a higher-force level (damping $\zeta_{1,1} = 0.0105$, force 0.08 N); the same damping ratio is assumed for all the generalized coordinates. Comparison of numerical and experimental results is good for all three cases; calculations

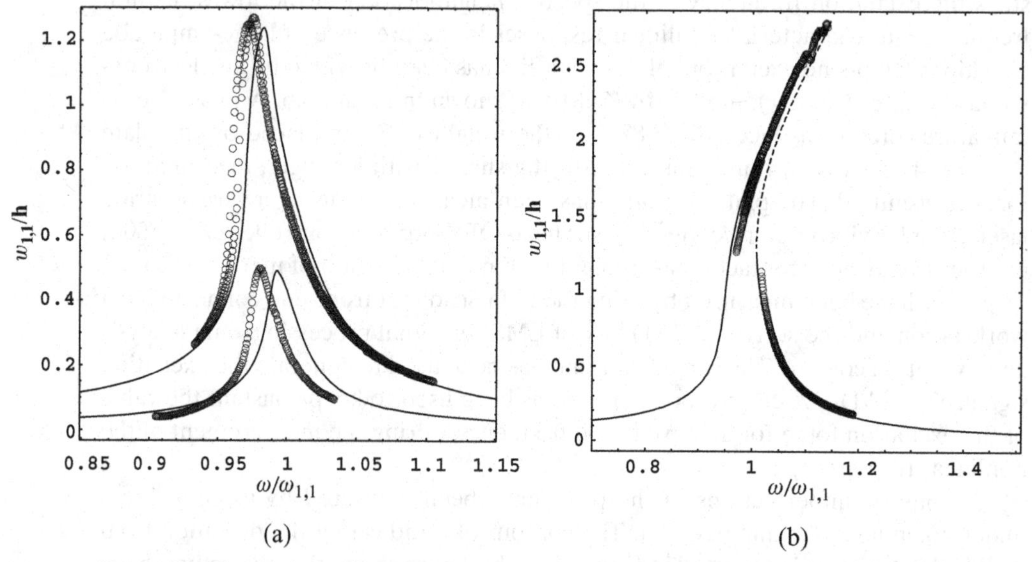

Figure 4.8. Comparison of numerical, case (c), and experimental results for the fundamental mode (1,1) at the center of the experimental plate; first harmonic only; $A_{1,1} = h$, $\zeta_{1,1} = 0.0105$; 22 dofs. ○, experimental data; —, stable theoretical solutions; - - -, unstable theoretical solutions. (a) $\tilde{f} = 0.01$ and 0.04 N. (b) $\tilde{f} = 0.08$ N.

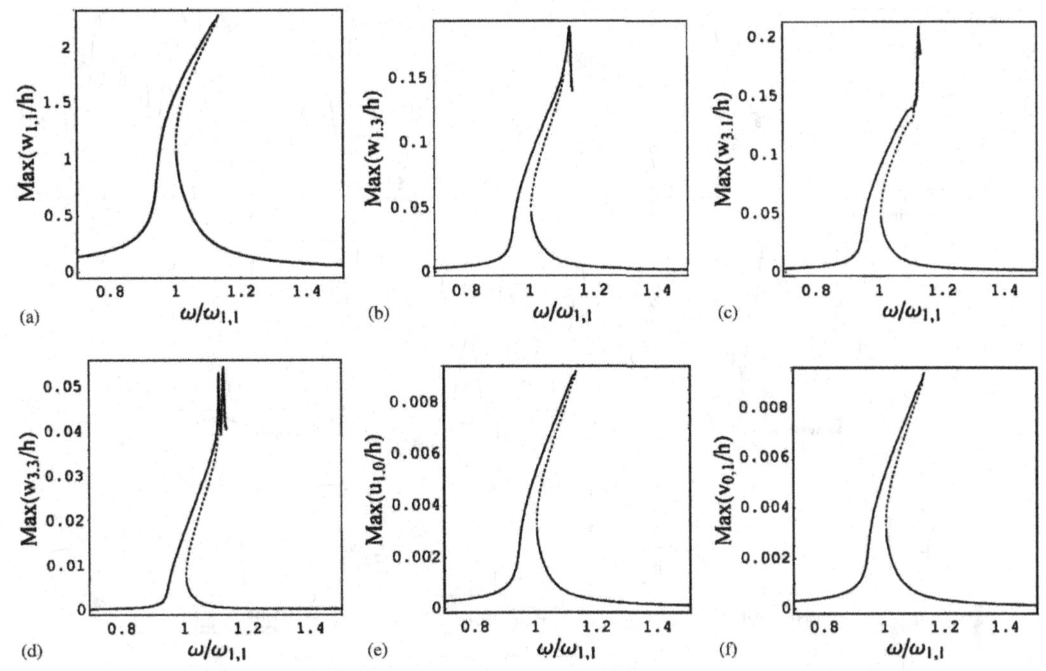

Figure 4.9. Response of the plate, case (c); fundamental mode (1,1), $A_{1,1} = h, \tilde{f} = 0.08$ N and $\zeta_{1,1} = 0.0105$; 22 dofs. —, stable periodic response; ---, unstable periodic response. (a) Maximum of the generalized coordinate $w_{1,1}$. (b) Maximum of the generalized coordinate $w_{1,3}$. (c) Maximum of the generalized coordinate $w_{3,1}$. (d) Maximum of the generalized coordinate $w_{3,3}$. (e) Maximum of the generalized coordinate $u_{1,0}$. (f) Maximum of the generalized coordinate $v_{0,1}$.

have been obtained introducing the geometric imperfection $A_{1,1} = h$, having the form of mode (1,1), which is in reasonable agreement with the measured actual plate surface, as reported in Figure 4.7 (measured imperfection at the center, about 0.41 mm, which would give $A_{1,1} = 1.37\,h$ if the imperfection had the shape of mode [1,1]). The model with the introduced imperfection is perfectly capable to reproduce qualitatively and quantitatively the nonlinear behavior of the imperfect plate, with the initial softening-type behavior, turning to hardening type for larger excitations and vibration amplitudes. It must be observed that, in Figure 4.8, the first harmonic of the oscillation amplitude of the generalized coordinate $w_{1,1}$ is reported (therefore the mean value is eliminated), which is practically coincident with the first harmonic of the vibration amplitude of the plate at the center.

The six main generalized coordinates associated with the plate response, given in Figure 4.8(b), are reported in Figure 4.9. In particular, $w_{1,3}$ and $w_{3,1}$ give a significant quantitative contribution to the plate response. In-plane vibration is smaller; however, even if small, the contribution of in-plane terms is fundamental in order to build an accurate model.

The time responses of the generalized coordinates close to the response peak (for excitation frequency $\omega = 1.135\ \omega_{1,1}$) are given in Figure 4.10 with their frequency spectra to show the contribution of superharmonics. In particular, the generalized coordinate $w_{1,1}$ shows a large asymmetric displacement of the plate inward and outward from the curvature created by the imperfection (a flat plate has symmetric

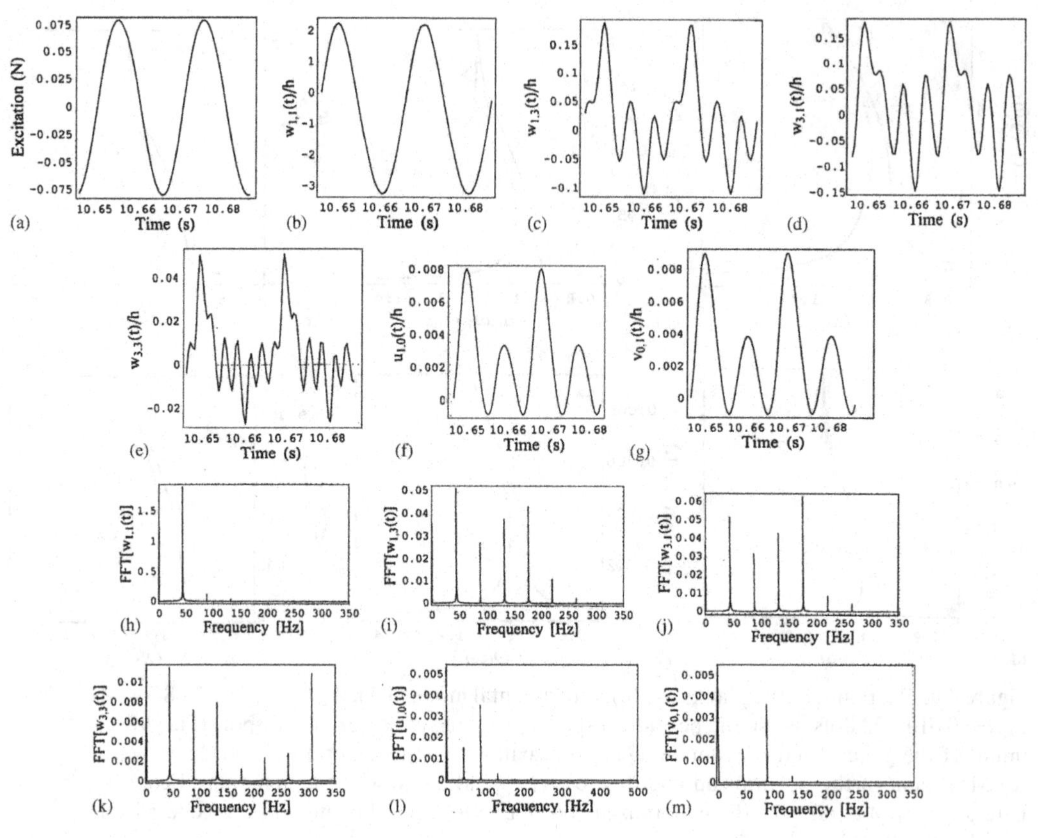

Figure 4.10. Computed time response and frequency spectrum of the generalized coordinates; mode (1,1), $\omega = 1.135\omega_{1,1}, A_{1,1} = h, \tilde{f} = 0.08$ N and $\zeta_{1,1} = 0.0105$; 22 dofs. (a) Force excitation. (b) Time response of $w_{1,1}$. (c) Time response of $w_{1,3}$. (d) Time response of $w_{3,1}$. (e) Time response of $w_{3,3}$. (f) Time response of $u_{1,0}$. (g) Time response of $v_{0,1}$. (h) Frequency spectrum of $w_{1,1}$. (i) Frequency spectrum of $w_{1,3}$. (j) Frequency spectrum of $w_{3,1}$. (k) Frequency spectrum of $w_{3,3}$. (l) Frequency spectrum of $u_{1,0}$. (m) Frequency spectrum of $v_{0,1}$.

displacement). This asymmetric displacement, resulting in a movement of the zero with respect to that of a sinusoidal oscillation (see Figure 4.10[b]), balances the imperfection around the resonance. In the frequency spectra, this zero frequency component is shown, as well as the second harmonic; these are due to quadratic nonlinearities in the equations of motion introduced by geometric imperfections.

The convergence of the solution, versus the number of generalized coordinates retained in the expansion, is shown in Figure 4.11. Although the solution with 13 dofs is isolated on the right-hand side, responses computed with 22 and 38 dofs are very close, indicating convergence. The model with 13 dofs has $w_{1,1}$, $u_{1,0}$, $u_{1,2}$, $u_{1,4}$, $u_{3,0}$, $u_{3,2}$, $u_{3,4}$, $v_{0,1}$, $v_{2,1}$, $v_{4,1}$, $v_{0,3}$, $v_{2,3}$, $v_{4,3}$. The model with 38 dofs has $w_{1,1}$, $w_{1,3}$, $w_{3,1}$, $w_{3,3}$, $w_{1,5}$, $w_{5,1}$, $u_{1,0}$, $u_{1,2}$, $u_{1,4}$, $u_{1,6}$, $u_{3,0}$, $u_{3,2}$, $u_{3,4}$, $u_{3,6}$, $u_{5,0}$, $u_{5,2}$, $u_{5,4}$, $u_{5,6}$, $u_{7,0}$, $u_{7,2}$, $u_{7,4}$, $u_{7,6}$, $v_{0,1}$, $v_{2,1}$, $v_{4,1}$, $v_{6,1}$, $v_{0,3}$, $v_{2,3}$, $v_{4,3}$, $v_{6,3}$, $v_{0,5}$, $v_{2,5}$, $v_{4,5}$, $v_{6,5}$, $v_{0,7}$, $v_{2,7}$, $v_{4,7}$, $v_{6,7}$.

The effect of the amplitude $A_{1,1}$ of geometric imperfection having the form of mode (1,1) on the nonlinear response is shown in Figure 4.12. Smaller amplitudes of $A_{1,1}$ with respect to h do not allow the peculiar softening-hardening behavior found in the experiments. Figure 4.12 shows that, after a while, all the curves for

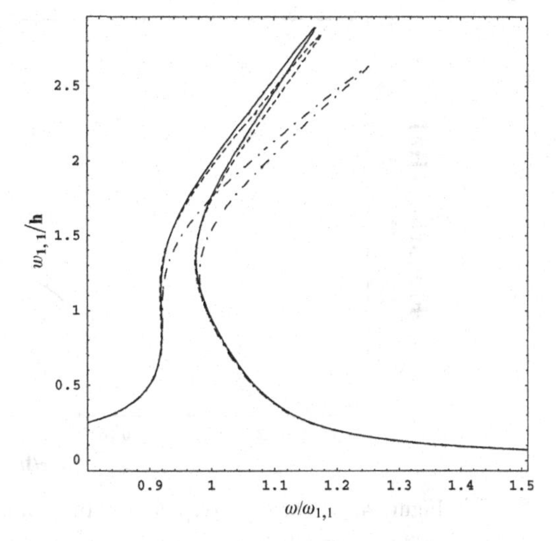

Figure 4.11. Convergence of results; fundamental mode (1,1), case (c), $A_{1,1} = 1.25\,h$, $\tilde{f} = 0.105$ N and $\xi_{1,1} = 0.0105$. —, 38 dofs; – –, 22 dofs; – - –, 13 dofs.

different initial imperfection become parallel to each other. This asymptotic result is of significant relevance in applications.

The effect of $A_{1,1}$ on the natural frequency of the fundamental mode (1,1) is shown in Figure 4.13. It is clearly shown that with $A_{1,1} = h$ the theoretical natural frequency raises from 32.7 to 38.5 Hz, becoming almost coincident to the experimental value of 38.7 Hz. Therefore, the imperfection introduced perfectly describes the linear and nonlinear behavior of the experimental plate.

4.6 Inertial Coupling in the Equations of Motion

For plates with added lumped or distributed masses and plates coupled to liquid, inertia coupling arises in the equations of motion, so that they have a different form with respect to equations (3.6) and cannot be immediately transformed in the

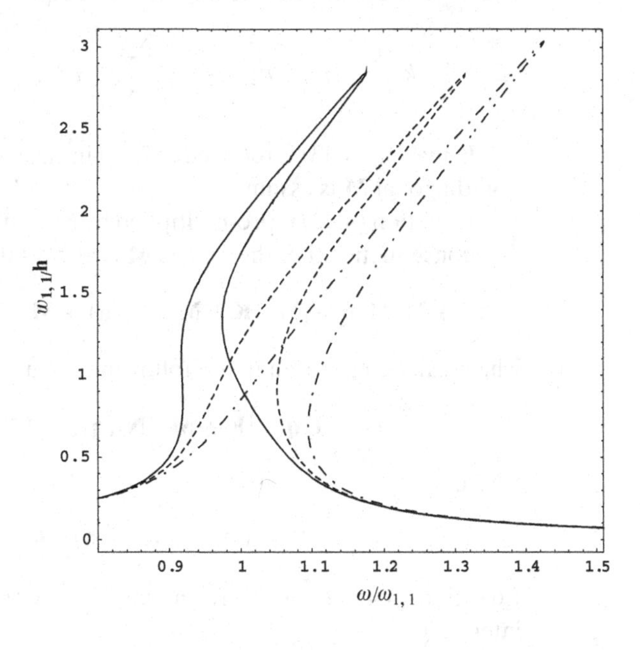

Figure 4.12. Effect of geometric imperfection $A_{1,1}$ on the nonlinear response of the fundamental mode (1,1), case (c); $\tilde{f} = 0.105$ N and $\zeta_{1,1} = 0.0105$. —, $A_{1,1} = 1.25\,h$; – –, $A_{1,1} = 0.75\,h$; – - –, $A_{1,1} = 0$.

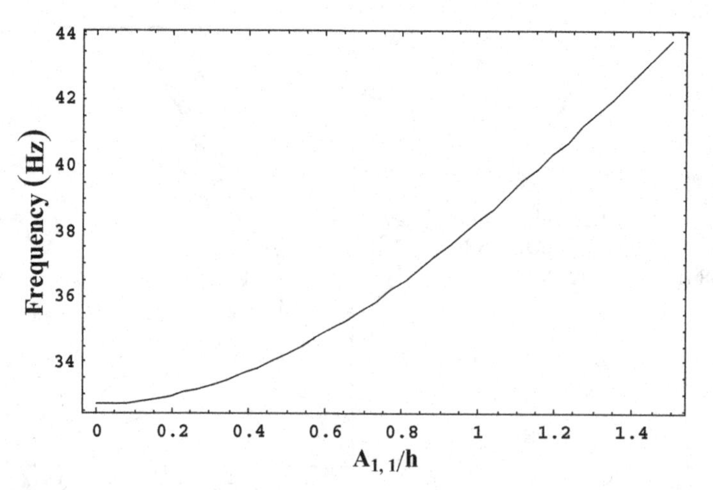

Figure 4.13. Natural frequency of the fundamental mode (1,1) versus the geometric imperfection $A_{1,1}$, case (c).

form (3.7), which is required for numerical integration. In particular, the equations of motion take the following form:

$$\mathbf{M}\ddot{\mathbf{q}} + \mathbf{C}\dot{\mathbf{q}} + [\mathbf{K} + \mathbf{N}_2(\mathbf{q}) + \mathbf{N}_3(\mathbf{q}, \mathbf{q})]\,\mathbf{q} = \mathbf{f}_0 \cos(\omega t), \tag{4.43}$$

where $\mathbf{M}$ is the mass matrix of dimension $\overline{N} \times \overline{N}$ ($\overline{N}$ being the number of degrees of freedom), $\mathbf{C}$ is the damping matrix, $\mathbf{K}$ is the linear stiffness matrix, which does not present terms involving $\mathbf{q}$, $\mathbf{N}_2$ is a matrix that involves linear terms in $\mathbf{q}$, therefore giving the quadratic nonlinear stiffness terms, $\mathbf{N}_3$ is a matrix that involves quadratic terms in $\mathbf{q}$, therefore giving the cubic nonlinear stiffness terms, $\mathbf{f}_0$ is the vector of excitation amplitudes and $\mathbf{q}$ is the vector of the $\overline{N}$ generalized coordinates, defined in equation (4.38). In particular, by using equation (4.42), the generic elements $k_{j,i}$, $n_{2_{j,i}}$ and $n_{3_{j,i}}$, of the matrices $\mathbf{K}$, $\mathbf{N}_2$ and $\mathbf{N}_3$, respectively, are given by

$$k_{j,i} = f_{j,i}, \quad n_{2_{j,i}}(\mathbf{q}) = \sum_{k=1}^{\overline{N}} f_{j,i,k}\, q_k, \quad n_{3_{j,i}}(\mathbf{q}, \mathbf{q}) = \sum_{k,l=1}^{\overline{N}} f_{j,i,k,l}\, q_k\, q_l.$$

Equation (4.43) is reduced to (3.6) in case that $\mathbf{M}$ is diagonal. In the present case, non-diagonal $\mathbf{M}$ is assumed.

Equation (4.43) is pre-multiplied by $\mathbf{M}^{-1}$ in order to diagonalize the mass matrix, as a consequence that the matrix $\mathbf{M}$ is always invertible; the result is

$$\mathbf{I}\ddot{\mathbf{q}} + \mathbf{M}^{-1}\mathbf{C}\,\dot{\mathbf{q}} + [\mathbf{M}^{-1}\mathbf{K} + \mathbf{M}^{-1}\mathbf{N}_2(\mathbf{q}) + \mathbf{M}^{-1}\mathbf{N}_3(\mathbf{q}, \mathbf{q})]\,\mathbf{q} = \mathbf{M}^{-1}\mathbf{f}_0 \cos(\omega t), \tag{4.44}$$

which can be rewritten in the following form

$$\mathbf{I}\ddot{\mathbf{q}} + \tilde{\mathbf{C}}\,\dot{\mathbf{q}} + [\tilde{\mathbf{K}} + \mathbf{M}^{-1}\mathbf{N}_2(\mathbf{q}) + \mathbf{M}^{-1}\mathbf{N}_3(\mathbf{q}, \mathbf{q})]\,\mathbf{q} = \tilde{\mathbf{f}}_0 \cos(\omega t), \tag{4.45}$$

where

$$\tilde{\mathbf{C}} = \mathbf{M}^{-1}\mathbf{C}, \quad \tilde{\mathbf{K}} = \mathbf{M}^{-1}\mathbf{K} \quad \text{and} \quad \tilde{\mathbf{f}}_0 = \mathbf{M}^{-1}\mathbf{f}_0. \tag{4.46a–c}$$

Equation (4.45) is in the form suitable for transformation in (3.7) for numerical integration.

4.7 Effect of Added Masses

A system of N lumped masses M_k at $x = x_k^*$, $y = y_k^*$ is considered. The reference translational kinetic energy of the added masses is given by

$$T_M = \frac{1}{2}\sum_{k=1}^{N} M_k \left[\dot{u}^2(x_k^*, y_k^*) + \dot{v}^2(x_k^*, y_k^*) + \dot{w}^2(x_k^*, y_k^*)\right]. \tag{4.47}$$

If a classical theory is used, so that only three variables u, v and w are used to describe the deformation and the rotary inertia of the plate is neglected, the rotary kinetic energy of the added masses is given by

$$T_{MR} = \frac{1}{2}\sum_{k=1}^{N}\left\{J_{y,k}\left[\frac{\partial \dot{w}}{\partial x}(x_k^*, y_k^*)\right]^2 + J_{x,k}\left[\frac{\partial \dot{w}}{\partial y}(x_k^*, y_k^*)\right]^2\right\}, \tag{4.48}$$

where $J_{x,k}$ and $J_{y,k}$ are the moment of inertia of the k-th mass with respect to an axis parallel to x and to y, respectively, passing by the middle plate surface at the mass location $x = x_k^*$, $y = y_k^*$.

Equations (4.47) and (4.48) must be added to (4.9) in order to have the kinetic energy of the plate with attached masses. In general, it gives equations of motion with inertia coupling that must be solved with the transformation of coordinates introduced in (4.44).

REFERENCES

A. Abe, Y. Kobayashi and G. Yamada 1998 *International Journal of Non-Linear Mechanics* **33**, 675–690. Two-mode response of simply supported, rectangular laminated plates.

M. Amabili 2004 *Computers and Structures* **82**, 2587–2605. Nonlinear vibrations of rectangular plates with different boundary conditions: theory and experiments.

M. Amabili 2006 *Journal of Sound and Vibration* **291**, 539–565. Theory and experiments for large-amplitude vibrations of rectangular plates with geometric imperfections.

M. Amabili and R. Garziera 1999 *Journal of Sound and Vibration* **224**, 519–539. A technique for the systematic choice of admissible functions in the Rayleigh-Ritz method.

S. I. Chang, A. K. Bajaj and C. M. Krousgrill 1993 *Nonlinear Dynamics* **4**, 433–460. Non-linear vibrations and chaos in harmonically excited rectangular plates with one-to-one internal resonance.

C.-Y. Chia 1980 *Nonlinear Analysis of Plates*. McGraw-Hill, New York, USA.

C.-Y. Chia 1988 *Applied Mechanics Reviews* **41**, 439–451. Geometrically nonlinear behavior of composite plates: a review.

H.-N. Chu and G. Herrmann 1956 *Journal of Applied Mechanics* **23**, 532–540. Influence of large amplitude on free flexural vibrations of rectangular elastic plates.

E. J. Doedel, A. R. Champneys, T. F. Fairgrieve, Y. A. Kuznetsov, B. Sandstede and X. Wang 1998 *AUTO 97: Continuation and Bifurcation Software for Ordinary Differential Equations (with HomCont)*. Concordia University, Montreal, Canada.

M. El Kadiri and R. Benamar 2003 *Journal of Sound and Vibration* **264**, 1–35. Improvement of the semi-analytical method, based on Hamilton's principle and spectral analysis, for determination of the geometrically non-linear response of thin straight structures. Part III: steady state periodic forced response of rectangular plates.

M. Ganapathi, T. K. Varadan and B. S. Sarma 1991 *Computers and Structures* **39**, 685–688. Nonlinear flexural vibrations of laminated orthotropic plates.

W. Han and M. Petyt 1997a *Computers and Structures* **63**, 295–308. Geometrically nonlinear vibration analysis of thin, rectangular plates using the hierarchical finite element method. I: the fundamental mode of isotropic plates.

W. Han and M. Petyt 1997b *Computers and Structures* **63**, 309–318. Geometrically non-linear vibration analysis of thin, rectangular plates using the hierarchical finite element method. II: 1st mode of laminated plates and higher modes of isotropic and laminated plates.

B. Harras, R. Benamar and R. G. White 2002 *Journal of Sound and Vibration* **251**, 579–619. Geometically non-linear free vibration of fully clamped symmetrically laminated rectangular composite plates.

D. Hui 1984 *Journal of Applied Mechanics* **51**, 216–220. Effects of geometric imperfections on large amplitude vibrations of rectangular plates with hysteresis damping.

A. W. Leissa 1969 *Vibration of Plates*. NASA SP-160, U.S. Government Printing Office, Washington, DC, USA (1993 reprinted by the Acoustical Society of America).

A. Y. T. Leung and S. G. Mao 1995 *Journal of Sound and Vibration* **183**, 475–491. A symplectic Galerkin method for non-linear vibration of beams and plates.

A. K. Noor, C. Andersen and J. M. Peters 1993 *Computer Methods in Applied Mechanics and Engineering* **103**, 175–186. Reduced basis technique for nonlinear vibration analysis of composite panels.

S. R. Rao, A. H. Sheikh and M. Mukhopadhyay 1993 *Journal of the Acoustical Society of America* **93**, 3250–3257. Large-amplitude finite element flexural vibration of plates/stiffened plates.

P. Ribeiro 2001 *Proceedings of the 42nd AIAA/ASME/ASCE/AHS/ASC Structures, Structural Dynamics, and Material Conference and Exhibit*, Seattle, WA, USA (paper A01-25098). Periodic vibration of plates with large displacements.

P. Ribeiro and M. Petyt 1999a *Journal of Sound and Vibration* **226**, 955–983. Geometrical non-linear, steady-state, forced, periodic vibration of plate. Part I: model and convergence study.

P. Ribeiro and M. Petyt 1999b *Journal of Sound and Vibration* **226**, 985–1010. Geometrical non-linear, steady-state, forced, periodic vibration of plate. Part II: stability study and analysis of multi-modal response.

P. Ribeiro and M. Petyt 2000 *International Journal of Non-Linear Mechanics* **35**, 263–278. Non-linear free vibration of isotropic plates with internal resonance.

M. Sathyamoorthy 1987 *Applied Mechanics Reviews* **40**, 1553–1561. Nonlinear vibration analysis of plates: a review and survey of current developments.

Y. Shi and C. Mei 1996 *Journal of Sound and Vibration* **193**, 453–464. A finite element time domain modal formulation for large amplitude free vibrations of beams and plates.

J. Yuan and S. M. Dickinson 1992 *Journal of Sound and Vibration* **152**, 203–216. On the use of artificial springs in the study of the free vibrations of systems comprised of straight and curved beams.

5 Vibrations of Empty and Fluid-Filled Circular Cylindrical Shells

5.1 Introduction

This chapter addresses the linear (small amplitude) and nonlinear (large amplitude) vibrations of circular cylindrical shells closed around the circumference (like a tube). Circular cylindrical shells are widely applied in aeronautical and aerospace engineering, in the petrol and chemical industry, in mechanical, civil, nuclear and naval engineering and in biomechanics. In fact, circular cylindrical shells are stiff structural elements that can be easily manufactured, for example, by folding and welding a metal sheet.

The linear vibrations of simply supported shells are studied by using the Donnell and the Flügge-Lur'e-Byrne theories. Fluid-structure interaction with still fluid is investigated. Numerical and experimental results are presented and compared.

The Rayleigh-Ritz method is introduced to study different boundary conditions and complicating effects, such as added masses and elastic bed.

The nonlinear vibrations of simply supported circular cylindrical shells under radial harmonic excitations are studied by using Donnell's nonlinear shallow-shell theory and the Galerkin method. For a periodic excitation applied to the circular cylindrical shell, a standing wave, symmetrical with respect to the point of application, is expected in the case of linear vibrations. This symmetrical standing wave is the *driven mode*. For large-amplitude vibrations, the response of the shell near resonance is given by circumferentially traveling waves, which can be in both directions (plus or minus θ). They can be described as the movement of the nodal lines of the driven mode. The traveling wave appears when a second standing wave (mode), the orientation of which is at $\pi/(2n)$ (where n is the number of nodal diameters) with respect to the previous one, is added to the driven mode. This second mode is called the *companion mode*. It has the same modal shape and frequency of the driven mode. The presence of this second mode arises because of the axial symmetry of the shell and is due to nonlinear coupling. The presence of a traveling-wave response near the resonance, which usually appears for vibration amplitudes smaller than the shell thickness for thin shells, is a fundamental difference with respect to linear vibrations.

The effect of geometric imperfections is investigated; they break the axial symmetry of the shell and change the nonlinear behavior. Numerical and experimental results are presented and satisfactorily compared.

5.1.1 Literature Review

Linear vibrations of circular cylindrical shells are extensively studied in the books by Leissa (1973) and Soedel (1993) for different boundary conditions. Among the possible boundary conditions, simply supported ends represent a particularly important case because this is achieved approximately in practical applications by connecting the shell to thin end plates and rings. Moreover, this condition is simple to treat mathematically.

Circular cylindrical shells containing or immersed in still and flowing fluid are extensively studied by Païdoussis (2003). Sloshing is deeply investigated by Ibrahim (2005).

A full literature review of work on the nonlinear vibrations and stability of shells in vacuo, filled with or surrounded by quiescent and flowing fluids, is given by Amabili and Païdoussis (2003). However, it is necessary to refer to some fundamental contributions.

The first study on nonlinear vibrations of circular cylindrical shells is attributed to Reissner (1955), who isolated a single half-wave (lobe) of the vibration mode and analyzed it for simply supported shells. By using Donnell's nonlinear shallow-shell theory for thin-walled shells, Reissner found that the nonlinearity could be either the hardening or softening type, depending on the geometry of the lobe. Almost at the same time, Grigolyuk (1955) studied large-amplitude free vibrations of circular cylindrical panels simply supported at all four edges. He used the same shell theory as Reissner (1955) and a two-mode expansion for the flexural displacement involving the first and third longitudinal modes. He also developed a single-mode approach. Results show a hardening-type nonlinearity. Chu (1961) continued with Reissner's work, extending the analysis to closed cylindrical shells. He found that nonlinearity in this case always leads to hardening-type characteristics, which, in some cases, can become quite strong. Nowinski (1963) confirmed Chu's results for closed circular shells. Evensen (1963) proved that Nowinski's analysis was not accurate, because it did not maintain a zero transverse deflection at the ends of the shell. Furthermore, Evensen found that Reissner's and Chu's theories did not satisfy the continuity of in-plane circumferential displacement for closed circular shells. Evensen (1963) noted in his experiments that the nonlinearity of closed shells is of the softening type and weak, as also observed by Olson (1965). Indeed, Olson (1965) observed a slight nonlinearity of the softening type in the experimental response of a thin seamless shell made of copper; the measured change in resonance frequency was only about 0.75%, for a vibration amplitude equal to 2.5 times the shell thickness.

In brief, it is possible to attribute to Evensen (1967) and Dowell and Ventres (1968) the original idea of mode expansions of the shell flexural displacement involving the two asymmetric modes with the same shape (sine and cosine functions around the shell circumference. One is directly driven by external excitation, the driven mode; the other is referred as the companion mode) and an axisymmetric term. Their intuition was supported by a few available experimental results.

The studies by Ginsberg (1973) and Chen and Babcock (1975) constitute fundamental contributions to the study of the influence of the companion mode on the nonlinear forced response of circular cylindrical shells.

Gonçalves and Batista (1988) and Amabili et al. (1998) studied the response of fluid-filled shells. In particular, Gonçalves and Batista (1988) neglected companion mode participation, the importance of which in the nonlinear response was investigated by Amabili et al. (1998). Ganapathi and Varadan (1996) developed a finite-element model (FEM) to study free nonlinear vibrations. In a series of papers, Amabili, Pellicano and Païdoussis (1999a,b; 2000a,b) systematically studied the nonlinear stability and large-amplitude forced vibrations of circular cylindrical shells with and without quiescent or flowing fluid by using a base of seven natural modes. The convergence of the solution was studied by Pellicano, Amabili and Païdoussis (2002) by using additional modes; a parametric study was also performed to investigate switches from softening-type to hardening-type nonlinearity.

A problem of one-to-one-to-one-to-two (1:1:1:2) internal resonances among driven, companion and axisymmetric modes for a water-filled circular cylindrical shell was studied by Amabili, Pellicano and Vakakis (2000c) and Pellicano, Amabili and Vakakis (2000). The phenomenon of saturation, with the energy being transferred from the mode directly excited to another mode, was observed. Extremely complex responses were found in the neighborhood of the primary resonance.

Kubenko and Lakiza (2000) studied experimentally the nonlinear vibrations of circular cylindrical shells made of transparent plastic, filled with (i) a liquid and (ii) a liquid containing gas. Both radial and axial excitations were used.

Autoparametric resonances for free flexural vibrations of infinitely long, circular cylindrical shells were investigated by Popov et al. (2001). A simple two-mode expansion, derived from Evensen (1967) but excluding the companion mode, was used. An energy approach was applied to obtain the equations of motion, which were transformed into action-angle coordinates and studied by averaging. Dynamics and stability were extensively investigated by using the methods of Hamiltonian dynamics, and chaotic motion was detected.

Jansen (2002) studied forced vibrations of simply supported, circular cylindrical shells under harmonic modal excitation and static axial load. Donnell's nonlinear shallow-shell theory was used, but the in-plane inertia of axisymmetric modes was taken into account in an approximate way. In-plane boundary conditions were satisfied on the average. A four-degree-of-freedom expansion of the radial displacement was used with the Galerkin method. Numerical results display a softening-type response with a traveling wave in the vicinity of resonance. The quasi-periodic response close to the resonance was deeply investigated. Jansen (2004) developed a further study by using various analytical-numerical models with different levels of complexity. Another theoretical study is by Kubenko et al. (2003).

Geometric imperfections of the shell geometry (out-of-roundness) were considered in buckling problems occurring since the end of the 1950s. However, there is no trace of their inclusion in studies of large-amplitude vibrations of shells until the beginning of the 1970s. Research on this topic was probably introduced by Vol'mir (1972). Kubenko et al. (1982) used a two-mode traveling-wave expansion, taking into account axisymmetric and asymmetric geometric imperfections. Only free vibrations were studied, but the effect of the number of circumferential waves on the nonlinearity was investigated. In particular, the results showed that the nonlinearity is of softening type and that it increases with the number of circumferential waves. The effect of axisymmetric and asymmetric imperfections was to increase the value of the natural frequency; this is in contrast with the results of other studies. Watawala and Nash (1983) studied the free and forced conservative vibrations of closed circular cylindrical shells by using Donnell's nonlinear shallow-shell theory. Empty and liquid-filled shells with a free surface and a rigid bottom were studied. A mode expansion that may be considered as a simple generalization of Evensen's analysis (1967) was introduced; therefore, it is not moment free at the ends of the shell.

A systematic analysis of geometric imperfections was carried out by Amabili (2003a) by using Donnell's nonlinear shallow-shell theory. In this study, the convergence of the solution was carefully verified and models of up to 16 degrees of freedom were built. A successful comparison with available results was performed. In particular, it was found that (i) axisymmetric imperfections do not split the double natural frequencies associated with each couple of asymmetric modes; outward (with respect to the center of curvature) axisymmetric imperfections increase natural frequencies; small inward (with respect to the center of curvature) axisymmetric imperfections decrease natural frequencies. (ii) Ovalization has a small effect on natural frequencies of modes with several circumferential waves and it does not split the double natural frequencies. (iii) Imperfections of the same shape as the resonant mode decrease both frequencies, but they decrease the frequency of the mode with the same angular orientation much more. (iv) Imperfections with twice the number of circumferential waves of the resonant mode decrease the frequency of the mode with the same angular orientation and increase the frequency of the other mode; they have a greater effect on natural frequencies than the imperfections of the same shape as the resonant mode. The split of the double natural frequencies, which is present in almost all real shells due to manufacturing imperfections, changes the traveling-wave response. Good agreement between theoretical and experimental results was obtained. All the modes investigated show a softening-type nonlinearity, which is much more accentuated for the water-filled shell, except in a case that presented a one-to-one internal resonance among modes with a different number of circumferential waves. Traveling-wave and amplitude-modulated responses were observed in the experiments.

Amabili (2003b) computed the large-amplitude response of perfect and imperfect, simply supported circular cylindrical shells subjected to harmonic excitation in the spectral neighborhood of the lowest natural frequency by using five different nonlinear shell theories: (i) Donnell's shallow-shell, (ii) Donnell's with in-plane inertia, (iii) Sanders-Koiter, (iv) Flügge-Lur'e-Byrne and (v) Novozhilov's theories. Except for the first theory, the Lagrange equations of motion were obtained by an energy approach, retaining damping through the Rayleigh dissipation function. Both empty and fluid-filled shells were investigated by using a potential fluid model. The effect of radial pressure and axial load was also studied. Results from the Sanders-Koiter, Flügge-Lur'e-Byrne and Novozhilov theories were practically coincident. A small difference was observed between the previous three theories and Donnell's theory with in-plane inertia. On the other hand, Donnell's nonlinear shallow-shell theory turned out to be the least accurate among the five theories compared. It gives excessive softening-type nonlinearity for empty shells. However, for water-filled shells, it gives sufficiently precise results for a large vibration amplitude. The different accuracies of Donnell's nonlinear shallow-shells theory for empty and water-filled shells can easily be explained by the fact that the in-plane inertia, which is neglected in Donnell's nonlinear shallow-shell theory, is much less important for a water-filled shell, which has a large radial inertia due to the liquid, than for an empty shell. Contained liquid, compressive axial loads and external pressure increase the softening-type nonlinearity of the shell. A minimum mode expansion necessary to capture the nonlinear response of the shell in the neighborhood of a resonance was determined and convergence of the solution was numerically investigated.

Experimental work is much less abundant than theoretical and numerical studies. Chen and Babcock (1975) measured the softening-type response for the fundamental mode of an empty shell with end rings (approximating simply supported boundary conditions) and detected traveling waves around the shell at an excitation frequency close to a resonance. A comparison of theoretical and experimental results is given by Amabili et al. (2002). A full series of comparisons of theoretical and experimental results has been conducted by Amabili (2003a).

5.2 Linear Vibrations of Simply Supported, Circular Cylindrical Shells

A thin circular cylindrical shell made of isotropic, homogeneous and linearly elastic material is considered, so that classic theories of shells are applicable. Many linear theories of shells have been developed. In this section, only the Donnell and Flügge-Lur'e-Byrne theories of shells are considered; these are obtained by neglecting nonlinear terms in the Donnell and Flügge-Lur'e-Byrne nonlinear shell theories. In particular, the Flügge-Lur'e-Byrne theory gives very accurate results for isotropic thin shells.

The shell is constrained at both ends by very thin diaphragms resistant to shear forces, but, at the same time, extremely flexible to loads orthogonal to their plane. This constraint is called simply support or "shear diaphragm." The boundary conditions, referring to Figure 1.12, are given as

$$N_x = M_x = v = w = 0 \quad \text{for}\, x = 0 \quad \text{and} \quad x = L, \tag{5.1}$$

where u, v, w are the displacements of a generic point in the longitudinal, circumferential and radial direction, respectively; N_x is the normal force, M_x is the bending moment per unit length, and L is the shell length.

5.2.1 Donnell's Theory of Shells

The equations of motion for linear vibrations of circular cylindrical shells are obtained from the nonlinear ones canceling nonlinear terms. For Donnell's theory of shells, equations (1.52–1.54) represent the nonlinear equations of motion. The equation of motion in the radial direction is given by substituting equation (1.142a–c) into (1.52) and canceling all the nonlinear terms. The in-plane equations are obtained by substituting $f_x = -\rho h \ddot{u}$ and $f_\theta = -\rho h \ddot{v}$, where ρ is the shell density and h is the shell thickness, deleting nonlinear terms; damping and external forcing are eliminated from the equations, as always in the study of natural vibrations. This gives the following equations of motion:

$$R^2 \frac{\partial^2 u}{\partial x^2} + \frac{1-\nu}{2} \frac{\partial^2 u}{\partial \theta^2} + R \frac{1+\nu}{2} \frac{\partial^2 v}{\partial x\, \partial \theta} - \nu R \frac{\partial w}{\partial x} = \rho \frac{1-\nu^2}{E} R^2 \frac{\partial^2 u}{\partial t^2}, \tag{5.2a}$$

$$\frac{\partial^2 v}{\partial \theta^2} + R^2 \frac{1-\nu}{2} \frac{\partial^2 v}{\partial x^2} + R \frac{1+\nu}{2} \frac{\partial^2 u}{\partial x\, \partial \theta} - \frac{\partial w}{\partial \theta} = \rho \frac{1-\nu^2}{E} R^2 \frac{\partial^2 v}{\partial t^2}, \tag{5.2b}$$

$$\nu R \frac{\partial u}{\partial x} + \frac{\partial v}{\partial \theta} - w - k \left(R^4 \frac{\partial^4 w}{\partial x^4} + 2R^2 \frac{\partial^4 w}{\partial x^2\, \partial \theta^2} + \frac{\partial^4 w}{\partial \theta^4} \right) = \rho \frac{1-\nu^2}{E} R^2 \frac{\partial^2 w}{\partial t^2}, \tag{5.2c}$$

where $k = (1/12)\,(h/R)^2$, x, r, θ are the cylindrical coordinates, R is the shell radius, ν is the Poisson's ratio and E is the Young's modulus.

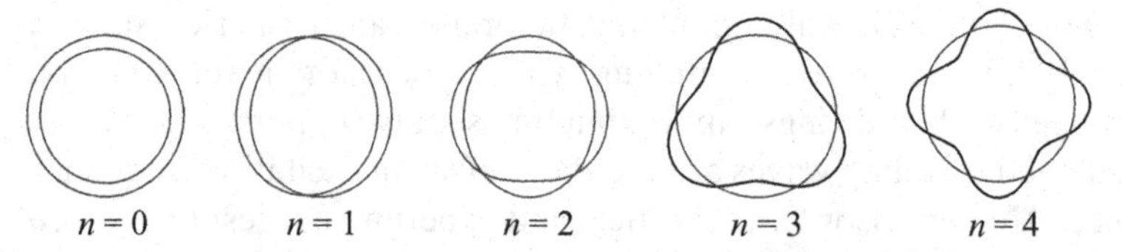

Figure 5.1. Number n of circumferential waves identifying mode shapes.

It is convenient to define the following differential operator:

$$\mathbf{L_D} = \begin{bmatrix} R^2\dfrac{\partial^2}{\partial x^2} + \dfrac{1-\nu}{2}\dfrac{\partial^2}{\partial \theta^2} & R\dfrac{1+\nu}{2}\dfrac{\partial^2}{\partial x\,\partial\theta} & -\nu R\dfrac{\partial}{\partial x} \\ R\dfrac{1+\nu}{2}\dfrac{\partial^2}{\partial x\,\partial\theta} & \dfrac{\partial^2}{\partial \theta^2} + R^2\dfrac{1-\nu}{2}\dfrac{\partial^2}{\partial x^2} & -\dfrac{\partial}{\partial\theta} \\ \nu R\dfrac{\partial}{\partial x} & \dfrac{\partial}{\partial\theta} & -1 - k\left(R^4\dfrac{\partial^4}{\partial x^4} + 2R^2\dfrac{\partial^4}{\partial x^2\,\partial\theta^2} + \dfrac{\partial^4}{\partial\theta^4}\right) \end{bmatrix}. \tag{5.3}$$

The differential operator $\mathbf{L_D}$ given in equation (5.3) becomes symmetric by changing the sign to the third column, which is equivalent to taking w positive outward. By using (5.3), equations (5.2) can be rewritten as

$$\mathbf{L_D}\begin{Bmatrix} u \\ v \\ w \end{Bmatrix} = \rho\frac{1-\nu^2}{E}R^2\frac{\partial^2}{\partial t^2}\begin{Bmatrix} u \\ v \\ w \end{Bmatrix}. \tag{5.4}$$

The solution of equations (5.2a–c), exactly satisfying all the boundary conditions (5.1), is given in the form:

$$u = C_1\cos(\lambda_m x)\cos(n\theta)\cos(\omega t), \tag{5.5a}$$

$$v = C_2\sin(\lambda_m x)\sin(n\theta)\cos(\omega t), \tag{5.5b}$$

$$w = C_3\sin(\lambda_m x)\cos(n\theta)\cos(\omega t), \tag{5.5c}$$

where $\lambda_m = m\pi/L$, $m = 1, 2, \ldots$ and $n = 0, 1, 2, \ldots$ are the number of axial half-waves and circumferential waves, respectively, as shown in Figures 5.1 and 5.2, ω is the circular frequency of natural vibration and t is time.

Equations (5.5a–c) describe the mode shape and C_i, $i = 1, 2, 3$ are the amplitude constants. However, for any asymmetric mode shape (i.e. for $n > 0$), there is a second mode, similar to the previous one, but orthogonal to it, with an angular shift of $\pi/2n$ (i.e. exchanging sine and cosine functions in the angular coordinate θ), as shown in Figure 5.3. This second mode presents the maximum vibration amplitude where the first mode has a node.

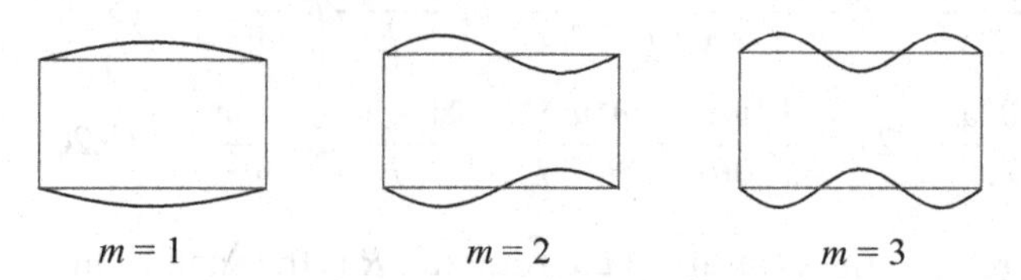

Figure 5.2. Number m of axial half-waves identifying mode shapes.

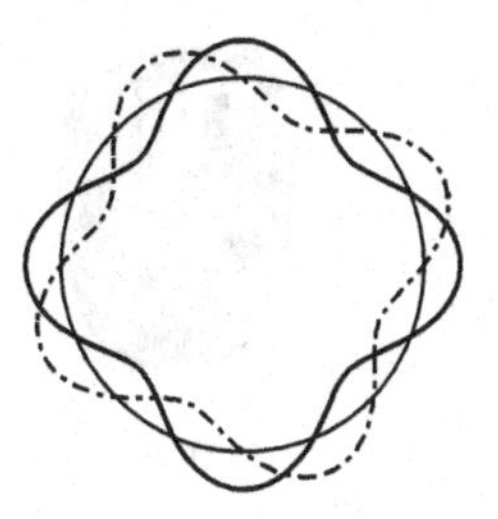

Figure 5.3. Couple of identical asymmetric modes; —, cos($n\theta$) mode; - - -, sin($n\theta$) mode; —, undeformed shell.

A frequency parameter Ω is defined as

$$\Omega^2 = (\rho/E)\,(1-\nu^2)\omega^2 R^2. \tag{5.6}$$

Substituting equations (5.5) into (5.2) yields

$$\Delta_{\mathbf{D}} \begin{Bmatrix} C_1 \\ C_2 \\ C_3 \end{Bmatrix} = 0, \tag{5.7}$$

where

$$\Delta_{\mathbf{D}} = \begin{bmatrix} \Omega^2 - H_1 & \dfrac{1+\nu}{2} n\tilde{\lambda} & -\nu\tilde{\lambda} \\ \dfrac{1+\nu}{2} n\tilde{\lambda} & \Omega^2 - H_2 & n \\ -\nu\tilde{\lambda} & n & \Omega^2 - H_3 \end{bmatrix}, \tag{5.8}$$

$\tilde{\lambda} = \lambda_m R$, $H_1 = \tilde{\lambda}^2 + \frac{1-\nu}{2} n^2$, $H_2 = \frac{1-\nu}{2}\tilde{\lambda}^2 + n^2$ and $H_3 = 1 + k(\tilde{\lambda}^2 + n^2)^2$. Equation (5.7) gives an eigenvalue problem, for which eigenvalues are given by the following bicubic characteristic equation:

$$\Omega^6 - K_2\,\Omega^4 + K_1\,\Omega^2 - K_0 = 0, \tag{5.9}$$

where the coefficients K_i are given by

$$K_2 = 1 + \frac{3-\nu}{2}(n^2 + \tilde{\lambda}^2) + k(n^2 + \tilde{\lambda}^2)^2, \tag{5.10a}$$

$$K_1 = \frac{1-\nu}{2}\left[(3+2\nu)\tilde{\lambda}^2 + n^2 + (n^2+\tilde{\lambda}^2)^2 + \frac{3-\nu}{1-\nu} k(n^2+\tilde{\lambda}^2)^3\right], \tag{5.10b}$$

$$K_0 = \frac{1-\nu}{2}\left[(1-\nu^2)\tilde{\lambda}^4 + k(n^2+\tilde{\lambda}^2)^4\right]. \tag{5.10c}$$

Equation (5.9) gives three real roots for any couple of m and n values. The corresponding eigenvectors give three different natural modes, one with a radial, one with a longitudinal and one with a circumferential prevalent displacement, each one having its own distinct natural frequency. The lowest natural frequency is usually associated with a mode with predominant radial displacement. The radial displacements of mode shapes with prevalent radial displacement are shown in Figure 5.4. In

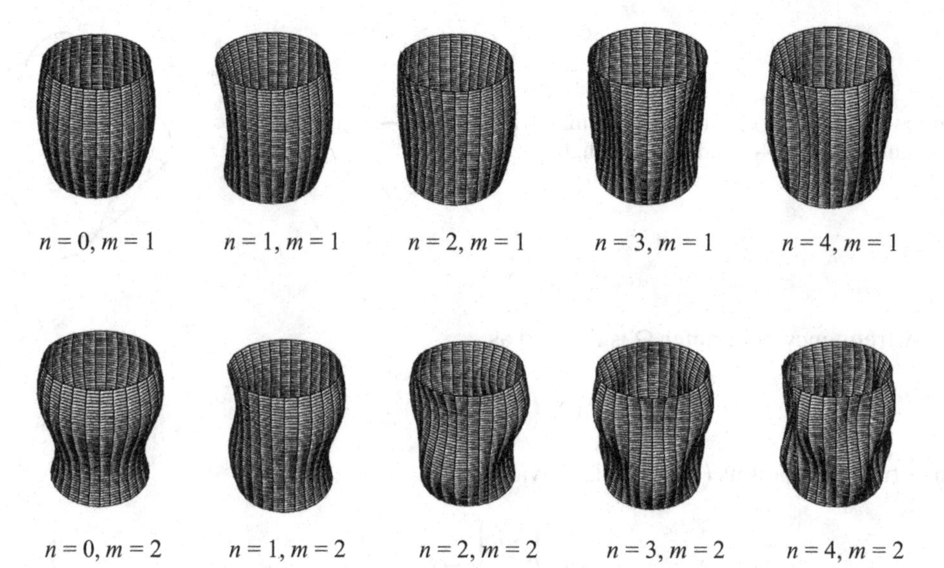

Figure 5.4. Mode shapes with prevalent radial displacement identified by the wavenumbers (n, m).

particular, mode shapes are obtained by the system of linear-dependent equations (5.7). It gives

$$\begin{bmatrix} \Omega^2 - H_1 & \dfrac{1+\nu}{2} n\tilde{\lambda} \\ \dfrac{1+\nu}{2} n\tilde{\lambda} & \Omega^2 - H_2 \end{bmatrix} \begin{Bmatrix} C_1/C_3 \\ C_2/C_3 \end{Bmatrix} = \begin{Bmatrix} \nu\tilde{\lambda} \\ -n \end{Bmatrix}. \tag{5.11}$$

In equation (5.11), the ratios C_1/C_3 and C_2/C_3 describe the mode shape; the remaining third coefficient, which cannot be determined by the linear-dependent equations (5.7), is a scale factor giving the vibration amplitude; this is determined by the initial conditions or by a normalization criterion.

5.2.2 Flügge-Lur'e-Byrne Theory of Shells

The equations of motion, according to the Flügge-Lur'e-Byrne theory of shells, can be obtained similarly to the Donnell equations of motion by using the stress-strain relationship (1.118–1.123), canceling nonlinear terms. Also in this case w is taken positive inward (see Figure 1.12). According to the Flügge-Lur'e-Byrne theory of shells (Leissa 1973),

$$(\mathbf{L_D} + k\,\mathbf{L_{MOD}}) \begin{Bmatrix} u \\ v \\ w \end{Bmatrix} = \rho \frac{1-\nu^2}{E} R^2 \frac{\partial^2}{\partial t^2} \begin{Bmatrix} u \\ v \\ w \end{Bmatrix}, \tag{5.12}$$

where

$$\mathbf{L_{MOD}} = \begin{bmatrix} \dfrac{1-\nu}{2}\dfrac{\partial^2}{\partial \theta^2} & 0 & R^3\dfrac{\partial^3}{\partial x^3} - \dfrac{1-\nu}{2} R \dfrac{\partial^3}{\partial x\,\partial \theta^2} \\ 0 & \dfrac{3(1-\nu)}{2} R^2 \dfrac{\partial^2}{\partial x^2} & \dfrac{3-\nu}{2} R^2 \dfrac{\partial^3}{\partial x^2\,\partial \theta} \\ -R^3\dfrac{\partial^3}{\partial x^3} + \dfrac{1-\nu}{2} R \dfrac{\partial^3}{\partial x\,\partial \theta^2} & -\dfrac{3-\nu}{2} R^2 \dfrac{\partial^3}{\partial x^2\,\partial \theta} & -1 - 2\dfrac{\partial^2}{\partial \theta^2} \end{bmatrix}. \tag{5.13}$$

Substituting equations (5.5a–c) into (5.12), the eigenvalue problem can be written as

$$\left(\Delta_{\mathbf{D}}+k\begin{bmatrix} -\frac{1-\nu}{2}n^2 & 0 & \tilde{\lambda}\left(-\tilde{\lambda}^2+\frac{1-\nu}{2}n^2\right) \\ 0 & -3\frac{1-\nu}{2}\tilde{\lambda}^2 & \frac{3-\nu}{2}n\tilde{\lambda}^2 \\ \tilde{\lambda}\left(-\tilde{\lambda}^2+\frac{1-\nu}{2}n^2\right) & \frac{3-\nu}{2}n\tilde{\lambda}^2 & -(1-2n^2) \end{bmatrix}\right)\begin{Bmatrix} C_1 \\ C_2 \\ C_3 \end{Bmatrix}=0. \tag{5.14}$$

By neglecting second-order and third-order terms in k, the eigenvalues are given by

$$\Omega^6-(K_2+k\Delta K_2)\,\Omega^4+(K_1+k\Delta K_1)\;\Omega^2-(K_0+k\Delta K_0)\;=0, \tag{5.15}$$

where the coefficients K_i are given by equations (5.10a–c) and

$$\Delta K_2=1-\frac{3+\nu}{2}n^2+\frac{3(1-\nu)}{2}\tilde{\lambda}^2, \tag{5.16a}$$

$$\Delta K_1=(3-2\nu)\tilde{\lambda}^2+(2-\nu)\,n^2+\frac{3-7\nu}{2}\tilde{\lambda}^4-(5-\nu)\tilde{\lambda}^2n^2-\frac{5-\nu}{2}n^4, \tag{5.16b}$$

$$\Delta K_0=\frac{1-\nu}{2}[(4-3\nu^2)\tilde{\lambda}^4+2\,(2-\nu)\,\tilde{\lambda}^2n^2+n^4-2\nu\tilde{\lambda}^6 \\ -6\tilde{\lambda}^4n^2-2\,(4-\nu)\,\tilde{\lambda}^2n^4-2n^6]. \tag{5.16c}$$

By neglecting k with respect to unity in the coefficients $K_2+k\Delta K_2$, $K_1+k\Delta K_1$ and $K_0+k\Delta K_0$, Flügge obtained the following simplified coefficients (Leissa 1973):

$$\Delta K_2=\Delta K_1=0, \tag{5.17a,b}$$

$$\Delta K_0=\frac{1-\nu}{2}[2\,(2-\nu)\,\tilde{\lambda}^2n^2+n^4-2\nu\tilde{\lambda}^6-6\tilde{\lambda}^4n^2-2\,(4-\nu)\,\tilde{\lambda}^2n^4-2n^6]. \tag{5.17c}$$

Mode shapes are given by

$$\begin{bmatrix} \Omega^2-H_1-k\frac{1-\nu}{2}n^2 & \frac{1+\nu}{2}n\tilde{\lambda} \\ \frac{1+\nu}{2}n\tilde{\lambda} & \Omega^2-H_2-3\frac{1-\nu}{2}k\tilde{\lambda}^2 \end{bmatrix}\begin{Bmatrix} C_1/C_3 \\ C_2/C_3 \end{Bmatrix} \\ =\begin{Bmatrix} \nu\tilde{\lambda}-k\tilde{\lambda}\left(-\tilde{\lambda}^2+\frac{1-\nu}{2}n^2\right) \\ -n-\frac{3-\nu}{2}kn\tilde{\lambda}^2 \end{Bmatrix}. \tag{5.18}$$

Figure 5.5 gives the natural frequencies, computed by using the Flügge theory with the coefficients given by equations (5.17a–c), of a simply supported circular cylindrical shell having the following dimensions and material properties: $L=520$ mm, $R=149.4$ mm, $h=0.519$ mm, $E=1.95\times10^{11}$ Pa, $\rho=7800$ kg/m^3 and $\nu=0.3$. Except for axisymmetric modes ($n=0$), the first natural frequency for any n is obtained for modes with predominant radial motion. For $n=0$ and $m=1$, the first mode is torsional, the second is axial; the radial mode has the largest frequency in this case. More specifically for ($n=0$, $m=1$), whereas the torsional mode is completely uncoupled presenting zero radial and axial displacement, the axial and radial

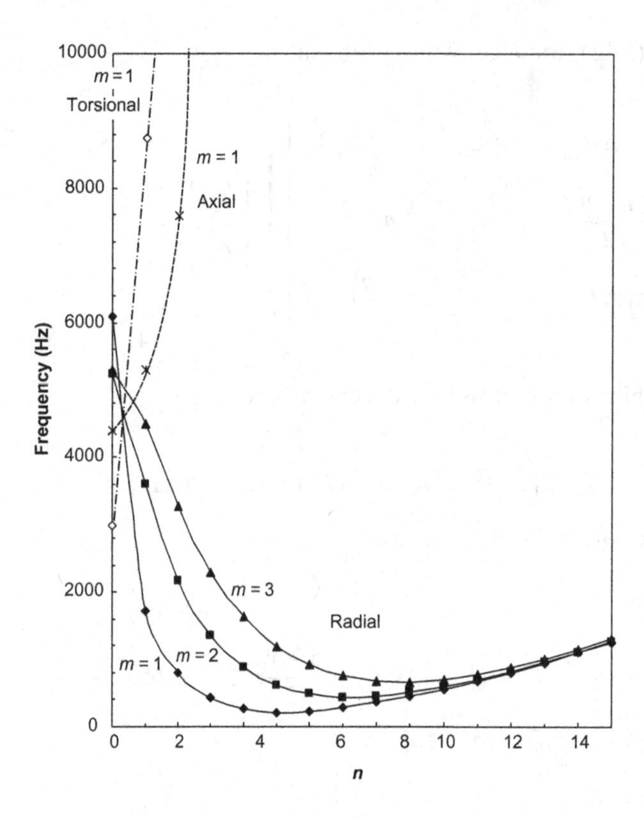

Figure 5.5. Natural frequencies of a simply supported circular cylindrical shell versus the number n of circumferential waves, for different number m of axial half-waves. ——, modes with prevalent radial motion; – –, modes with prevalent axial motion; –·–, modes with prevalent torsional motion.

modes are largely coupled presenting large contribution of both displacements. The contribution of axial and torsional motion to radial modes with $m = 1$ is shown in Figure 5.6.

Experiments have been conducted on a commercial circular cylindrical shell made of stainless steel with the same dimensions and material properties used for calculation in Figures 5.5 and 5.6. Two stainless steel annular plates with external and internal radii of 149.4 and 60 mm, respectively, and a thickness of 0.25 mm have been welded to the shell ends to approximate the simply supported boundary condition of the shell. A comparison of theoretical and experimental natural frequencies is presented in Figure 5.7, showing a good agreement. For some modes, two distinct experimental natural frequencies are reported for given n and m values. This is due to small imperfections that separate the natural frequencies of the couple of identical sine and cosine modes.

5.3 Circular Cylindrical Shells Containing or Immersed in Still Fluid

An inviscid, irrotational and compressible fluid coupled to a vibrating shell is considered (Morand and Ohayon 1992). The fluid is driven into oscillation by the shell and has no mean flow. The fluid can be internal or external to the shell and, in the longitudinal direction, is confined in the shell length. In the case of external fluid, this condition is equivalent to that of a shell of infinite length surrounded by fluid and infinite supports, each one at distance L from the next one; therefore, for external fluid, L is the shell length between two consecutive supports. The shell is completely filled in the case of internal fluid; in the case of external fluid, the fluid domain is assumed to be infinite in the radial direction. Therefore, the fluid does not present a

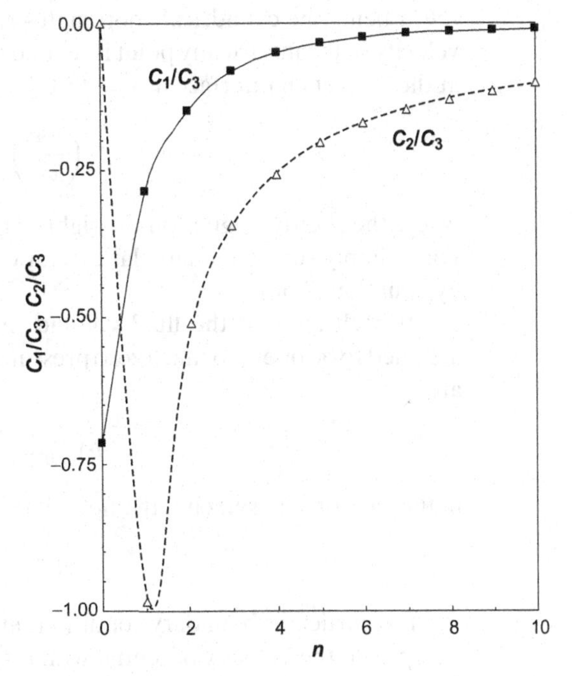

Figure 5.6. Contribution of axial (C_1/C_3) and torsional (C_2/C_3) vibration to radial modes versus n for $m = 1$.

free surface and sloshing is not present. The velocity potential Φ of the fluid satisfies the Helmholtz equation

$$\nabla^2 \Phi = \frac{1}{c^2} \frac{\partial^2 \Phi}{\partial t^2}, \tag{5.19}$$

where the Laplacian operator ∇^2 in cylindrical coordinates is given by

$$\nabla^2 = \frac{\partial^2}{\partial r^2} + \frac{1}{r} \frac{\partial}{\partial r} + \frac{1}{r^2} \frac{\partial^2}{\partial \theta^2} + \frac{\partial^2}{\partial x^2}, \tag{5.20}$$

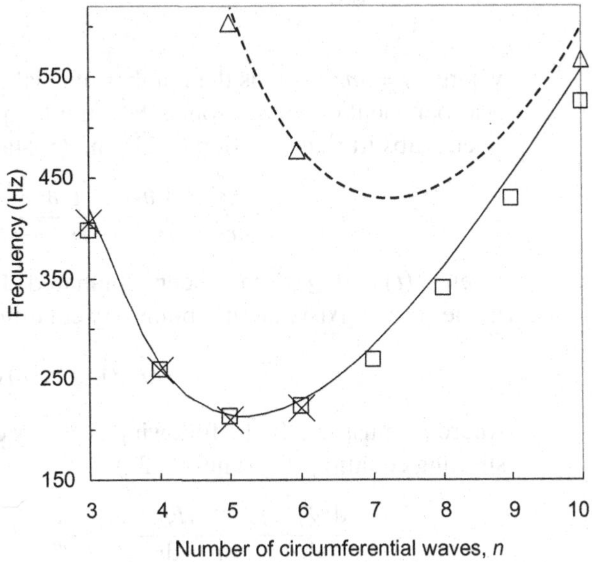

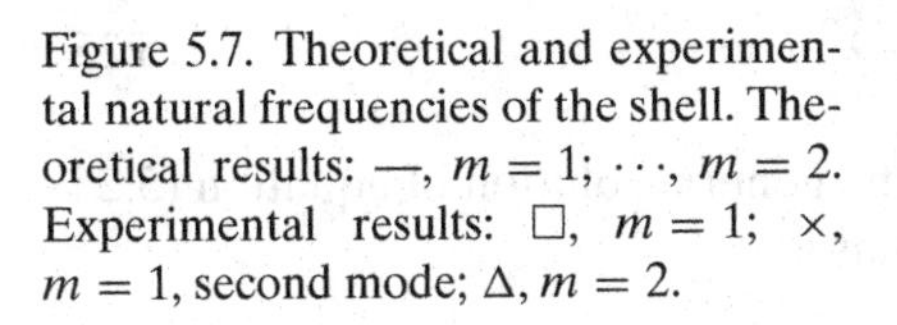
Figure 5.7. Theoretical and experimental natural frequencies of the shell. Theoretical results: —, $m = 1$; $\cdots$, $m = 2$. Experimental results: □, $m = 1$; ×, $m = 1$, second mode; Δ, $m = 2$.

r, θ, x being the cylindrical coordinates and c the velocity of sound in the fluid. The velocity of the fluid at any point is related to Φ by $\mathbf{v} = -\nabla\Phi$. No cavitation is assumed at the fluid-shell interface at $r = R$:

$$\left(\frac{\partial \Phi}{\partial r}\right)_{r=R} = \dot{w}, \tag{5.21}$$

where the positive sign is on the right-hand side if the w is assumed positive inward. For a simply supported circular cylindrical shell, the radial displacement w is given by equation (5.5c).

If both ends of the fluid volume (in correspondence of the shell edges) are assumed to be open, so that a zero pressure is assumed there, the boundary conditions are

$$(\Phi)_{x=0} = (\Phi)_{x=L} = 0; \tag{5.22a}$$

in the case of a closed end, the boundary condition becomes

$$\partial\Phi/\partial x = 0. \tag{5.22b}$$

For particular boundary conditions at the ends of the fluid domain, it is possible to separate the velocity potential with respect to x

$$\Phi(r, \theta, x, t) = \phi(r, \theta)\, Q(x)\, \dot{f}(t), \tag{5.23}$$

where $f(t) = \cos(\omega t)$ is the harmonic function describing the shell vibrations and ω is the circular frequency of free vibration.

The longitudinal mode shape $Q(x)$, in the following special cases, is given by

$$\begin{aligned} Q(x) &= \sin(\lambda_m x) && \text{if both ends are open,} && (5.24a)\\ &= \sum_{m=1}^{\infty} \sigma_m \cos(\lambda_m x) && \text{if both ends are closed,} && (5.24b)\\ &= \sum_{m=1}^{\infty} \sigma_m \cos\left[\frac{(2m-1)\,\pi x}{2L}\right] && \text{if open at } x = L \text{ and closed at } x = 0, && (5.24c) \end{aligned}$$

where $\lambda_m = m\pi/L$, m is the number of axial half-waves, L is the shell length and σ_m is a coefficient depending on m. For simplicity, both the fluid ends are assumed to be open. Substituting equation (5.23) and (5.24a) into (5.19), yields

$$\frac{\partial^2\phi}{\partial r^2} + \frac{1}{r}\frac{\partial\phi}{\partial r} + \frac{1}{r^2}\frac{\partial^2\phi}{\partial\theta^2} - \lambda_m^2\phi = -\frac{\omega^2}{c^2}\phi, \tag{5.25}$$

where $\dot{f}(t)$ and $Q(x)$ have been eliminated. The general solution of equation (5.25) in the case of axisymmetric boundary conditions is

$$\phi(r, \theta) = F_n(r)\cos(n\theta), \tag{5.26}$$

where F_n must satisfy the following ordinary differential equation, obtained by substituting equation (5.26) into (5.25)

$$\frac{\mathrm{d}^2 F_n(r)}{\mathrm{d}r^2} + \frac{1}{r}\frac{\mathrm{d}F_n(r)}{\mathrm{d}r} - \left[\lambda_m^2 - \left(\frac{\omega}{c}\right)^2 + \left(\frac{n}{r}\right)^2\right] F_n(r) = 0. \tag{5.27}$$

The function F_n takes one of the following expressions:

$$F_n(r) = A_{mn}\mathrm{J}_n(kr) + B_{mn}\mathrm{Y}_n(kr) \qquad \text{if } \omega/c > \lambda_m, \tag{5.28}$$

where $k^2 = (\omega^2/c^2) - \lambda_m^2$, or

$$F_n(r) = C_{mn}\mathrm{I}_n(k_1 r) + D_{mn}\mathrm{K}_n(k_1 r) \qquad \text{if } \omega/c < \lambda_m, \tag{5.29}$$

where $k_1^2 = \lambda_m^2 - (\omega^2/c^2)$, J_n and Y_n are the Bessel functions of the first and second kind and I_n and K_n are the modified Bessel function of the first and second kind, respectively.

For internal fluid, the function F_n must be regular at $r = 0$. For external fluid, the spatial velocity potential Φ must satisfy the radiation condition

$$\lim_{r\to\infty} \Phi = \lim_{r\to\infty} \frac{\partial \Phi}{\partial r} = \lim_{r\to\infty} \frac{\partial \Phi}{\partial x} = \lim_{r\to\infty} \frac{\partial \Phi}{\partial \theta} = 0.$$

Therefore, four different expressions of $F_n(r)$ are possible; if $\omega/c > \lambda_m$, they are

$$F_n(r) = A_{mn}\mathrm{J}_n(kr) \qquad \text{for internal fluid,} \tag{5.30a}$$
$$F_n(r) = B_{mn}\mathrm{Y}_n(kr) \qquad \text{for external fluid.} \tag{5.30b}$$

If $\omega/c < \lambda_m$, they are

$$F_n(r) = C_{mn}\mathrm{I}_n(k_1 r) \qquad \text{for internal fluid,} \tag{5.31a}$$
$$F_n(r) = D_{mn}\mathrm{K}_n(k_1 r) \qquad \text{for external fluid.} \tag{5.31b}$$

In the following discussion, it is assumed that $\omega/c < \lambda_m$, which is the condition of low compressibility of the fluid. The case $\omega/c > \lambda_m$ can be easily obtained substituting k_1 with k and I and K with J and Y, respectively.

Substituting (5.5a), (5.31a), (5.26), (5.24a) and (5.23) into the boundary condition (5.21) yields for a simply supported shell with internal fluid

$$C_{mn} = \frac{C_3}{k_1 \mathrm{I}_n'(k_1 R)}, \tag{5.32}$$

where the apex indicates the derivative of the Bessel function with respect to its argument; in particular, $\mathrm{I}_n'(k_1 R) = [\mathrm{I}_{n-1}(k_1 R) + \mathrm{I}_{n+1}(k_1 R)]/2$. Equations (5.32), (5.31a), (5.26), (5.24a) and (5.23) finally give

$$\Phi(r, \theta, x, t) = C_3 \frac{\mathrm{I}_n(k_1 r)}{k_1 \mathrm{I}_n'(k_1 R)} \sin(\lambda_m x)\cos(n\theta)\dot{f}(t). \tag{5.33a}$$

For external fluid, equation (5.33a) is transformed into

$$\Phi(r, \theta, x, t) = C_3 \frac{\mathrm{K}_n(k_1 r)}{k_1 \mathrm{K}_n'(k_1 R)} \sin(\lambda_m x)\cos(n\theta)\dot{f}(t), \tag{5.33b}$$

where $\mathrm{K}_n'(k_1 R) = -[\mathrm{K}_{n-1}(k_1 R) + \mathrm{K}_{n+1}(k_1 R)]/2$. In the case of incompressible fluid, k_1 is replaced by λ_m.

The unsteady pressure p, positive outward on the shell, exerted by the fluid on the shell wall, is obtained by using Bernoulli's equation and simply yields

$$p = \rho_F \frac{\partial \Phi_{r=R}}{\partial t}, \tag{5.34}$$

where ρ_F is the mass density of the fluid.

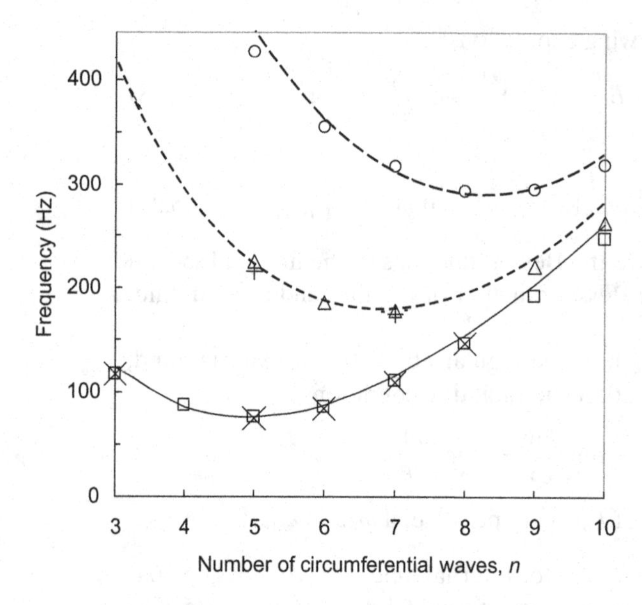

Figure 5.8. Theoretical and experimental natural frequencies of the water-filled shell. Theoretical results: —, $m = 1$; $\cdots$, $m = 2$; – – –, $m = 3$. Experimental results: □, $m = 1$; ×, $m = 1$, second mode; Δ, $m = 2$; +, $m = 2$, second mode; ○, $m = 3$.

In the presence of an internal pressure p (positive in direction $-w$) acting on the shell surface, equation (5.12) is modified adding the term $[(1-\nu^2)/(Eh)]R^2 p$ to the third row of the equation on the right-hand side. Therefore, by adding the pressure p, the following equation of motion is obtained:

$$(\mathbf{L_D} + k\mathbf{L_{MOD}})\begin{Bmatrix} u \\ v \\ w \end{Bmatrix} = \rho\frac{1-\nu^2}{E}R^2\frac{\partial^2}{\partial t^2}\begin{Bmatrix} u \\ v \\ w \end{Bmatrix} + \rho_V\frac{1-\nu^2}{E}R^2\frac{\partial^2}{\partial t^2}\begin{Bmatrix} 0 \\ 0 \\ w \end{Bmatrix}, \quad (5.35)$$

where ρ_v is the added virtual density factor, given by

$$\rho_v = \frac{\rho_F}{k_1 h}\frac{\mathrm{I}_n(k_1 R)}{\mathrm{I}'_n(k_1 R)}, \quad \text{for internal fluid} \quad (5.36a)$$

$$\rho_v = \frac{\rho_F}{k_1 h}\frac{\mathrm{K}_n(k_1 R)}{\mathrm{K}'_n(k_1 R)}, \quad \text{for external fluid.} \quad (5.36b)$$

Figure 5.8 shows the theoretical and experimentally measured natural frequencies of the simply supported, water-filled shell, previously studied in Figures 5.5–5.7; the agreement is very good. In the calculations of natural frequencies of the water-filled shell, it has been assumed that water is incompressible; therefore, $k_1 = \lambda_m$ and $\rho_F = 1000$ kg/m^3. The added mass effect due to the contained water is evident comparing Figures 5.7 (for empty shell) and 5.8 (water-filled shell); a very large decrement of natural frequencies is obtained. In the shell used in the experiments, the ends were closed by flexible rubber circular plates in order to simulate open ends. A smaller decrement of frequencies is obtained for thicker shells.

5.4 Rayleigh-Ritz Method for Linear Vibrations

The case of simply supported circular cylindrical shells is very special, because the closed-form analytical solution of mode shapes (5.5a–c) is obtained. For different boundary conditions, or by complicating the system by attaching lumped masses or springs, a closed-form analytical solution cannot be obtained. In these cases, it is necessary to use an approximate method to discretize the system. For this purpose, a powerful tool is the Rayleigh-Ritz method.

The equation of motion for free harmonic vibrations of an undamped structure can be written in the following form:

$$\mathbf{L}(\mathbf{u}) = \omega^2 \mu \mathbf{u}, \tag{5.37}$$

where $\mathbf{L}$ is a self-adjoint differential operator, $\mathbf{u}$ is the displacement vector of the middle surface (for plates and shells) of the structure, ω is the corresponding radian frequency and μ is the mass per unit length or area ($\mu = \rho h$ for plates and shells). For the system considered, it is possible to write the Rayleigh quotient

$$\omega^2 = \frac{\int_\Gamma \mathbf{u} \cdot \mathbf{L}(\mathbf{u})\, \mathrm{d}S}{\int_\Gamma \mu\, \mathbf{u} \cdot \mathbf{u}\, \mathrm{d}S}, \tag{5.38}$$

Γ being the middle surface of the structure. The numerator on the right-hand side of equation (5.38) equals twice the maximum potential energy of the system

$$U = (1/2) \int_\Gamma \mathbf{u} \cdot \mathbf{L}(\mathbf{u})\, \mathrm{d}S, \tag{5.39}$$

and the denominator is twice the reference kinetic energy, that is, the maximum kinetic energy divided by ω^2 of the system

$$\overline{T} = (1/2) \int_\Gamma \mu\, \mathbf{u} \cdot \mathbf{u}\, \mathrm{d}S. \tag{5.40}$$

The Rayleigh-Ritz method is used to find natural frequencies and mode shapes. In particular, $\mathbf{u}$ is expanded by using a sum of admissible vectorial functions $\mathbf{x}_i$ and appropriate unknown coefficients a_i

$$\mathbf{u} = \sum_{i=1}^{\infty} a_i\, \mathbf{x}_i. \tag{5.41}$$

The infinite sum in equation (5.41) is truncated to N terms in the applications. The choice of the admissible functions is very important to simplify the calculations and to guarantee convergence to the actual solution. The choice of admissible functions, which are all the eigenfunctions, eventually including rigid body modes, of the closest, "less-constrained" problem extracted from the one considered is useful in many cases. In particular, it is applicable in all the cases where it is possible to extract a less-constrained problem having mode shapes expressed by analytical expression in closed form.

In fact, the expansion theorem states that any function (or vectorial function) $\mathbf{u}$, defined in the structure, satisfying the homogeneous boundary conditions of the system, and for which $\mathbf{L}(\mathbf{u})$ is continuous, can be represented by an absolutely and convergent series of the eigenfunctions of the system (Meirovitch 1967, see pages 211–233). This theorem can be applied to the present case where $\mathbf{u}$ is a mode of the considered system and the eigenfunctions $\mathbf{x}_i$ are those of the less-constrained system. In fact, $\mathbf{u}$ satisfies the same boundary conditions as the less-constrained system, if all the additional constraints are replaced by translational and rotational springs. The nature of the convergence of the method and its rate in the case of a beam are discussed by Amabili and Garziera (1999).

In the Rayleigh-Ritz method, the unknown coefficients a_i are determined by minimizing the quantity $(U - \omega^2\overline{T})$ with respect to each of the coefficients a_i, for $i = 1, \ldots, N$. This operation leads to the following eigenvalue problem:

$$\frac{\partial}{\partial a_1}(U - \omega^2\overline{T}) = 0, \quad \ldots, \quad \frac{\partial}{\partial a_N}(U - \omega^2\overline{T}) = 0, \tag{5.42}$$

which determines a set of N linear algebraic homogeneous equations in the N unknowns a_i. The nontrivial solution is obtained by solving the associated eigenvalue problem, with eigenvalues ω^2 and eigenvectors that give the coefficients a_i of mode shapes. It can be proved that by increasing the number N of terms in the expansion (5.41), the computed natural frequencies converge to the actual values from above. In practical applications, not more than the lowest $N/2$ natural frequencies can be estimated with sufficient accuracy by using an expansion with N, accurately chosen, terms in equation (5.41). The most accurate estimation is for the lowest natural frequency, usually referred to as the fundamental frequency.

The use of admissible functions $\mathbf{x}_i$, which are the natural modes of a less-constrained problem, allows an interesting simplification. In fact, the maximum potential energy of the system can be obtained as the multiplication of the reference kinetic energy of natural mode in the less-constrained problem by the corresponding eigenvalue ω_i^2 (the squared radian frequency) of the same problem and by the coefficient a_i, and then adding all the products. For each term of the expansion it is possible to write

$$\int_\Omega \mathbf{x}_i \cdot \mathbf{L}(\mathbf{x}_i)\, \mathrm{d}S = \omega_i^2 \int_\Omega \mu\, \mathbf{x}_i \cdot \mathbf{x}_i\, \mathrm{d}S. \tag{5.43}$$

Equation (5.43) is obtained by equalizing the maximum potential and the maximum kinetic energies of each natural mode of the less-constrained problem. Therefore, twice the maximum potential energy of the system can be written as

$$\int_\Omega \mathbf{u} \cdot \mathbf{L}(\mathbf{u})\, \mathrm{d}S = \sum_{i=1}^{N} a_i^2 \omega_i^2 \int_\Omega \mu\, \mathbf{x}_i \cdot \mathbf{x}_i\, \mathrm{d}S. \tag{5.44}$$

In equation (5.44), the orthogonality of the eigenfunctions of the less-constrained problem has been used.

5.5 Nonlinear Vibrations of Empty and Fluid-Filled, Simply Supported, Circular Cylindrical Shells with Donnell's Nonlinear Shallow-Shell Theory

The problem is approached by using Donnell's nonlinear shallow-shell theory, including geometric imperfections, given by equations (1.162) and (1.163). The Galerkin method is used here, differently from the energy approach used in Chapter 4 for nonlinear vibrations of rectangular plates, and is applied directly to the partial differential equations describing the dynamics in order to discretize them, that is, to transform them into a finite set of ordinary differential equations. The Galerkin method is less flexible than the Lagrange equations of motion (energy approach) used in Chapter 4. However, as a consequence of Donnell's nonlinear shallow-shell theory being used, it allows us to obtain models of smaller size (i.e. with less degrees of freedom) with a reduced effort to find the appropriate expansions of the shell displacements. In fact, only w is explicitly used. The energy approach will be used in Chapter 7 to solve the same problem by using different nonlinear shell theories.

An external modal excitation f of unspecified physical origin is considered, which has the form:

$$f = f_{1,n} \cos(n\theta) \sin(\pi x/L) \cos(\omega t), \tag{5.45}$$

where $f_{1,n}$ is a coefficient having the dimension of pressure. Excitations with frequency close to the natural frequency of the lowest modes are considered; low-frequency modes have a predominant radial motion and are identified by the pair (n, m), where n is the number of circumferential waves and m is the number of axial half-waves.

This kind of external excitation is unrealistic. In practice, one or more forces are usually applied to the system. More realistic is the case of a harmonic point excitation, modeling, for instance, the excitation by an electrodynamical exciter (shaker); the considerations above are important in the analysis of the experimental results presented in the following sections. The point force excitation $\tilde{f}$ is the resultant of the following pressure distribution:

$$f = \tilde{f}\,\delta(R\theta - R\tilde{\theta})\,\delta(x - \tilde{x}) \cos(\omega t), \tag{5.46}$$

where δ is the Dirac delta function, $\tilde{f}$ is the magnitude of the localized force, $\tilde{\theta}$ and $\tilde{x}$ give the angular and axial coordinates of the point of application of the force, respectively.

The viscous damping model introduced in equation (1.162) is unrealistic and will be substituted by modal damping coefficients experimentally identified in the equations of motion used to perform numerical calculations.

The forces per unit length in the axial and circumferential directions, as well as the shear force, are given by equations (1.155a–c), where $f_x = f_\theta = 0$. The force-displacement relations, inserting geometric imperfection w_0 into equations (1.142a–c), are

$$(1-\nu^2)\frac{N_x}{Eh} = -\frac{\nu w}{R} + \frac{1}{2}\left(\frac{\partial w}{\partial x}\right)^2 + \frac{\partial w}{\partial x}\frac{\partial w_0}{\partial x} + \frac{\nu}{2}\left(\frac{\partial w}{R\partial\theta}\right)^2 + \nu\frac{\partial w}{R\partial\theta}\frac{\partial w_0}{R\partial\theta} + \frac{\partial u}{\partial x} + \frac{\nu}{R}\frac{\partial v}{\partial\theta}, \tag{5.47a}$$

$$(1-\nu^2)\frac{N_\theta}{Eh} = -\frac{w}{R} + \frac{\nu}{2}\left(\frac{\partial w}{\partial x}\right)^2 + \nu\frac{\partial w}{\partial x}\frac{\partial w_0}{\partial x} + \frac{1}{2}\left(\frac{\partial w}{R\partial\theta}\right)^2 + \frac{\partial w}{R\partial\theta}\frac{\partial w_0}{R\partial\theta} + \nu\frac{\partial u}{\partial x} + \frac{1}{R}\frac{\partial v}{\partial\theta}, \tag{5.47b}$$

$$(1-\nu^2)\frac{N_{x\theta}}{Eh} = 2(1-\nu)\left[\frac{1}{R}\frac{\partial w}{\partial x}\frac{\partial w}{\partial\theta} + \frac{1}{R}\frac{\partial w}{\partial x}\frac{\partial w_0}{\partial\theta} + \frac{1}{R}\frac{\partial w_0}{\partial x}\frac{\partial w}{\partial\theta} + \frac{1}{R}\frac{\partial u}{\partial\theta} + \frac{\partial v}{\partial x}\right]. \tag{5.47c}$$

In this study, attention is focused on a finite, simply supported, circumferentially closed circular cylindrical shell of length L. The following out-of-plane boundary conditions are imposed:

$$w = w_0 = 0, \tag{5.48a}$$

$$M_x = -D\{(\partial^2 w/\partial x^2) + \nu[\partial^2 w/(R^2\,\partial\theta^2)]\} = 0, \tag{5.48b}$$

and

$$\partial^2 w_0/\partial x^2 = 0 \quad \text{at } x = 0,\, L, \tag{5.48c}$$

where M_x is the bending moment per unit length. The in-plane boundary conditions are

$$N_x = 0 \quad \text{and} \quad v = 0 \quad \text{at } x = 0,\; L. \tag{5.49a,b}$$

Moreover, u, v and w must be continuous in θ.

Past studies show that a linear modal base is the simplest choice to discretize the simply supported shell. In particular, in order to reduce the number of degrees of freedom, it is important to use only the most significant modes. It is necessary to consider, in addition to the asymmetric mode directly driven into vibration by the excitation (driven mode): (i) the orthogonal mode having the same shape and natural frequency but rotated by $\pi/(2n)$ (companion mode), (ii) additional asymmetric modes, and (iii) axisymmetric modes. In fact, it has been clearly established that, for large-amplitude shell vibrations, the deformation of the shell involves significant axisymmetric oscillations inward. In particular, the shell presents an inward axisymmetric dynamic contraction with twice the excitation frequency to guarantee the in-plane quasi-inextensibility of the shell. In fact, without this axisymmetric contraction, the shell increases its length along the circumference during large-amplitude radial vibrations (see Figure 5.1 for $n \geq 2$), which is against the shell mechanics: a shell bends more easily than it stretches.

According to these considerations, the radial displacement w is expanded by using the eigenmodes of the empty shell (which are unchanged for the completely filled shell with open ends):

$$w(x,\theta,t) = \sum_{m=1}^{M_1}\sum_{n=1}^{N} [A_{m,n}(t)\cos(n\theta) + B_{m,n}(t)\sin(n\theta)]\sin(\lambda_m x) + \sum_{m=1}^{M_2} A_{m,0}(t)\sin(\lambda_m x), \tag{5.50}$$

where n is the number of circumferential waves, m is the number of longitudinal half-waves, $\lambda_m = m\pi/L$ and t is the time; $A_{m,n}(t)$, $B_{m,n}(t)$ and $A_{m,0}(t)$ are the generalized coordinates that are unknown functions of t. The integers N, M_1 and M_2 must be selected with care in order to obtain the required accuracy and acceptable dimensions of the nonlinear problem. The number of degrees of freedom used in the present numerical calculations is 16 or more. It is observed that, for symmetry reasons, the nonlinear interaction among linear modes of the chosen base involves (i) the asymmetric modes ($n > 0$) having a given n value (the resonant mode), (ii) the asymmetric modes having a multiple of this value of circumferential waves ($k \times n$, where k is an integer) and (iii) axisymmetric modes ($n = 0$). Asymmetric modes with a different number of circumferential waves that do not satisfy the relationship $k \times n$ have interaction only if their natural frequencies are very close to a relationship of the type 1:1, 1:2 or 1:3 with the frequency of the resonant mode (n, m), that is, giving rise to internal resonances. Only modes with an odd m value of longitudinal half-waves can be considered for symmetry reasons in the study of resonance of modes with $m = 1$ (if imperfections with an even m value are not introduced). In particular, asymmetric modes having up to three longitudinal half-waves ($M_1 = 3$, only odd m values) and modes having n, $2 \times n$ and $3 \times n$ circumferential waves have been considered in the numerical calculations. For axisymmetric modes, $M_2 = 7$ has been used (only odd m values). Actually, $M_1 = M_2 = 3$ is sufficient to guarantee good accuracy. However, as it has been shown by Amabili et al. (2000a) and Pellicano et al. (2002), additional axisymmetric modes give a small contribution to the shell response, whereas additional asymmetric modes with more than three longitudinal half-waves are practically negligible. This is the reason for having M_2 larger than M_1. The form of the radial displacement used in the numerical calculation is

$$w(x,\theta,t) = \sum_{m=1}^{3}\sum_{k=1}^{3}\left[A_{m,kn}(t)\cos(kn\theta) + B_{m,kn}(t)\sin(kn\theta)\right]\sin(\lambda_m x) + \sum_{m=1}^{4} A_{(2m-1),0}(t)\sin(\lambda_{(2m-1)}x). \quad (5.51)$$

However, expansions involving additional asymmetric modes that do not satisfy the relationship $k \times n$, but having frequency close to relationship 1:1, 1:2 or 1:3 with the driven mode, have to be introduced in the case of internal resonances. The expansion in equation (5.51) has 22 degrees of freedom (dof), which are reduced to 16 if resonances of modes with $m = 1$ are studied and imperfections with $m = 2$ are neglected; in fact, modes with $m = 2$ do not give contributions for reasons of symmetry.

The presence of couples of modes having the same shape but different angular orientations, the first one described by $\cos(n\theta)$ (driven mode for the excitation given by equation [5.45]) and the other by $\sin(n\theta)$ (companion mode), in the periodic response of the shell leads to the appearance of traveling-wave vibration around the shell in the angular direction. This phenomenon is related to the axial symmetry of the system.

When the excitation has a frequency close to the resonance of mode ($n, m = 1$), results show that the generalized coordinates $A_{1,n}(t)$ and $B_{1,n}(t)$ have the same frequency of the excitation. The coordinates $A_{1,2n}(t)$, $B_{1,2n}(t)$, $A_{3,2n}(t)$, $B_{3,2n}(t)$ and all the coordinates associated with axisymmetric modes have twice the frequency of the excitation. The coordinates $A_{3,n}(t)$, $B_{3,n}(t)$, $A_{1,3n}(t)$, $B_{1,3n}(t)$, $A_{3,3n}(t)$ and $B_{3,3n}(t)$ have three times the frequency of the excitation.

The initial radial imperfection w_0 is expanded in the same form of w, that is, in a double Fourier series satisfying the boundary conditions (5.48a,c) at the shell edges:

$$w_0(x,\theta) = \sum_{m=1}^{\tilde{M}_1}\sum_{n=1}^{\tilde{N}}\left[\tilde{A}_{m,n}\cos(n\theta) + \tilde{B}_{m,n}\sin(n\theta)\right]\sin(\lambda_m x) + \sum_{m=1}^{\tilde{M}_2}\tilde{A}_{m,0}\sin(\lambda_m x), \quad (5.52)$$

where $\tilde{A}_{m,n}$, $\tilde{B}_{m,n}$ and $\tilde{A}_{m,0}$ are the modal amplitudes of imperfections; $\tilde{N}$, $\tilde{M}_1$ and $\tilde{M}_2$ are integers indicating the number of terms in the expansion.

5.5.1 Fluid-Structure Interaction

The contained fluid is assumed to be incompressible and inviscid. Liquid-filled shells vibrating in a low-frequency range satisfy the incompressible hypothesis very well. The nonlinear effects in the dynamic pressure and in the boundary conditions at the fluid-structure interface are neglected. These nonlinear effects are negligible because displacements of the order of the shell thickness have significant nonlinear effects on the shell, but the deformation of the flow remains small from the hydrodynamic and aerodynamic point of view since h/R is small. The shell prestress due to the fluid weight is also neglected. By using an extension of equation (5.33a) for the expression of w given by (5.51), and by using (5.34), the dynamic pressure p exerted by the contained fluid on the shell is given by

$$p = \rho_F(\dot{\Phi})_{r=R} = \sum_{m=1}^{M}\sum_{n=0}^{N} \rho_F[\ddot{A}_{mn}(t)\cos(n\theta) + \ddot{B}_{mn}(t)\sin(n\theta)]\frac{\mathrm{I}_n(\lambda_m R)}{\lambda_m \mathrm{I}_n'(\lambda_m R)}\sin(\lambda_m x), \tag{5.53}$$

where ρ_F is the mass density of the internal fluid. Equation (5.53) shows that the inertial effects due to the fluid are different for the asymmetric and the axisymmetric terms of the mode expansion. Hence, the fluid is expected to change the nonlinear behavior of the fluid-filled shell. Usually the inertial effect of the fluid is larger for axisymmetric modes, thus enhancing the nonlinear behavior of the shell.

5.5.2 Stress Function and Galerkin Method

Expansions (5.50) and (5.51) used for the radial displacement w satisfy identically the boundary conditions given by equations (5.48a,b). Moreover, they satisfy exactly the continuity of the circumferential displacement, which is a fundamental condition,

$$\int_0^{2\pi} (\partial v/\partial \theta)\, d\theta = v|_{\theta=2\pi} - v|_{\theta=0} = 0. \tag{5.54}$$

Equation (5.54) is proved in Section 5.5.4.

The boundary conditions for the in-plane displacements, equations (5.49a,b), give very complex expressions when transformed into equations involving w. Therefore, they are modified into simpler integral expressions that satisfy equations (5.49a,b) on the average, as originally introduced by Dowell and Ventres (1968). Specifically, the following conditions are imposed:

$$\int_0^{2\pi} N_x\, R\, d\theta = 0, \quad \text{at } x = 0, L, \tag{5.55}$$

$$\int_0^{2\pi}\int_0^{L} N_{x\theta}\, \mathrm{d}x\; R\, d\theta = 0. \tag{5.56}$$

Equation (5.55) ensures a zero axial force N_x on the average at $x = 0, L$; equation (5.56) is satisfied when $v = 0$ on the average at $x = 0, L$ and u is continuous in θ on the average. Substitution of equations (5.49a,b) by equations (5.55) and (5.56) simplifies computations, although it introduces an approximation (boundary conditions [5.49a,b] are exactly satisfied at n discrete points, where n is the number of circumferential waves).

When the expansions of w and w_0, equations (5.50) and (5.52), are substituted in the right-hand side of equation (1.163), a partial differential equation for the stress function F is obtained. The solution may be written as

$$F = F_h + F_p, \tag{5.57}$$

where F_h is the homogeneous solution and F_p is the particular solution. The particular solution is given by

$$F_p = \sum_{m=1}^{2M}\sum_{n=1}^{2N}[F_{mn1}\sin(m\eta)\ \sin(n\theta) + F_{mn2}\sin(m\eta)\ \cos(n\theta)$$
$$+ F_{mn3}\cos(m\eta)\ \sin(n\theta) + F_{mn4}\cos(m\eta)\ \cos(n\theta)]$$
$$+\sum_{n=1}^{2N}[F_{0n3}\sin(n\theta) + F_{0n4}\cos(n\theta)] + \sum_{m=1}^{2M}[F_{m02}\sin(m\eta) + F_{m04}\cos(m\eta)], \tag{5.58}$$

where N is the same as in equation (5.50), M is the largest between M_1 and M_2, $\eta = \pi x/L$ and the functions of time F_{mnj}, $j = 1, \ldots, 4$, have a long expression not reported here. They have been identified by using the *Mathematica* computer program (Wolfram 1999) for symbolic manipulations. If equation (5.50) is substituted into the right-hand side of equation (1.163), after some algebra, the following expressions are obtained:

$$\frac{1}{R}\frac{\partial^2 w}{\partial x^2} = \frac{\pi^2}{RL^2}\frac{\partial^2 w}{\partial \eta^2} = -\frac{\pi^2}{RL^2}\sum_{m=1}^{M}\sum_{n=0}^{N}[A_{mn}\cos(n\theta) + B_{mn}\sin(n\theta)]m^2\sin(m\eta), \tag{5.59}$$

$$\left(\frac{1}{R}\frac{\partial^2 w}{\partial x\,\partial\theta}\right)^2 = \frac{\pi^2}{R^2L^2}\sum_{m_1=1}^{M}\sum_{m_2=1}^{M}\sum_{n_1=0}^{N}\sum_{n_2=0}^{N} m_1m_2n_1n_2\cos(m_1\eta)\cos(m_2\eta)$$
$$\times[-A_{m_1n_1}\sin(n_1\theta) + B_{m_1n_1}\cos(n_1\theta)]$$
$$\times[-A_{m_2n_2}\sin(n_2\theta) + B_{m_2n_2}\cos(n_2\theta)]\ , \tag{5.60}$$

$$\frac{\partial^2 w}{\partial x^2}\frac{\partial^2 w}{R^2\,\partial\theta^2} = \frac{\pi^2}{R^2L^2}\sum_{m_1=1}^{M}\sum_{m_2=1}^{M}\sum_{n_1=0}^{N}\sum_{n_2=0}^{N} m_1^2n_2^2\sin(m_1\eta)\sin(m_2\eta)$$
$$\times[A_{m_1n_1}\cos(n_1\theta) + B_{m_1n_1}\sin(n_1\theta)]$$
$$\times[A_{m_2n_2}\cos(n_2\theta) + B_{m_2n_2}\sin(n_2\theta)], \tag{5.61}$$

$$\frac{2\,\partial^2 w}{R\partial x\,\partial\theta}\frac{\partial^2 w_0}{R\partial x\,\partial\theta} = \frac{2\pi^2}{R^2L^2}\sum_{m_1=1}^{M}\sum_{m_2=1}^{\tilde{M}}\sum_{n_1=0}^{N}\sum_{n_2=0}^{\tilde{N}} m_1m_2n_1n_2\cos(m_1\eta)\cos(m_2\eta)$$
$$\times[-A_{m_1n_1}\sin(n_1\theta) + B_{m_1n_1}\cos(n_1\theta)]$$
$$\times\left[-\tilde{A}_{m_2n_2}\sin(n_2\theta) + \tilde{B}_{m_2n_2}\cos(n_2\theta)\right], \tag{5.62}$$

$$-\left(\frac{\partial^2 w}{\partial x^2}\frac{\partial^2 w_0}{R^2\,\partial\theta^2} + \frac{\partial^2 w_0}{\partial x^2}\frac{\partial^2 w}{R^2\,\partial\theta^2}\right) = \frac{-\pi^2}{R^2L^2}\sum_{m_1=1}^{M}\sum_{m_2=1}^{\tilde{M}}\sum_{n_1=0}^{N}\sum_{n_2=0}^{\tilde{N}}(m_1^2n_2^2 + m_2^2n_1^2)$$
$$\times\sin(m_1\eta)\sin(m_2\eta)[A_{m_1n_1}\cos(n_1\theta) + B_{m_1n_1}$$
$$\times\sin(n_1\theta)][\tilde{A}_{m_2n_2}\cos(n_2\theta) + \tilde{B}_{m_2n_2}\sin(n_2\theta)], \tag{5.63}$$

where $\tilde{M}$ is the largest between $\tilde{M}_1$ and $\tilde{M}_2$; the time dependence has been suppressed here for the sake of brevity. By substituting equations (5.58–5.63) into equation (1.163), the unknown functions F_{mni} can be identified by using the *Mathematica* computer program.

The homogeneous solution may be assumed to have the form:

$$F_h = \frac{1}{2}\overline{N}_x\, R^2\theta^2 + \frac{1}{2}x^2\left\{\overline{N}_\theta - \frac{1}{2\pi RL}\int_0^L\int_0^{2\pi}\left[\frac{\partial^2 F_p}{\partial x^2}\right] R\,\mathrm{d}\theta\,\mathrm{d}x\right\} - \overline{N}_{x\theta}\, x\, R\theta, \quad (5.64)$$

where $\overline{N}_x$, $\overline{N}_\theta$ and $\overline{N}_{x\theta}$ are the average in-plane restraint stresses (forces per unit length) generated at the shell ends, defined as

$$\overline{N}_\# = [1/(2\pi L)]\int_0^{2\pi}\int_0^L N_\#\,\mathrm{d}x\,\mathrm{d}\theta, \quad (5.65)$$

where the symbol # must be replaced by x, θ and $x\theta$. Boundary conditions (5.55, 5.56) allow us to express $\overline{N}_x$, $\overline{N}_\theta$ and $\overline{N}_{x\theta}$, see equations (5.47a–c), in terms of w, w_0 and their derivatives. Equation (5.64) is chosen in order to satisfy the boundary conditions imposed. Moreover, it satisfies equations (1.155a–c) on the average.

By using equations (5.47a–c) and (5.65) and taking into account that, from the boundary conditions (5.48) and (5.49), $\overline{N}_x = \text{constant}$ (in particular, $\overline{N}_x = 0$) and $\overline{N}_{x\theta} = 0$, after simple calculations it is obtained:

$$\overline{N}_\theta = \nu\overline{N}_x + \frac{Eh}{2\pi RL}\int_0^{2\pi}\int_0^L\left[-\frac{w}{R} + \frac{1}{2}\left(\frac{\partial w}{R\partial\theta}\right)^2 + \frac{\partial w}{R\partial\theta}\frac{\partial w_0}{R\partial\theta}\right]\mathrm{d}x\, R\mathrm{d}\theta. \quad (5.66)$$

By inserting equations (5.50) and (5.52) into equation (5.66), the following expression is obtained:

$$\overline{N}_\theta = \frac{Eh}{2\pi R}\left\{-2\sum_{m=1}^{M_2}\frac{A_{m,0}(t)}{m}\left[1-(-1)^m\right] + \frac{\pi}{4R}\sum_{n=1}^{N}\sum_{m=1}^{M_1} n^2\left(A_{m,n}^2(t) + B_{m,n}^2(t)\right)\right.$$
$$\left. + \frac{\pi}{2R}\sum_{n=1}^{\overline{N}}\sum_{m=1}^{\overline{M}} n^2(\tilde{A}_{m,n}A_{m,n}(t) + \tilde{B}_{m,n}B_{m,n}(t))\right\}, \quad (5.67)$$

where $\overline{N}$ is the smallest between N and $\tilde{N}$ and $\overline{M}$ is the smallest between M_1 and $\tilde{M}_1$.

By use of the Galerkin method, 16 (or more) second-order, ordinary, coupled nonlinear differential equations are obtained for the variables $A_{m,kn}(t)$, $B_{m,kn}(t)$ and $A_{m,0}(t)$, for $m = 1, \ldots, M$ and $k = 1, \ldots, 3$, by successively weighting the single original equation (1.162) with the functions that describe the shape of the modes retained in equation (5.51); the weighting functions ϕ_j, for $j = 1, \ldots, 16$, are defined as

$$\varphi_j(x,\theta) = \begin{cases} \cos(n\theta)\,\sin(\lambda_m x) & \text{for } j=1 \\ \sin(n\theta)\,\sin(\lambda_m x) & \text{for } j=2 \\ \vdots & \vdots \end{cases}. \quad (5.68)$$

The Galerkin projection, in this case, can be defined as

$$\langle D\nabla^4 w + ch\dot{w} + \rho h\ddot{w} = \ldots,\ \varphi_j\rangle = \int_0^{2\pi}\int_0^L (D\nabla^4 w + ch\dot{w} + \rho h\ddot{w} = \ldots) \times \varphi_j(x,\theta)\ \mathrm{d}x\ R\mathrm{d}\theta. \quad (5.69)$$

The Galerkin projections of the equation of motion (1.162) have been performed analytically by using the *Mathematica* computer software (Wolfram 1999) in order to avoid numerical errors, which arise from numerical calculations of surface integrals of trigonometric functions.

The Galerkin projection of the modal excitation f on the weighting functions z_j, gives

$$\langle f, z_j\rangle = \int_0^{2\pi}\int_0^L f z_j\ \mathrm{d}x\ \mathrm{R}\ \mathrm{d}\theta = \frac{\pi RL}{2} f_{1,n}\cos(\omega t) \quad \text{for } j = 1$$
$$= 0 \quad \text{for } j \neq 1. \quad (5.70)$$

Therefore, the modal excitation considered gives a nonzero contribution only to mode $A_{1,n}(t)\cos(n\theta)\sin(\pi x/L)$, that is, on the driven mode. The mode having the shape of the driven mode, but rotated by $\pi/(2n)$, that is, with form $B_{1,n}(t)\sin(n\theta)$ $\sin(\pi x/L)$, is referred to as the companion mode and is not directly excited in this case.

If the point excitation is located at $\tilde{\theta} = 0$, $\tilde{x} = L/2$, the Galerkin projection of the point excitation f on the weighing functions z_j gives

$$\langle f, z_j\rangle = \tilde{f}\cos(\omega t) \quad \text{for } z_j \text{ axisymmetric or containing } \cos(n\theta),$$
$$= 0 \quad \text{for } z_j \text{ containing } \sin(n\theta). \quad (5.71)$$

In this case, setting $\tilde{f} = f_{1,n}\pi RL/2$, the only difference between modal excitation and point excitation is that the point excitation directly drives all modes described by a cosine function in angular direction θ as well as the axisymmetric modes.

The generic j-th equation of motion takes the form of equation (3.6) after dividing it by the modal mass associated with $\ddot{A}_{m,n}(t)$ or $\ddot{B}_{m,n}(t)$. The equations of motion resulting after the Galerking projections have very long expressions, containing quadratic and cubic nonlinear terms, and are studied by (i) use of the software AUTO 97 (Doedel et al. 1998) for continuation and bifurcation analysis of nonlinear ordinary differential equations and (ii) direct integration of the equations of motion by using the DIVPAG routine of the Fortran library IMSL. The software AUTO 97 is capable of continuation of the solution, bifurcation analysis and branch switching by using the pseudo-arclength continuation method. In particular, the shell response under harmonic excitation has been studied by using an analysis in two steps: (i) the excitation frequency has been fixed far enough from resonance and the magnitude of the excitation has been used as bifurcation parameter. The solution has been started at zero force, corresponding to the trivial undisturbed configuration of the shell, and has been continued up to reach the desired force magnitude. (ii) When the desired magnitude of excitation has been reached, the solution has been continued by using the excitation frequency as bifurcation parameter.

5.5.3 Traveling-Wave Response

Away from resonance, the companion mode solution disappears ($B_{1,n}(t) = 0$) and the generalized coordinates are nearly in phase or in opposite phase. The presence of the companion mode in the shell response leads to the appearance of a traveling wave and to more complex phase relationships among the generalized coordinates. The flexural mode shapes are represented by equation (5.51). Supposing that the response of the driven and companion modes have the same frequency of oscillation, that is, $A_{1,n}(t) = a_{1,n}\cos(\omega t + \theta_1)$ and $B_{1,n}(t) = b_{1,n}\cos(\omega t + \theta_2)$, and considering the other coordinates having smaller amplitude, equation (5.51) can be rearranged as

$$w = \{[a_{1,n}\cos(\omega t + \theta_1) + b_{1,n}\sin(\omega t + \theta_2)]\cos(n\theta) + b_{1,n}\sin(n\theta - \omega t - \theta_2)\} \times \sin(\pi x/L) + \mathrm{O}\left(a_{1,n}^2,\, b_{1,n}^2,\, a_{3,n}, \ldots a_{1,0},\, a_{3,0} \ldots\right), \tag{5.72}$$

where $a_{1,n}$ and $b_{1,n}$ are the amplitudes of driven and companion modes, respectively, θ_1 and θ_2 are the phases and O are negligible terms. Equation (5.72) gives a combined solution consisting of a standing wave plus a traveling wave of amplitude $b_{1,n}$ moving in an angular direction around the shell with angular velocity ω/n. The resulting standing wave is given by the sum of the two standing waves, one of amplitude $a_{1,n}$ and one of amplitude $b_{1,n}$, having the same circular frequency ω and the same shape, but having a phase difference of $\theta_2 - \theta_1 - \pi/2$. When $\theta_2 - \theta_1 \cong \pi/2$, as is generally observed in calculations and experiments, the amplitude of the resulting standing wave is almost $a_{1,n} - b_{1,n}$; therefore, it becomes zero for $a_{1,n} = b_{1,n}$ and $\theta_2 - \theta_1 = \pi/2$, when a pure traveling wave appears. The amplitude and frequency of the traveling-wave solution are not affected by the phase relationship between driven and companion modes. For reasons of symmetry, traveling waves can appear in one direction or in the opposite angular direction; both are solutions of the equations of motion.

It is important to observe that the companion mode arises as a consequence of the symmetry of the system. This phenomenon represents a fundamental difference vis-à-vis linear vibrations.

5.5.4 Proof of the Continuity of the Circumferential Displacement

The continuity condition of the circumferential displacement v, equation (5.54), is satisfied exactly by the assumed mode expansion. For the assumed boundary conditions considered at the shell ends, by using equations (1.142a–c) and (1.155a–c), equation (5.54) is transformed into

$$\int_0^{2\pi} \left[\frac{1}{Eh}\left(\frac{\partial^2 F}{\partial x^2} - \nu\frac{\partial^2 F}{R^2\,\partial\theta^2}\right) + \frac{w}{R} - \frac{1}{2}\left(\frac{\partial w}{R\,\partial\theta}\right)^2\right] \mathrm{d}\theta = 0. \tag{5.73}$$

For simplicity, only seven terms are considered in the expansion of w and the geometric imperfection is set to zero; in particular, $m = 1, 2$ and $k = 1$ for asymmetric terms and $m = 1, 2, 3$ for axisymmetric terms in equation (5.51). The terms involved in equation (5.73), after some calculations, are

$$\frac{1}{Eh}\int_0^{2\pi} \frac{\partial^2 F}{\partial x^2}\,\mathrm{d}\theta = \frac{2\pi}{Eh}\,\overline{N}_\theta + \frac{4}{R}\left(A_{1,0}(t) + \frac{A_{3,0}(t)}{3} + \frac{A_{5,0}(t)}{5}\right) - \frac{2\pi}{R}\,[A_{1,0}(t)$$

$$\times \sin(\pi x/L) + A_{3,0}(t)\sin(3\pi x/L) + A_{5,0}(t)\sin(5\pi x/L)] - \frac{\pi n^2}{R^2}$$
$$\times\left[-\frac{A_{1,n}(t)\,A_{2,n}(t) + B_{1,n}(t)\,B_{2,n}(t)}{2}\cos(\pi x/L) + \frac{A_{1,n}^2(t) + B_{1,n}^2(t)}{4}\cos(2\pi x/L)\right.$$
$$\left. + \frac{A_{1,n}(t)\,A_{2,n}(t) + B_{1,n}(t)\,B_{2,n}(t)}{2}\cos(3\pi x/L) + \frac{A_{2,n}^2(t) + B_{2,n}^2(t)}{4}\cos(4\pi x/L)\right], \tag{5.74}$$

$$\frac{1}{Eh}\int_0^{2\pi} \frac{\partial^2 F}{R^2\,\partial\theta^2}\,d\theta = \frac{2\pi}{Eh}\,\overline{N}_x, \tag{5.75}$$

$$\frac{1}{Eh}\int_0^{2\pi} \frac{w}{R}\,d\theta = \frac{2\pi}{EhR}[A_{1,0}(t)\sin(\pi x/L) + A_{3,0}(t)\sin(3\pi x/L) + A_{5,0}(t)\sin(5\pi x/L)], \tag{5.76}$$

$$\frac{1}{Eh}\int_0^{2\pi}\left(\frac{\partial w}{R\partial\theta}\right)^2 d\theta = \frac{\pi n^2}{EhR^2}\left\{\left(A_{1,n}^2(t) + B_{1,n}^2(t)\right)\left[\frac{1-\cos(2\pi x/L)}{2}\right]\right.$$
$$+ \left(A_{2,n}^2(t) + B_{2,n}^2(t)\right)\left[\frac{1-\cos(4\pi x/L)}{2}\right] + (A_{1,n}(t)\,A_{2,n}(t)$$
$$\left. + B_{1,n}(t)\,B_{2,n}(t))\,[\cos(\pi x/L) - \cos(3\pi x/L)]\right\} \tag{5.77}$$

and

$$\overline{N}_\theta - \nu\overline{N}_x = Eh\left[-\frac{2}{\pi R}(A_{1,0}(t) + A_{3,0}(t) + A_{5,0}(t))\right.$$
$$\left. + \frac{n^2}{8R^2}\left(A_{1,n}^2(t) + B_{1,n}^2(t) + A_{2,n}^2(t) + B_{2,n}^2(t)\right)\right]. \tag{5.78}$$

Substituting equations (5.74–5.78) into equation (5.73), it is found that the left-hand part of this equation is identically zero for all x values. This proves that the continuity condition is exactly satisfied and the extension to geometric imperfections and to the 16 terms in equation (5.51) can be easily obtained.

5.6 Numerical Results for Nonlinear Vibrations of Simply Supported Shells

5.6.1 Empty Shell

A test case of a simply supported circular cylindrical shell is analyzed. The shell characteristics are: $L = 0.2$ m, $R = 0.1$ m, $h = 0.247 \times 10^{-3}$ m, $E = 71.02 \times 10^9$ Pa, $\rho = 2796$ kg/m^3 and $\nu = 0.31$; the driven mode is $n = 6$ and $m = 1$, which corresponds to a case studied by several authors (Chen and Babcock 1975; Ganapathi and Varadan 1996; Amabili et al. 1998; Amabili et al. 2000b; Pellicano et al. 2002). The software AUTO 97 (Doedel et al. 1998) has been used for continuation and bifurcation analysis of the nonlinear equations of motion.

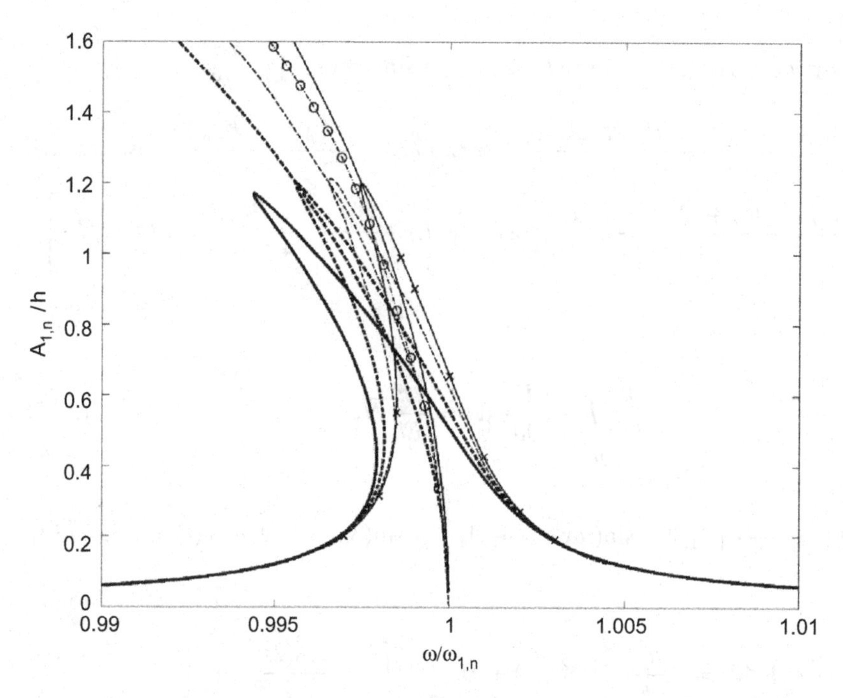

Figure 5.9. Frequency-response curve without companion modes participation. —, 16 dofs model; – –, 4 dofs model; —, 3 dofs model (Amabili et al. 1998); – –, Chen and Babkock (1975); – o -, backbone curve only (Ganapathi and Varadan 1996).

Figure 5.9 shows the frequency-response curve (computed by using the model with 16 degrees of freedom [dofs]) of the driven mode ($A_{m=1,n=6}(t)$) without companion mode participation ($B_{m,(nn1)}=0$); this is an artificial constraint that is introduced in order to simplify the nonlinear response, which becomes similar to that studied in Section 3.2, and to identify the trend of nonlinearity, which is softening in this case. The present results (continuous thick line) are compared to those obtained analytically by Chen and Babcock (1975), Ganapathi and Varadan (1996) (only backbone curve), Amabili et al. (1998) and the 4 dofs model including $A_{1,6}$, $B_{1,6}$, $A_{1,0}$, $A_{3,0}$ (obtained by removing the artificial kinematic constraint between $A_{1,0}$ and $A_{3,0}$ introduced by Amabili et al. [1998] in order to reduce to three the number of degrees of freedom). The amplitude of the external modal excitation is $f_{1,6}=0.0012\,h^2\rho\omega_{1,6}^2$ and the damping ratio is $2\zeta_{1,6}=0.001$ (with $\zeta_{j,k}=\zeta_{1,6}\,\omega_{j,k}/\omega_{1,6}$ for the additional generalized coordinates). The linear circular frequency of the driven and companion modes is $\omega_{1,6}=2\pi\times 564.2$ rad/s. The backbone curves (pertaining to undamped free vibrations) are also shown, with the exception of the present result (for clarity). Figure 5.9 shows reasonably good agreement between the present results and those obtained previously. However, the complete model used here shows that the actual response is more strongly softening. In particular, the 3 dofs model (Amabili et al. 1998) gives a stiffer response with respect to the present system using a fuller mode expansion. Note that, after removing the kinematic constraint of the 3 dofs model, that is, using the 4 dofs model, the response approaches the complete model, demonstrating a rapid convergence of the series expansion of the radial displacement w.

In order to better investigate the convergence of the expansion (5.50), the 16 dofs response is compared with reduced models; only the solution branch with no companion mode participation is shown in Figure 5.10, where various expansions are compared. In all models, both $A_{1,6}$, $B_{1,6}$ are used, plus: (case i), the pair $A_{1,12}$, $B_{1,12}$

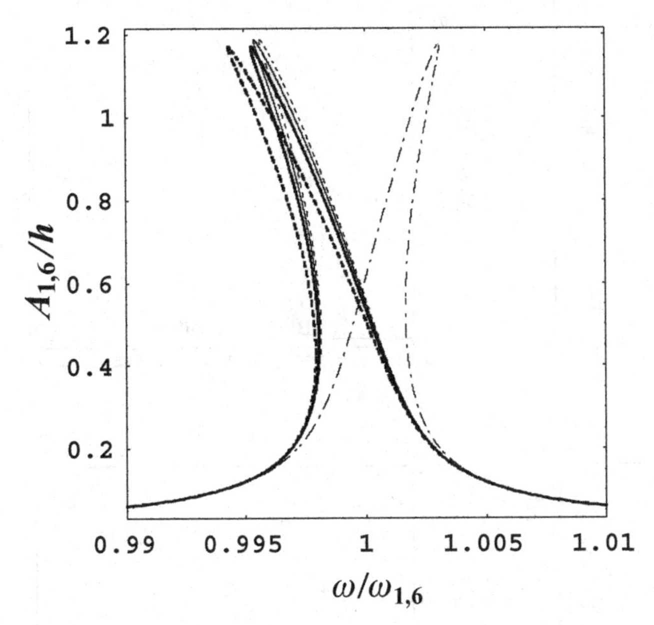

Figure 5.10. Frequency-response curves without companion modes participation. —, case (i), 6 dofs model; —, case (ii); – –, case (iii), 4 dofs model; –·–, case (iv); – –, case (v), 16 dofs model.

of modes having $2n$ circumferential waves plus axisymmetric modes $A_{1,0}, A_{3,0}$ with 1 and 3 longitudinal half-waves (continuous thick line); (case ii) axisymmetric modes $A_{1,0}, A_{3,0}, A_{5,0}$ having 1, 3 and 5 longitudinal half-waves (continuous thin line); (case iii) axisymmetric modes $A_{1,0}, A_{3,0}$ having 1 and 3 longitudinal half-waves (dashed thin line); (case iv) the first axisymmetric mode $A_{1,0}$ (dashed dotted line); (case v) the 16 dofs model (dashed thick line). Figure 5.10 proves that if only the first axisymmetric mode, case (iv), is used in the expansion, a hardening nonlinearity is erroneously obtained. However, the convergence of the series is still not completely reached with the 4 dofs expansion, case (iii); the 16 dofs model, case (v), gives more strongly softening results, even if the difference is not large.

Figure 5.11 shows the frequency-response relationship with companion mode participation (i.e. the actual response of the shell) for the 6 dofs model including modes with $2n$ circumferential waves. The main branch 1 in Figure 5.11 corresponds to vibration with zero amplitude of the companion mode $B_{1,6}(t)$. This branch has pitchfork bifurcations (BP) at $\omega/\omega_{1,6} = 0.9953$ and at 1.0011, where branch 2 appears. This new branch corresponds to participation of both $A_{1,6}(t)$ and $B_{1,6}(t)$, giving a traveling-wave response. The companion mode presents a node at the location of the excitation force and therefore it is not directly excited; its amplitude is different than zero only for large-amplitude vibrations, due to nonlinear coupling. In the narrow frequency region where both $A_{1,6}(t)$ and $B_{1,6}(t)$ are different from zero, they give rise to a traveling wave around the shell; phase shift between the two coordinates is almost $\pi/2$. The appearance of branch 2 is related to the 1:1 internal resonance of $A_{1,6}(t)$ and $B_{1,6}(t)$, which is due to the axial symmetry of the circular cylindrical shell. This branch appears for sufficiently large excitation, and it can be observed for vibration amplitude of the order of 1/10 of the shell thickness.

Branch 2 undergoes two Neimark-Sacker (torus) bifurcations (TR), at $\omega/\omega_{1,6} = 0.99534$ and 0.9998. Amplitude-modulated (quasi-periodic) response is indicated

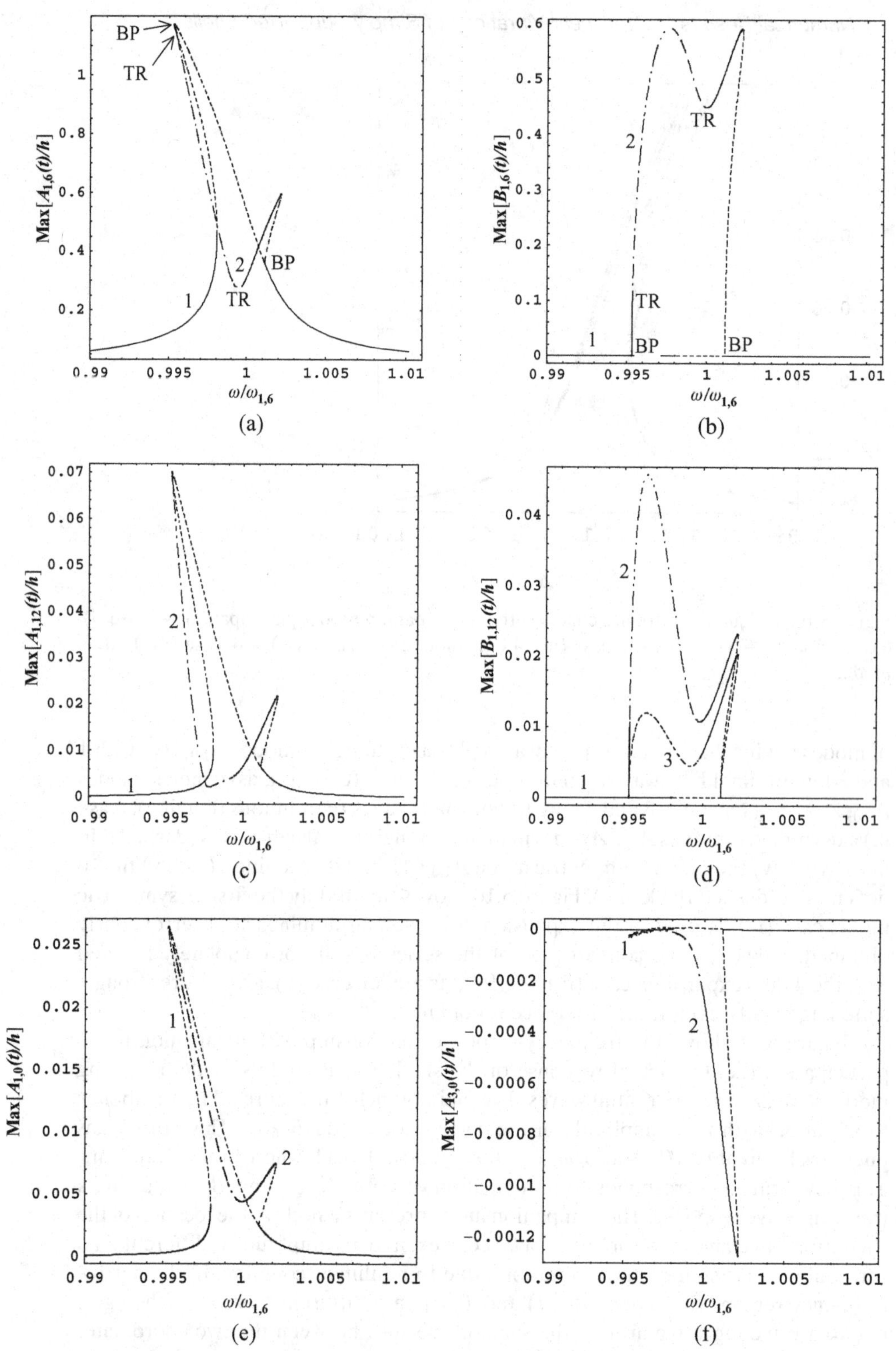

Figure 5.11. Frequency-response curve with companion mode participation for the model including modes with $2n$ circumferential waves (6 dofs). —, Stable periodic solution; –·–, stable quasi-periodic solution; – –, unstable solutions; BP, pitchfork bifurcation; TR, Neimark-Sacker bifurcation. (a) Maximum of $A_{1,6}(t)/h$. (b) Maximum of $B_{1,6}(t)/h$. (c) Maximum of $A_{1,12}(t)/h$. (d) Maximum of $B_{1,12}(t)/h$. (e) Maximum of $A_{1,0}(t)/h$. (f) Maximum of $A_{3,0}(t)/h$.

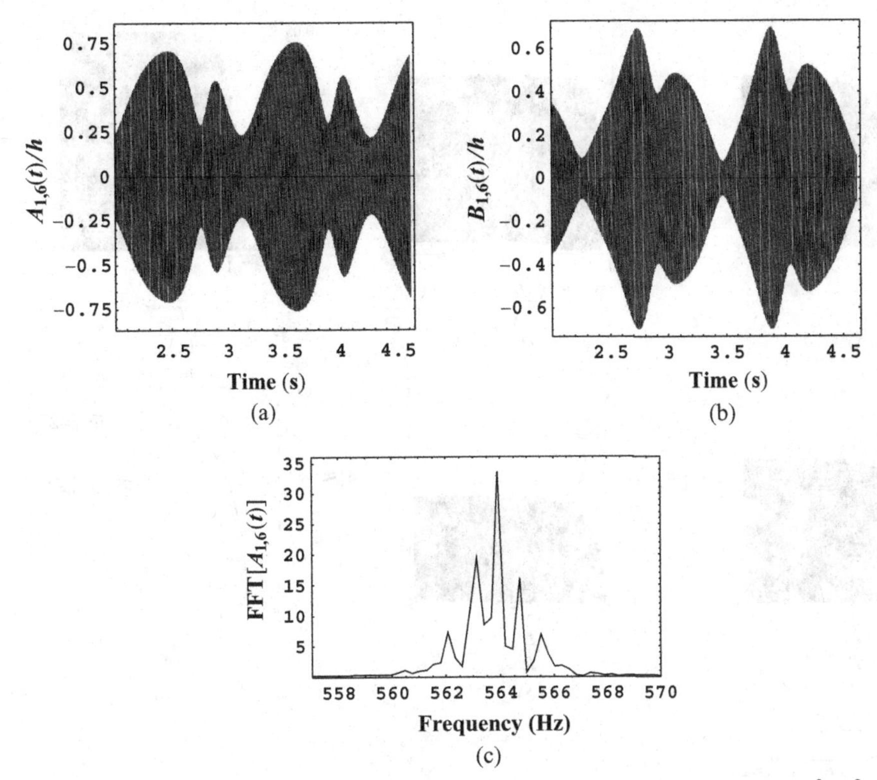

Figure 5.12. Quasi-periodic response for $\omega/\omega_{1,6} = 0.999$ and $f_{1,6} = 0.0012\,h^2\rho\omega_{1,6}^2$; 6 dofs model. (a) Time history $A_{1,6}(t)/h$. (b) Time history $B_{1,6}(t)/h$. (c) Frequency spectrum of $A_{1,6}(t)$.

in Figure 5.11 on branch 2 for $0.99534 < \omega/\omega_{1,6} < 0.9998$, that is, bracketed by the two Neimark-Sacker bifurcations. A further complication is shown in Figure 5.11(d), related to $B_{1,12}(t)$, where a third branch appears. The amplitude of axisymmetric modes in Figures 5.11(e,f) is small if compared with $A_{1,6}(t)$ and $B_{1,6}(t)$; however, the presence of $A_{1,0}(t)$ and $A_{3,0}(t)$ in the model is fundamental to predicting the correct nonlinearity of the shell.

It can been observed in Figure 5.11 that no stable periodic solutions exist for $0.9981 < \omega/\omega_{1,6} < 0.9998$, for the 6 dofs model. Therefore, the response of the system to harmonic excitation has been investigated further in this region. Figure 5.12(a,b) shows the quasi-periodic time response of the generalized coordinates $A_{1,6}(t)$ and $B_{1,6}(t)$, respectively. This figure also shows that the maximum of $A_{1,6}(t)$ at any time is approximately indicated by points for $\omega/\omega_{1,6} = 0.999$ (i.e. on a vertical line) in the region contained between the two branches 1 and 2. The response is not chaotic as can be observed from the frequency spectrum given in Figure 5.12(c). It shows that the single-frequency response obtained in the stable region is divided into several closely spaced frequencies that give a beating phenomenon. The whole area where no stable periodic solutions exist is associated with beating phenomena that give quasi-periodic response with modulations in the oscillation amplitude. Note that the small portion of the spectrum given in Figure 5.12(c) contains most of the energy in the time signal, that is, no harmonic components are present at lower and higher frequencies.

In order to clarify definitively the role of the fundamental ratios h/R and L/R on the characterization of nonlinearity (softening or hardening), a two-parameter study has been performed on a simple model including $A_{1,6}$, $A_{1,0}$ and $A_{3,0}$, because the trend of nonlinearity is not affected by the companion mode $B_{1,6}$. Note that

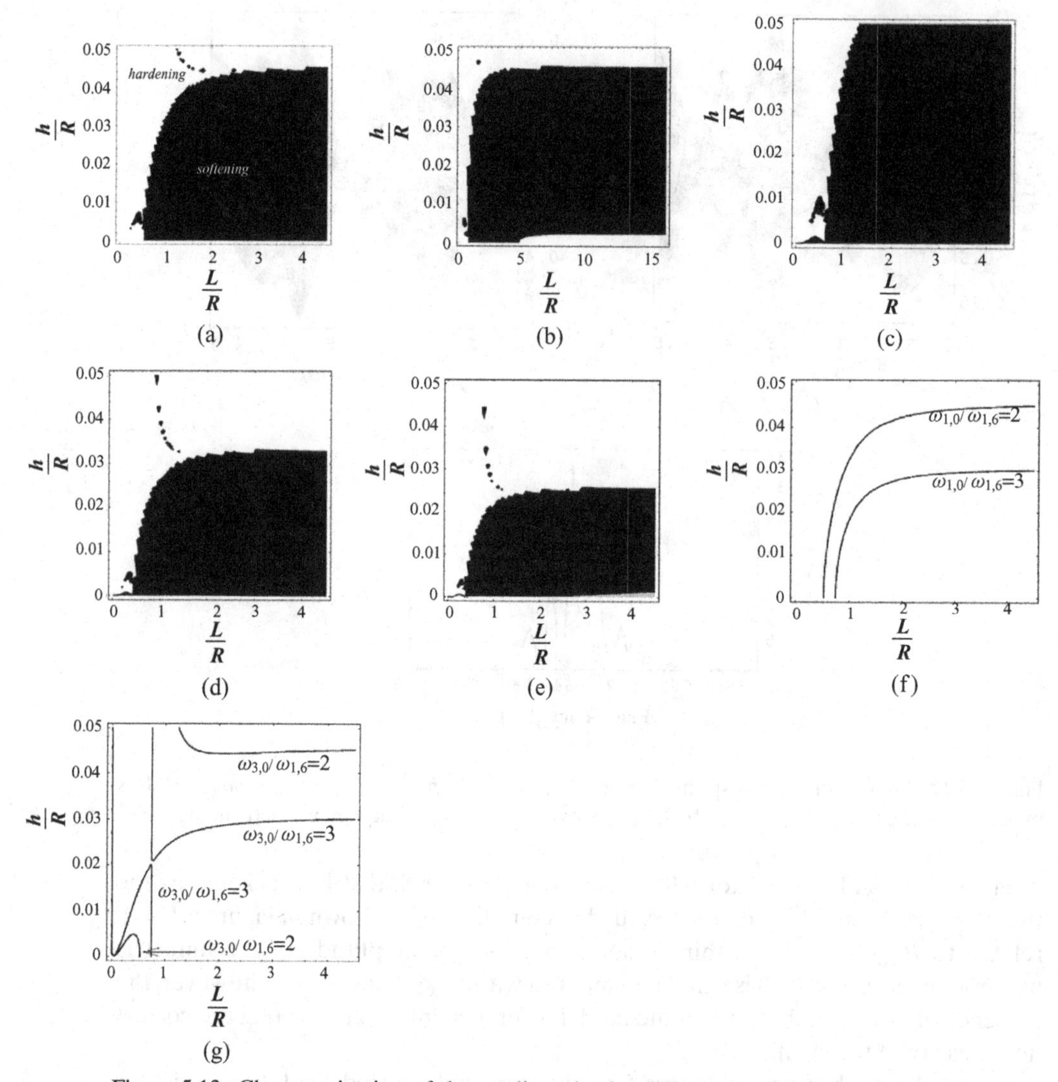

Figure 5.13. Characterization of the nonlinearity for different nodal diameters n with $m = 1$. White region, hardening nonlinearity; black region, softening nonlinearity. (a) $n = 6$. (b) $n = 6$ with enlarged L/R axis. (c) $n = 5$. (d) $n = 7$. (e) $n = 8$. (f) Internal resonances between modes (1,6) and (1,0). (g) Internal resonances between modes (1,6) and (3,0).

the mode ($m = 1, n = 6$) is not generally the fundamental mode (lower natural frequency) when ratios h/R and L/R vary; the fundamental mode can have a different value n. In Figures 5.13(a–e) maps in the $(h/R, L/R)$-plane are shown. The dark region represents softening behavior, and the white region represents hardening-type nonlinearity. Figures 5.13(a,b), which are for $n = 6$, show that, for small values of L/R or large values of h/R, shells have hardening behavior, but the greater part of the plane is dominated by softening behavior. However, for sufficiently long shells having small h/R ratios, the system can behave in a hardening way. Some other, irregularly shaped softening regions, are present within the hardening region. The passage from hardening to softening is quite interesting. Indeed, it may be observed that the main boundary in Figure 5.13(a) is due to the presence of an internal resonance. This correspondence is analyzed in detail by following the presence of the

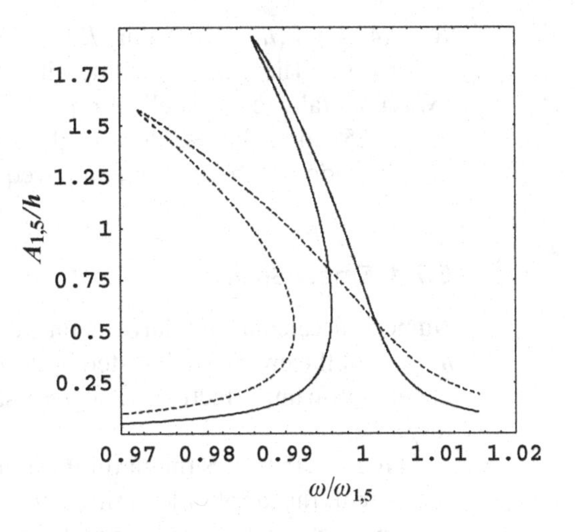

Figure 5.14. Theoretical frequency-response curve for an empty and water-filled, simply supported shell. —, water-filled shell, excitation $\tilde{f} = 3$ N, damping $\zeta_{1,n} = 0.0017$; – – –, empty shell, excitation $\tilde{f} = 1.5$ N, damping $\zeta_{1,n} = 0.0008$.

1:2 and 1:3 internal resonances between (1,6) and (1,0) modes, Figure 5.13(f), and between (1,6) and (3,0) modes, Figure 5.13(g), in the $(h/R, L/R)$-plane. Analysis of these internal resonances suggests that the main boundary, between hardening and softening nonlinearity, is due to a 1:2 internal resonance between modes (1,6) and (1,0) (upper part of the main softening region). The additional irregular regions present in Figure 5.13(a) are due to 1:3 and 1:2 internal resonances between modes (1,6) and (3,0), as shown in Figure 5.13(g). The small hardening region in Figure 5.13(b) for long shells ($L/R > 5$) with small h/R ratios does not seem to be due to an internal resonance. The results presented in Figure 5.13 depend on the considered mode $(1, n)$; for this reason Figures 5.13(c,d,e) show the results for $n = 5, 7, 8$, respectively. The softening region is reduced as n is increased; nevertheless, it must be pointed out that the number of nodal diameters n characterizing the fundamental mode increases as the ratio h/R decreases. Therefore, the present analysis confirms that circular cylindrical shells present "generally" a softening-type nonlinearity.

5.6.2 Water-Filled Shell

The dimensions and material properties of a simply supported, water-filled shell are $L = 520$ mm, $R = 149.4$ mm, $h = 0.519$ mm, $E = 1.98 \times 10^{11}$ Pa, $\rho = 7800$ kg/m^3, $\rho_F = 1000$ kg/m^3 and $\nu = 0.3$; the mode ($m = 1, n = 5$) is considered, with natural frequency of 79.21 Hz. Figure 5.14 presents the frequency-response curve for forced vibrations of the same empty and water-filled shell. A modal damping coefficient $\zeta_{1,5} = 0.0008$ is used here for the empty shell (natural frequency, 210.9 Hz), whereas $\zeta_{1,5} = 0.0017$ is used for the water-filled shell. Excitation of 1.5 N is used for the empty shell and 3 N for the water-filled shell. Figure 5.14 shows that added mass due to the water contained in the shell generates a much stronger softening-type nonlinearity than for the same empty shell, which shows a weak nonlinearity. In Figure 5.14, the empty and water-filled shells have a different linear frequency $\omega_{1,5}$, which is used to nondimensionalize the excitation frequency ω.

5.7 Effect of Geometric Imperfections

Calculations have been performed for circular cylindrical shells made of stainless steel with the following dimensions and material properties: $L = 520$ mm,

$R = 149.4\,\text{mm}, h = 0.519\,\text{mm}, E = 1.98 \times 10^{11}\,\text{Pa}, \rho = 7800\,\text{kg/m}^3, \rho_F = 1000\,\text{kg/m}^3$ and $\nu = 0.3$. These data coincide with those used in Section 5.6.2 and with those of an experimentally tested shell, as reported in Section 5.8. An harmonic point excitation at $x = 251$ mm, that is, close to the middle of the shell, is applied in the spectral neighborhood of the fundamental frequency.

5.7.1 Empty Shell

Numerical calculations have been performed for the fundamental mode ($n = 5$, $m = 1$) of the empty shell tested in the experiments in order to investigate the effect of geometric imperfections of a given shape on the natural frequency and nonlinear response.

Figure 5.15 shows the natural frequency of the fundamental mode of the empty shell versus the amplitude of three different geometric imperfections: (i) axisymmetric imperfection $\tilde{A}_{1,0}$, (ii) asymmetric imperfection $\tilde{A}_{1,5}$ having the same shape of the fundamental mode and (iii) asymmetric imperfection $\tilde{A}_{1,10}$ having twice the number of circumferential waves of the fundamental mode. It is important to observe that modes with $3n$ circumferential waves must be included in equation (5.51) in order to study the effect of imperfection $\tilde{A}_{1,10}$. The axisymmetric imperfection does not divide the double eigenvalue associated with the fundamental mode; this is only obtained by asymmetric imperfections. Small positive axisymmetric imperfections (inward gaussian curvature) decrease the natural frequency, negative axisymmetric imperfections (outward gaussian curvature) increase it. Moreover, imperfections having twice the number of circumferential waves with respect to the resonant mode have a very large effect on its natural frequency, as indicated in Figure 5.15(c). Imperfections with the same number of circumferential waves have a smaller effect, but still significant, as shown in Figure 5.15(b). Imperfections with a number of circumferential waves that is not a multiple of n play a very small role.

The frequency-response relationship of the perfect, empty shell under harmonic point excitation of magnitude 1.5 N is given in Figure 5.16 assuming modal damping on the fundamental mode $\zeta_{1,5} = 0.0008$. Only the resonant generalized coordinates $A_{1,5}(t)$ and $B_{1,5}(t)$ are plotted. The coordinate $A_{1,5}(t)$ is the driven mode because it has an antinode at the location of the excitation force and the coordinate $B_{1,5}(t)$ is the companion mode.

Asymmetric imperfections with the same shape of the resonant mode have a large effect on the nonlinear response. Figure 5.17 shows the shell response along with responses of the imperfect shell with $\tilde{A}_{1,5} = 0.5h$ and $\tilde{B}_{1,5} = 0.5h$ (positive or negative asymmetric imperfections with n waves give the same result; this is not true for axisymmetric imperfections). Results show that the trend of nonlinearity is minimally affected in this case, but the traveling-wave response is largely modified, because natural frequencies of the driven and companion modes do not coincide anymore. It is important to observe that the linear frequency $\omega_{1,5}$, which is used to nondimensionalize the excitation frequency ω, is changed by the geometric imperfection; therefore, each curve represented in Figure 5.17 has been nondimensionalized by using a different linear frequency $\omega_{1,5}$.

The effect of geometric imperfections with 10 circumferential waves, which are the most important, on the nonlinear response are analyzed in Section 5.8.

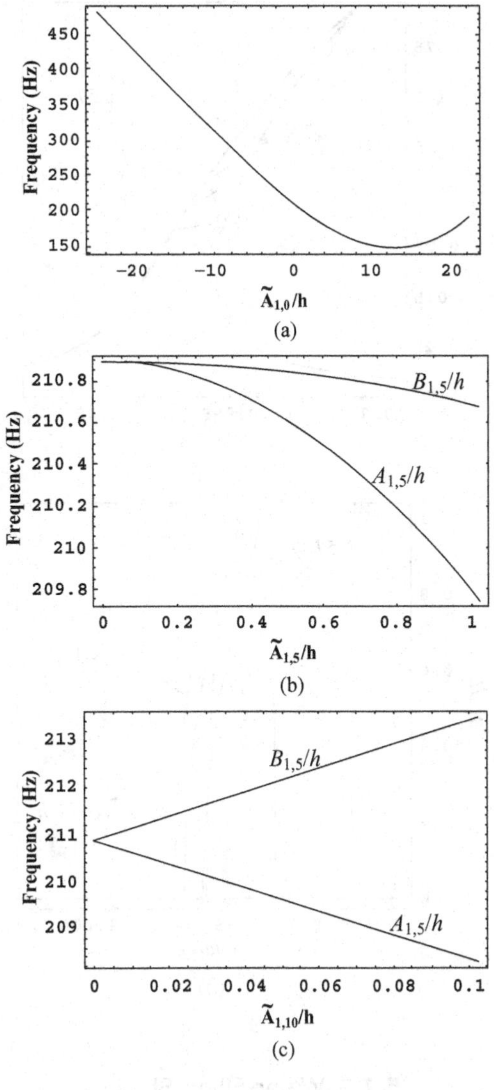

Figure 5.15. Natural frequency of the fundamental mode ($n = 5, m = 1$) of the empty shell versus the amplitude of geometric imperfections. (a) Axisymmetric imperfection $\tilde{A}_{1,0}$. (b) Asymmetric imperfection $\tilde{A}_{1,5}$. (c) Asymmetric imperfection $\tilde{A}_{1,10}$.

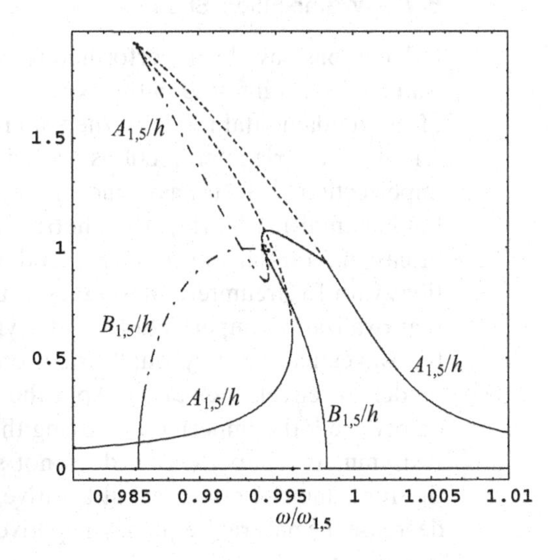

Figure 5.16. Maximum amplitude of vibration of $A_{1,5}(t)$, driven mode, and $B_{1,5}(t)$, companion mode, versus excitation frequency, empty shell, excitation 1.5 N, damping $\zeta_{1,5} = 0.0008$, 16 dofs. —, stable periodic solutions; –·–, stable quasi-periodic solutions; – –, unstable periodic solutions.

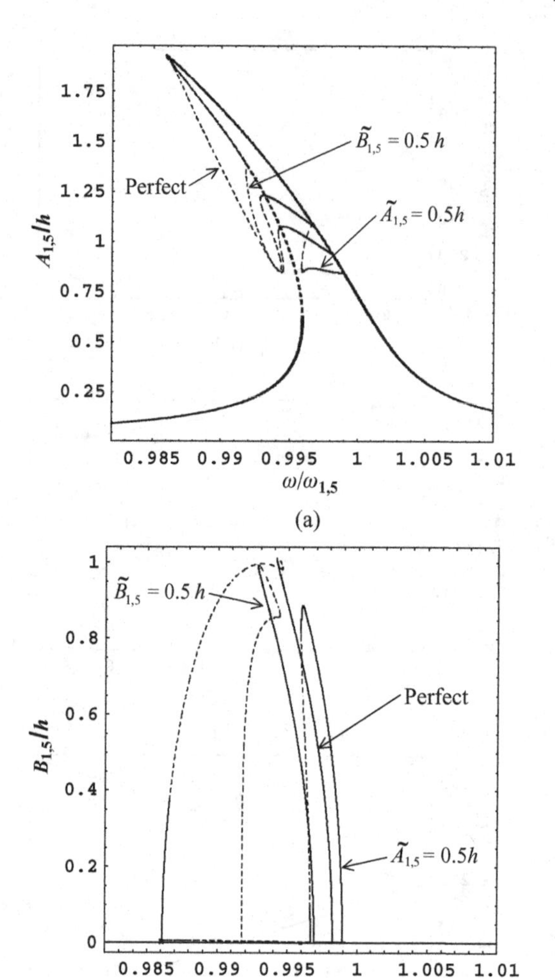

Figure 5.17. Frequency-response curve of the resonant generalized coordinates for the fundamental mode of the empty shell; perfect shell and shell with geometric imperfections $\tilde{A}_{1,5}$ and $\tilde{B}_{1,5}$. (a) Driven mode $A_{1,5}(t)$. (b) Companion mode $B_{1,5}(t)$.

5.7.2 Water-Filled Shell

Calculations have been performed for the fundamental mode ($n = 5, m = 1$) of the water-filled shell tested in the experiments. Figure 5.18 shows the natural frequency of the fundamental mode of the water-filled shell versus the amplitude of five different geometric imperfections: (i) axisymmetric imperfection $\tilde{A}_{1,0}$, (ii) ovalization imperfection $\tilde{A}_{1,2}$, (iii) asymmetric imperfection $\tilde{A}_{1,5}$ having the same shape of the fundamental mode, (iv) asymmetric imperfection $\tilde{A}_{1,10}$ having twice the number of circumferential waves of the fundamental mode and (v) asymmetric imperfections with 16 circumferential waves, which are not a multiple of $n = 5$. Results show that ovalization imperfections and asymmetric imperfections with 16 circumferential waves have a very small effect on the natural frequency of the fundamental mode; moreover, they do not split the double eigenvalue. Similar to the case of an empty shell, the imperfection giving the larger effect on natural frequency is $\tilde{A}_{1,10}$. Axisymmetric imperfection does not split the double eigenvalue associated with the fundamental mode; small positive imperfections (inward gaussian curvature) decrease the natural frequency, negative (outward gaussian curvature) imperfections increase it.

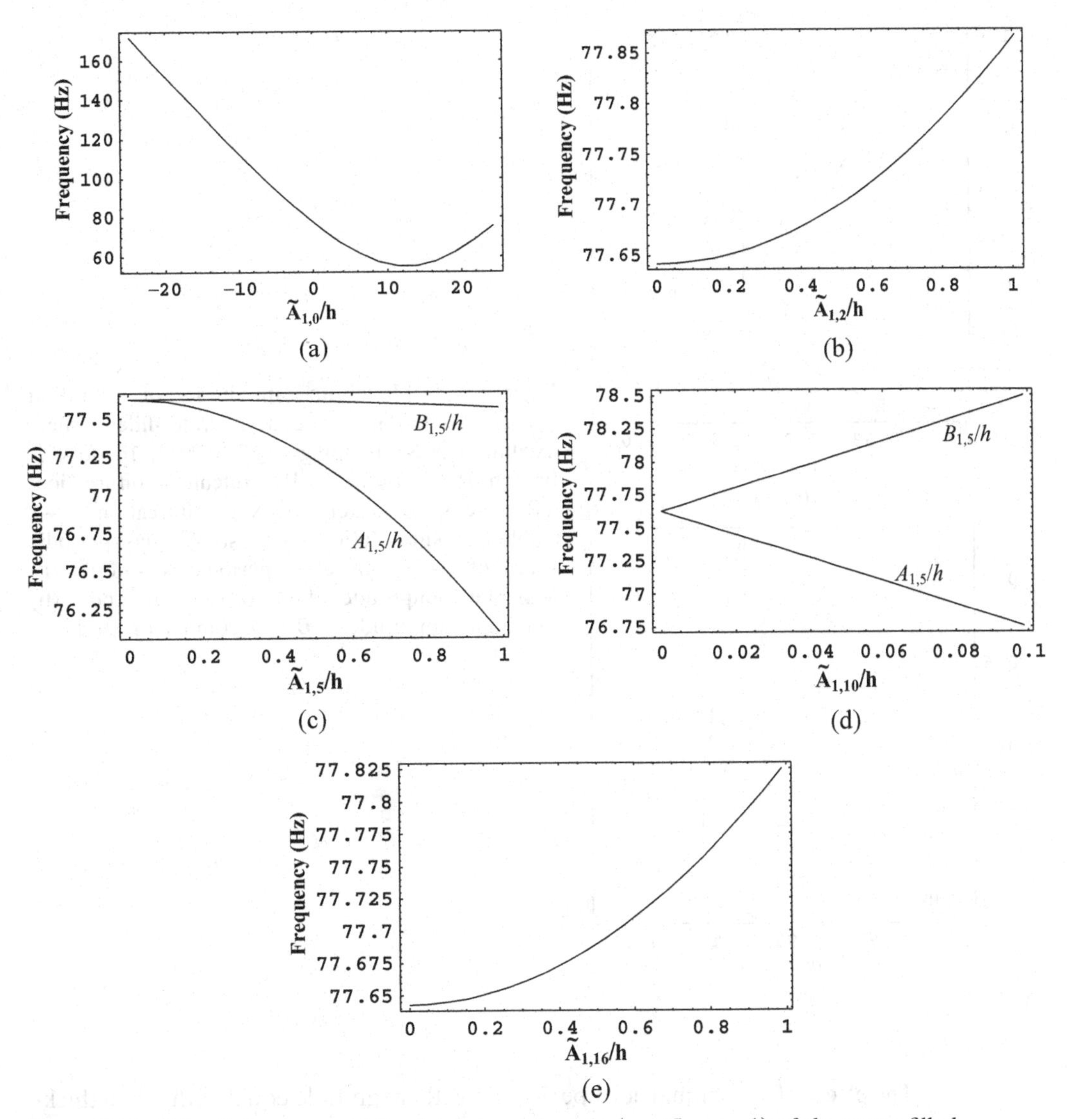

Figure 5.18. Natural frequency of the fundamental mode ($n = 5$, $m = 1$) of the water-filled shell versus the amplitude of geometric imperfections. (a) Axisymmetric imperfection $\tilde{A}_{1,0}$. (b) Ovalization imperfection $\tilde{A}_{1,2}$. (c) Asymmetric imperfection $\tilde{A}_{1,5}$. (d) Asymmetric imperfection $\tilde{A}_{1,10}$. (e) Asymmetric imperfection $\tilde{A}_{1,16}$.

The frequency-response relationship of the fundamental mode of the perfect, water-filled shell under harmonic point excitation of magnitude 3 N is given in Figure 5.19, assuming modal damping $\zeta_{1,5} = 0.0017$. The frequency-response relationship of the shell in Figure 5.19 shows a largely increased softening behavior, due to the contained water, with respect to the response of the empty shell given in Figure 5.16. Figure 5.19 presents a main branch 1 corresponding to zero amplitude of the companion mode $B_{1,5}(t)$. This branch has pitchfork bifurcations (BP) at $\omega/\omega_{1,5} = 0.9714$ and 1.0018 where branch 2 appears. This new branch corresponds to participation of both $A_{1,5}(t)$ and $B_{1,5}(t)$ that gives the traveling-wave response. Branch 2 undergoes two Neimark-Sacker (torus) bifurcations (TR), at $\omega/\omega_{1,5} = 0.9716$ and 0.9949. The response of the shell on branch 2 for $0.9716 < \omega/\omega_{1,5} < 0.9949$ is modulated in amplitude (quasi-periodic).

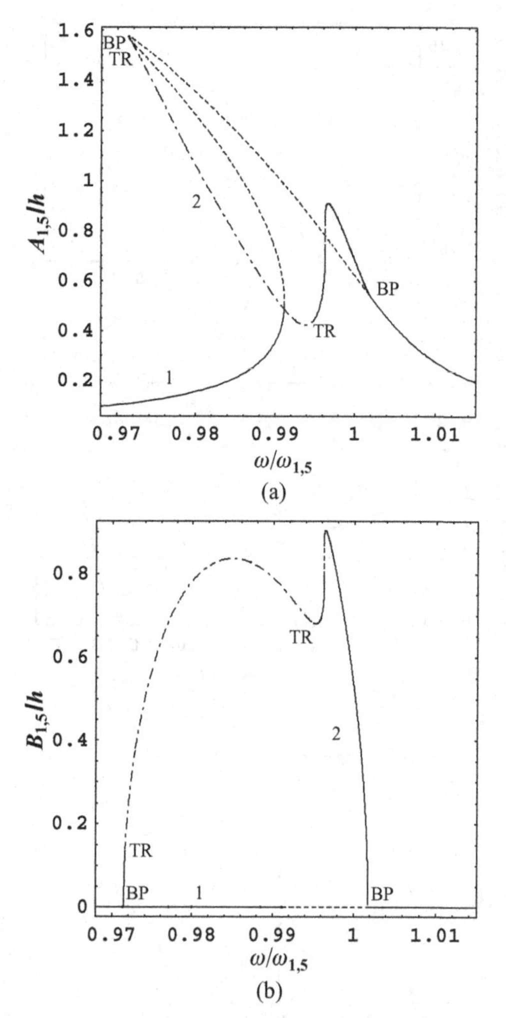

Figure 5.19. Maximum amplitude of vibration versus excitation frequency, water-filled shell, excitation 3 N, damping $\zeta_{1,5} = 0.0017$, 16 dofs. 1, branch 1; 2, branch 2; BP, pitchfork bifurcation; TR, Neimark-Sacker (torus) bifurcations. —, stable periodic solutions; –·–, stable quasi-periodic solutions; – –, unstable periodic solutions. (a) Maximum amplitude of $A_{1,5}(t)$, driven mode. (b) Maximum amplitude of $B_{1,5}(t)$, companion mode.

The effect of axisymmetric imperfection with magnitude equal to the shell thickness on the nonlinear response is limited, as shown in Figure 5.20. The ovalizaton of an amplitude twice as big as the shell thickness is even less important. However, asymmetric imperfections with the same shape of the resonant mode have a large effect on the nonlinear response, as shown in Figures 5.21 and 5.22. Similarly to what is observed in Figure 5.17 for the empty shell, the trend of nonlinearity is minimally affected in this case, but the traveling-wave response is completely modified. It is easy to conclude that imperfections having the same shape as the mode excited and with the same order of magnitude as the shell thickness (the magnitude $0.5h$ has been simulated here) are able to change the shell response around a resonance almost completely.

5.8 Comparison of Numerical and Experimental Results

Tests have been conducted on a commercial circular cylindrical shell made of stainless steel and having a longitudinal seam weld; dimensions and material properties are those given at the beginning of Section 5.7. Two stainless-steel annular plates with

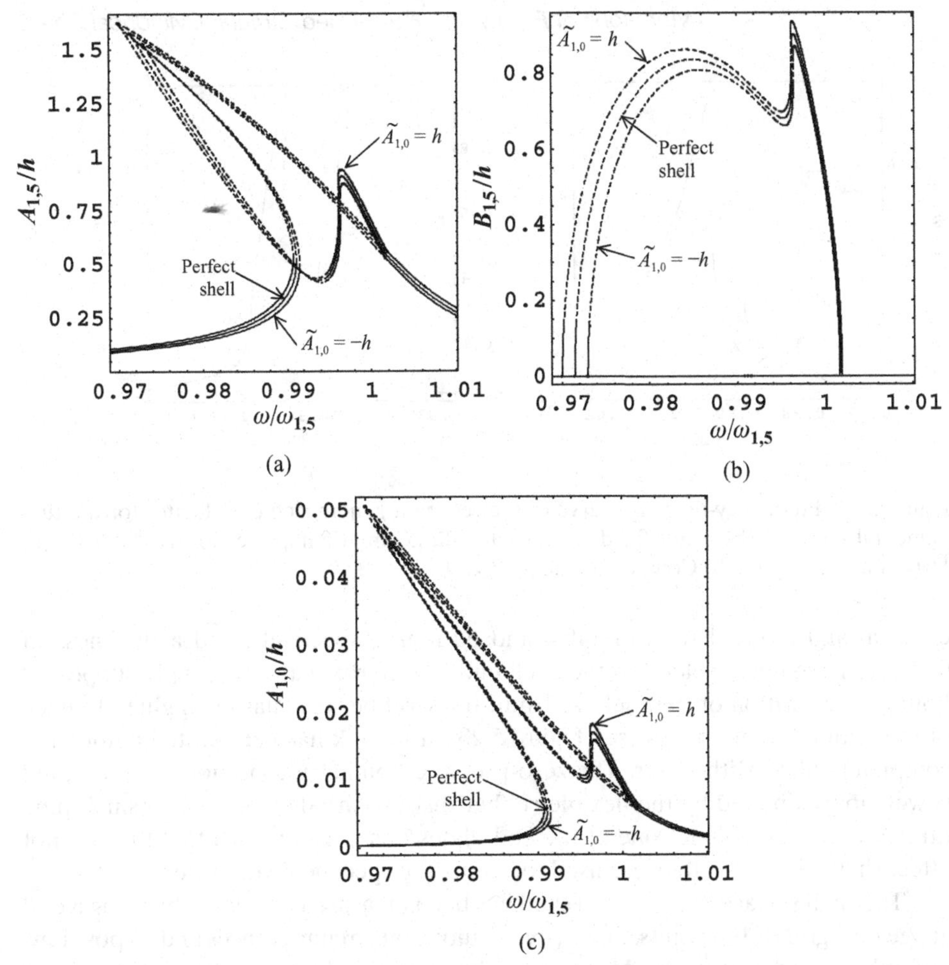

Figure 5.20. Frequency-response curve of the main generalized coordinates for the fundamental mode of the water-filled shell; perfect shell and shell with axisymmetric geometric imperfections $\tilde{A}_{1,0} = \pm h$. (a) Driven mode $A_{1,5}(t)$. (b) Companion mode $B_{1,5}(t)$. (c) First axisymmetric mode $A_{1,0}(t)$.

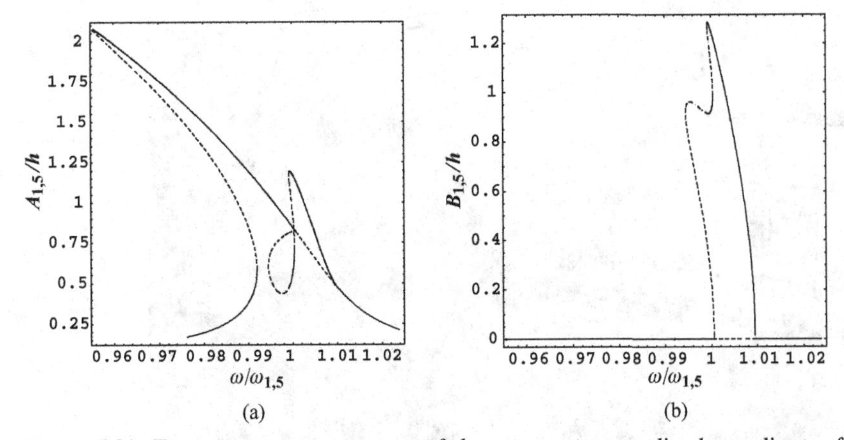

Figure 5.21. Frequency-response curve of the resonant generalized coordinates for the fundamental mode of the water-filled shell; shell with geometric imperfection $\tilde{A}_{1,5} = 0.5h$. (a) Driven mode $A_{1,5}(t)$. (b) Companion mode $B_{1,5}(t)$.

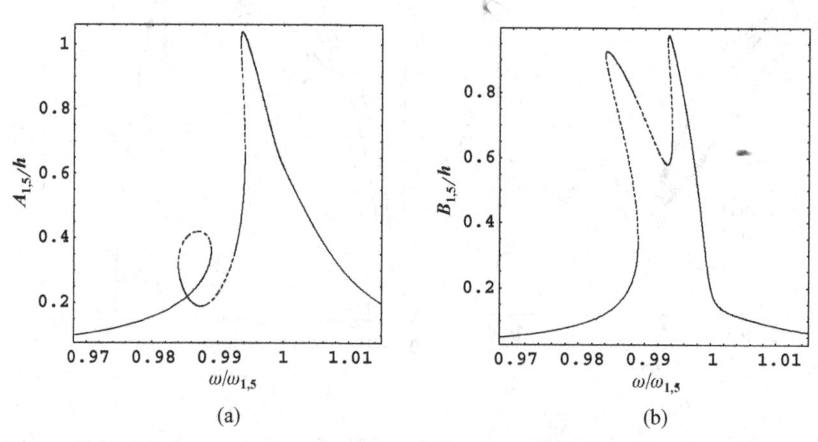

Figure 5.22. Frequency-response curve of the resonant generalized coordinates for the fundamental mode of the water-filled shell; shell with geometric imperfection $\tilde{B}_{1,5} = 0.5h$. (a) Driven mode $A_{1,5}(t)$. (b) Companion mode $B_{1,5}(t)$.

external and internal radii of 149.4 and 60 mm, respectively, and a thickness of 0.25 mm have been welded to the shell ends to approximate the simply supported boundary condition of the shell. A 1-mm-thick rubber disk has been glued to each of these annular end plates (see Figure 5.23). The tank has been tested empty and completely filled with water. The zero-pressure boundary condition for the liquid is well approximated by the flexible rubber disks at the shell ends. Two small pipe fittings were welded onto one of the end plates in a position such that they do not affect shell vibrations; they are used for filling the specimen with water.

The shell surface of the tested shell has been measured on a lathe by using a dial gauge on a grid of 100 points, that is, five equidistant circumferences and 20 positions around each circumference. Moreover, a fine grid of additional 68 points has been

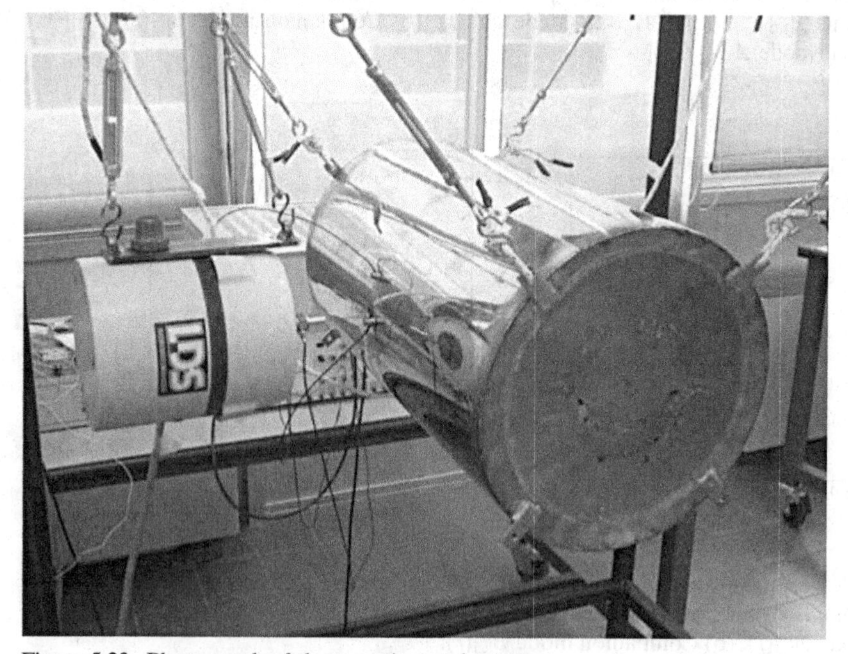

Figure 5.23. Photograph of the experimental setup.

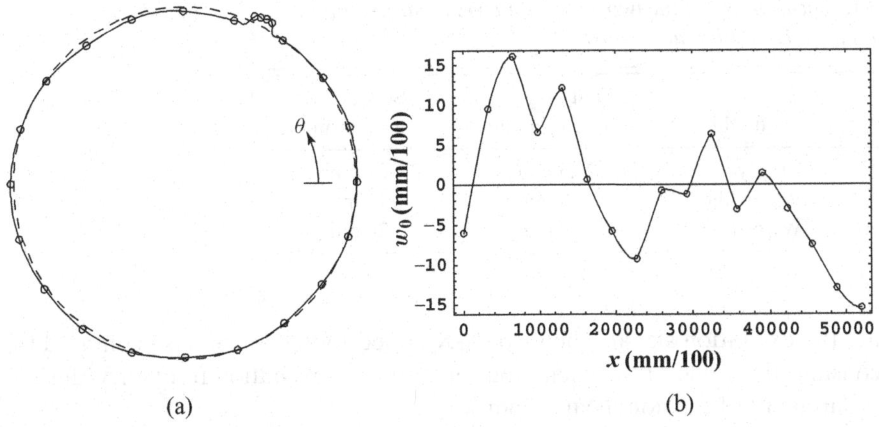

Figure 5.24. Geometric imperfections of the shell surface. (a) Central circumference (imperfections magnified 10 times). (b) Line at $\theta = 60°$. ∘, measured point; —, Fourier interpolation; – –, perfect geometry.

measured around the longitudinal weld where small deformations of the shell are present. Figure 5.24(a) shows the geometry of the measured central circumference with the magnitude of detected imperfection magnified by a factor 10; in particular, both measured points and their associated Fourier interpolation are given. The origin of the angular coordinate θ is taken at the excitation point. The geometric imperfections in the central circumference are mainly localized around the weld at $\theta = 63°$, as shown in Figure 5.24(a), with a protuberance of about 0.15 mm. The Fourier series interpolating the measured points of Figure 5.24(a) have been computed. Obviously, Fourier coefficients from measurements on other circumferences are slightly different. Geometric imperfections in the longitudinal direction are plotted in Figure 5.24(b) at $\theta = 60°$, that is, 3° ahead of the longitudinal welding. Results show that deformations are mainly concentrated at the longitudinal weld and at the shell ends, where the annular end plates have been welded to the shell.

The shell has been suspended horizontally with cables to a box-type frame and has been subjected to (i) burst-random excitation to identify the natural frequencies and perform a modal analysis by measuring the shell response on a grid of points, and to (ii) harmonic excitation, increasing or decreasing by very small steps the excitation frequency in the spectral neighborhood of the lowest natural frequencies to characterize nonlinear responses in presence of large-amplitude vibrations. The excitation has been provided by an electrodynamical exciter (shaker), model LDS V406 with power amplifier LDS PA100E, connected to the shell by a stinger glued in a position close to the middle of the shell, specifically at $x = 251$ mm. A piezoelectric force transducer, model B&K 8200, of mass 21 g, placed on the shaker and connected to the shell by a stinger, measured the force transmitted. The shell response has been measured by using two accelerometers, model B&K 4393, of mass 2.4 g. For all nonlinear tests, the two accelerometers have been glued close to the middle of the shell length, specifically at $x = 264$ mm, at different angular positions corresponding to an antinode and a node of the excited driven mode to measure the nonlinear response. The specific locations of the accelerometers are given in Table 5.1 for the different modes investigated. The time responses have been measured by using the Difa Scadas II front-end connected to a workstation with the software CADA-X of LMS for signal processing and data analysis; the same front-end has been used to

Table 5.1. *Location $R\theta$ of the two accelerometers in large-amplitude vibration tests. Excitation at origin $\theta = 0$*

Mode	Shell filling	First accelerometer	Second accelerometer
$n = 5$	Empty	−26.5 mm	21 mm
$n = 5$	Water-filled	30 mm	−17 mm
$n = 10$	Water-filled	0 mm	24 mm

generate the excitation signal. The CADA-X closed-loop control has been used to keep constant the value of the excitation force for any excitation frequency, during the measurement of the nonlinear response.

5.8.1 Empty Shell

The fundamental mode of the empty shell has been measured to be divided to a couple of orthogonal modes having the same shape ($n = 5$, $m = 1$) but slightly different frequency, 209.33 and 212.6 Hz. The small difference in the frequency of the two modes is attributed to imperfections of the test specimen, previously reported, and to the mass of the sensors glued to the shell. In addition, imperfections affect the position of the mode shapes on the circumference with respect to the excitation point, making it more difficult to find nodes. Without imperfections, the first of the couple of modes presents an antinode at the excitation point; the second mode presents a node there. The node closest to the excitation point has been experimentally determined for each of these two modes, and an accelerometer has been placed there. Therefore, each one of the two accelerometers is at a node of one mode, corresponding to an antinode of the other, orthogonal mode. The location of the accelerometers is given in Table 5.1. It can be immediately observed that both accelerometers are considerably far from the excitation point. Therefore, both modes are directly driven by the excitation. The projections of the excitation force on $A_{1,5}(t)$ and $B_{1,5}(t)$ give $0.64\tilde{f}$ and $0.768\tilde{f}$, respectively.

Figures 5.25(a,b) show the experimental accelerations measured by the two accelerometers around the fundamental frequency versus the excitation frequency for four different force levels: 0.025, 0.25, 0.5 and 0.75 N. The closed-loop control used in the experiments keeps the excitation force constant after filtering the signal from the load cell in order to use only the harmonic component with the given excitation frequency. The measured accelerations reported in Figure 5.25 have been filtered in order to eliminate any frequency except the excitation frequency. Experiments have been performed both increasing and decreasing the excitation frequency; the frequency resolution used in this case is 0.025 Hz. The hysteresis between the two curves (up = increasing frequency; down = decreasing frequency) is clearly visible. Sudden increments (jumps) of the vibration amplitude are observed increasing and decreasing the excitation frequency. These are characteristic of a weak softening-type nonlinearity. When the vibration amplitude is equal to 0.7 times the shell thickness, the peak of the response appears for a frequency that is lower by about 0.37% with respect to the linear one (i.e. the one measured with force 0.025 N). The traveling-wave response around the shell, associated with significant acceleration measured by both accelerometers at the same frequency, is not observed in this case because

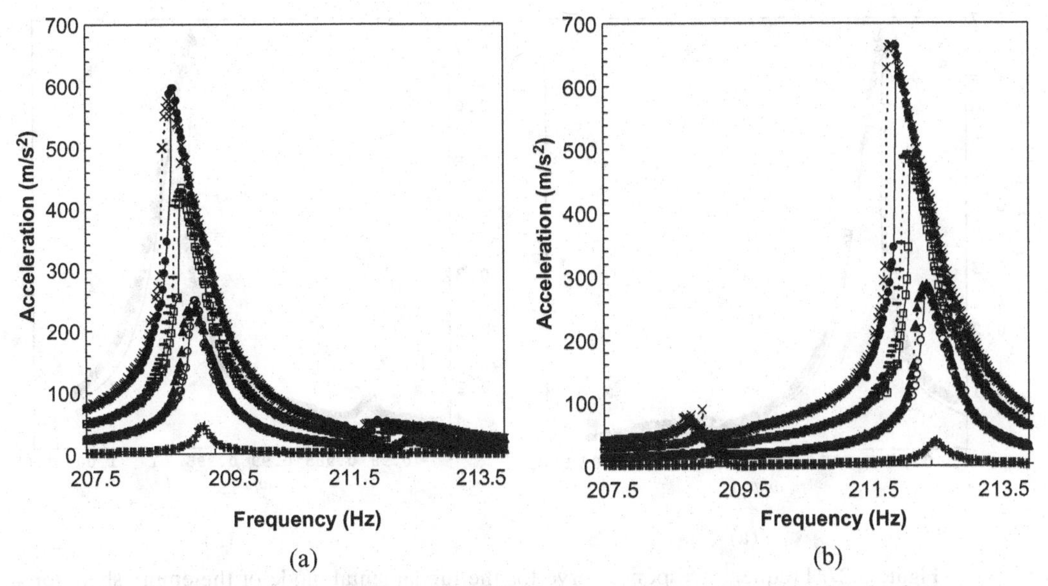

Figure 5.25. Experimentally measured acceleration versus excitation frequency for the fundamental mode of the empty shell; (a) first accelerometer. (b) Second accelerometer. – + –, force 0.025 N; –○–, force 0.25 N up (i.e. increasing the excitation frequency); --▲--, force 0.25 N down (i.e. decreasing the excitation frequency); –□–, force 0.5 N up; -----, force 0.5 N down; –•–, force 0.75 N up; – – × – –, force 0.75 N down.

the separation of the two modes with the same shape is too large for the maximum force level reached in the experiments.

The phase relationships between the two accelerations (measured positive outward) and the force input (measured positive inward) are given in Figure 5.26 for excitation of 0.75 N.

The measured accelerations have been converted to displacements, dividing by the excitation circular frequency squared, and have been plotted in Figure 5.27 together with the theoretical responses for the case with a force level of 0.75 N; normalization of frequency with respect to $\omega_{1,5} = 212.6 \times 2\pi$ has been performed. When comparing filtered experimental results to theoretical results, only coordinates $A_{1,5}(t)$ and $B_{1,5}(t)$ must be considered, because the others, which have higher

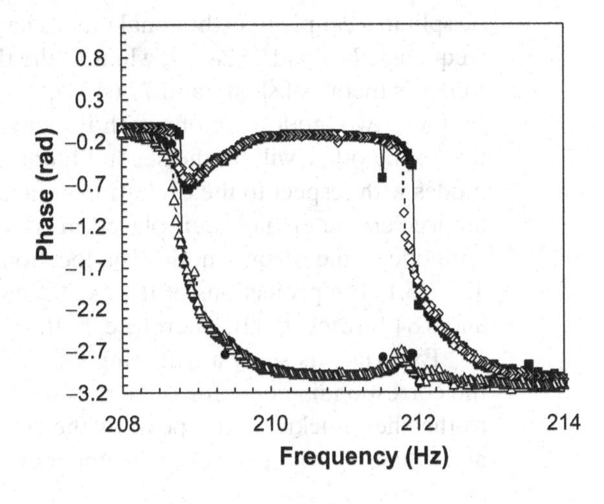

Figure 5.26. Experimentally measured response phase-frequency curves for the fundamental mode of the empty shell; force 0.75 N. –•–, first accelerometer up; –■–, second accelerometer up; –Δ–, first accelerometer down; –◇–, second accelerometer down.

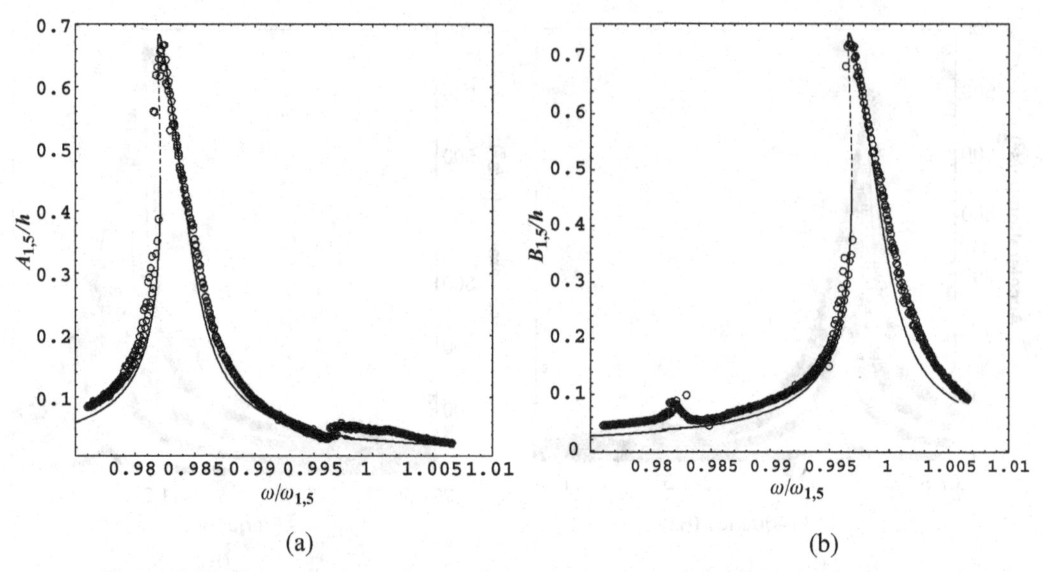

Figure 5.27. Frequency-response curve for the fundamental mode of the empty shell; force 0.75 N. ○, experimental data; —, stable theoretical solutions; - - -, unstable theoretical solutions. (a) Displacement/h from the first accelerometer. (b) Displacement/h from the second accelerometer.

frequencies, are eliminated by the filter. The theoretical curves have been computed by including the geometric imperfection $\tilde{A}_{1,10} = 0.072h$, which is the value needed to reproduce the frequency separation between the two modes ($n = 5, m = 1$); this value is not too far from the geometric imperfection with 10 circumferential waves measured on the central circumference, which is $0.11h$. The modal damping used to calculate the theoretical curves is $\zeta_{1,5} = 0.00077$ (0.077%) for $A_{1,5}$ and $\zeta_{1,5} = 0.00084$ (0.084%) for $B_{1,5}$ and have been identified by matching the maximum value of the measured response. The agreement between the theoretical curves and the experimental results is particularly good.

5.8.2 Water-Filled Shell

The fundamental mode ($n = 5, m = 1$) of the water-filled shell has been measured to be split in a couple of orthogonal modes having the same shape but slightly different frequency, 74.9 and 76.28 Hz, whereas the theoretical value is 76.16 Hz, according to Flügge's theory of shells, and 77.64 Hz, according to Donnell's shallow-shell theory. Of these two modes, the one with the lowest frequency will be referred as the first mode; the other will be the second mode. The location of the shapes of these two modes with respect to the excitation point has been experimentally determined and an accelerometer has been placed at a node of each mode, corresponding to an antinode of the second mode. The locations of the two accelerometers are given in Table 5.1. The projections of the excitation force on $A_{1,5}(t)$ and $B_{1,5}(t)$ give $0.543\tilde{f}$ and $0.84\tilde{f}$, respectively. Therefore, both modes are directly excited.

Experiments show a softening-type nonlinearity that is much stronger than in the corresponding case for the empty shell. When the vibration amplitude is equal to the shell thickness, the peak of the response appears for a frequency lower by more than 1% with respect to the linear one (i.e. the one measured with force 0.1 N).

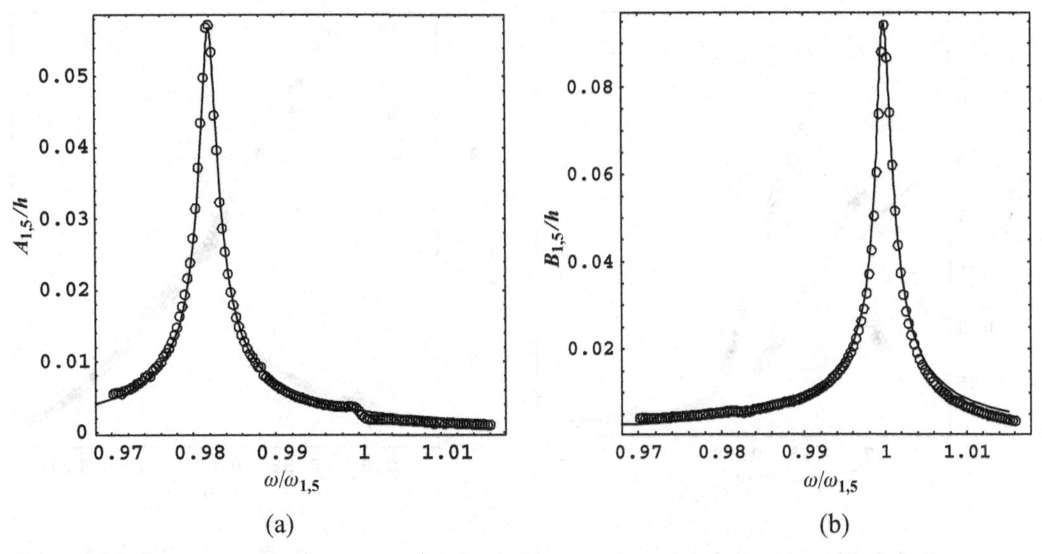

(a) (b)

Figure 5.28. Frequency-response curve for the fundamental mode of the water-filled shell; force 0.1 N, which gives linear behavior. ∘, experimental data; —, stable theoretical solutions. (a) Displacement/h from the first accelerometer. (b) Displacement/h from the second accelerometer.

The traveling-wave response around the shell has been detected in a relatively large frequency range around the resonance, especially for the excitation force of 6 N.

The measured accelerations have been converted to displacements, dividing by the excitation circular frequency squared, and have been plotted in Figures 5.28–5.30 together with the theoretical responses for the case with force levels of 0.1 (linear case), 3 and 6 N, respectively; normalization of frequency with respect to

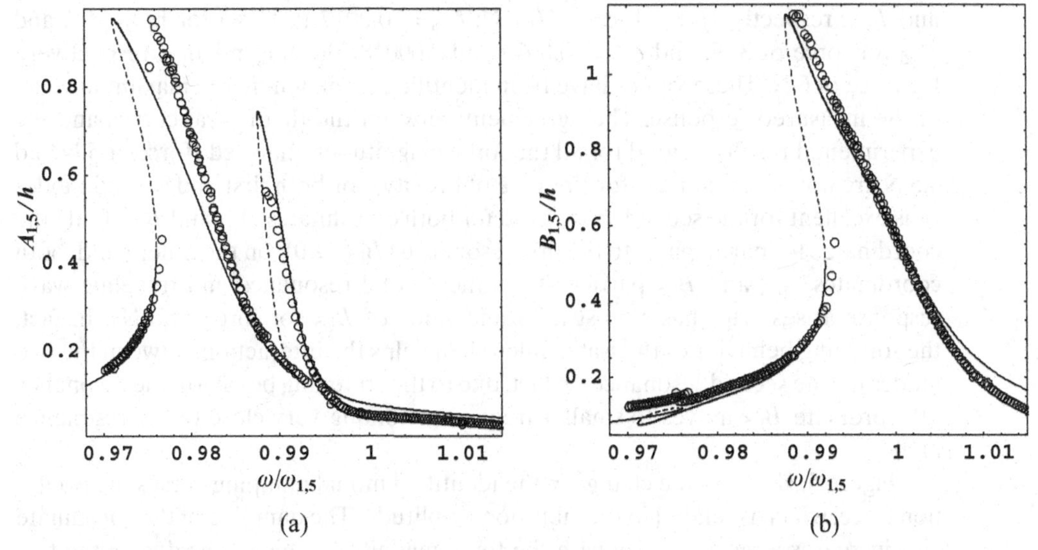

(a) (b)

Figure 5.29. Frequency-response curve for the fundamental mode of the water-filled shell; force 3 N. ∘, experimental data; —, stable theoretical solutions; – – –, unstable theoretical solutions. (a) Displacement/h from the first accelerometer. (b) Displacement/h from the second accelerometer.

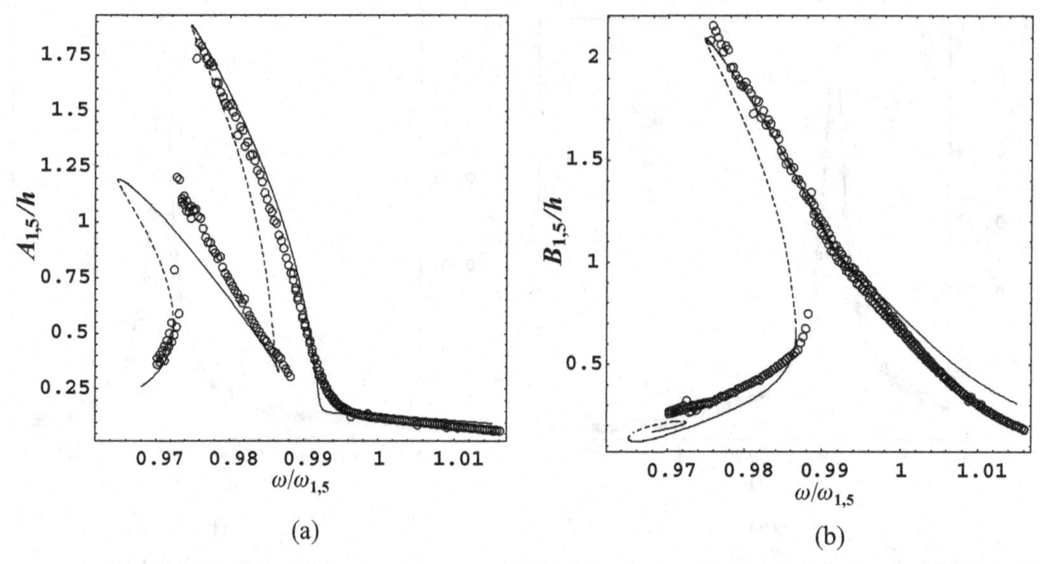

Figure 5.30. Frequency-response curve for the fundamental mode of the water-filled shell; force 6 N. ∘, experimental data; —, stable theoretical solutions; – – –, unstable theoretical solutions. (a) Displacement/h from the first accelerometer. (b) Displacement/h from the second accelerometer.

$\omega_{1,n} = 76.28 \times 2\pi$ has been performed. The coordinate $A_{1,5}$ has resonance for the first mode; $B_{1,5}$ has resonance for the second mode. The theoretical curves have been computed by including the geometric imperfection $\tilde{A}_{1,10} = 0.0817h$, which is the value needed to reproduce the frequency difference between the two modes ($n = 5$, $m = 1$). This value is close enough to the geometric imperfection with 10 circumferential waves measured on the central circumference, which is $0.11h$. The modal damping used to calculate the theoretical curves is $\zeta_{1,5} = 0.001$ and 0.00091 for $A_{1,5}$ and $B_{1,5}$, respectively, for force of 0.1 N; $\zeta_{1,5} = 0.0017$ (0.17%) for both $A_{1,5}$ and $B_{1,5}$ for force of 3 N; and $\zeta_{1,5} = 0.0029$ and 0.00058 for $A_{1,5}$ and $B_{1,5}$, respectively, for force of 6 N. These values have been identified by matching the maximum value of the measured response. The agreement between the theoretical curves and the experimental results is good for all the force magnitudes analyzed (forces 1.5 N and 4.5 N are not reported here for the sake of brevity) for both first and second modes (it is excellent for the second mode) and for both coordinates $A_{1,5}$ and $B_{1,5}$. Only the coordinate $A_{1,5}$ participates to the first resonance ($B_{1,5} \cong 0$); on the other hand, both coordinates $A_{1,5}$ and $B_{1,5}$ participate to the second resonance and traveling-wave response arises ($A_{1,5}$ has almost the same value of $B_{1,5}$ for force of 6 N). In fact, the softening behavior of the water-filled shell helps the interaction between the two modes for the second resonance. In fact, due to the softening behavior, the resonance of coordinate $B_{1,5}$ moves to smaller frequency, coming very close to the resonance of $A_{1,5}$.

Figure 5.31 shows the change in the identified modal damping versus the excitation force, which is related to the vibration amplitude. The damping of the coordinate $A_{1,5}$ increases almost linearly with the force magnitude; this seems due to the fact that no traveling-wave response is observed for the resonance of $A_{1,5}$. Damping coefficients of coordinates $A_{1,5}$ and $B_{1,5}$ have similar values for forces of 0.1 and 1.5 N

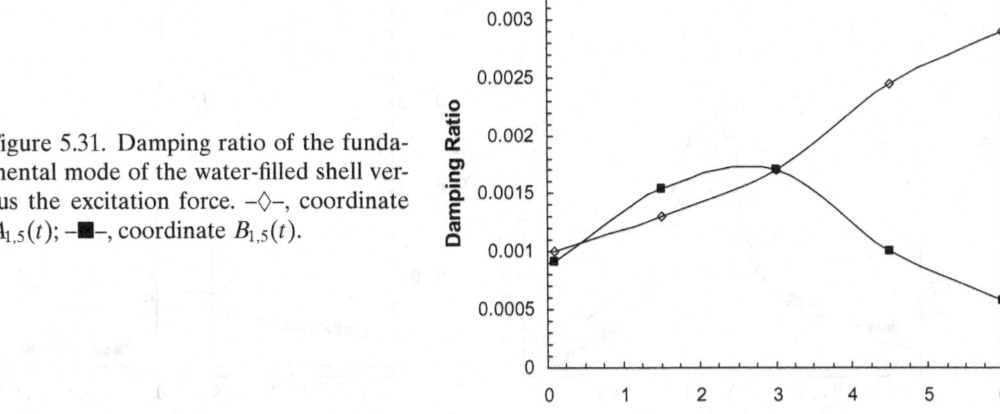

Figure 5.31. Damping ratio of the fundamental mode of the water-filled shell versus the excitation force. -◇-, coordinate $A_{1,5}(t)$; -■-, coordinate $B_{1,5}(t)$.

and have exactly the same value for the force of 3 N. For this force magnitude the traveling-wave response becomes significant (the amplitude of $A_{1,5}$ becomes about one-half of $B_{1,5}$ around resonance of the second mode) and damping on the coordinate $B_{1,5}$ largely decreases for larger forces.

Then, mode ($n = 10, m = 1$) is investigated; results for additional modes are presented by Amabili (2003a). This case is interesting because the measured frequency separation between the couple of modes is very small. Moreover, the positions of nodes and antinodes where the two accelerometers have been placed (see Table 5.1) show that $A_{1,5}(t)$ is directly excited (driven mode) and $B_{1,5}(t)$ is not excited (companion mode). Therefore, instead of using small geometric imperfections to perfectly reproduce the experimental results, calculations have been performed for a perfect shell. The measured accelerations have been converted to displacements, dividing by the excitation circular frequency squared, and have been plotted in Figure 5.32 together with the theoretical responses for the case with force level 1.5 N. Normalization of frequency with respect to $\omega_{1,10} = 246.8 \times 2\pi$ has been performed. The modal damping used to calculate the theoretical curves is $\zeta_{1,10} = 0.0023$ for $A_{1,10}$ and $\zeta_{1,10} = 0.0025$ for $B_{1,10}$. The agreement between the theoretical and experimental results is strong. Around the resonance the amplitudes of $A_{1,10}(t)$ and $B_{1,10}(t)$ are close, with $B_{1,10}(t)$ smaller than $A_{1,10}(t)$; in this frequency range, the shell response is a traveling wave around the circumference. Before the appearance of the pitchfork bifurcation giving rise to branch 2, the measured acceleration on the second accelerometer, corresponding to $B_{1,10}(t)$, should be zero. However, because of the finite diameter of the sensor, a small contribution of $A_{1,10}(t)$ is recorded so that the measured vibration level is small but not exactly zero.

The measured response of the shell for force of 3 N is shown in Figure 5.33. It is particularly interesting to note that amplitudes of $A_{1,10}(t)$ and $B_{1,10}(t)$ are coincident in this case in a frequency range of almost 2 Hz around the resonance. The companion mode participation grows with the vibration amplitude (and the magnitude of the excitation) from zero to the driven mode curve. After this point, driven and companion mode amplitudes increase simultaneously, giving rise to a pure traveling-wave response. The measured accelerations, converted to displacements, have been plotted

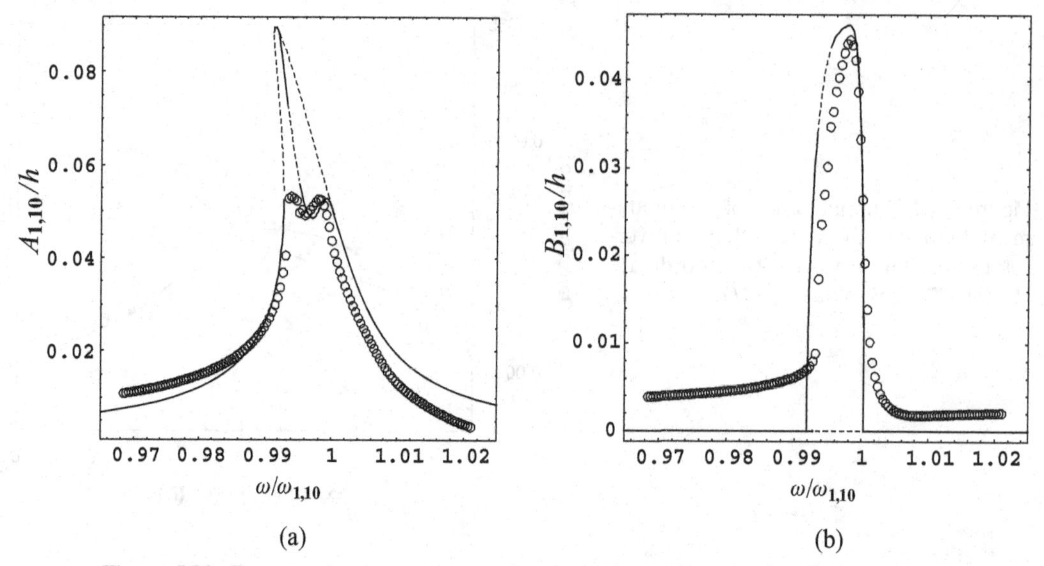

Figure 5.32. Frequency-response curve for mode ($n = 10$, $m = 1$) of the water-filled shell; force 1.5 N. o, experimental data; —, stable theoretical solutions; – – –, unstable theoretical solutions. (a) Displacement/h from the first accelerometer. (b) Displacement/h from the second accelerometer.

in Figure 5.34 together with the theoretical responses for the case with a force level of 3 N. The modal damping used to calculate the theoretical curves is $\zeta_{1,10} = 0.003$ for both $A_{1,10}$ and $B_{1,10}$. The agreement between the theoretical and experimental results is good. The main branch 1 corresponds to zero vibration amplitude of the companion mode $B_{1,10}(t)$. This branch has pitchfork bifurcations at $\omega/\omega_{1,10} = 0.9791$ and 1.0043 where branch 2 appears. Branch 2 loses stability through two Neimark-Sacker (torus) bifurcations at $\omega/\omega_{1,10} = 0.9795$ and 0.9944. No stable periodic response is indicated

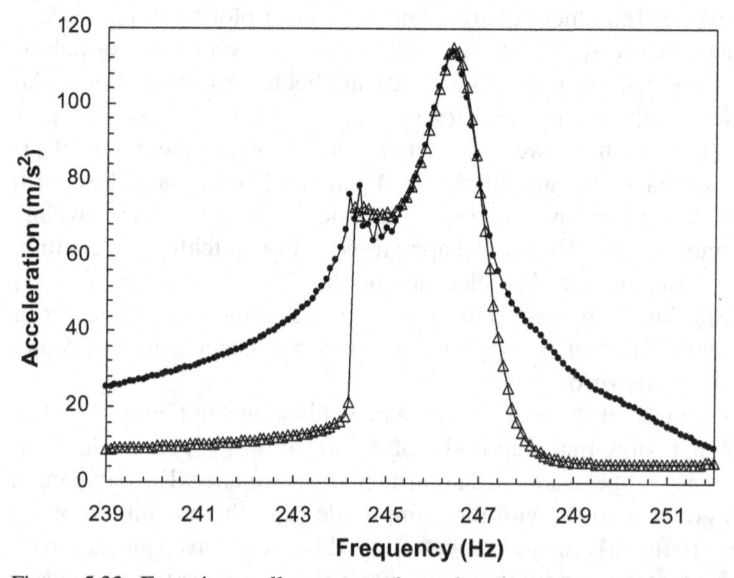

Figure 5.33. Experimentally measured acceleration versus excitation frequency for modes ($n = 10$, $m = 1$) of the water-filled shell; force 3 N. -•-, first accelerometer; -Δ-, second accelerometer.

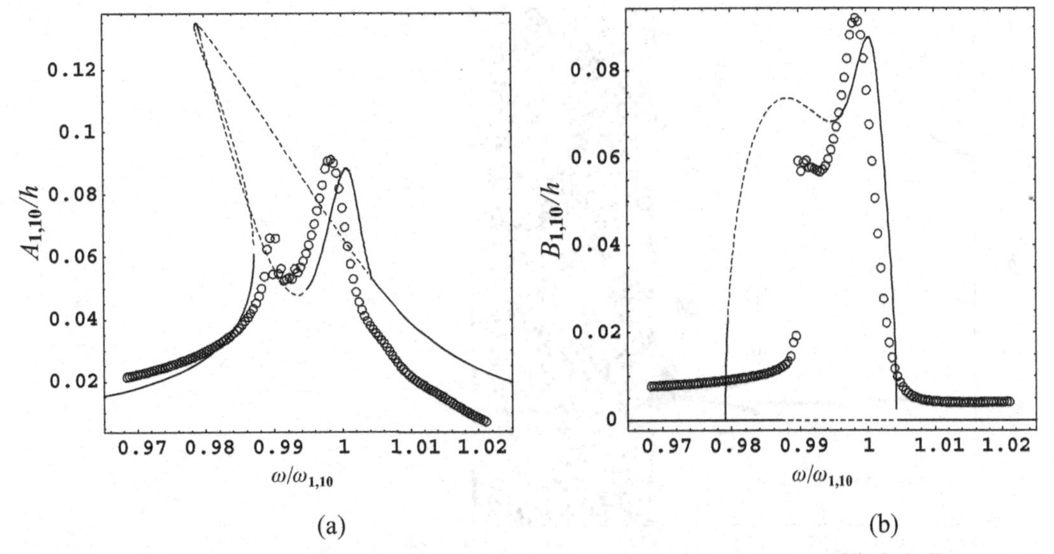

Figure 5.34. Frequency-response curve for mode ($n = 10$, $m = 1$) of the water-filled shell; force 3 N. ∘, experimental data; —, stable theoretical solutions; – – –, unstable theoretical solutions. (a) Displacement/h from the first accelerometer. (b) Displacement/h from the second accelerometer.

in Figure 5.34 in the frequency range between the two Neimark-Saker bifurcations; in fact, the response is quasi-periodic.

5.9 Chaotic Vibrations of a Water-Filled Shell

The same shell described at the beginning of Section 5.7 is considered. The shell is assumed to be perfect, simply supported and water-filled with open ends. The 16 dofs model has been used.

Poincaré maps have been computed by direct integration of the equations of motion. The excitation frequency ω has been kept constant, $\omega = 0.92\,\omega_{1,5}$ (the shell displays softening-type response; therefore, for large excitation, the resonance is obtained for $\omega < \omega_{1,5}$), and the excitation amplitude has been varied between 0 and 600 N. The force range has been divided into 500 steps, so that the force is varied in steps of 1.2 N. Each time the force is changed by a step, 500 periods have been allowed to elapse in order to eliminate the transient motion. The initial condition at the first step is zero displacement and velocity for all the variables. In the following steps, the solution at the previous step, with addition of a small perturbation in order to find a stable solution, is used as the initial condition. The bifurcation diagrams obtained by all these Poincaré maps by using the conventional Galerkin model are shown in Figures 5.35 and 5.36. In particular, in Figure 5.35 the load is increased from 0 to 600 N; in Figure 5.36 the load is decreased from 600 N to 0. Simple periodic motion, a period-doubling bifurcation, subharmonic response, amplitude modulations and chaotic response have been detected, as shown in Figures 5.35(a,b) and 5.36(a,b). This indicates the existence of complex nonlinear dynamics for the circular cylindrical shell subject to large harmonic excitation. Comparison of Figures 5.35 and 5.36, where only stable solutions are reported, shows that only in some regions the shell exhibits the same behavior when the load (bifurcation parameter) is increased or decreased. This

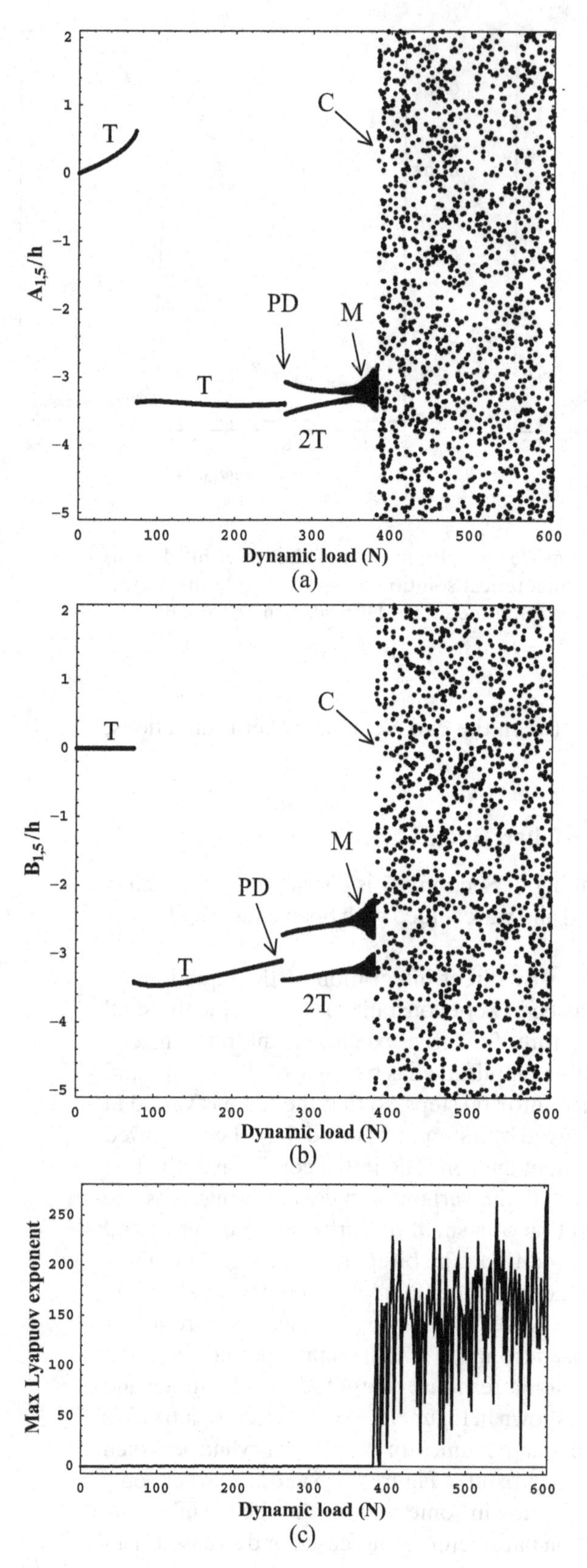

Figure 5.35. Bifurcation diagram of Poincaré maps and maximum Lyapunov exponent for the water-filled shell under increasing harmonic load with frequency $\omega = 0.92\,\omega_{1,5}$; conventional Galerkin model, 16 dofs. (a) Generalized coordinate $A_{1,5}(t)$, driven mode; T, response period equal to excitation period; PD, period-doubling bifurcation; 2T, periodic response with two times the excitation period; M, amplitude modulations; C, chaos. (b) Generalized coordinate $B_{1,5}(t)$, companion mode. (c) Maximum Lyapunov exponent.

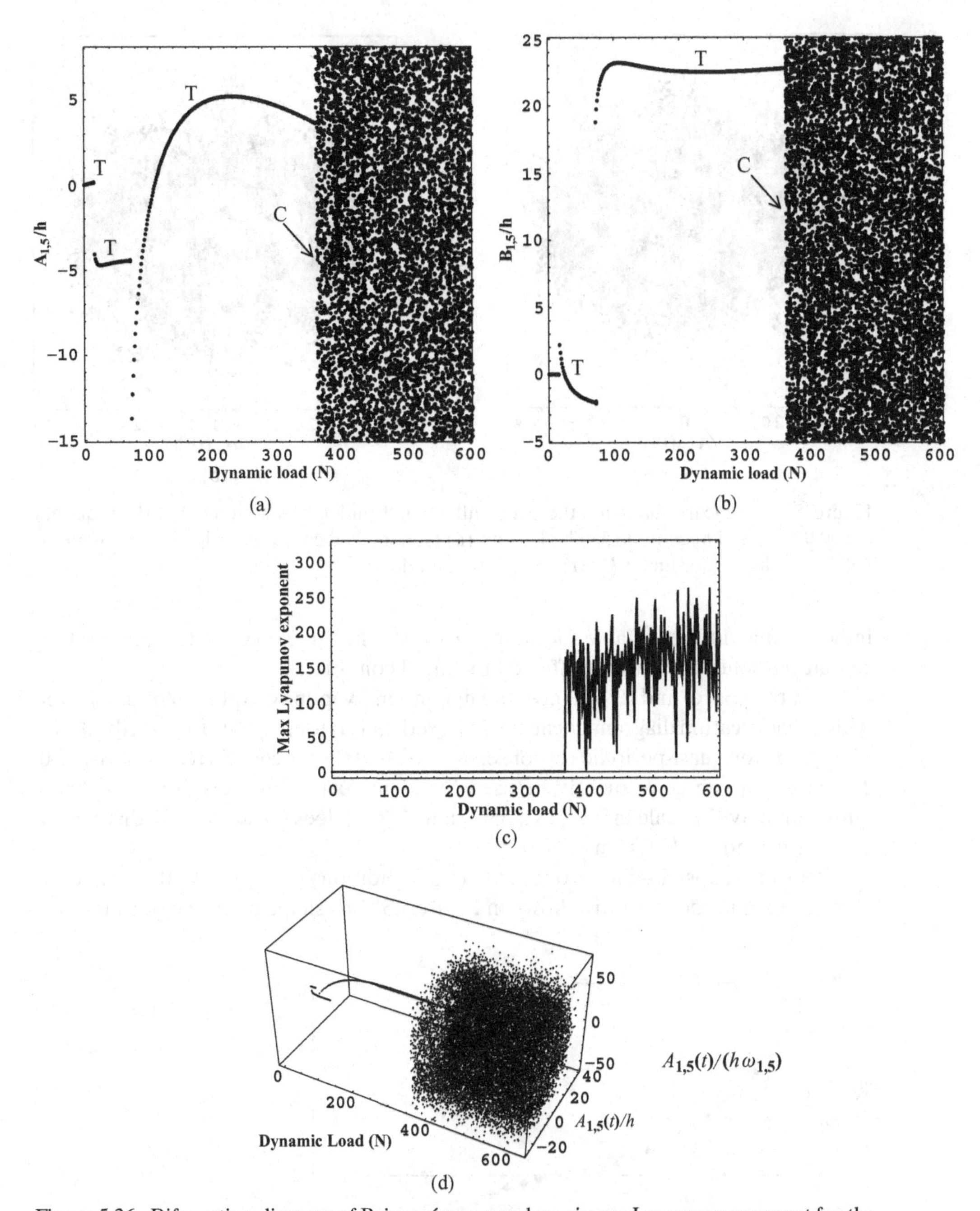

Figure 5.36. Bifurcation diagram of Poincaré maps and maximum Lyapunov exponent for the water-filled shell under decreasing harmonic load with frequency $\omega = 0.92\,\omega_{1,5}$; 16 dofs. (a) Generalized coordinate $A_{1,5}(t)$, driven mode; T, response period equal to excitation period; C, chaos. (b) Generalized coordinate $B_{1,5}(t)$, companion mode. (c) Maximum Lyapunov exponent. (d) Three-dimensional representation of the bifurcation diagram: generalized coordinate $A_{1,5}(t)$, enlarged scale.

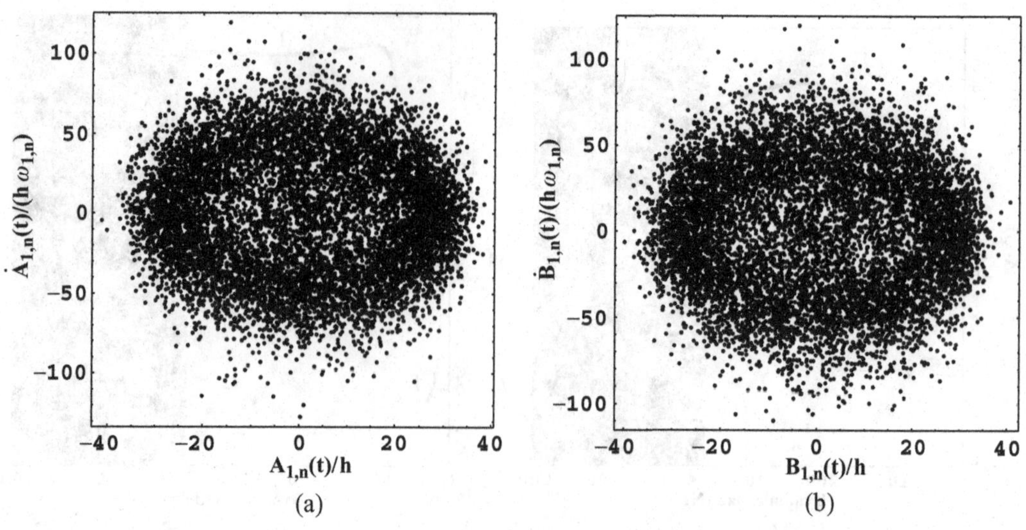

Figure 5.37. Poincaré maps for the water-filled shell under harmonic load with frequency $\omega = 0.92\,\omega_{1,n}$ and magnitude 550 N; 16 dofs. (a) Generalized coordinate $A_{1,5}(t)$, driven mode. (b) Generalized coordinate $B_{1,5}(t)$, companion mode.

indicates that different stable solutions coexist for the same set of system parameters, so that the solution is largely affected by initial conditions.

Figures 5.35(c) and 5.36(c) give the maximum Lyapunov exponent σ_1 associated with the bifurcation diagram. It can be observed that (i) for periodic forced vibrations, $\sigma_1 < 0$, (ii) for quasi-periodic response, $\sigma_1 = 0$ and (iii) for chaotic response, $\sigma_1 > 0$. Therefore, σ_1 can conveniently be used for identification of the system dynamics. Unfortunately, the scale in Figures 5.35(c) and 5.36(c) does not allow us to distinguish $\sigma_1 < 0$ (around -0.13) from $\sigma_1 = 0$.

Poincaré maps obtained from zero initial conditions for $\omega/\omega_{1,n} = 0.92$ and excitation of magnitude 550 N are shown in Figure 5.37; the shape of the chaotic attractor

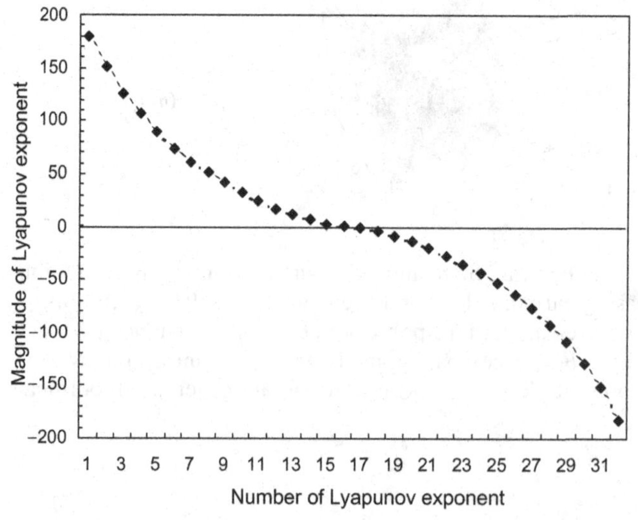

Figure 5.38. All the Lyapunov exponents for the water-filled shell under harmonic load with frequency $\omega = 0.92\,\omega_{1,n}$ and magnitude 550 N; 16 dofs.

is clear. All the Lyapunov exponents have been computed for this case and are given in Figure 5.38. It is evident that the response is associated with hyperchaos with 16 positive Lyapunov exponents, which is a very-high-dimensional chaos, as indicated by the Lyapunov dimension, which is 31.8. Such a large number of positive Lyapunov exponents for chaotic behavior of a structure (in the present model the fluid gives only an added inertia) do not seem to have been reported in the literature before. This indicates that the 16 dofs model is already at its limit for describing the chaotic response.

REFERENCES

M. Amabili 2003a *Journal of Sound and Vibration* **262**, 921–975. Theory and experiments for large-amplitude vibrations of empty and fluid-filled circular cylindrical shells with imperfections.

M. Amabili 2003b *Journal of Sound and Vibration* **264**, 1091–1125. Comparison of shell theories for large-amplitude vibrations of circular cylindrical shells: Lagrangian approach.

M. Amabili and R. Garziera 1999 *Journal of Sound and Vibration* **224**, 519–539. A technique for the systematic choice of admissible functions in the Rayleigh-Ritz method.

M. Amabili, R. Garziera and A. Negri 2002 *Journal of Fluids and Structures* **16**, 213–227. Experimental study on large-amplitude vibrations of water-filled circular cylindrical shells.

M. Amabili and M. P. Païdoussis 2003 *Applied Mechanics Reviews* **56**, 349–381. Review of studies on geometrically nonlinear vibrations and dynamics of circular cylindrical shells and panels, with and without fluid-structure interaction.

M. Amabili, F. Pellicano and M. P. Païdoussis 1998 *Journal of Fluids and Structures* **12**, 883–918. Nonlinear vibrations of simply supported, circular cylindrical shells, coupled to quiescent fluid.

M. Amabili, F. Pellicano and M. P. Païdoussis 1999a *Journal of Sound and Vibration* **225**, 655–699. Non-linear dynamics and stability of circular cylindrical shells containing flowing fluid. Part I: stability.

M. Amabili, F. Pellicano and M. P. Païdoussis 1999b *Journal of Sound and Vibration* **228**, 1103–1124. Non-linear dynamics and stability of circular cylindrical shells containing flowing fluid. Part II: large-amplitude vibrations without flow.

M. Amabili, F. Pellicano and M. P. Païdoussis 2000a *Journal of Sound and Vibration* **237**, 617–640. Non-linear dynamics and stability of circular cylindrical shells containing flowing fluid. Part III: truncation effect without flow and experiments.

M. Amabili, F. Pellicano and M. P. Païdoussis 2000b *Journal of Sound and Vibration* **237**, 641–666. Non-linear dynamics and stability of circular cylindrical shells containing flowing fluid. Part IV: large-amplitude vibrations with flow.

M. Amabili, F. Pellicano and A. F. Vakakis 2000c *ASME Journal of Vibration and Acoustics* **122**, 346–354. Nonlinear vibrations and multiple resonances of fluid-filled, circular shells. Part I: equations of motion and numerical results.

J. C. Chen and C. D. Babcock 1975 *AIAA Journal* **13**, 868–876. Nonlinear vibration of cylindrical shells.

H.-N. Chu 1961 *Journal of Aerospace Science* **28**, 602–609. Influence of large-amplitudes on flexural vibrations of a thin circular cylindrical shell.

E. J. Doedel, A. R. Champneys, T. F. Fairgrieve, Y. A. Kuznetsov, B. Sandstede and X. Wang 1998 *AUTO 97: Continuation and Bifurcation Software for Ordinary Differential Equations (with HomCont)*. Concordia University, Montreal, Canada.

E. H. Dowell and C. S. Ventres 1968 *International Journal of Solids and Structures* **4**, 975–991. Modal equations for the nonlinear flexural vibrations of a cylindrical shell.

D. A. Evensen 1963 *AIAA Journal* **1**, 2857–2858. Some observations on the nonlinear vibration of thin cylindrical shells.

D. A. Evensen 1967 *Nonlinear Flexural Vibrations of Thin-Walled Circular Cylinders*. NASA TN D-4090. U.S. Government Printing Office, Washington, DC, USA.

M. Ganapathi and T. K. Varadan 1996 *Journal of Sound and Vibration* **192**, 1–14. Large amplitude vibrations of circular cylindrical shells.

J. H. Ginsberg 1973 *Journal of Applied Mechanics* **40**, 471–477. Large amplitude forced vibrations of simply supported thin cylindrical shells.

P. B. Gonçalves and R. C. Batista 1988 *Journal of Sound and Vibration* **127**, 133–143. Non-linear vibration analysis of fluid-filled cylindrical shells.

E. I. Grigolyuk 1955 *Prikladnaya Matematika i Mekhanika* **19**, 376–382. Vibrations of circular cylindrical panels subjected to finite deflections (in Russian).

R. A. Ibrahim 2005 *Liquid Sloshing Dynamics. Theory and Applications*. Cambridge University Press, Cambridge, UK.

E. L. Jansen 2002 *International Journal of Non-Linear Mechanics* **37**, 937–949. Non-stationary flexural vibration behaviour of a cylindrical shell.

E. L. Jansen 2004 *Computers and Structures* **82**, 2647–2658. A comparison of analytical-numerical models for nonlinear vibrations of cylindrical shells.

V. D. Kubenko, P. S. Koval'chuk and T. S. Krasnopol'skaya 1982 *Soviet Applied Mechanics* **18**, 34–39. Effect of initial camber on natural nonlinear vibrations of cylindrical shells.

V. D. Kubenko, P. S. Koval'chuk and L. A. Kruk 2003 *Journal of Sound and Vibration* **265**, 245–268. Non-linear interaction of bending deformations of free-oscillating cylindrical shells.

V. D. Kubenko and V. D. Lakiza 2000 *International Applied Mechanics* **36**, 896–902. Vibration effects of the dynamic interaction between a gas-liquid medium and elastic shells.

A. W. Leissa 1973 *Vibrations of Shells*. NASA SP-288, U.S. Government Printing Office, Washington, DC, USA (1993 reprinted by the Acoustical Society of America).

L. Meirovitch 1967 *Analytical Methods in Vibrations*. Macmillan, New York, USA, 211–233.

H. J.-P. Morand and R. Ohayon 1992 *Interactions Fluides-Structures*. Masson, Paris. (English edition 1995 *Fluid Structure Interaction*. Wiley, New York, USA).

J. Nowinski 1963 *AIAA Journal* **1**, 617–620. Nonlinear transverse vibrations of orthotropic cylindrical shells.

M. D. Olson 1965 *AIAA Journal* **3**, 1775–1777. Some experimental observations on the nonlinear vibration of cylindrical shells.

M. P. Païdoussis 2003 *Fluid-Structure Interactions: Slender Structures and Axial Flow*, Vol. 2. Elsevier, London, UK.

F. Pellicano, M. Amabili and M. P. Païdoussis 2002 *International Journal of Non-Linear Mechanics* **37**, 1181–1198. Effect of the geometry on the nonlinear vibrations analysis of circular cylindrical shells.

F. Pellicano, M. Amabili and A. F. Vakakis 2000 *ASME Journal of Vibration and Acoustics* **122**, 355–364. Nonlinear vibrations and multiple resonances of fluid-filled, circular shells. 2: perturbation analysis.

A. A. Popov, J. M. T. Thompson and F. A. McRobie 2001 *Journal of Sound and Vibration* **248**, 395–411. Chaotic energy exchange through auto-parametric resonance in cylindrical shells.

E. Reissner 1955 Nonlinear effects in vibrations of cylindrical shells. Ramo-Wooldridge Corporation Report AM5-6.

W. Soedel 1993 *Vibrations of Shells and Plates*, 2nd edition. Marcel Dekker, New York, USA.

A. S. Vol'mir 1972 *Nonlinear Dynamics of Plates and Shells* (in Russian). Nauka, Moscow.

L. Watawala and W. A. Nash 1983 *Computers and Structures* **16**, 125–130. Influence of initial geometric imperfections on vibrations of thin circular cylindrical shells.

S. Wolfram 1999 *The Mathematica Book*, 4th edition. Cambridge University Press, Cambridge, UK.

6 Reduced-Order Models: Proper Orthogonal Decomposition and Nonlinear Normal Modes

6.1 Introduction

Reduced-order models are very useful to study nonlinear dynamics of fluid and solid systems. In fact, the study of nonlinear vibrations and dynamics of systems described by too many degrees of freedom, like those obtained by using commercial finite-element programs, is practically impossible. In fact, not only is the computational time needed to obtain a solution by changing a system parameter too long, as the excitation frequency around a resonance, but also spurious solutions and lack of convergence are easily obtained for large-dimension systems. For this reason, techniques to reduce the number of degrees of freedom in nonlinear problems are an important and challenging research area.

The two most popular methods used to build reduced-order models (ROMs) are the proper orthogonal decomposition (POD) and the nonlinear normal modes (NNMs) methods. The first method (POD, also referred to as the Karhunen-Loève method) uses a cloud of points in the phase space, obtained from simulations or from experiments, in order to build the reduced subspace that will contain the most information (Zahorian and Rothenberg 1981; Sirovich 1987; Aubry et al. 1988; Breuer and Sirovich 1991; Georgiou et al. 1999; Azeez and Vakakis 2001; Amabili et al. 2003; Kerschen et al. 2003, 2005; Sarkar and Païdoussis 2003, 2004; Georgiou 2005; Amabili et al. 2006). This method is, in essence, linear, as it furnishes the best orthogonal basis, which decorrelates the signal components and maximizes variance.

Amabili et al. (2003, 2006) compared the Galerkin and POD models of a water-filled circular cylindrical shell from moderate to extremely large vibration amplitudes. Accurate POD models can be built by using only POD modes with significant energy. In particular, Amabili et al. (2006) found that more proper orthogonal modes are necessary to reach energy convergence using time series extracted from more complex responses (chaotic or quasi-periodic) than from the periodic ones. Therefore, by using complex responses it is possible to build models with larger dimensions, suitable to accurately describe large variations of the system parameters.

The second method (NNMs) constructs and defines the researched subspaces from specific properties of the dynamical systems, by adapting the reduction theorems provided by the mathematics: center manifold theorem (Carr 1981; Guckenheimer and Holmes 1983) and normal form theory (Poincaré 1892; Elphick et al. 1987; Iooss and Adelmeyer 1998). Their application to vibratory systems led to two definitions of NNMs, which are equivalent in a conservative framework: either a family of periodic orbits in the vicinity of the equilibrium point (Rosenberg 1966; Mikhlin 1995; Vakakis et al. 1996) or an invariant manifold containing these periodic orbits (Shaw and Pierre 1991). Numerous asymptotic methods have been proposed for their computation: the application of the center manifold theorem (Shaw and Pierre 1993), the normal form theory (Jézéquel and Lamarque 1991; Touzé et al. 2004; Touzé and Amabili 2006; Amabili and Touzé 2007), the conservation of energy for conservative systems (King and Vakakis 1994) or the method of multiple scales (Lacarbonara et al. 2003). Numerical procedures have also been proposed recently by Jiang et al. (2005a,b), who extended the method proposed by Pesheck et al. (2002) for conservative cases. Bellizzi and Bouc (2005) propose a numerical resolution of an extended KBM (Krylov-Bogoliubov-Mitropolsky) method, whereas Slater (1996) used continuation techniques to generate the NNMs.

Application of the POD and NNM methods to reduced-order modeling enabled us to show that a few degrees of freedom in general are enough to catch the nonlinear behavior of many structures, versus the several degrees of freedom necessary in the corresponding Galerkin models. A comparison of these two reduction techniques has been proposed by Amabili and Touzé (2007).

6.2 Reference Solution

The Galerkin method, employing any set of basis functions ϕ_i, approximates the nonlinear partial differential equation (PDE) by transforming it into a finite set of coupled ordinary differential equations (ODEs), with the solution being expressed as:

$$w(\boldsymbol{\xi}, t) = \sum_{i=1}^{N} q_i(t)\varphi_i(\boldsymbol{\xi}), \tag{6.1}$$

where t is time, $\boldsymbol{\xi}$ is the vector of spatial coordinate (x, θ) describing the shell middle surface Ω, $q_i(t)$ are the generalized coordinates and N is the number of generalized coordinates (degrees of freedom), that is, the number of basis functions assumed. The linear modal base is the best choice for discretizing the shell, as these are the eigenfunctions of the linear operator of the PDE. The orthogonality property of the eigenmodes allows decoupling the ODEs at the linear stage. Other sets of basis functions may be used, with the consequence that the ODEs are linearly coupled, and more functions are needed to attain convergence. The key question in the Galerkin method is the convergence of the solution.

6.3 Proper Orthogonal Decomposition (POD) Method

The proper orthogonal modes (also referred to as spatially coherent modes) obtained by the POD method are used as a basis in conjunction with the Galerkin approach. The POD method optimally extracts the spatial information necessary to characterize the spatio-temporal complexity and inherent dimension of a system, from a set of temporal snapshots of the response, gathered from either numerical simulations or experimental data (Zahorian and Rothenberg 1981; Sirovich 1987; Aubry et al. 1988; Breuer and Sirovich 1991; Azeez and Vakakis 2001; Sarkar and Païdoussis 2003). In the present context, the temporal responses are obtained via conventional Galerkin simulations. It can be observed here that, for large-amplitude experimentally measured vibration, responses can be highly noise polluted.

The solution can be expressed by using the base of the proper orthogonal modes $\psi_i(\boldsymbol{\xi})$,

$$w(\boldsymbol{\xi}, t) = \sum_{i=1}^{\tilde{K}} a_i(t)\, \psi_i(\boldsymbol{\xi}), \tag{6.2}$$

where a_i are the proper orthogonal coordinates and $\tilde{K}$ is the number of degrees of freedom of the POD solution (in general, significantly lower than N).

The displacement field w is divided into its time-mean value $\overline{w}(\boldsymbol{\xi})$ and the zero-mean response $\tilde{w}(\boldsymbol{\xi}, t) = (w(\boldsymbol{\xi}, t) - \overline{w}(\boldsymbol{\xi}))$. In the POD method, the proper orthogonal modes are obtained by minimizing the objective function,

$$\tilde{\lambda} = \langle(\psi(\boldsymbol{\xi}) - \tilde{w}(\boldsymbol{\xi}, t))^2\rangle \qquad \forall \boldsymbol{\xi} \in \Omega, \tag{6.3}$$

with $\langle\,\rangle$ denoting the time-averaging operation. If the temporal snapshots of $\tilde{w}$ are denoted by $\{\tilde{w}_n\}$, the time-averaging operation of a series of $\overline{N}$ snapshots is $\langle\tilde{w}(\boldsymbol{\xi}, t)\rangle = (1/\overline{N}) \sum_{n=1}^{\overline{N}} \tilde{w}_n(\boldsymbol{\xi})$. By assuming that $\tilde{w}$ is a random field, imposing the normalization $\int_\Omega \psi^2(\boldsymbol{\xi})\, d\boldsymbol{\xi} = 1$ and developing the squared expression in equation (6.3), the proper orthogonal modes are obtained by maximizing the quantity $\langle\int_\Omega \psi(\boldsymbol{\xi})\tilde{w}(\boldsymbol{\xi}, t)\, d\boldsymbol{\xi}\rangle$ for any $\boldsymbol{\xi}$ of Ω. To ensure that this quantity is positive, maximization of $\langle\int_\Omega (\psi(\boldsymbol{\xi})\tilde{w}(\boldsymbol{\xi}, t))^2\, d\boldsymbol{\xi}\rangle$ is performed. Therefore, the minimization of the objective function in equation (6.3) can be replaced by the maximization of the objective function:

$$\lambda = \frac{\left\langle\int_\Omega (\psi(\boldsymbol{\xi})\, \tilde{w}(\boldsymbol{\xi}, t))^2\, d\boldsymbol{\xi}\right\rangle}{\int_\Omega \psi^2(\boldsymbol{\xi})\, d\boldsymbol{\xi}}, \tag{6.4}$$

with respect to $\psi(\boldsymbol{\xi})$. In equation (6.4), the denominator is equal to unity. Maximization of the objective function (6.4) is obtained by solving the eigenvalue problem:

$$\int_\Omega \langle\tilde{w}(\boldsymbol{\xi}, t)\, \tilde{w}(\boldsymbol{\xi}', t)\rangle\psi(\boldsymbol{\xi}')d\boldsymbol{\xi}' = \lambda\, \psi(\boldsymbol{\xi}), \tag{6.5}$$

where $\langle\tilde{w}(\boldsymbol{\xi}, t)\, \tilde{w}(\boldsymbol{\xi}', t)\rangle$ is the time-averaged spatial autocorrelation function.

A Galerkin projection scheme for determining proper orthogonal modes semi-analytically, and in parallel to approximate the solution of the PDE, is presented next. The proper orthogonal modes are projected on the eigenmodes $\varphi(\boldsymbol{\xi})$ of the empty shell as

$$\psi(\boldsymbol{\xi}) = \sum_{i=1}^{N} \alpha_i \varphi_i(\boldsymbol{\xi}), \tag{6.6}$$

where α_i are unknown coefficients. By substituting equations (6.2) and (6.6) into (6.5), the following expression is obtained:

$$\sum_{i=1}^{N} \varphi_i(\boldsymbol{\xi}) \sum_{j=1}^{N}\sum_{k=1}^{N} \langle\tilde{q}_i(t)\tilde{q}_j(t)\rangle\, \alpha_k \int_\Omega \varphi_j(\boldsymbol{\xi}')\varphi_k(\boldsymbol{\xi}')d\boldsymbol{\xi}' = \lambda \sum_{i=1}^{N} \alpha_i\, \varphi_i(\boldsymbol{\xi}), \tag{6.7}$$

where $\tilde{q}_i = (q_i - \overline{q}_i)$ is the zero-mean response of the i-th generalized coordinate, with $\overline{q}_i$ being its mean. Equation (6.7) is multiplied by $\varphi_m(\boldsymbol{\xi})$ and integrated over Ω

for any m from 1 to N. By using the orthogonality relationships of the basis functions $\varphi_m(\boldsymbol{\xi})$, the following eigenvalue problem is finally obtained:

$$\mathbf{A}\,\boldsymbol{\alpha} = \lambda\,\mathbf{B}\boldsymbol{\alpha}, \tag{6.8}$$

where

$$A_{ij} = \tau_i\,\tau_j\langle\tilde{q}_i(t)\,\tilde{q}_j(t)\rangle, \qquad B_{ij} = \tau_i\,\delta_{ij}, \qquad \tau_i = \int_\Omega \varphi_i^2(\boldsymbol{\xi})\mathrm{d}\boldsymbol{\xi}, \quad (6.9\text{–}6.11)$$

and δ_{ij} is the Kronecker delta. The norm of the basis functions τ_i in the present case is $\pi RL/2$ for asymmetric modes and πRL for axisymmetric modes; without effect on the results, they can be assumed to be 0.5 and 1, respectively. In equation (6.8), $\mathbf{A}$ and $\mathbf{B}$ are symmetric and positive definite matrices of dimension $N \times N$, and $\boldsymbol{\alpha}$ is a vector containing the N unknown coefficients of the proper orthogonal modes. The eigenvectors $\boldsymbol{\alpha}$ corresponding to the largest eigenvalues (known as dominant proper orthogonal modes) in equation (6.8) can now be inserted in equation (6.6) that gives a basis for the approximate solution of the PDE using the Galerkin approach. This will be referred to as the POD-Galerkin scheme hereafter. The optimal number of terms $\tilde{K}$ to be retained can be estimated by $\sum_{i=1}^{\tilde{K}} \lambda_i / \sum_{i=1}^{N} \lambda_i \geq 0.999$ in equation (6.8), but for each problem this cut-off value can be different. It can be useful to check the convergence of the solution by increasing the value $\tilde{K}$. Over a certain value, results become less accurate, because the additional terms introduced in the expansion are highly noise polluted. As mentioned previously, the order of the POD-Galerkin model necessary to capture the salient dynamical features of the original PDE is significantly lower than that of the conventional Gakerkin model.

In some applications, it may be better to use time responses obtained for different system parameters in order to produce better proper orthogonal modes. For example, in the case of two such responses, equation (6.8) is still used, but equation (6.9) is replaced by

$$\mathbf{A} = \frac{p_1\,\mathbf{A}^{(1)}}{\|\mathbf{A}^{(1)}\|} + \frac{p_2\,\mathbf{A}^{(2)}}{\|\mathbf{A}^{(2)}\|}, \tag{6.12}$$

where $\|\mathbf{A}\| = \sqrt{\mathrm{Tr}\,(\mathbf{A}\,\mathbf{A}^{\mathrm{T}})}$ is the Frobenius norm of $\mathbf{A}$, Tr gives the trace of the matrix,

$$A_{ij}^{(1)} = \tau_i\,\tau_j\left\langle\tilde{q}_i^{(1)}(t)\,\tilde{q}_j^{(1)}(t)\right\rangle, \tag{6.13}$$

$$A_{ij}^{(2)} = \tau_i\,\tau_j\left\langle\tilde{q}_i^{(2)}(t)\,\tilde{q}_j^{(2)}(t)\right\rangle, \tag{6.14}$$

and $\tilde{q}_i^{(1)}(t)$ and $\tilde{q}_i^{(2)}(t)$ are the zero-mean first and second responses, respectively. In equation (6.12) p_1 and p_2 are two coefficients used to give a different weight to the first and second response in the calculation of the proper orthogonal modes. The Frobenius norm is introduced just to normalize the matrices, as the amplitude of the generalized coordinates $q_i(t)$ can be very different for the two responses.

For simply supported, circular cylindrical shells, by using equations (6.2), (6.6) and (5.50), the expansion used for the POD solution is given by

$$w(\boldsymbol{\xi}, t) = \sum_{i=1}^{\tilde{K}} a_i(t) \sum_{j=1}^{N} \alpha_{j,i}\,\varphi_j(\boldsymbol{\xi}) = \sum_{i=1}^{\tilde{K}} a_i(t) \sum_{m=1}^{M} \sum_{n=0}^{\tilde{N}} [\alpha_{m,n,i}\,\cos(n\theta) + \beta_{m,n,i}\,\sin(n\theta)]\,\sin(\lambda_m x), \tag{6.15}$$

where, on the right-hand side, two different symbols $\alpha_{m,n,i}$ and $\beta_{m,n,i}$ have been introduced to differentiate the coefficients of the proper orthogonal modes for cosine and sine terms in θ and are given by the corresponding $\alpha_{j,i}$. Equation (6.15) is used to solve equations (1.160) and (1.161) with the Galerkin method to find the equations of motion of the reduced-order model. Moreover equation (6.15) has still the same shape over the shell surface of equation (5.50). Therefore, the fluid-structure interaction can be treated with the same approach used for the Galerkin method. This is not surprising, because the POD modes have been projected on the eigenmodes.

6.4 Asymptotic Nonlinear Normal Modes (NNMs) Method

Nonlinear normal modes are here defined as an invariant manifold of the state space. They are moreover chosen tangent at the origin, which corresponds to the position of the structure at rest. An asymptotic procedure, based on the normal form theory, is used to compute the NNMs of the system. The method is here recalled in brief, and the interested reader is referred to Touzé and Amabili (2006) for a complete description. In particular, the present formulation of the NNM method allows us to take modal damping into account in the derivations, hence extending previous results obtained for conservative systems (Touzé et al. 2004). A third-order asymptotic development is applied in order to perform a nonlinear change of coordinates for the system of the damped unforced equation of motion, corresponding to equations (3.6) with the right-hand side equal to zero. A real formulation is used, so that normal forms are expressed with oscillators; the formulation is applicable to systems with diagonalized linear matrices, for example, obtained by using the technique introduced in Section 4.6. The dummy variable $y_j = \dot{x}_j$ for the nondimensional velocity permits recasting of the system of equations into first order. The nonlinear change of coordinates is

$$x_j = r_j + \sum_{i=1}^{K}\sum_{p\geq i}^{K}\left(a_{ip}^{(j)} r_i r_p + b_{ip}^{(j)} s_i s_p\right) + \sum_{i=1}^{K}\sum_{p=1}^{K} c_{ip}^{(j)} r_i s_p + \sum_{i=1}^{K}\sum_{p\geq i}^{K}\sum_{k\geq p}^{K}\left(d_{ipk}^{(j)} r_i r_p r_k + e_{ipk}^{(j)} s_i s_p s_k\right) + \sum_{i=1}^{K}\sum_{p=1}^{K}\sum_{k\geq p}^{K}\left(t_{ipk}^{(j)} s_i r_p r_k + u_{ipk}^{(j)} r_i s_p s_k\right), \tag{6.16a}$$

$$y_j = s_j + \sum_{i=1}^{K}\sum_{p\geq i}^{K}\left(\alpha_{ip}^{(j)} r_i r_p + \beta_{ip}^{(j)} s_i s_p\right) + \sum_{i=1}^{K}\sum_{p=1}^{K} \gamma_{ip}^{(j)} r_i s_p + \sum_{i=1}^{K}\sum_{p\geq i}^{K}\sum_{k\geq p}^{K}\left(\lambda_{ipk}^{(j)} r_i r_p r_k + \mu_{ipk}^{(j)} s_i s_p s_k\right) + \sum_{i=1}^{K}\sum_{p=1}^{K}\sum_{k\geq p}^{K}\left(\nu_{ipk}^{(j)} s_i r_p r_k + \xi_{ipk}^{(j)} r_i s_p s_k\right), \tag{6.16b}$$

for $j = 1, \ldots, N$, where N and K are the number of degrees of freedom in the original and the reduced-order model, respectively, r_j is the transformed nondimensional displacement and s_j is the transformed nondimensional velocity. Other symbols are the transformation coefficients. After substitution of (6.16) into (3.6), the dynamics, written with the newly introduced variables (r_j, s_j), is expressed in an invariant-based span of the state space. As a result, proper truncations can now be realized, as all invariant-breaking terms between oscillators in equation (3.6) have been canceled. The reduction can now be applied by simply selecting the most important normal coordinates for simulation (master coordinates) and canceling all the others.

The nonlinear change of coordinates given by equations (6.16) leads to the cancelation of all the quadratic terms in the original nonlinear equations of motion, because these terms are nonresonant as long as no low-order internal resonance relationship exists. On the other hand, several of the cubic coefficients introduced in the nonlinear transformation (6.16) vanish, because they correspond to cubic resonant terms in the original equations of motion; these cubic resonant terms stay in the normal forms. The dynamics in the normal coordinates is written as

$$\dot{r}_j = s_j, \tag{6.17a}$$

$$\begin{aligned}\dot{s}_j = &-\omega_j^2 r_j - 2\zeta_j\omega_j s_j - \left(h_{jjj}^{(j)} + A_{jjj}^{(j)}\right) r_j^3 - B_{jjj}^{(j)} r_j s_j^2 - C_{jjj}^{(j)} r_j^2 s_j \\ &- r_j \left[\sum_{p>j}^{K} \left(h_{jpp}^{(j)} + A_{jpp}^{(j)} + A_{pjp}^{(j)}\right) r_p^2 + B_{jpp}^{(j)} s_p^2 + \left(C_{jpp}^{(j)} + C_{pjp}^{(j)}\right) r_p s_p\right] \\ &+ \sum_{i=1}^{i<j} \left[\left(h_{iij}^{(j)} + A_{iij}^{(j)} + A_{jii}^{(j)}\right) r_i^2 + B_{jii}^{(j)} s_i^2 + \left(C_{jii}^{(j)} + C_{iji}^{(j)}\right) r_i s_i\right] \\ &- s_j \left[\sum_{p>j}^{K} \left(B_{pjp}^{(j)} r_p s_p + C_{ppj}^{(j)} r_p^2\right) + \sum_{i=1}^{i<j} \left(B_{iij}^{(j)} r_i s_i + C_{iij}^{(j)} r_i^2\right)\right],\end{aligned} \tag{6.17b}$$

for $j = 1, \ldots, K$, where $h_{ipk}^{(j)} = z_{jipk}$ are the coefficients of cubic terms in equation (3.6), denoted here with a different symbol in order to avoid confusion with coefficients of quadratic terms, $A_{ipk}^{(j)}$, $B_{ipk}^{(j)}$ and $C_{ipk}^{(j)}$ arise from the cancelation of the quadratic terms, and are expressed by

$$A_{ipk}^{(j)} = \sum_{l \geq i}^{K} z_{jil} a_{pk}^{(l)} + \sum_{l=1}^{l \leq i} z_{jli} a_{pk}^{(l)}, \tag{6.18a}$$

$$B_{ipk}^{(j)} = \sum_{l \geq i}^{K} z_{jil} b_{pk}^{(l)} + \sum_{l=1}^{l \leq i} z_{jli} b_{pk}^{(l)}, \tag{6.18b}$$

$$C_{ipk}^{(j)} = \sum_{l \geq i}^{K} z_{jil} c_{pk}^{(l)} + \sum_{l=1}^{l \leq i} z_{jli} c_{pk}^{(l)}, \tag{6.18c}$$

and z_{jil} are the coefficients of quadratic terms in equation (3.6).

For the simply supported, fluid-filled, circular cylindrical shells considered here, the minimum model must retain the NNMs corresponding to the driven mode (r_1, s_1, that are the transformations of $A_{1,5}$ and $\dot{A}_{1,5}$) and the companion mode (r_2, s_2, that are the transformations of $B_{1,5}$ and $\dot{B}_{1,5}$), as these two modes have the same eigenfrequency (1:1 internal resonance). Finally, the reduced-order model (ROM) built by selecting these two pairs of coordinates takes the form

$$\begin{aligned}&\ddot{r}_1 + \omega_1^2 r_1 + 2\zeta_1\omega_1\dot{r}_1 + \left(A_{111}^{(1)} + h_{111}^{(1)}\right) r_1^3 + B_{111}^{(1)} r_1 \dot{r}_1^2 + \left(A_{212}^{(1)} + A_{122}^{(1)} + h_{122}^{(1)}\right) r_1 r_2^2 \\ &\quad + B_{122}^{(1)} r_1 \dot{r}_2^2 + B_{212}^{(1)} r_2 \dot{r}_1 \dot{r}_2 + C_{111}^{(1)} r_1^2 \dot{r}_1 + \left(C_{122}^{(1)} + C_{212}^{(1)}\right) r_1 r_2 \dot{r}_2 + C_{221}^{(1)} r_2^2 \dot{r}_1 = \hat{f}\cos(\omega t),\end{aligned} \tag{6.19a}$$

$$\begin{aligned}&\ddot{r}_2 + \omega_2^2 r_2 + 2\zeta_2\omega_2\dot{r}_2 + \left(A_{222}^{(2)} + h_{222}^{(2)}\right) r_2^3 + B_{222}^{2} r_2 \dot{r}_2^2 + \left(A_{112}^{(2)} + A_{211}^{(2)} + h_{112}^{(2)}\right) r_2 r_1^2 \\ &\quad + B_{211}^{(2)} r_2 \dot{r}_1^2 + B_{112}^{(2)} r_1 \dot{r}_1 \dot{r}_2 + C_{222}^{(2)} r_2^2 \dot{r}_2 + \left(C_{121}^{(2)} + C_{211}^{(2)}\right) r_1 r_2 \dot{r}_1 + C_{112}^{(2)} r_1^2 \dot{r}_2 = 0.\end{aligned} \tag{6.19b}$$

A forcing term has been added at the end of the process and now appears in equation (6.19a). This is the second approximation used for building this ROM, as a time-invariant manifold is used. The most accurate solution would have consisted of constructing a periodically forced invariant manifold; see, for example, Jiang et al. (2005b). However, this results in a very complicated formulation and time-consuming numerical calculations for constructing the ROM. The proposed method has the advantage of simplicity and quickness of computation. It allows the derivation of a differential model that could be used easily for parametric studies. However, it is valid, strictly speaking, only for small values of $\hat{f}$.

With the NNMs method, the original 16 degrees of freedom of the conventional Galerkin model obtained in Chapter 5 have been reduced to two in equations (6.19). However, differently than the POD method, the structure of the equations of motion is changed. In fact, quadratic nonlinear terms have been canceled, but cubic terms involve both the transformed nondimensional displacement and the transformed nondimensional velocity.

6.5 Discussion on POD and NNMs

After presentation of the two reduction methods, a first discussion on their theoretical settings is here provided in order to underline their abilities and limitations.

The POD method, which consists in finding the best orthogonal hyperplanes that contain the most information, is essentially a linear method. This can be seen as an advantage because few manipulations involving linear algebra only are needed to construct the ROM. The key formula of the method, equation (6.8), is an eigenvalue problem. On the other hand, the linear essence of the method may be a drawback, as curved subspaces are generally more suitable to capture clouds of points with complicated shapes. An NNM, being an invariant manifold in state space, is a curved subspace, so that the NNM reduction method is essentially nonlinear. The invariance property is the key that allows finding the lowest dimensional subspaces that contain dynamical properties, because dynamical motions do not exist within any other subspace that does not share this invariance property. This is the main advantage of the NNM method as compared with the POD. It is expected that fewer NNMs are necessary than POD modes.

The POD method is global in the sense that it is able to capture any motion in state space and furnishes the adapted basis for decomposing it. This is an advantage compared with the asymptotic NNM method used here, which relies on a local theory. The third-order development, equation (6.16), is valid for small values of the modal amplitudes. The use of a time-invariant manifold in NNMs can give unreliable results for large values of the amplitude of the external forcing; in fact, the oscillations of the manifold will be too large, and the time-independent transformation will become too crude an approximation. When increasing the nonlinearities by feeding more energy into the system, the results provided by the asymptotic NNM method are expected to deteriorate. This is not the case for the POD, if one has taken care to construct its POD-based ROM with clouds of points that are significant for a large range of values of the nonlinearity. In this context, it has already been argued (Kerschen et al. 2003; Amabili et al. 2006) that a chaotic response is the best candidate for building the POD.

Finally, the two methods differ radically in the way the ROM is built. For the POD, it is mandatory to have a response of the system to build the ROM. In the

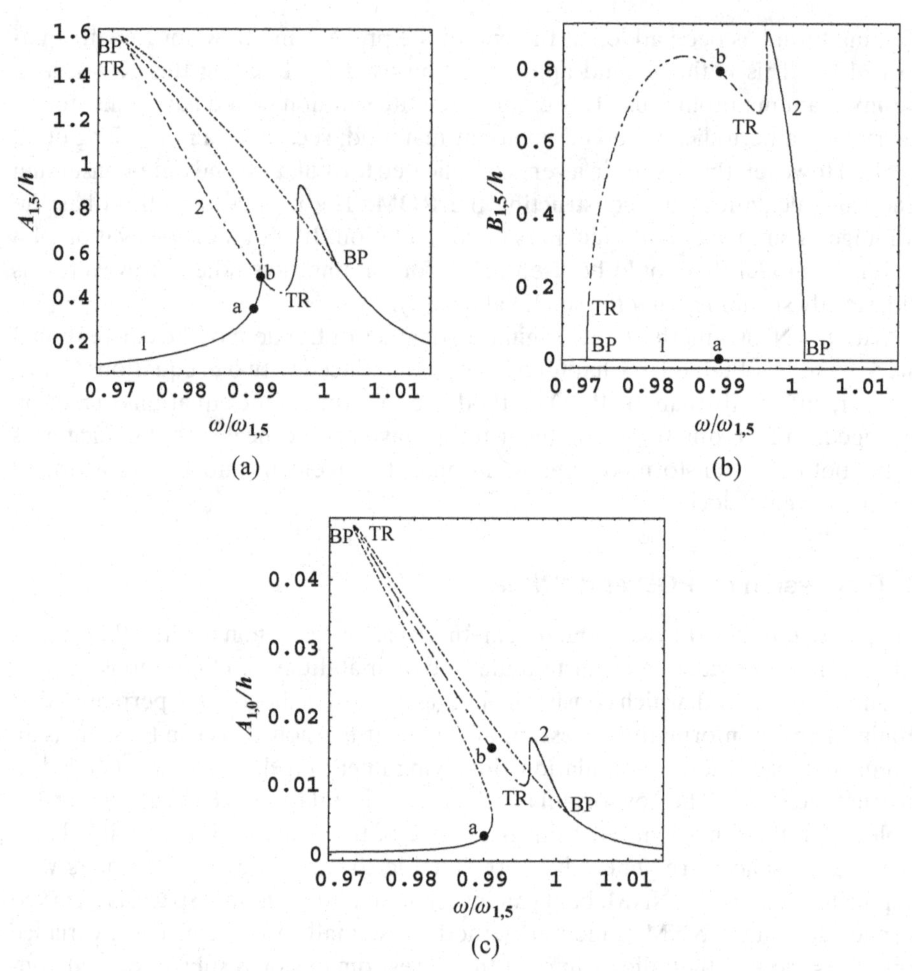

Figure 6.1. Maximum amplitude of vibration versus excitation frequency, excitation 3 N; conventional Galerkin model, 16 dofs. (a) Maximum amplitude of $A_{1,5}(t)$, driven mode. (b) Maximum amplitude of $B_{1,5}(t)$, companion mode. (c) Maximum amplitude of $A_{1,0}(t)$, first axisymmetric mode. 1, branch 1; 2, branch 2; BP, pitchfork bifurcation; TR, Neimark-Sacker (torus) bifurcations; a, response "a"; b, response "b." —, stable periodic solutions; – • –, stable quasi-periodic solutions; – –, unstable periodic solutions.

present context of a completely theoretical model, this is a drawback, because one must compute time responses to be in a position of reducing the order. Moreover, it has been emphasized by Amabili et al. (2003, 2006) that the choice of these time responses is not an easy task that could be done blindly. By comparison, the asymptotic NNM method does not need any response of the system, but dynamical properties only, which are the discretized equations of motion, that is, after projection of the PDE with the Galerkin method. With the eigenvalues of the linear part (eigenfrequencies ω_j and damping coefficients ζ_j) in hand and the nonlinear coefficients z_{jil} and z_{jikl}, the nonlinear change of coordinates, equation (6.16), can be applied directly to obtain the ROM. As the coefficients in equation (6.16) are computed once and for all, application of the method is easy and not too demanding in terms of computation time. In Section 6.6, all these conclusions will be illustrated with the numerical results.

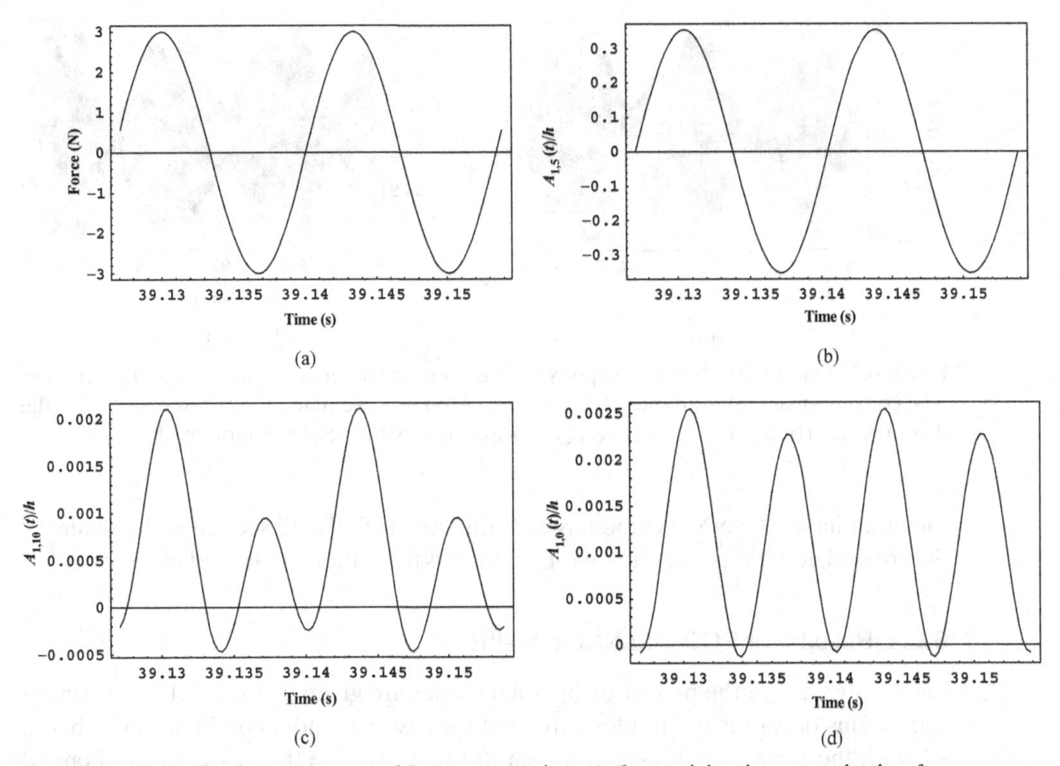

Figure 6.2. Periodic time response without companion mode participation at excitation frequency $\omega/\omega_{1,5}$= 0.99, excitation 3 N; conventional Galerkin model, 16 dofs. (a) Excitation. (b) Modal coordinate $A_{1,5}(t)$ associated to the driven mode. (c) Modal coordinate $A_{1,10}(t)$. (d) Modal coordinate $A_{1,0}(t)$ associated with the first axisymmetric mode.

6.6 Numerical Results

The simply supported, water-filled circular cylindrical shell (without imperfections) investigated in Chapter 5 is considered, with the following dimensions and material properties: $L = 520\,\text{mm}$, $R = 149.4\,\text{mm}, h = 0.519\,\text{mm}$, $E = 2.06 \times 10^{11}$ Pa, $\rho = 7800$ kg/m^3, $\rho_F = 1000$ kg/m^3 and $\nu = 0.3$. Numerical calculations have been performed for the fundamental mode ($m = 1$, $n = 5$) of the water-filled shell. The natural frequency $\omega_{1,5}$ of this mode is 79.21 Hz, according to Donnell's theory of shells, and modal damping $\zeta_{1,5} = 0.0017$ is assumed in the Galerkin model for the fundamental mode. For additional modes in the Galerkin model, the following modal damping has been assumed: $\zeta_{m,n} = \zeta_{1,5}\, \omega_{m,n}/\omega_{1,5}$. The same values of modal damping have been used in all the reduced-order models.

The response of the fundamental mode of the water-filled shell to harmonic point excitation of 3 N at $\tilde{x} = L/2$ and $\tilde{\theta} = 0$ has been computed by using the conventional Galerkin model with 16 degrees of freedom (dofs). The results were given in Figure 5.19 and are also reported in Figure 6.1 with the indication of two particular time responses: a on branch 1 for $\omega = 0.99\, \omega_{1,5}$, and b on branch 2 for $\omega = 0.991\, \omega_{1,5}$. Time response in case a is shown in Figure 6.2; it has zero amplitude of the companion mode $B_{1,n}(t)$. The quasi-periodic time response b is reported in Figure 6.3 for the most significant generalized coordinates. The Galerkin model is used as reference solution, and the time responses a and b are used in the POD method. On

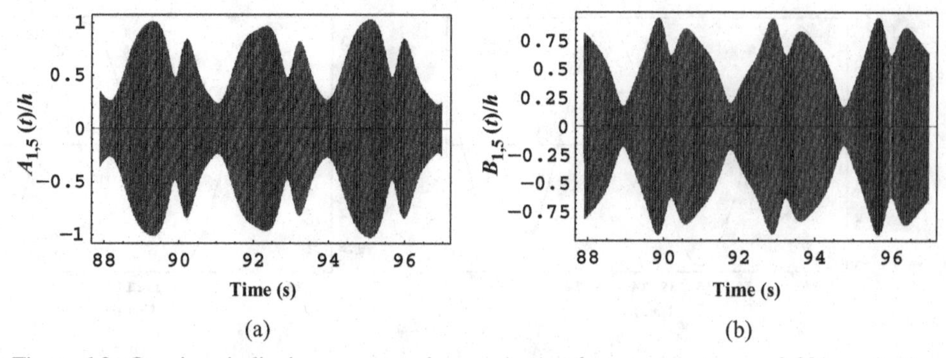

Figure 6.3. Quasi-periodic time response b at excitation frequency $\omega/\omega_{1,5} = 0.991$, excitation 3 N; conventional Galerkin model, 16 dofs. (a) Modal coordinate $A_{1,5}(t)$ associated with the driven mode. (b) Modal coordinate $B_{1,5}(t)$ associated with the companion mode.

the other hand, the NNMs model is built directly from the 16 second-order ordinary differential equations obtained with the Galerkin method presented in Chapter 5.

6.6.1 Results for POD and NNMs Methods

The coefficients of the proper orthogonal modes are given in Table 6.1 (most significant terms only) for the modes extracted for cases: a, and a combination of b and −b with the same weight (see equation [6.12], $p_1 = p_2 = 0.5$); case −b is obtained by changing in case "b" the sign to the generalized coordinates associated to $\sin(n\theta)$ terms in equation (5.50). In fact, associated with the response b there is always an identical response −b traveling in the opposite angular direction around the shell. By taking into account both responses and combining them with identical weights, the symmetry of the system is recovered. For simplicity, the combination of cases b and −b is indicated with case b hereafter.

The optimal number of proper orthogonal modes $\tilde{K}$ to be retained can be estimated by plotting $\sum_{i=1}^{\tilde{K}} \lambda_i / \sum_{i=1}^{N} \lambda_i$ as a function of $\tilde{K}$, as has been done in Figures 6.4(a,b) for cases a and b, respectively. For case a, the first mode has almost all the energy of the system, which is completely attained with two modes. An additional mode is necessary for case b.

The first results analyzed are those with proper orthogonal modes extracted from time series a. The maximum amplitude of the shell response for all the frequency range in the spectral neighborhood of the fundamental frequency is shown in Figure 6.5(a), calculated by using two proper orthogonal modes. The reference Galerkin

Table 6.1. *Coefficients of the main proper orthogonal modes (POMs)*

Time response	i-th POM	$\alpha_{1,5,i}$	$\beta_{1,5,i}$	$\alpha_{1,10,i}$	$\beta_{1,10,i}$	$\alpha_{1,0,i}$	$\alpha_{3,0,i}$
a	1	−1	0	−0.00163	0	−0.000391	−0.00006
	2	0.001658	0	−0.5819	0	−0.7914	0.1490
0.5 b	1	1	0	0.000213	0	0.0000434	8.85×10^{-6}
+0.5 −b	2	0	1	0	−0.000291	0	0
	3	0.000123	0	−0.1847	0	−0.9641	0.1855

Figure 6.4. Significance of POD eigenvalues versus the number of proper orthogonal modes. (a) POD applied to time response a. (b) POD applied to time response b combined with −b.

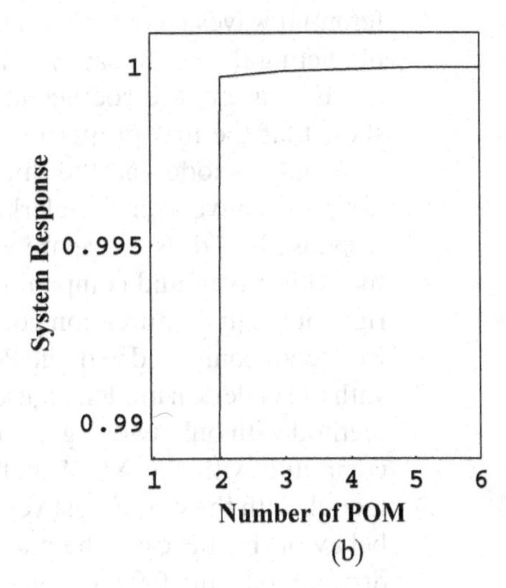

(b)

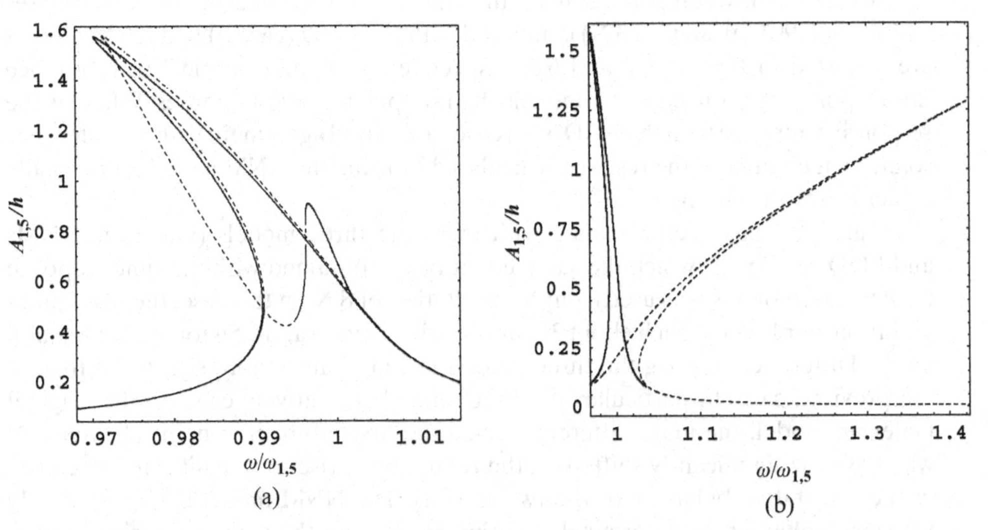

(a) (b)

Figure 6.5. Maximum amplitude of vibration versus excitation frequency; POD model versus conventional Galerkin model, case a. Maximum amplitude of $A_{1,5}(t)$, driven mode. (a) —, POD model (two modes); —, stable conventional Galerkin solutions; – – –, unstable conventional Galerkin solutions. (b) —, POD model with three modes; —, POD model with two modes (coincident with model with three modes); – – –, POD model with one mode.

solution is also plotted for comparison. Results are converted from proper orthogonal coordinates to the more intuitive modal coordinate. Results for branch 1 are extremely close, even though the dimension of the system has been reduced from 16 to 2. In this case, the proper orthogonal modes are not able to detect the bifurcation points and branch 2 because they have been computed by using a time response with zero companion mode participation. The convergence of the solution is investigated in Figure 6.5(b), where the response computed with one, two and three proper orthogonal modes is shown. Results show that the additional third mode gives zero contribution to the shell response and that only two modes are necessary; results with only one mode are completely wrong. The first of the two necessary modes is associated with the driven mode ($m = 1, n = 5$), whereas the second is associated mainly with axisymmetric oscillation involving contribution of modes with $2n$ circumferential waves (see Table 6.1). The axisymmetric term is fundamental for correctly predicting the nonlinearity of the shell, as discussed in Chapter 5.

For case b, the coefficients of the main proper orthogonal modes in Table 6.1 show that the first proper orthogonal mode is the driven mode, the second is the companion mode and the third is the axisymmetric mode. Responses obtained by using the conventional Galerkin method (16 dofs) and the POD method from time response b (3 dofs) compare very well for excitation of 3 N, as shown in Figure 6.6 for both driven and companion modes. The main difference is a slight shift on the right of the first bifurcation point of branch 1. It can also be observed that the natural frequency computed with the POD model is practically identical to the one computed with the Galerkin model. Figure 6.6 also shows the response computed with the NNM method with only two degrees of freedom. It can be observed that also the response computed with the NNM method compares very well with the original Galerkin model, with the curves just very slightly shifted to the left and with exact qualitative behavior. In this case, the maximum vibration amplitudes reach about $1.5h$ for the driven mode and $0.9h$ for the companion mode.

In order to also compare results in the time domain, the quasi-periodic responses ($\omega/\omega_{1,5} = 0.991$, excitation 3 N) computed with the POD (case b) and NNM models are reported in Figures 6.7 and 6.8, respectively; it is more critical to reproduce this response by reduced-order models than simple periodic responses. Whereas the response computed with the POD is in reasonably good agreement with the reference solution in Figure 6.3, the response calculated by using the NNM model is practically coincident with this one.

Figure 6.9 has been obtained with the same three models (Galerkin, NNMs and POD case b, for which the same equations as obtained with the time response for excitation of 3 N was used), but for excitation of 8 N. In this case, the maximum-vibration amplitudes reach about $3h$ for the driven mode and $2.5h$ for the companion mode. Differences among the three models become much more significant than in the previous case. In particular, the POD model is relatively close to the original Galerkin model, the main difference being the first bifurcation point of branch 1, which is now significantly shifted on the right, giving rise to a significant difference in the qualitative behavior of the two models. The NNM model has qualitatively the same behavior as the original Galerkin model, but the response is significantly shifted to the left, giving rise to the model overestimating the softening nonlinearity of the system.

It can be observed here that the POD model could be improved by using a time response computed for an excitation of 8 N to find the proper orthogonal modes.

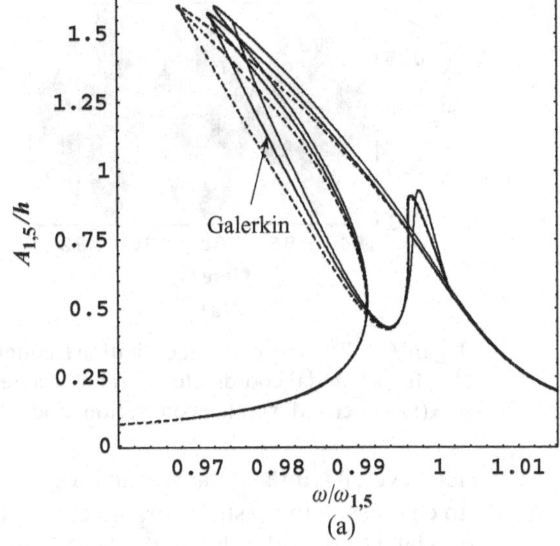

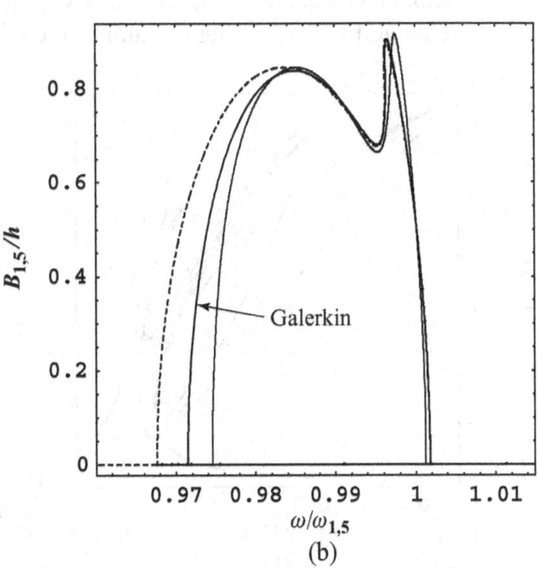

Figure 6.6. Maximum amplitude of vibration versus excitation frequency, excitation 3 N; conventional Galerkin model, POD model (Case b) with three modes and NNMs model with two modes. (a) Maximum amplitude of $A_{1,5}(t)$, driven mode. (b) Maximum amplitude of $B_{1,5}(t)$, companion mode. —, conventional Galerkin model (16 dofs); —, POD model (3 dofs); – –, NNM model (2 dofs).

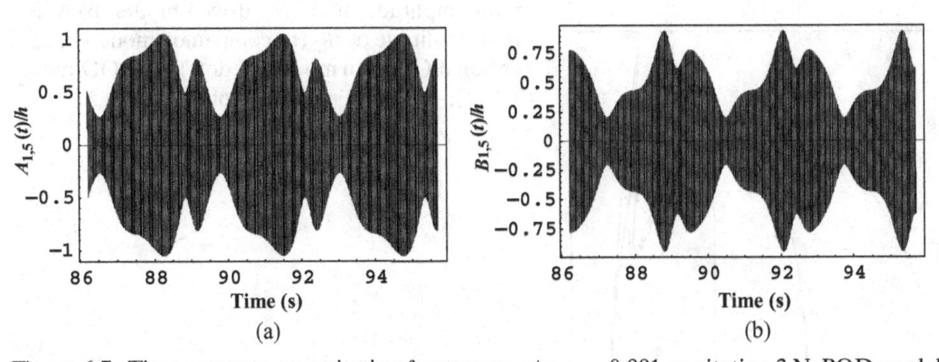

Figure 6.7. Time response at excitation frequency $\omega/\omega_{1,5} = 0.991$, excitation 3 N; POD model from time response b, 3 dofs. (a) Modal coordinate $A_{1,5}(t)$ associated with the driven mode. (b) Modal coordinate $B_{1,5}(t)$ associated with the companion mode.

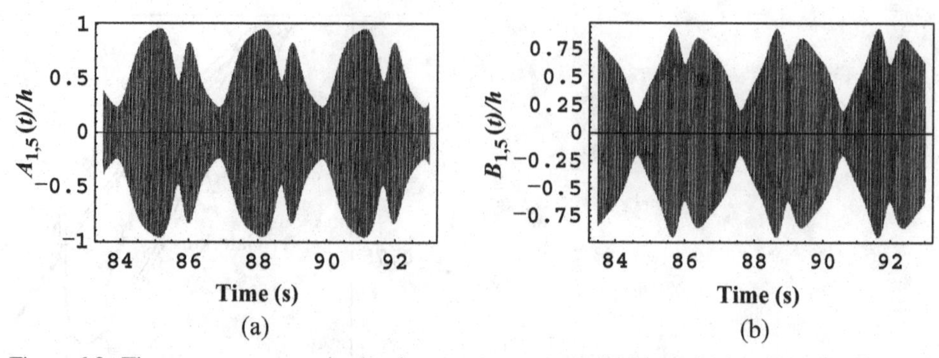

Figure 6.8. Time response at excitation frequency $\omega/\omega_{1,5} = 0.991$, excitation 3 N; NNM model, 2 dofs. (a) Modal coordinate $A_{1,5}(t)$ associated with the driven mode. (b) Modal coordinate $B_{1,5}(t)$ associated with the companion mode.

However, it is interesting here to investigate the robustness of a reduced-order model to changes in the system parameters, and it is therefore convenient to use the same model. On the other hand, the NNM model is built once and for all and may not be changed when varying the amplitude of the forcing. The observed differences with

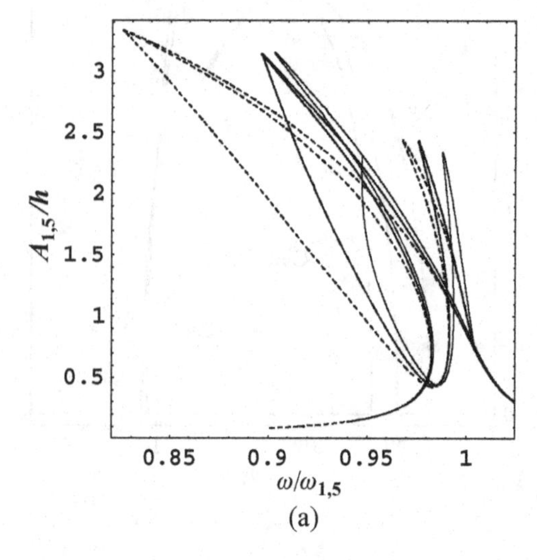

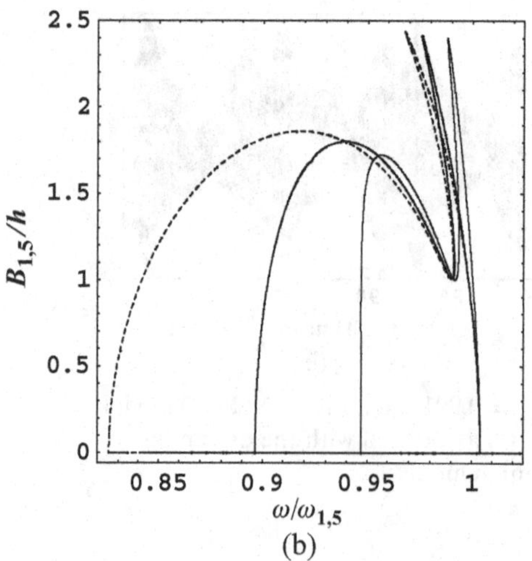

Figure 6.9. Maximum amplitude of vibration versus excitation frequency, excitation 8 N; conventional Galerkin model, POD model (case b) with three modes and NNM model with two modes. (a) Maximum amplitude of $A_{1,5}(t)$, driven mode. (b) Maximum amplitude of $B_{1,5}(t)$, companion mode. —, conventional Galerkin model (16 dofs); —, POD model (3 dofs); – –, NNM model (2 dofs).

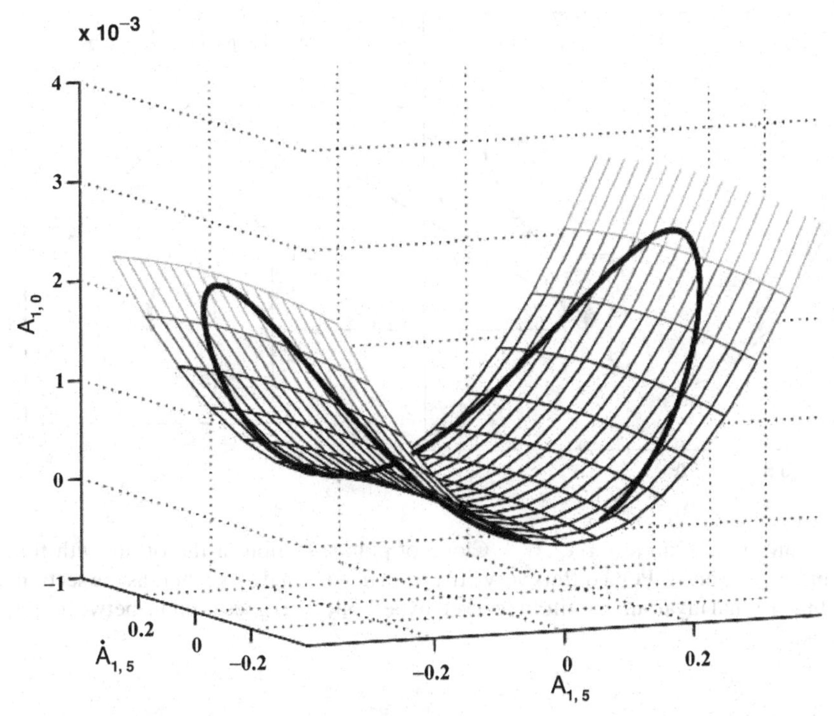

Figure 6.10. Projection of the 32-dimensional state space onto the three-dimensional subspace spanned by $(A_{1,5}, \dot{A}_{1,5}, A_{1,0})$. Closed orbit: forced motion for $\omega/\omega_{1,5} = 0.99$ and excitation 3 N, case a. Curved surface: invariant manifold corresponding to the first NNM (driven mode).

the reference solution are the consequences of the two approximations used to build it: asymptotic development and time-invariant manifold.

It is well known that the contribution of the axisymmetric mode, even if it is small compared with the oscillation of the fundamental mode, is fundamental to predicting the correct nonlinear behavior of the shell. It can be interesting to observe that both reduced-order models do not need this knowledge a priori, because they act on time responses, in the case of POD, or on the equations of motion, in the case of NNMs, where this information on the system dynamics is included.

6.6.2 Geometrical Interpretation

In order to get a geometrical interpretation of the frequency-response curves shown in the previous section, projections and Poincaré sections of the solutions are here proposed. The state space of the reference solution is 32-dimensional (16 dofs selected with displacement and velocity as independent variables for each degree of freedom), plus one for the external forcing. Two different time responses are considered for excitation of 3 N: case a, $\omega/\omega_{1,5} = 0.99$ on branch 1 with no companion mode participation; case b, $\omega/\omega_{1,5} = 0.991$ on branch 2 with companion mode participation and quasi-periodic response with amplitude modulations.

First, a two-dimensional invariant manifold, corresponding to a single NNM, is shown in Figure 6.10. This NNM, defined by the coordinates (r_1, s_1), corresponds to the continuation of the driven asymmetric mode: $(A_{1,5}, \dot{A}_{1,5})$. The geometry of the manifold is given by equations (6.19). Keeping K master coordinates (here $K = 1$)

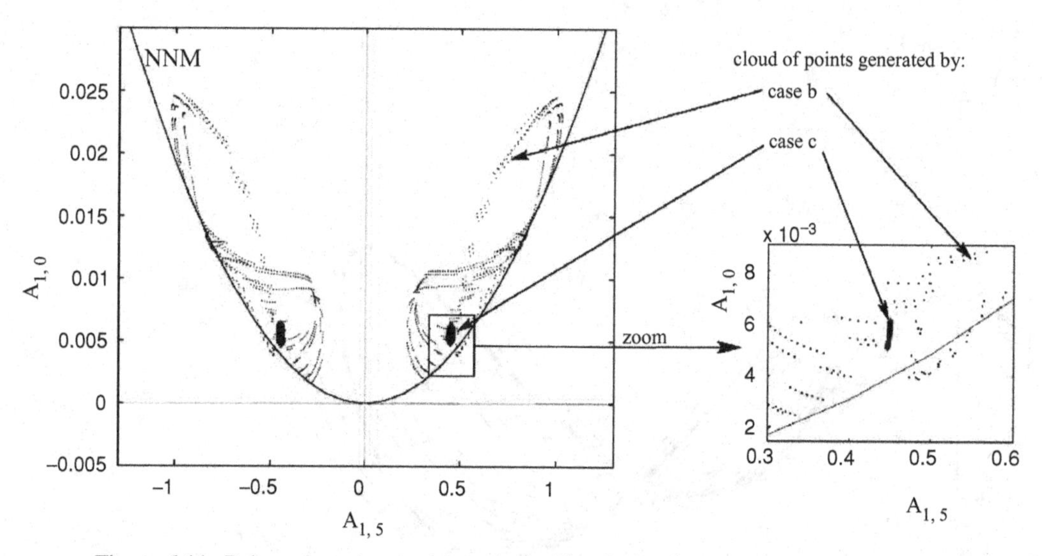

Figure 6.11. Poincaré section in ($A_{1,5}$, $A_{1,0}$). Cloud of points: section of the orbits with the Poincaré plane, cases b and c. POD differs very little from the original axis, whereas the section of the four-dimensional invariant manifold (curved hyperbolic line) goes right in-between the points (NNMs).

and canceling the others, equations (6.19) define a $2K$-dimensional invariant manifold in state space. The closed orbit is computed by time integration of the reference solution and corresponds to case a, where the companion mode is not excited, ensuring a two-dimensional motion. One can observe that the forced motion occurs in the vicinity of the time-invariant manifold, illustrating the quality of the proposed approximation. Once again, the small differences (the orbit goes before the manifold for $A_{1,5} > 0$, and behind for $A_{1,5} < 0$) are due to the asymptotic approximations used for constructing the subspace.

The dynamics in the vicinity of the first eigenfrequency is essentially governed by the 1:1 internal resonance and the coupling between driven and companion mode. Hence, the dynamics is essentially four-dimensional, so that only Poincaré sections will allow illustration of the geometry of the subspaces used for reduction.

Figure 6.11 shows such a Poincaré map, where the selected section is the plane ($A_{1,5}$, $A_{1,0}$), chosen in order to show the important contribution of the first axisymmetric mode on the asymmetric motion. Two clouds of points are represented, corresponding to time series computed by the reference model at points b and a new point c corresponding to a periodic motion on branch 2, whereas at point b the stable solution is quasi-periodic, so that the cloud of points occupies a larger part of the state space. This difference explains, in particular, why the POD constructed with point b is better than the one built with point c (Amabili et al. 2003), because the problem defined by equation (6.8) is numerically better posed when the cloud of points spans a large portion of the state space. Continuous lines in Figure 6.11 are the sections of the subspaces provided by POD and NNM methods, respectively. The new POD axes are given by the α_i defined in equation (6.6), whose numerical values are recalled in Table 6.1. This figure clearly shows why three POD modes are necessary to recover the dynamics correctly. The new POD axes are extremely close to the original ones, and the cloud of points lies precisely in between them, with a significant contribution on $A_{1,0}$. Hence, the third POD mode, which corresponds to $A_{1,0}$, must mandatorily be kept in the reduced POD model, otherwise significant dynamical information is discarded.

On the other hand, the section of the four-dimensional invariant manifold (corresponding to two NNMs) is given by the parabolic curve, which goes exactly where the computed points are located. This explains why only two NNMs are necessary to recover the dynamics, as the four-dimensional manifold displays an important curvature in the direction of $A_{1,0}$.

Both the proper orthogonal decomposition (POD) and the nonlinear normal mode (NNM) methods have been verified to be suitable for building reduced-order models of a water-filled shell. In particular, a larger reduction of the model is possible by using the NNM method. However, the asymptotic formulation used here in the NNM method does not make it suitable for studying very large vibration amplitudes, where the POD model performs better.

The nonlinear character of the invariant manifold that defines NNMs allows better reduction than the POD method, which is a linear decomposition. On the other hand, the global nature of the POD provides very robust results even for complex dynamics, whereas the asymptotic NNMs has been found to fail in recovering these motions. However, it has been found that, despite being an approximation, the qualitative behavior of the NNMs compares very well with the original solution, until the validity limits are attained. It has been found numerically that these limits are not that small: amplitude of vibration up to $3h$.

Construction of the NNM-based reduced-order model with the asymptotic method is direct and does not need intensive computations, because a single nonlinear change of coordinates, computed once and for all, is required. The method can thus be blindly applied, provided the Galerkin projection has been performed on a large number of modes. On the other hand, for the POD, particular care must be taken in the choice of the time responses used to build it.

To conclude, the investigations conducted here show that for moderate vibration amplitudes, the asymptotic NNM method provides more reduced equations that always recover the qualitative behavior. However, the method must be modified in order to bypass its main limitation, which is due to the asymptotic development. Unfortunately, only numerical solutions are possible, thus leading to an intense numerical effort in order to build the reduced-order model. Hence, for very large vibration amplitude and a large range of parameter variations, the POD method still performs better because of its global nature.

REFERENCES

M. Amabili, A. Sarkar and M. P. Païdoussis 2003 *Journal of Fluids and Structures* **18**, 227–250. Reduced-order models for nonlinear vibrations of cylindrical shells via the proper orthogonal decomposition method.

M. Amabili, A. Sarkar and M. P. Païdoussis 2006 *Journal of Sound and Vibration* **290**, 736–762. Chaotic vibrations of circular cylindrical shells: Galerkin versus reduced-order models via the proper orthogonal decomposition method.

M. Amabili and C. Touzé 2007 *Journal of Fluids and Structures* **23**, 885–903. Reduced-order models for nonlinear vibrations of fluid-filled circular cylindrical shells: comparison of POD and asymptotic nonlinear normal modes methods.

N. Aubry, P. Holmes, J. L. Lumley and E. Stone 1988 *Journal of Fluid Mechanics* **192**, 115–173. The dynamics of coherent structures in the wall region of a turbulent boundary layer.

M. F. Azeez and A. F. Vakakis 2001 *Journal of Sound and Vibration* **240**, 859–889. Proper orthogonal decomposition (POD) of a class of vibroimpact oscillations.

S. Bellizzi and R. Bouc 2005 *Journal of Sound and Vibration* **287**, 545–569. A new formulation for the existence and calculation of nonlinear normal modes.

K. S. Breuer and L. Sirovich 1991 *Journal of Computational Physics* **96**, 277–296. The use of the Karhunen-Loève procedure for the calculation of linear eigenfunctions.

J. Carr 1981 *Applications of Centre Manifold Theory*. Springer-Verlag, New York, USA.

C. Elphick, E. Tirapegui, M. Brachet, P. Coullet and G. Iooss 1987 *Physica D* **29**, 95–127. A simple global characterization for normal forms of singular vector fields.

I. T. Georgiou 2005 *Nonlinear Dynamics* **41**, 69–110. Advanced proper orthogonal decomposition tools: using reduced order models to identify normal modes of vibration of slow invariant manifolds in the dynamics of planar nonlinear rods.

I. T. Georgiou, I. Schwartz, E. Emaci and A. Vakakis 1999 *Journal of Applied Mechanics* **66**, 448–459. Interaction between slow and fast oscillations in an infinite degree-of-freedom linear system coupled to nonlinear subsystem: theory and experiment.

J. Guckenheimer and P. Holmes 1983 *Non-linear Oscillations, Dynamical Systems and Bifurcations of Vector Fields*. Springer, New York, USA.

G. Iooss and M. Adelmeyer 1998 *Topics in Bifurcation Theory*, 2nd edition. World Scientific, New York, USA.

L. Jézéquel and C. H. Lamarque 1991 *Journal of Sound and Vibration* **149**, 429–459. Analysis of non-linear dynamical systems by the normal form theory.

D. Jiang, C. Pierre and S. Shaw 2005a *International Journal of Non-linear Mechanics* **40**, 729–746. The construction of non-linear normal modes for systems with internal resonance.

D. Jiang, C. Pierre and S. Shaw 2005b *Journal of Sound and Vibration* **288**, 791–812. Nonlinear normal modes for vibratory systems under harmonic excitation.

G. Kerschen, B. F. Feeny, J.-C. Golinval 2003. *Computer Methods in Applied Mechanics and Engineering* **192**, 1785–1795. On the exploitation of chaos to build reduced-order models.

G. Kerschen, J.-C. Golinval, A. F. Vakakis and L. A. Bergman 2005 *Nonlinear Dynamics* **41**, 147–169. The method of proper orthogonal decomposition for dynamical characterization and order reduction of mechanical systems: an overview.

M. E. King and A. F. Vakakis 1994 *Journal of Vibration and Acoustics* **116**, 332–340. Energy-based formulation for computing nonlinear normal modes in undamped continuous systems.

W. Lacarbonara, G. Rega and A. H. Nayfeh 2003 *International Journal of Non-linear Mechanics* **38**, 851–872. Resonant non-linear normal modes. Part I: analytical treatment for structural one-dimensional systems.

Y. V. Mikhlin 1995 *Journal of Sound and Vibration* **182**, 577–588. Matching of local expansions in the theory of non-linear vibrations.

E. Pesheck, C. Pierre and S. Shaw 2002 *Journal of Sound and Vibration* **249**, 971–993. A new Galerkin-based approach for accurate non-linear normal modes through invariant manifolds.

H. Poincaré 1892 *Les Méthodes Nouvelles de la Mécanique Céleste*. Gauthiers-Villars, Paris, France.

R. M. Rosenberg 1966 *Advances in Applied Mechanics* **9**, 155–242. On non-linear vibrations of systems with many degrees of freedom.

A. Sarkar and M. P. Païdoussis 2003 *Journal of Fluids and Structures* **17**, 525–539. A compact limit-cycle oscillation model of a cantilever conveying fluid.

A. Sarkar and M. P. Païdoussis 2004 *International Journal of Non-Linear Mechanics* **39**, 467–481. A cantilever conveying fluid: coherent modes versus beam modes.

S. Shaw and C. Pierre 1991 *Journal of Sound and Vibration* **150**, 170–173. Non-linear normal modes and invariant manifolds.

S. Shaw and C. Pierre 1993 *Journal of Sound and Vibration* **164**, 85–124. Normal modes for non-linear vibratory systems.

L. Sirovich 1987 *Quarterly Journal of Applied Mathematics* **45**, 561–571. Turbulence and dynamics of coherent structures. Part I: coherent structures.

J. C. Slater 1996 *Nonlinear Dynamics* **10**, 19–30. A numerical method for determining nonlinear normal modes.

C. Touzé and M. Amabili 2006 *Journal of Sound and Vibration* **298**, 958–981. Non-linear normal modes for damped geometrically non-linear systems: application to reduced-order modeling of harmonically forced structures.

C. Touzé, O. Thomas and A. Chaigne 2004 *Journal of Sound and Vibration* **273**, 77–101. Hardening/softening behaviour in nonlinear oscillations of structural systems using non-linear normal modes.

A. F. Vakakis, L. I. Manevich, Y. V. Mikhlin, V. N. Pilipchuk and A. A. Zevin 1996 *Normal Modes and Localization in Non-linear Systems*. Wiley, New York, USA.

S. A. Zahorian and M. Rothenberg 1981 *Journal of the Acoustical Society of America* **69**, 519–524. Principal component analysis for low-redundancy encoding of speech spectra.

7 Comparison of Different Shell Theories for Nonlinear Vibrations and Stability of Circular Cylindrical Shells

7.1 Introduction

Most of studies on large-amplitude (geometrically nonlinear) vibrations of circular cylindrical shells used Donnell's nonlinear shallow-shell theory to obtain the equations of motion, as shown in Chapter 5. Only a few used the more refined Sanders-Koiter or Flügge-Lur'e-Byrne nonlinear shell theories. The majority of these studies do not include geometric imperfections, and some of them use a single-mode approximation to describe the shell dynamics.

This chapter presents a comparison of shell responses to radial harmonic excitation in the spectral neighborhood of the lowest natural frequency computed by using five different nonlinear shell theories: (i) Donnell's shallow-shell, (ii) Donnell's with in-plane inertia, (iii) Sanders-Koiter, (iv) Flügge-Lur'e-Byrne and (v) Novozhilov theories. These five shell theories are practically the only ones applied to geometrically nonlinear problems among the theories that neglect shear deformation. Donnell's shallow-shell theory has already been used in Chapter 5, and the numerical results presented there are used for comparison. Shell theories including shear deformation and rotary inertia are not considered in this chapter. The results presented are based on the study by Amabili (2003).

7.2 Energy Approach

The elastic strain energy of the shell is given by equation (1.141), in which the expressions of the middle surface strain-displacement relationships and changes in curvature and torsion must be inserted according to the selected nonlinear shell theory.

The kinetic energy T_S of a circular cylindrical shell, by neglecting rotary inertia but retaining in-plane inertia, is given by

$$T_S = \frac{1}{2}\rho h \int_0^{2\pi} \int_0^L (\dot{u}^2 + \dot{v}^2 + \dot{w}^2)\,\mathrm{d}x\,R\,\mathrm{d}\theta, \tag{7.1}$$

where ρ is the mass density of the shell. In equation (7.1), the overdot denotes time derivative.

The virtual work W done by the external forces is written as

$$W = \int_0^{2\pi}\int_0^L (q_x u + q_\theta v + q_r w)\,\mathrm{d}x\,R\,\mathrm{d}\theta, \tag{7.2}$$

where q_x, q_θ and q_r are the distributed forces per unit area acting in axial, circumferential and radial directions, respectively. Initially, only a single harmonic radial force is considered; therefore, $q_x = q_\theta = 0$. The external radial distributed load q_r applied to the shell, because of the radial concentrated force $\tilde{f}$, is given by equation (5.46). Here, the point excitation is located at $\tilde{x} = L/2$, $\tilde{\theta} = 0$. Equation (7.2) can be rewritten in the following form:

$$W = \tilde{f}\cos(\omega t)\,(w)_{x=L/2,\,\theta=0}\,. \tag{7.3}$$

In order to reduce the system to finite dimensions, the middle surface displacements u, v and w are expanded by using trial functions. The boundary conditions imposed at the shell ends, $x = 0, L$, are given by equations (5.48) and (5.49).

The study in Chapter 5 shows that it is necessary to consider, in addition to the asymmetric mode directly driven into vibration by the excitation given in equation (5.46) (driven mode), (i) the orthogonal mode having the same shape and natural frequency but rotated by $\pi/(2n)$ (companion mode), (ii) additional asymmetric modes and (iii) axisymmetric modes. According to these considerations, the displacements u, v and w are expanded by using the eigenmodes of the simply supported, empty shell:

$$u(x,\theta,t) = \sum_{m=1}^{M_1}\sum_{j=1}^{N}\left[u_{m,j,c}(t)\cos(j\theta) + u_{m,j,s}(t)\sin(j\theta)\right]\cos(\lambda_m x) + \sum_{m=1}^{M_2} u_{m,0}(t)\cos(\lambda_m x), \tag{7.4a}$$

$$v(x,\theta,t) = \sum_{m=1}^{M_1}\sum_{j=1}^{N}\left[v_{m,j,c}(t)\sin(j\theta) + v_{m,j,s}(t)\cos(j\theta)\right]\sin(\lambda_m x) + \sum_{m=1}^{M_2} v_{m,0}(t)\sin(\lambda_m x), \tag{7.4b}$$

$$w(x,\theta,t) = \sum_{m=1}^{M_1}\sum_{j=1}^{N}\left[w_{m,j,c}(t)\cos(j\theta) + w_{m,j,s}(t)\sin(j\theta)\right]\sin(\lambda_m x) + \sum_{m=1}^{M_2} w_{m,0}(t)\sin(\lambda_m x), \tag{7.4c}$$

where j is the number of circumferential waves, m is the number of longitudinal half-waves, $\lambda_m = m\pi/L$ and t is time; $u_{m,j}(t)$, $v_{m,j}(t)$ and $w_{m,j}(t)$ are the generalized coordinates that are unknown functions of t. The additional subscript c or s indicates if the generalized coordinate is associated with cosine or sine function in θ, except for v, for which the notation is reversed (no additional subscript is used for axisymmetric terms). The integers N, M_1 and M_2 must be selected with care in order to obtain the required accuracy and acceptable dimension of the nonlinear problem.

Excitation in the neighborhood of resonance of mode with $m = 1$ longitudinal half-wave and n circumferential waves, indicated as resonant mode (m, n) for simplicity, is considered. The minimum number of degrees of freedom that has been found to be necessary to predict the nonlinear response with good accuracy is 14. In particular, asymmetric modes having up to three longitudinal half-waves ($M_1 = 3$, only odd m values) and modes having $n, 2 \times n, 3 \times n$ and $4 \times n$ circumferential waves have been considered in the numerical calculations. For axisymmetric modes, up to $M_2 = 9$ have been used (only odd m values).

The minimal expansion used in the numerical calculation for excitation in the neighborhood of resonance of mode $(m = 1, n)$ is

$$u(x, \theta, t) = [u_{1,n,c}(t)\cos(n\theta) + u_{1,n,s}(t)\sin(n\theta)]\ \cos(\lambda_1 x) + \sum_{m=1}^{2} u_{2m-1,0}(t)\cos(\lambda_{2m-1}x), \tag{7.5a}$$

$$v(x, \theta, t) = \sum_{k=1}^{2}[v_{1,kn,c}(t)\sin(kn\theta) + v_{1,kn,s}(t)\cos(kn\theta)]\ \sin(\lambda_1 x) + [v_{3,2n,c}(t)\sin(2n\theta) + v_{3,2n,s}(t)\cos(2n\theta)]\sin(\lambda_3 x), \tag{7.5b}$$

$$w(x, \theta, t) = [w_{1,n,c}(t)\cos(n\theta) + w_{1,n,s}(t)\sin(n\theta)]\ \sin(\lambda_1 x) + \sum_{m=1}^{2} w_{2m-1,0}(t)\sin(\lambda_{2m-1}x). \tag{7.5c}$$

This expansion has 14 generalized coordinates (degrees of freedom) and guarantees good accuracy for the calculation performed here. Accuracy and convergence of the solution have been checked numerically. It has been verified that no terms can be eliminated from the expansion given in equations (7.5) without introducing large errors in the shell response. However, in most calculations, 16 or more (26, 36 and 48) generalized coordinates have been used. In the expansion with 16 coordinates, which has been largely used, terms $[u_{1,2n,c}(t)\cos(2n\theta) + u_{1,2n,s}(t)\sin(2n\theta)]\ \cos(\lambda_1 x)$ have been added to equation (7.5a); however, they do not give a significant improvement to the accuracy of the solution with respect to equations (7.5). It can be observed that, if a different choice for the mode expansion with respect to equation (7.4) is used, a different number of generalized coordinates is necessary to obtain good accuracy.

The point excitation considered gives a direct excitation only to modes described by a cosine function in an angular direction in equation (7.5c), referred to as the driven modes, and to axisymmetric modes. The modes having the shape of the driven modes, but rotated by $\pi/(2n)$, that is, with the form described by the sine function in the angular direction θ, are referred to as the companion modes.

It is extremely interesting to observe that in equations (7.5a) and (7.5c) only the resonant driven and companion modes plus the first and third axisymmetric modes are retained in the minimal expansion. This is in perfect agreement with the result obtained in Chapter 5 for the expansion of the radial displacement w used with Donnell's nonlinear shallow-shell theory. For the circumferential displacement v, no axisymmetric modes are necessary. In fact, the first axisymmetric modes with prevalent radial displacement have contribution of axial displacement but zero contribution of circumferential displacement, as shown in Figure 5.6. However, it is necessary to retain in the expansion of v terms with $2n$ circumferential waves and three longitudinal half-waves. It must be clarified that Donnell's nonlinear shallow-shell theory used in Chapter 5 takes into account in-plane displacements by using the stress function, which has terms having twice the axial and angular wavenumbers of radial terms.

Initial geometric imperfections of the circular cylindrical shell are considered only in the radial direction. They are associated with zero initial stress. The radial imperfection w_0 is expanded in the form given by equation (5.52).

7.2.1 Additional Terms to Satisfy the Boundary Conditions

Equations (5.48a,c) and (5.49b) are identically satisfied by the expansions of u, v, w and w_0. Moreover, the continuity in θ of all the displacements is also satisfied. On the other hand, equations (5.48b) and (5.49a) can be rewritten in the following form:

$$M_x = \frac{Eh^3}{12(1-\nu^2)}(k_x + \nu k_\theta) = 0, \tag{7.6}$$

$$N_x = \frac{Eh}{1-\nu^2}(\varepsilon_{x,0} + \nu \varepsilon_{\theta,0}) = 0. \tag{7.7}$$

Equation (7.6) is identically satisfied for the expressions of k_x and k_θ given in equations (1.106, 1.107) and (1.136, 1.137) for Donnell's and the Sanders-Koiter nonlinear shell theories, respectively. For the Flügge-Lur'e-Byrne, modified in order to retain nonlinear terms in changes of curvature and torsion, and Novozhilov nonlinear shell theories, equation (7.6) is not satisfied only by nonlinear terms that are extremely small for thin shells. For this reason, equation (7.6) can be considered identically satisfied also for these two shell theories.

Equation (7.7) is not identically satisfied by all the nonlinear shell theories. According to the Sanders-Koiter theory (see equations [1.133] and [1.134]), and eliminating null terms at the shell edges, equation (7.7) can be rewritten as

$$\left[\frac{\partial \hat{u}}{\partial x} + \frac{1}{8}(1+\nu)\left(\frac{\partial v}{\partial x} - \frac{\partial u}{R\partial\theta}\right)^2 + \frac{1}{2}\left(\frac{\partial w}{\partial x}\right)^2 + \frac{\partial w}{\partial x}\frac{\partial w_0}{\partial x}\right]_{x=0,\,L} = 0, \tag{7.8}$$

where $\hat{u}$ is a term added to the expansion of u, given in equation (7.4a) or (7.5a), in order to satisfy exactly the axial boundary conditions $N_x = 0$ (note that $\partial u/\partial x = 0$ at $x = 0$, L according to equations [7.4a] and [7.5a]). Because $\hat{u}$ is a second-order term in the shell displacement, it has not been inserted in the second-order terms that involve u in equation (7.8).

All the generalized coordinates, except the six coordinates associated with the resonant mode (m, n), which are $u_{m,n,c}(t)$, $v_{m,n,c}(t)$, $w_{m,n,c}(t)$, $u_{m,n,s}(t)$, $v_{m,n,s}(t)$, $w_{m,n,s}(t)$, are neglected because they are an infinitesimal of higher order. Calculations give

$$\begin{aligned}\hat{u}(t) = &-\frac{1}{32}[a(t) + b(t)\cos(2n\theta) + c(t)\sin(2n\theta)]\sin(2m\pi x/L) - (m\pi/L)\\ &\times [w_{m,n,c}(t)\cos(n\theta) + w_{m,n,s}(t)\sin(n\theta)]\\ &\times \sum_{j=0}^{\tilde{N}}\sum_{i=1}^{\tilde{M}} \frac{i}{m+i}\left[\tilde{A}_{i,j}\cos(j\theta) + \tilde{B}_{i,j}\sin(j\theta)\right]\sin\left[(m+i)\pi x/L\right],\end{aligned} \tag{7.9}$$

where $\tilde{M}$ is the largest between $\tilde{M}_1$ and $\tilde{M}_2$ and

$$\begin{aligned} a(t) = {} & (4m\pi/L)\left(w_{m,n,c}^2 + w_{m,n,s}^2\right) + (1+\nu)(m\pi/L)\left(v_{m,n,c}^2 + v_{m,n,s}^2\right) \\ & + (1+\nu)[Ln^2/(m\pi R^2)]\left(u_{m,n,c}^2 + u_{m,n,s}^2\right) \\ & - 2(1+\nu)(n/R)\left(v_{m,n,s}u_{m,n,s} - v_{m,n,c}u_{m,n,c}\right), \end{aligned} \tag{7.10}$$

$$\begin{aligned} b(t) = {} & (4m\pi/L)\left(w_{m,n,c}^2 - w_{m,n,s}^2\right) + (1+\nu)(m\pi/L)\left(v_{m,n,s}^2 - v_{m,n,c}^2\right) \\ & + (1+\nu)[Ln^2/(m\pi R^2)]\left(u_{m,n,s}^2 - u_{m,n,c}^2\right) \\ & - 2(1+\nu)(n/R)\left(v_{m,n,s}u_{m,n,s} + v_{m,n,c}u_{m,n,c}\right), \end{aligned} \tag{7.11}$$

$$\begin{aligned} c(t) = {} & (8m\pi/L)w_{m,n,c}w_{m,n,s} + 2(1+\nu)(m\pi/L)v_{m,n,c}v_{m,n,s} \\ & - 2(1+\nu)[Ln^2/(m\pi R^2)]u_{m,n,c}u_{m,n,s} \\ & - 2(1+\nu)(n/R)\left(v_{m,n,c}u_{m,n,s} - v_{m,n,s}u_{m,n,c}\right). \end{aligned} \tag{7.12}$$

For the Donnell, Flügge-Lur'e-Byrne and Novozhilov nonlinear shell theories, equation (7.9) is still correct, but equations (7.10–7.12) are simplified. For the Donnell nonlinear shell theory,

$$a(t) = (4m\pi/L)\left(w_{m,n,c}^2 + w_{m,n,s}^2\right), \tag{7.13}$$

$$b(t) = (4m\pi/L)\left(w_{m,n,c}^2 - w_{m,n,s}^2\right), \tag{7.14}$$

$$c(t) = (8m\pi/L)\, w_{m,n,c}w_{m,n,s}. \tag{7.15}$$

For the Flügge-Lur'e-Byrne and Novozhilov theories,

$$a(t) = (4m\pi/L)\left(w_{m,n,c}^2 + w_{m,n,s}^2 + v_{m,n,c}^2 + v_{m,n,s}^2\right) + 4\nu[Ln^2/(m\pi R^2)]\left(u_{m,n,c}^2 + u_{m,n,s}^2\right), \tag{7.16}$$

$$b(t) = (4m\pi/L)\left(w_{m,n,c}^2 - w_{m,n,s}^2 + v_{m,n,s}^2 - v_{m,n,c}^2\right) + 4\nu[Ln^2/(m\pi R^2)]\left(u_{m,n,s}^2 - u_{m,n,c}^2\right), \tag{7.17}$$

$$c(t) = (8m\pi/L)\left(w_{m,n,c}w_{m,n,s} + v_{m,n,c}v_{m,n,s}\right) - 8\nu\left[Ln^2/\left(m\pi R^2\right)\right]u_{m,n,c}u_{m,n,s}. \tag{7.18}$$

7.2.2 Fluid-Structure Interaction

The shell is assumed to be fluid filled and to have open ends. In addition, in this case, nonlinear mechanics in the fluid deformation can be neglected, and the stresses due to the fluid weight in the shell are also neglected. Only the kinetic energy T_F is associated with inviscid and incompressible fluid:

$$T_F = \frac{1}{2}\rho_F, \iiint_v \nabla\Phi\cdot\nabla\Phi\, dV, \tag{7.19}$$

where V is the fluid volume and the potential function Φ is given in the form of equation (5.33a) with the simplification $k_1 = \lambda_m$ for incompressible fluid. As introduced in Chapter 5, the velocity of the fluid at any point is related to Φ by $\mathbf{v} = \nabla\Phi$. Here the sign has been changed since w is positive outward.

By using the Green's theorem (Amabili 2000),

$$\iiint_V \nabla\Phi \cdot \nabla\Phi \, \mathrm{d}V = \iint_{\partial V} \Phi \frac{\partial \Phi}{\partial \eta} \, \mathrm{d}S = \iint_S \Phi \frac{\partial \Phi}{\partial \eta} \, \mathrm{d}S + \iint_{S_E} \Phi \frac{\partial \Phi}{\partial \eta} \, \mathrm{d}S, \quad (7.20)$$

where ∂V is the boundary of the fluid volume, which is given by the shell internal surface S and the two circular surfaces (globally indicated with S_E) at the shell edges where the condition $\Phi = 0$ is imposed for open ends; η indicates the normal to ∂V directed outward the fluid domain. The last term in equation (7.20) is identically zero for open ends. By using equations (7.20) and (5.21), and considering that the normal η coincides with r on the surface S, equation (7.19) can be transformed into

$$T_F = \frac{1}{2} \rho_F \int_0^{2\pi} \int_0^L (\Phi)_{r=R} \, \dot{w} \, \mathrm{d}x \, R \, \mathrm{d}\theta. \quad (7.21)$$

As proved by Amabili (2000) in the case of harmonic oscillation, equation (7.21) is also valid in the case of compressible fluid.

7.2.3 Lagrange Equations of Motion

The nonconservative damping forces are assumed to be of viscous type and are taken into account by using Rayleigh's dissipation function:

$$F = \frac{1}{2} c \int_0^{2\pi} \int_0^L (\dot{u}^2 + \dot{v}^2 + \dot{w}^2) \mathrm{d}x \, R \, \mathrm{d}\theta, \quad (7.22)$$

where c has a different value for each term of the mode expansion. Simple calculations give

$$F = \frac{1}{2}(L/2) R \sum_{n=0}^{N} \sum_{m=1}^{M} \psi_n \left[c_{m,n,c} \left(\dot{u}_{m,n,c}^2 + \dot{v}_{m,n,c}^2 + \dot{w}_{m,n,c}^2 \right) + c_{m,n,s} \left(\dot{u}_{m,n,s}^2 + \dot{v}_{m,n,s}^2 + \dot{w}_{m,n,s}^2 \right) \right], \quad (7.23)$$

where

$$\psi_n = \begin{cases} 2\pi & \text{if } n = 0 \\ \pi & \text{if } n > 0. \end{cases} \quad (7.24)$$

The damping coefficient $c_{m,n,c \text{ or } s}$ is related to the modal damping ratio, which can be evaluated from experiments, by $\zeta_{m,n,c \text{ or } s} = c_{m,n,c \text{ or } s} / (2\mu_{m,n}\omega_{m,n})$, where $\omega_{m,n}$ is the natural circular frequency of mode (m, n) and $\mu_{m,n}$ is the modal mass of this mode, given by $\mu_{m,n} = \psi_n (\rho + \rho_V) h (L/2) R$, and the virtual mass due to contained fluid is

$$\rho_V = \frac{\rho_F}{\lambda_m h} \frac{\mathrm{I}_n(\lambda_m R)}{\mathrm{I}'_n(\lambda_m R)}. \quad (7.25)$$

The total kinetic energy of the system is

$$T = T_S + T_F. \quad (7.26)$$

The potential energy of the system is only the elastic strain energy of the shell:

$$U = U_S. \quad (7.27)$$

The virtual work done by the concentrated radial force $\tilde{f}$, given by equation (7.3), is specialized for the expression of w given in equation (7.4c):

$$W = \tilde{f}\cos(\omega t)\,(w)_{x=L/2,\,\theta=0} = \tilde{f}\cos(\omega t)\left[\sum_{m=1}^{M_1}\sum_{j=1}^{N} w_{m,j,c}(t) + \sum_{m=1}^{M_2} w_{m,0}(t)\right]. \quad (7.28)$$

In the presence of axial loads and radial pressure acting on the shell, additional virtual work is done by the external forces. Let us consider a time-dependent axial load $N(t)$ applied at both shell ends; $N(t)$ is positive in the x direction. In particular, $-N(t)$ is applied at $x = 0$ and $N(t)$ is applied at $x = L$. The axial distributed force q_x has the following expression:

$$q_x = N(t)\,[-\delta(x) + \delta(x - L)], \quad (7.29)$$

where δ is the Dirac delta function. The virtual work done by the axial load is

$$W = \int_0^{2\pi}\int_0^{L} N(t)\,[-\delta(x) + \delta(x - L)]\,u\,\mathrm{d}x\,R\,\mathrm{d}\theta = -4\pi RN(t)\sum_{m=1}^{M_2} u_{m,0}(t). \quad (7.30)$$

In the case of uniform internal time-varying pressure $p_r(t)$, the radial distributed force q_r is obviously

$$q_r = p_r(t). \quad (7.31)$$

The virtual work done by radial pressure is

$$W = \int_0^{2\pi}\int_0^{L} p_r(t) w\,\mathrm{d}x\,R\,\mathrm{d}\theta = 4Rp_r(t)L\sum_{m=1}^{M_2} w_{m,0}(t)/m. \quad (7.32)$$

Equations (7.30) and (7.32) show that only axisymmetric vibrations are directly excited by uniform axial loads and radial pressure.

The following notation is introduced:

$$\mathbf{q} = \{u_{m,n,c}, u_{m,n,s}, v_{m,n,c}, v_{m,n,s}, w_{m,n,c}, w_{m,n,s}\}^{\mathrm{T}}, \quad m = 1, \ldots, M \text{ and } n = 0, \ldots, N. \quad (7.33)$$

The generic element of the time-dependent vector $\mathbf{q}$ is referred to as q_j. The dimension of $\mathbf{q}$ is $\overline{N}$, which is the number of degrees of freedom (dofs) used in the mode expansion.

The generalized forces Q_j are obtained by differentiation of the Rayleigh dissipation function and of the virtual work done by external forces:

$$Q_j = -\frac{\partial F}{\partial \dot{q}_j} + \frac{\partial W}{\partial q_j} = -c_j\dot{q}_j + \begin{cases} 0 & \text{if } q_j = u_{m,n,c/s},\ v_{m,n,c/s} \text{ or } w_{m,n,s}, \\ \tilde{f}\cos(\omega t) & \text{if } q_j = w_{m,n,c}, \end{cases} \quad (7.34)$$

where the subscript c/s indicates c or s.

The Lagrange equations of motion for the fluid-filled shell are

$$\frac{\mathrm{d}}{\mathrm{d}t}\left(\frac{\partial T}{\partial \dot{q}_j}\right) - \frac{\partial T}{\partial q_j} + \frac{\partial U}{\partial q_j} = Q_j, \quad j = 1, \ldots, \overline{N}, \quad (7.35)$$

where $\partial T/\partial q_j = 0$. These second-order equations have very long expressions containing quadratic and cubic nonlinear terms. In particular,

$$\frac{\mathrm{d}}{\mathrm{d}t}\left(\frac{\partial T}{\partial \dot{q}_j}\right) = \begin{cases} \rho h(L/2)\psi_n R\ddot{q}_j & \text{if } q_j = u_{m,n,c/s} \text{ or } v_{m,n,c/s}, \\ (\rho + \rho_V)h(L/2)\psi_n R\ddot{q}_j & \text{if } q_j = w_{m,n,c/s}, \end{cases} \tag{7.36}$$

which shows that no inertial coupling among the Lagrange equations exists for the shell with simply supported edges with the mode expansion used.

The very long term containing quadratic and cubic nonlinearities can be written in the form:

$$\frac{\partial U}{\partial q_j} = \sum_{i=1}^{\bar{N}} f_{j,i} q_i + \sum_{i,k=1}^{\bar{N}} f_{j,i,k} q_i q_k + \sum_{i,k,l=1}^{\bar{N}} f_{j,i,k,l} q_i q_k q_l, \tag{7.37}$$

where coefficients f have long expressions that also include geometric imperfections.

7.3 Numerical Results for Nonlinear Vibrations

All calculations, with the exception of those in Section 7.3.1.1, have been performed for a shell having the following dimensions and material properties: $L = 520$ mm, $R = 149.4$ mm, $h = 0.519$ mm, $E = 1.98 \times 10^{11}$ Pa, $\rho = 7800$ kg/m^3 and $\nu = 0.3$. The shell has been considered empty or completely water filled with zero pressure on the fluid at the shell ends ($\rho_F = 1000$ kg/m^3), and water filled with geometric imperfections in Section 6.3. Comparison of present results with those available in the literature for an empty shell is given in Section 7.3.1.1. More results are presented for the water-filled shell than for the empty shell because the presence of a liquid largely enhances the softening-type nonlinearity of the response, as observed theoretically and experimentally. Therefore, a nonlinear study is much more important for the water-filled shell than for the empty one.

7.3.1 Empty Shell

The response of the circular cylindrical shell subjected to a harmonic point excitation of 2 N applied at the middle of the shell in the neighborhood of the lowest (fundamental) resonance $\omega_{1,n} = 2\pi \times 215.3$ rad/s, corresponding to mode ($m = 1$, $n = 5$), is given in Figure 7.1; only the principal (radial) coordinates, corresponding to driven and companion modes, are shown for brevity. The Flügge-Lur'e-Byrne nonlinear theory of shells has been used in the calculation, with modal damping $\zeta_{1,n} = 0.001(\zeta_{1,n} = \zeta_{1,n,c \text{ or } s})$. All the calculations reported in this section, if not diversely specified, have been performed by using an expansion involving 16 generalized coordinates, which are the ones given in equations (7.5) with the additional term $[u_{1,2n,c}(t)\cos(2n\theta) + u_{1,2n,s}(t)\sin(2n\theta)]\ \cos(\lambda_1 x)$ in equation (7.5a). The solution initially presents a single branch 1 with a folding and a typical softening-type behavior and corresponds to a driven mode vibration with zero amplitude of the companion mode. Branch 1 presents a pitchfork bifurcation around the peak of the response where branch 2 arises and where branch 1 loses stability. Branch 2 is the solution with participation of both driven and companion modes, giving a standing wave plus a traveling-wave response around the shell. Branch 2 loses stability at a Neimark-Sacker bifurcation where amplitude modulations of the solution arise; modulations end at a second Neimark-Sacker bifurcation. The response between the two

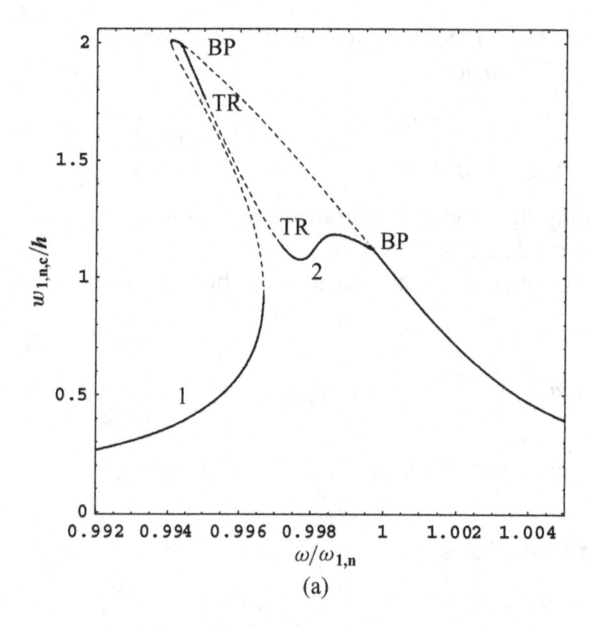

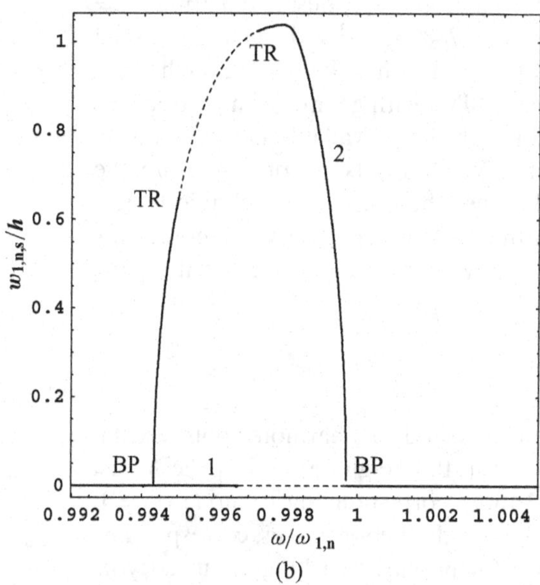

Figure 7.1. Frequency-response curve for the fundamental mode of the perfect, empty shell; $\zeta_{1,n} = 0.001$; Flügge-Lur'e-Byrne theory. Only resonant generalized coordinates are reported. (a) Amplitude of $w_{1,n,c}(t)$, driven mode. (b) Amplitude of $w_{1,n,s}(t)$, companion mode. 1, branch 1; 2, branch 2; BP, pitchfork bifurcation; TR, Neimark-Sacker bifurcation.

Neimark-Sacker bifurcations is quasi-periodic. Branch 2 ends at a second pitchfork bifurcation where it merges with branch 1 that regains stability. A similar behavior was found in Chapter 5 based on Donnell's nonlinear shallow-shell theory and the Galerkin method.

The response of the shell in the time domain is given in Figures 7.2 and 7.3 for excitation frequency $\omega/\omega_{1,n} = 0.998$ (periodic response on branch 2) and 0.9969 (quasi-periodic response), respectively. In Figures 7.2(a–c) the generalized coordinates associated with driven, companion and first axisymmetric modes are reported. It is clearly shown that the phase difference between driven and companion modes is $\theta_2 - \theta_1 \cong \pi/2$, giving a traveling-wave response. The relative amplitudes among the generalized coordinates $u_{1,n,c/s}$, $v_{1,n,c/s}$ and $w_{1,n,c/s}$ associated with driven and companion modes are almost the same as the ones observed for linear free vibrations

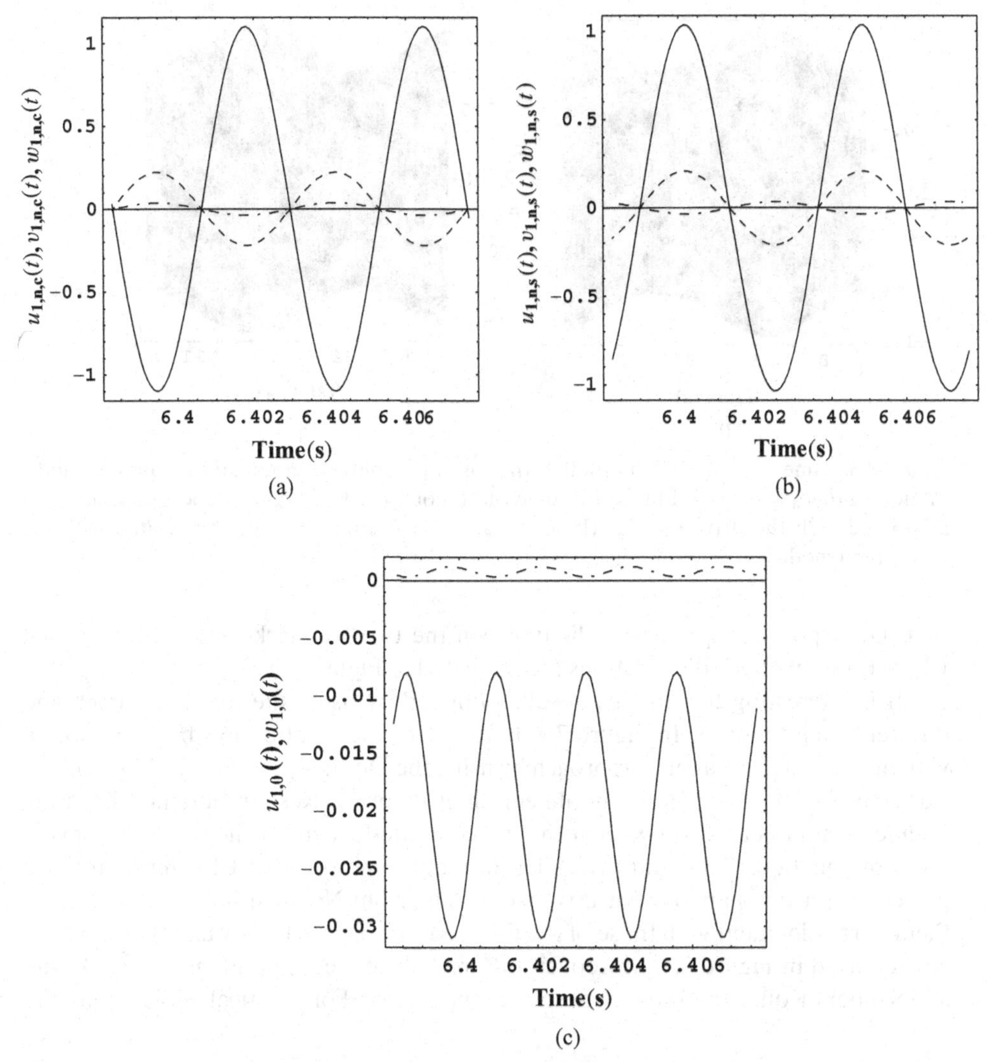

Figure 7.2. Time response of the shell in the case of traveling-wave response; $\omega/\omega_{1,n} = 0.998$; Flügge-Lur'e-Byrne theory. (a) Generalized coordinates associated with the driven mode: —, $w_{1,n,c}(t)$; – –, $v_{1,n,c}(t)$; –·–, $u_{1,n,c}(t)$. (b) Generalized coordinates associated with the companion mode: —, $w_{1,n,s}(t)$; – –, $v_{1,n,s}(t)$; –·–, $u_{1,n,s}(t)$. (c) Generalized coordinates associated with the first axisymmetric mode: —, $w_{1,0}(t)$; –·–, $u_{1,0}(t)$.

(eigenvectors of the linear solution). Only very small variations of the relative amplitudes are observed for these modes in the neighborhood of the resonance. However, this is not true for all the other generalized coordinates. When the excitation has a frequency close to the resonance of mode $(m=1, n)$, results show that the generalized coordinates associated with driven and companion modes $(m=1, n)$ have the same frequency as the excitation. The coordinates associated with modes $(m=1, 2n)$ and $(m=3, 2n)$ and all the coordinates associated with axisymmetric modes, as shown in Figure 7.2(c), have twice the frequency of the excitation; the coordinates associated with modes $(m=3, n)$, $(m=1, 3n)$ and $(m=3, 3n)$ have three times the frequency of the excitation. Figure 7.2(c) also shows a negative (inward) axisymmetric oscillation of the shell associated with large-amplitude asymmetric vibration. This is a characteristic of nonlinear vibrations of circular cylindrical shells. Figures 7.3(a,b) confirm

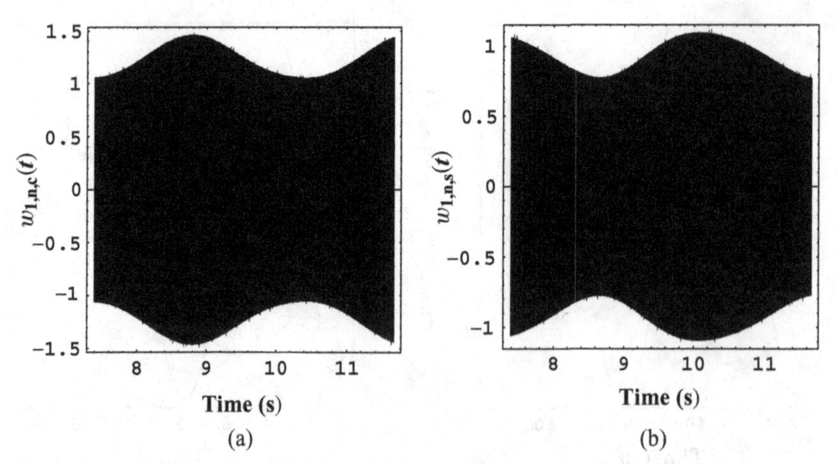

Figure 7.3. Time response of the shell in the case of amplitude modulations (quasi-periodic motion); $\omega/\omega_{1,n} = 0.9969$; Flügge-Lur'e-Byrne theory. (a) Generalized coordinate $w_{1,n,c}(t)$ associated with the driven mode. (b) Generalized coordinate $w_{1,n,s}(t)$ associated with the companion mode.

that the response is quasi-periodic between the two Neimark-Sacker bifurcations, where no stable periodic solutions are indicated in Figure 7.1.

It is interesting to compare results obtained by using the same approach and different shell theories. In Figure 7.4, branch 1 of the solution has been computed with the present Lagrangian approach by using the Donnell, Sanders-Koiter, Flügge-Lur'e-Byrne and Novozhilov nonlinear shell theories. Results obtained by using Donnell's nonlinear shallow-shell theory, computed by using the mode expansion given in equation (7.5c) and the Galerkin method reported in Chapter 5, are also presented for comparison. For this case, results from Novozhilov's nonlinear shell theory are coincident with those of the Flügge-Lur'e-Byrne theory and therefore are not reported in Figure 7.4. Moreover, differences between the Flügge-Lur'e-Byrne and Sanders-Koiter theories are absolutely negligible. For practical applications, the

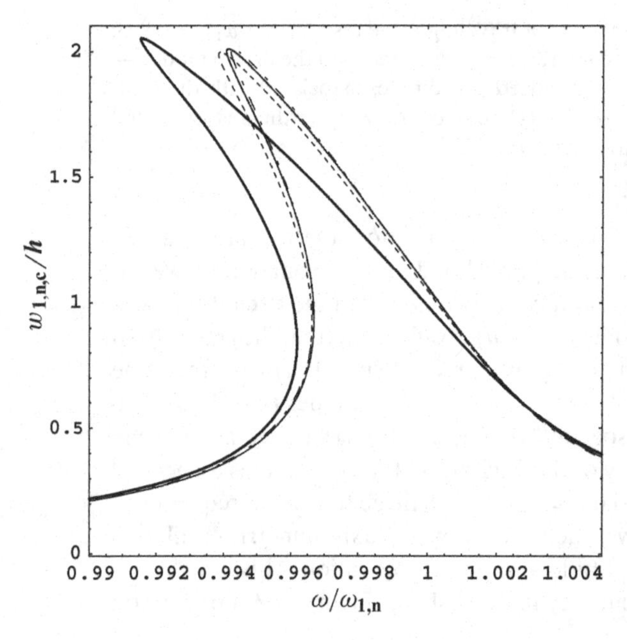

Figure 7.4. Frequency-response curve for the fundamental mode of the perfect, empty shell; branch 1 only; $\zeta_{1,n} = 0.001$. Only the generalized coordinate $w_{1,n,c}(t)$ is reported. —, Donnell's nonlinear shallow-shell theory; – –, Donnell's theory (Lagrangian approach); –·–, Sanders-Koiter theory; —, Flügge-Lur'e-Byrne theory (coincident with Novozhilov).

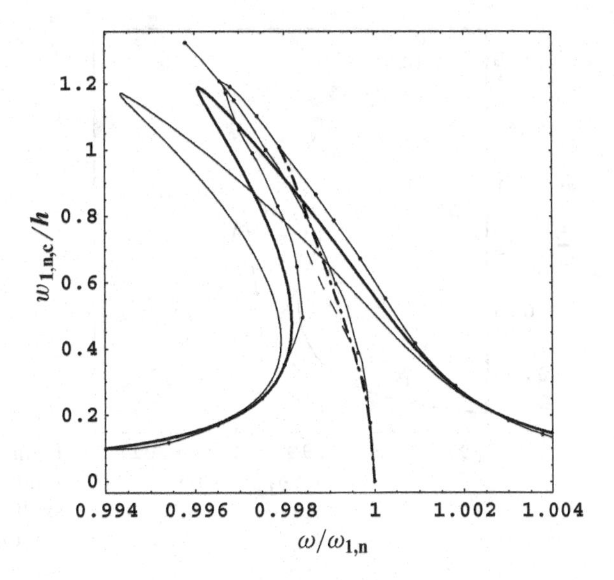

Figure 7.5. Frequency-response curve for the fundamental mode of the perfect, empty shell studied by Chen and Babcock (1975); branch 1 only. Only the generalized coordinate $w_{1,n,c}(t)$ is reported. ——, present study with 36 dofs (Donnell's theory, Lagrangian approach); —, shell response from Pellicano et al. (2002), also given in Figure 5.9 in Chapter 5, by using the Donnell's nonlinear shallow-shell theory with 16 dofs; —•—, backbone curve and response from Chen and Babcock (1975); – –, backbone curve from Varadan et al. (1989); — · —, babckbone curve from Ganapathi and Varadan (1996).

difference between the Flügge-Lur'e-Byrne and Donnell (with an energy approach that includes in-plane inertia) theories can be considered very small as well. However, although all the results obtained by using the present Lagrangian approach are very close to each other, differences from Donnell's nonlinear shallow-shell theory are quite significant; in particular, an excessive softening nonlinearity is predicted by Donnell's nonlinear shallow-shell theory. In fact, this theory neglects in-plane inertia, which plays a relevant role in axisymmetric modes. In this case, Donnell's nonlinear shallow-shell theory gives a wrong evaluation of the linear natural frequency of the first axisymmetric mode, and it explains the difference with all the other theories. It is interesting to observe that the mode investigated has $n = 5$ circumferential waves, which is a number almost at the limit of applicability of Donnell's shallow-shell theory.

7.3.1.1 Comparison with Results Available in the Literature

In this subsection only, numerical results for a different shell are carried out in order to compare present results to those available in the literature. Calculations have been performed for a simply supported, empty shell previously studied by Chen and Babcock (1975), Pellicano, Amabili and Païdoussis (2002), Varadan et al. (1989) and Ganapathi and Varadan (1996). Dimensions and material properties are: $L = 0.2$ m, $R = 0.1$ m, $h = 0.247$ mm, $E = 71.02 \times 10^9$ Pa, $\rho = 2796$ kg/m^3 and $\nu = 0.31$. A harmonic force excitation $\tilde{f} = 0.0785$ N and modal damping $\zeta_{1,n} = 0.0005$ are assumed. Donnell's nonlinear theory, retaining in-plane inertia, has been used with 36 degrees of freedom (dofs). In particular, the 36 generalized coordinates used are given by equations (7.4a–c) by inserting $j = k \times n$, $n = 6$ for the fundamental mode, $k = 1, \dots 4$; $M_1 = 3$ if $k = 2$, $M_1 = 1$ otherwise (odd m only); $M_2 = 5$ (odd m only). Only the fundamental mode ($n = 6, m = 1$) is investigated; it has a natural frequency of 555.9 Hz.

The shell response is shown in Figure 7.5. It is compared to (i) the response obtained in Chapter 5 and reported in Figure 5.9 by using Donnell's nonlinear shallow-shell theory with a mode expansion involving 16 dofs, originally obtained by Pellicano et al. (2002); (ii) the response calculated by Chen and Babcock (1975), who used the perturbation method to solve the nonlinear equations obtained by

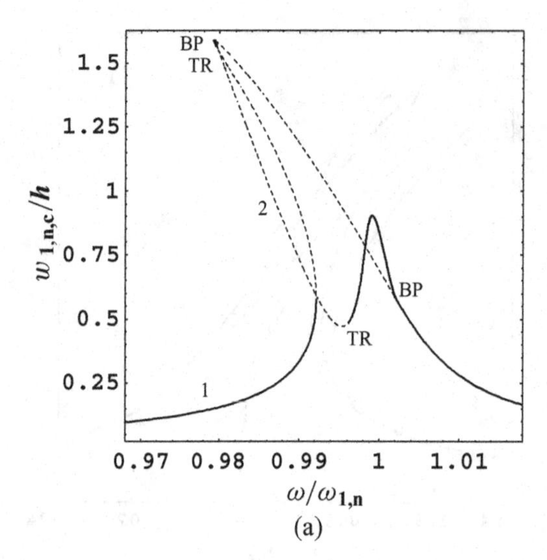

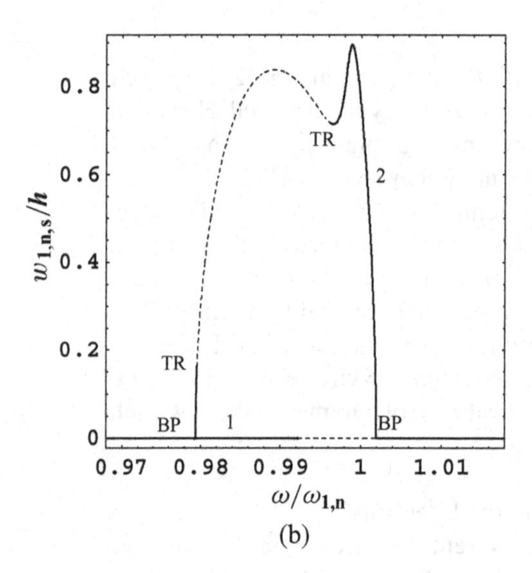

Figure 7.6. Frequency-response curve for the fundamental mode of the perfect, water-filled shell; $\zeta_{1,n} = 0.0017$; $\tilde{f} = 3$ N; Sanders-Koiter theory. Only the resonant generalized coordinates are reported. (a) Amplitude of $w_{1,n,c}(t)$, driven mode. (b) Amplitude of $w_{1,n,s}(t)$, companion mode. 1, branch 1; 2, branch 2: BP, pitchfork bifurcation; TR, Neimark-Sacker bifurcation.

Donnell's nonlinear shallow-shell theory; (iii) the backbone curve (indicating only the resonance, i.e. the peak of the response) computed by Varadan et al. (1989), who used a simple three-mode expansion and Donnell's nonlinear shallow-shell theory and (iv) the backbone curve obtained by Ganapathi and Varadan (1996), who used the finite-element method with the nonlinear terms of the Donnell shell theory. Present results are intermediate between those obtained by Pellicano et al. (2002) and those obtained by Ganapathi and Varadan (1996) and are close to those of Chen and Babcock (1975). This can be considered a good validation of the present model.

7.3.2 Water-Filled Shell

The response of the water-filled, circular cylindrical shell subjected to a harmonic point excitation of 3 N applied at the middle of the shell in the neighborhood of the lowest (fundamental) resonance $\omega_{1,n} = 2\pi \times 76.15$ rad/s, corresponding to mode $(m = 1, n = 5)$, is given in Figure 7.6. Only the principal coordinates, corresponding to driven and companion modes, are shown for brevity. The Sanders-Koiter nonlinear

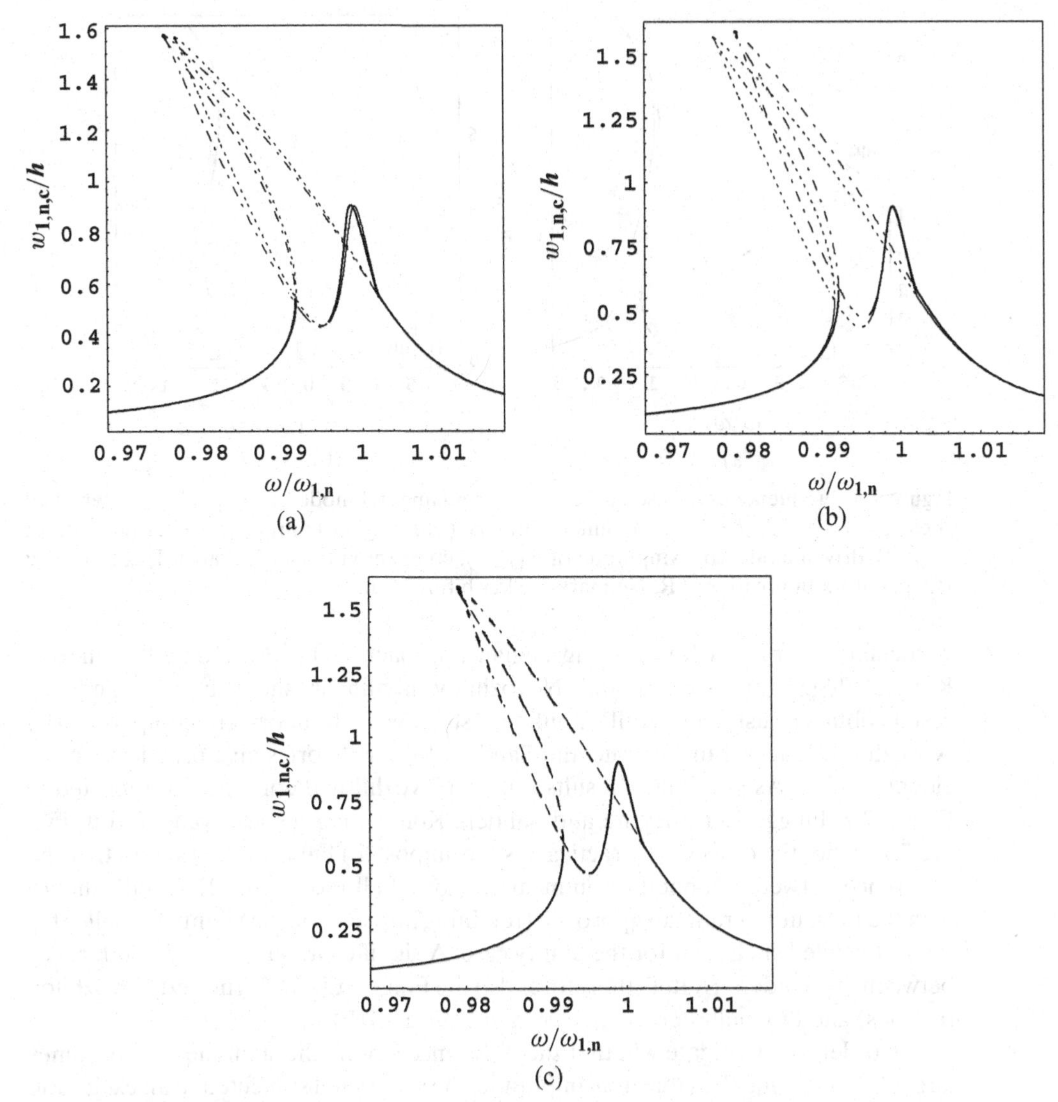

Figure 7.7. Frequency-response curve for the fundamental mode of the perfect, water-filled shell; branch 1 only; $\zeta_{1,n} = 0.0017$; $\tilde{f} = 3$ N. Only $w_{1,n,c}(t)$ is reported. (a) –·–, Donnell's nonlinear shallow-shell theory; – –, Donnell theory (Lagrangian approach). (b) – –, Donnell's nonlinear shallow-shell theory; –·–, Sanders-Koiter theory. (c) –·–, Sanders-Koiter theory; – –, Flügge-Lur'e-Byrne theory (coincident with Novozhilov).

theory of shells has been used in the calculation, with modal damping $\zeta_{1,n} = 0.0017$. All the calculations reported in this section, if not diversely specified, have been performed by using the expansion involving the 16 generalized coordinates previously specified. The response is qualitatively close to the response of the fundamental mode of the empty shell (see Figure 7.1). The main difference is the peak on branch 2 close to the second pitchfork bifurcation in Figure 7.6. From the quantitative point of view, the change is much more important because the frequency scale is greatly enlarged; it means that the water-filled shell presents a greatly enhanced nonlinear behavior of the softening type with respect to the same empty shell.

The comparison among different nonlinear shell theories is given in Figure 7.7, where both branches 1 and 2 are reported for the main coordinate $w_{1,n,c}(t)$. In

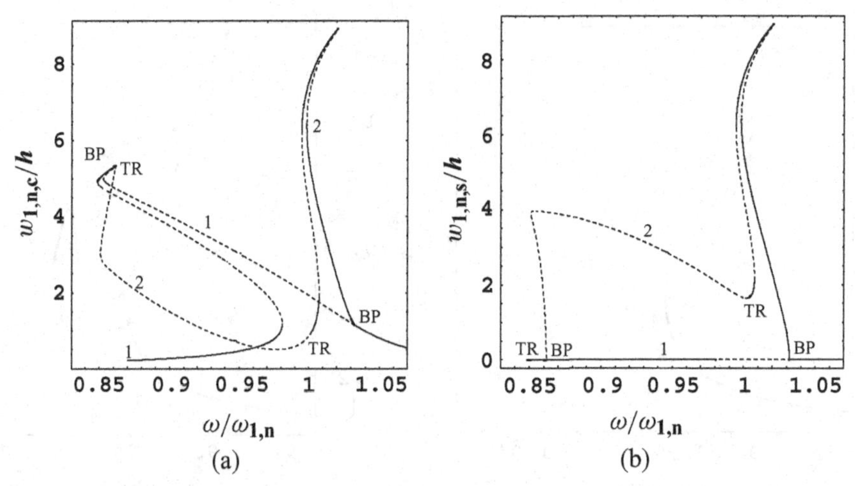

Figure 7.8. Frequency-response curve for the fundamental mode of the perfect, water-filled shell; $\zeta_{1,n} = 0.0017$; $\tilde{f} = 30$ N; Donnell's theory (Lagrangian approach). (a) Amplitude of $w_{1,n,c}(t)$, driven mode. (b) Amplitude of $w_{1,n,s}(t)$, companion mode. 1, branch 1; 2, branch 2; BP, pitchfork bifurcation; TR, Neimark-Sacker bifurcation.

particular, the results for the Lagrangian approach using the Donnell, Sanders-Koiter, Flügge-Lur'e-Byrne and Novozhilov nonlinear shell theories and the results obtained using Donnell's nonlinear shallow-shell theory are compared. The Novozhilov, Flügge-Lur'e-Byrne and Sanders-Koiter theories give practically coincident results. (As previously, results for the Novozhilov theory are not reported in Figure 7.7. Flügge-Lur'e-Byrne and Sanders-Koiter theories are compared in Figure 7.7[c], but the curves are practically superimposed.) Figure 7.7(a) shows that the difference between Donnell's nonlinear shallow-shell theory and Donnell's theory with the present Lagrangian approach (retaining in-plane inertia) is much smaller for the water-filled shell than for the empty case. A significant, but not large, difference between the Sanders-Koiter (therefore, also the Flügge-Lur'e-Byrne and Novozhilov theories) and Donnell theories is shown in Figure 7.7(b).

In order to investigate whether the difference among the shell theories becomes larger by increasing the vibration amplitude, the same shell subjected to an excitation of 30 N has been investigated. Figure 7.8 presents the response computed by using Donnell's nonlinear theory with the present Lagrangian approach. Large qualitative differences between Figures 7.6 and 7.8 are evident. In particular, the peak on branch 2 close to the second bifurcation is greatly increased with respect to the other part of the response and the frequency range with quasi-periodic responses is enlarged. The branch 1 of the response shows a softening-type nonlinearity and presents a second folding for a vibration amplitude of about $5h$.

The comparison among the different shell theories is given in Figure 7.9. In this case as well, differences among Donnell's nonlinear shallow-shell, the Donnell with Lagrangian approach and the Sanders-Koiter theories are significant, but not excessive.

It can be concluded from these results that Donnell's nonlinear shallow-shell theory is much more accurate for water-filled than for empty shells. This can be easily explained because the in-plane inertia, which is neglected by Donnell's nonlinear shallow-shell theory, is much less important for a water-filled shell that has a large radial inertia due to the liquid associated with radial deflection.

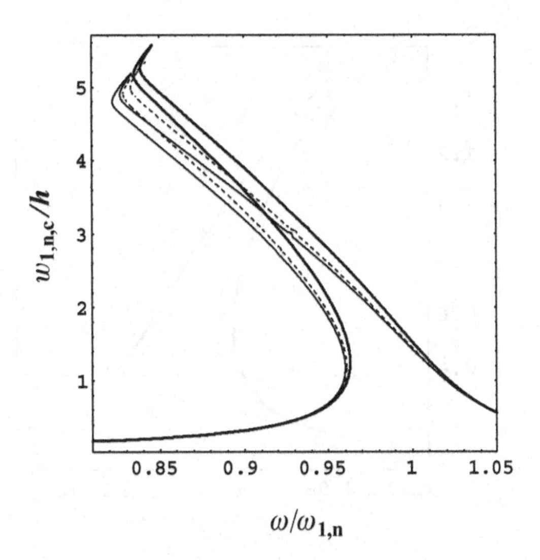

Figure 7.9. Frequency-response curve for the fundamental mode of the perfect, water-filled shell; branch 1 only; $\zeta_{1,n} = 0.0017$; $\tilde{f} = 30$ N. Only $w_{1,n,c}(t)$ is reported. —, Donnell's nonlinear shallow-shell theory; – –, Donnell's theory (Lagrangian approach); —, Sanders-Koiter theory.

The effect of neglecting the additional term $\hat{u}$, see equations (7.8–7.18), is investigated in Figure 7.10 for Donnell's theory with Lagrangian approach. The difference is small and the additional term $\hat{u}$ can be neglected without a significant difference in the computed response. A similar effect of $\hat{u}$ has been found for the other shell theories.

A crucial point of the present Lagrangian approach is the expansion of the middle surface displacements. The expansion given by equations (7.5a–c) has been compared with reduced and enlarged expansions in Figure 7.11 for Donnell's theory with the Lagrangian approach. In particular, results show that the expansion given by equations (7.5a–c) is the smallest one necessary to obtain correct results. The expansion with 16 dofs used in all the previous calculations gives results very close to this one. A largely increased expansion with 36 dofs, which can be considered close to convergence, gives a shell response relatively close to the one computed with 14 dofs. In particular, the 36 generalized coordinates used in the largest model are given by equations (7.4a–c) by inserting $j = k \times n$, $n = 5$ for the fundamental mode, $k = 1, \ldots, 4$; $M_1 = 3$ if $k = 2$, $M_1 = 1$ otherwise (odd m only); $M_2 = 5$ (odd m only). On the other hand, as shown by the 12 dofs model, if the generalized coordinates

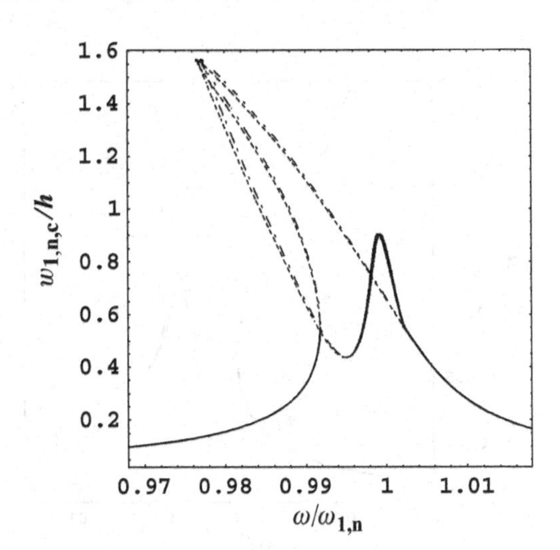

Figure 7.10. Frequency-response curve for the fundamental mode of the perfect, water-filled shell; $\zeta_{1,n} = 0.0017$; $\tilde{f} = 3$ N; Donnell's theory (Lagrangian approach). – –, solution with $\hat{u}$; –·–, solution neglecting $\hat{u}$. 1, branch 1; 2, branch 2.

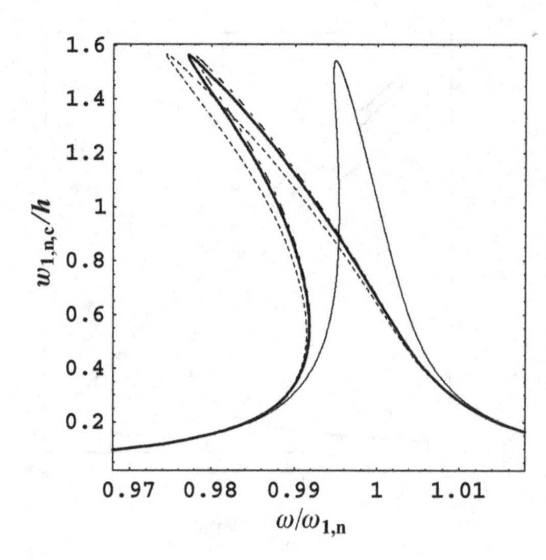

Figure 7.11. Convergence of the frequency-response curve for the fundamental mode of the perfect, water-filled shell; branch 1 only; $\zeta_{1,n} = 0.0017$; $\tilde{f} = 3$ N; Donnell's theory (Lagrangian approach). —, 12 dofs model; –·–, 14 dofs model, equation (19a–c); **—**, 16 dofs model; – –, 36 dofs model.

$v_{3,2n,c}$ and $v_{3,2n,s}$ are removed from equations (7.5a–c), a completely wrong response is obtained.

The convergence of the solution for other shell theories is very similar. In particular, Figure 7.12 shows the shell response computed by using the Sanders-Koiter shell theory and expansions with 16 and 26 dofs. The 26 generalized coordinates are given by equations (7.4a–c) by inserting $j = k \times n$, $n = 5$, $k = 1, \ldots, 3$, $M_1 = 1$ and $M_2 = 5$ with odd m only.

7.3.2.1 Water-Filled Shell with Imperfections

The effect of geometric imperfections on the natural frequency of the fundamental mode of the water-filled shell is investigated in Figures 7.13(a–c) for both the Donnell (Lagrangian approach) and Sanders-Koiter theories with expansion involving 48 dofs. Natural frequencies are evaluated by eliminating all the nonlinear terms in the generalized coordinates (but not nonlinear terms associated with imperfections) from the Lagrange equations. The 48 generalized coordinates are given by equations (7.4a–c) inserting $j = k \times n$, $n = 5$, $k = 1, \ldots, 4$; $M_1 = 1$ if $k = 4$, $M_1 = 3$ otherwise (odd m only); $M_2 = 5$ (odd m only). In particular, three different geometric

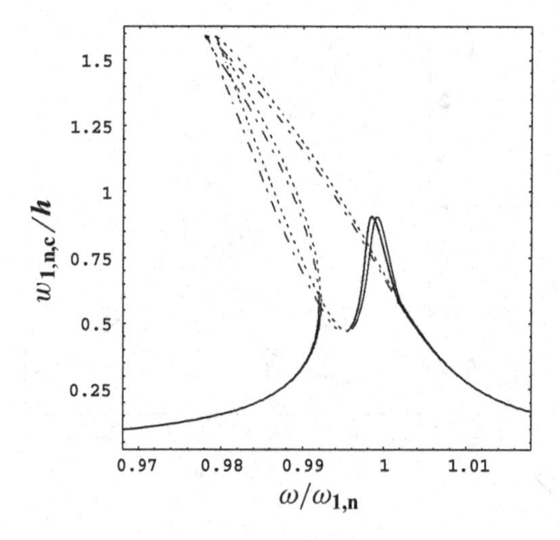

Figure 7.12. Convergence of the frequency-response curve for the fundamental mode of the perfect, water-filled shell; $\zeta_{1,n} = 0.0017$; $\tilde{f} = 3$ N; Sanders-Koiter theory. – –, 16 dofs model; –·–, 26 dofs model. 1, branch 1; 2, branch 2.

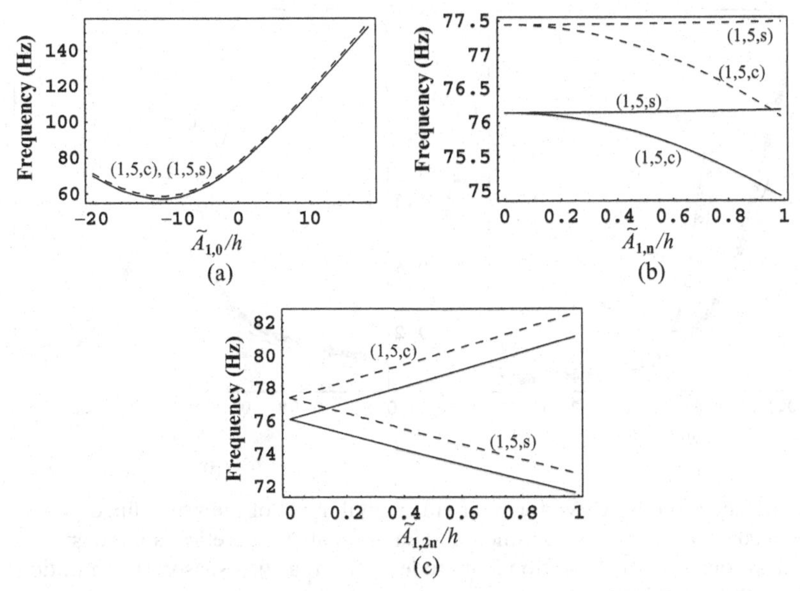

Figure 7.13. Natural frequency of the fundamental mode ($n = 5$, $m = 1$) of the water-filled shell versus the amplitude of geometric imperfections. (a) Axisymmetric imperfection $\tilde{A}_{1,0}$. (b) Asymmetric imperfection $\tilde{A}_{1,n}$. (c) Asymmetric imperfection $\tilde{A}_{1,2n}$. – –, Donnell's theory (Lagrangian approach); —, Sanders-Koiter theory.

imperfections are considered (i) axisymmetric imperfection $\tilde{A}_{1,0}$, (ii) asymmetric imperfection $\tilde{A}_{1,n}$ having the same shape of the fundamental mode and (iii) asymmetric imperfection $\tilde{A}_{1,2n}$ having twice the number of circumferential waves of the fundamental mode. The axisymmetric imperfection does not split the double eigenvalue associated with the fundamental mode; this is only obtained by asymmetric imperfections. Positive axisymmetric imperfections (outward Gaussian curvature) increase the natural frequency; small negative axisymmetric imperfections (inward Gaussian curvature) decrease it, but the trend is inverted around $\tilde{A}_{1,0} \cong -10h$. Imperfections having twice the number of circumferential waves with respect to the resonant mode have a very large effect on its natural frequency, as indicated in Figure 7.13(c) (note that curves cross at zero imperfection in this case and that the position of modes [1, 5, *c*] and [1, 5, *s*] is inverted for negative imperfections); imperfections with the same number of circumferential waves of the fundamental mode have a smaller effect, but still quite large. Imperfections with a number of circumferential waves that is not a multiple of n play a very small role, as shown in Chapter 5. The main difference between the Donnell and Sanders-Koiter results is related to the calculation of the natural frequency of the fundamental mode for zero imperfection. The effect of imperfections on the frequency is almost the same for both theories.

The response of a simply supported, water-filled stainless steel shell experimentally tested by Amabili (2003), and studied in Chapter 5, is computed by using Donnell's nonlinear shell theory (Lagrangian approach) with the 36 dofs model previously specified. Results are compared to the experimental results reported by Amabili (2003) for validation purposes. The fundamental mode ($m=1$, $n=5$) is investigated. The modal damping used in the calculation is $\zeta_{1,n} = 0.0017$ and the excitation is given by a harmonic radial force of 3 N applied at $x = L/2$ and $\theta = 0$. A geometric imperfection $\tilde{A}_{1,2n} = -0.147h$ has been used to reproduce the difference between the natural frequency of driven and companion modes. The comparison is

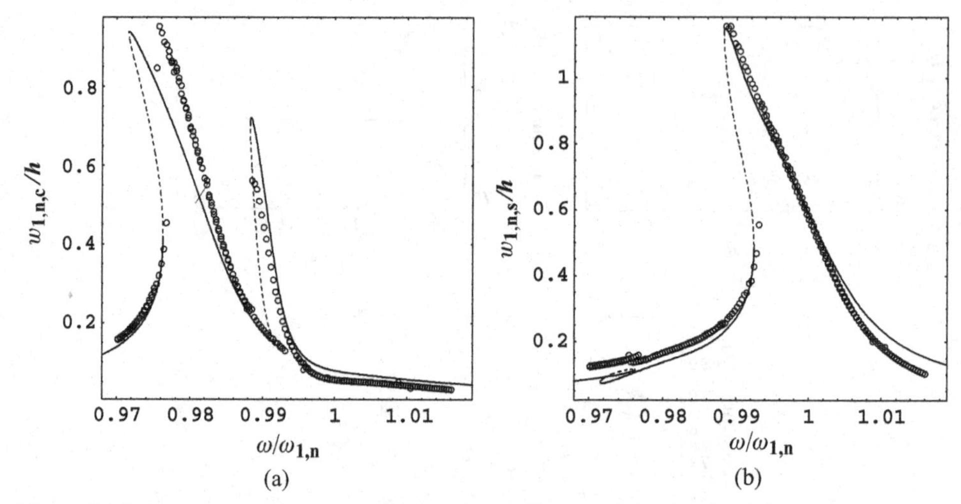

Figure 7.14. Frequency-response curve for the fundamental mode of the water-filled imperfect shell; $\zeta_{1,n} = 0.0017$; $\tilde{f} = 3$ N. ○, experimental data; —, stable theoretical solutions; – –, unstable theoretical solutions. (a) Vibration amplitude / h from the first sensor. (b) Vibration amplitude / h from the second sensor.

shown in Figure 7.14, where a good agreement between theoretical and experimental results is shown. In particular, the agreement is excellent for the second peak.

7.3.3 Discussion

Results from the Sanders-Koiter, Flügge-Lur'e-Byrne and Novozhilov theories are extremely close, for both empty and water-filled shells. For the thin shell numerically investigated in this study, for which $h/R \cong 288$, there is almost no difference among them. Small difference has been observed between the previous three theories and Donnell's theory with in-plane inertia. On the other hand, Donnell's nonlinear shallow-shell theory is the least accurate among the five theories here compared. It gives excessive softening-type nonlinearity for empty shells. However, for water-filled shells, it gives sufficiently precise results, also for quite a large vibration amplitude. The different accuracy of Donnell's nonlinear shallow-shell theory for empty and water-filled shells can easily be explained by the fact that the in-plane inertia, which is neglected in Donnell's nonlinear shallow-shell theory, is much less important for a water-filled shell, which has a large radial inertia due to the liquid, than for an empty shell. Contained liquid, compressive axial loads and external pressure increase the softening-type nonlinearity of the shell.

The Lagrangian approach developed has the advantage of being suitable to be applied to different nonlinear shell theories, of exactly satisfying the boundary conditions and of being very flexible in structural modifications without complicating the solution procedure.

7.4 Effect of Axial Load and Pressure on the Nonlinear Stability and Response of the Empty Shell

The effects of a uniform axial load per unit length N on the stability and the response to harmonic radial excitation of the fundamental mode ($m = 1, n = 5$) of the empty

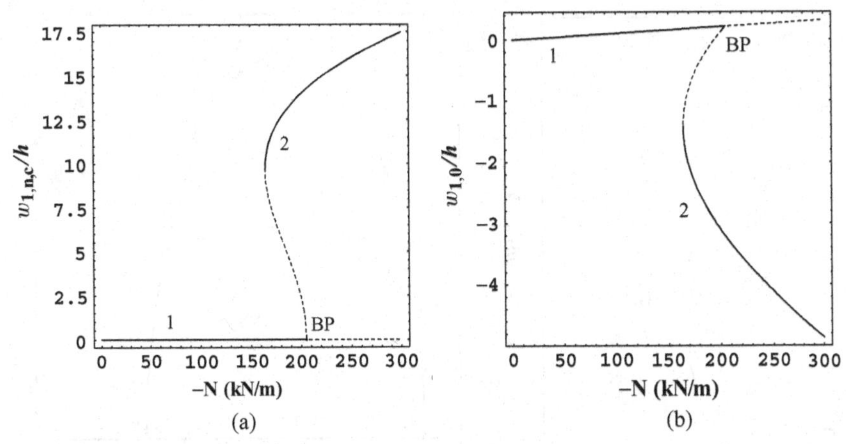

Figure 7.15. Buckling and post-buckling of the empty shell under static axial load per unit length N. (a) Generalized coordinate $w_{1,n,c}$. (b) Generalized coordinate $w_{1,0}$. 1, branch 1; 2, branch 2; BP, pitchfork bifurcation.

shell have been investigated. The Sanders-Koiter nonlinear theory has been used. The mode expansion involves 22 dofs; they are those used in the previous model with 16 dofs but with $M_2 = 9$ (odd m only). This addition of axisymmetric modes is necessary in the case of pressurization or axial loads in order to predict with accuracy the static deformation and prestress of the shell.

The stability of the shell under static axial compressive load N is shown in Figure 7.15, where the post-buckling behavior of the shell (without companion mode participation) corresponds to branch 2, which starts at a pitchfork bifurcation point, indicating buckling, at $N = -205$ kN/m. The bifurcation of the equilibrium is slightly subcritical; this behavior is characteristic of softening-type systems. The mode that goes first in instability is the fundamental mode of vibration ($m = 1, n = 5$) in this case. The simple formula $N_{cr} = Eh^2/(R\sqrt{3(1-\nu^2)})$ (Yamaki 1984) gives a buckling load per unit length $N_{cr} = 216$ kN/m, which is very close to the bifurcation point found in Figure 7.15.

The nonlinear response of the shell under a harmonic radial excitation of 2 N (applied at $L/2$), damping $\zeta_{1,n} = 0.001$ and compressive axial load $N = -150$ kN/m is shown in Figure 7.16. The axial load decreases the natural frequency of the shell to about 0.51 $\omega_{1,n}$. By comparing Figures 7.16 and 7.1, which has been obtained for the same case without static axial load, it is evident that a compressive axial load largely increases the softening-type nonlinearity of the shell.

The effect of external uniform pressure has been also investigated with the same model. The stability of the shell under pressure p_r is shown in Figure 7.17 for mode ($m = 1, n = 5$). Buckling is reached for $p_r = -49.9$ kPa (external pressure). Comparison of Figures 7.15(b) and 7.17(b) shows that a larger pre-buckling deformation (specifically, in the opposite direction) is obtained under axial load than under pressure. The mode that is first in instability is the mode ($m = 1, n = 6$) for p_r of about −38 kPa in this case; the fundamental mode of vibration ($m = 1, n = 5$) reaches instability just after this one.

The nonlinear response of the shell under a harmonic radial excitation of 2 N, damping $\zeta_{1,n} = 0.001$ and pressure $p_r = -37$ kPa is shown in Figure 7.18 for mode

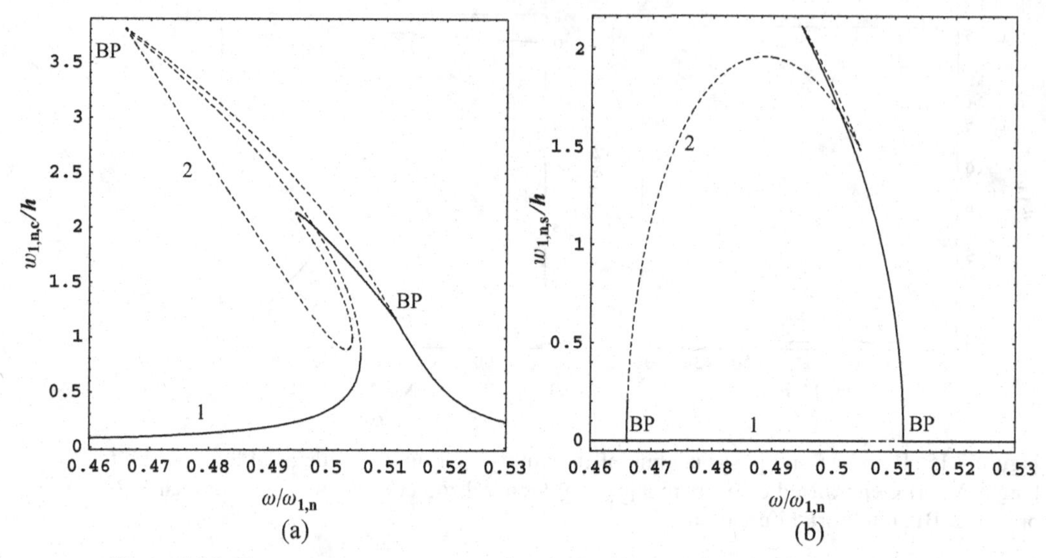

Figure 7.16. Frequency-response curve for the fundamental mode of the perfect, empty shell under axial load $N = -150$ kN/m; $\zeta_{1,n} = 0.001$; Sanders-Koiter theory. (a) Amplitude of $w_{1,n,c}(t)$, driven mode. (b) Amplitude of $w_{1,n,s}(t)$, companion mode. 1, branch 1; 2, branch 2; BP, pitchfork bifurcation.

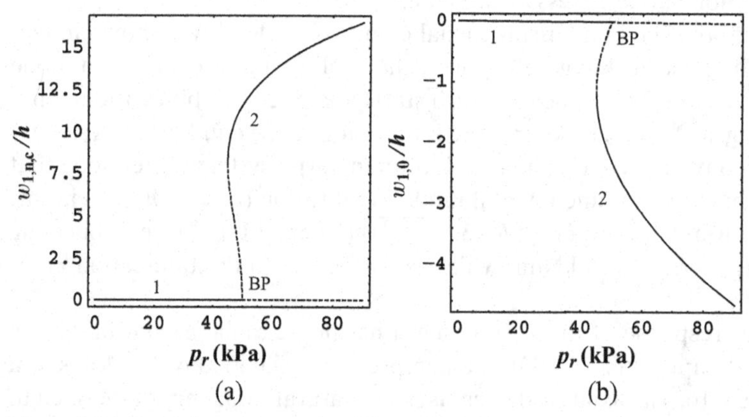

Figure 7.17. Buckling and post-buckling of the empty shell under pressure p_r. (a) Generalized coordinate $w_{1,n,c}$. (b) Generalized coordinate $w_{1,0}$. 1, branch 1; 2, branch 2; BP, pitchfork bifurcation.

($m = 1$, $n = 5$). The imposed external pressure decreases the natural frequency of the shell to about 0.51 $\omega_{1,n}$, as in the previous case. By comparing Figures 7.18 and 7.1, it is evident that an external pressure largely increases the softening-type nonlinearity of the shell, as observed in Figure 7.16 for the compressive axial load.

Geometric imperfections are not considered here. As is very well known, geometric imperfections have a large effect on the buckling of axially compressed shells.

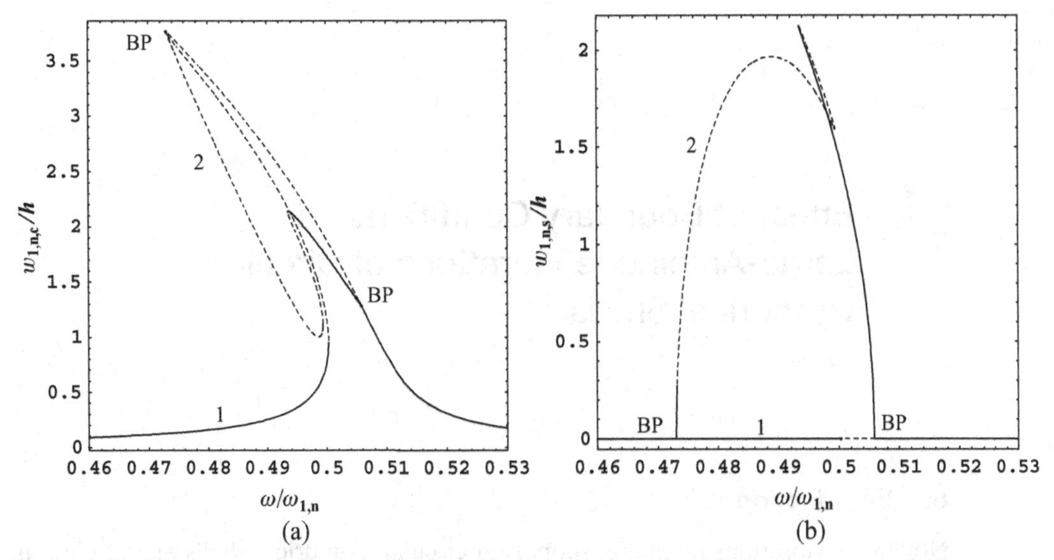

Figure 7.18. Frequency-response curve for the fundamental mode of the perfect, empty shell under pressure $p_r = -37$ kPa; $\zeta_{1,n} = 0.001$; the Sanders-Koiter theory. (a) Amplitude of $w_{1,n,c}(t)$, driven mode. (b) Amplitude of $w_{1,n,s}(t)$, companion mode. 1, branch 1; 2, branch 2; BP, pitchfork bifurcation.

REFERENCES

M. Amabili 2000 *Journal of Sound and Vibration* **231**, 79–97. Eigenvalue problems for vibrating structures coupled with quiescent fluids with free surface.

M. Amabili 2003 *Journal of Sound and Vibration* **264**, 1091–1125. Comparison of shell theories for large-amplitude vibrations of circular cylindrical shells: Lagrangian approach.

J. C. Chen and C. D. Babcock 1975 *AIAA Journal* **13**, 868–876. Nonlinear vibration of cylindrical shells.

M. Ganapathi and T. K. Varadan 1996 *Journal of Sound and Vibration* **192**, 1–14. Large-amplitude vibrations of circular cylindrical shells.

F. Pellicano, M. Amabili and M. P. Païdoussis 2002 *International Journal of Non-Linear Mechanics* **37**, 1181–1198. Effect of the geometry on the nonlinear vibrations analysis of circular cylindrical shells.

T. K. Varadan, G. Prathap and H. V. Ramani 1989 *AIAA Journal* **27**, 1303–1304. Nonlinear free flexural vibration of thin circular cylindrical shells.

N. Yamaki 1984 *Elastic Stability of Circular Cylindrical Shells*. North-Holland, Amsterdam, The Netherlands.

8 Effect of Boundary Conditions on Large-Amplitude Vibrations of Circular Cylindrical Shells

8.1 Introduction

Nonlinear vibrations of simply supported circular cylindrical shells are studied in Chapters 5 and 7. However, shells can have different boundary conditions in practical applications.

In this chapter, the effect of boundary conditions on the nonlinear forced vibrations of circular cylindrical shells is investigated. Numerical results show that, for the case analyzed, the axial constraint largely increases the softening-type nonlinearity of the shell with respect to the simply supported shell. On the other end, for the studied thin shell, the effect of the rotational constraint is very small.

8.1.1 Literature Review

Studies comparing the results for nonlinear vibrations of circular cylindrical shells with different constraints are very scarce. In fact, most of the literature deals with simply supported shells. Not many studies on shells with different boundary conditions are available. In particular, Matsuzaki and Kobayashi (1969) studied large-amplitude vibrations of clamped circular cylindrical shells theoretically and experimentally. They based their analysis on Donnell's nonlinear shallow-shell theory and used a simple mode expansion with two degrees of freedom (dofs). The analysis found a softening-type nonlinearity for clamped shells, in agreement with their own experimental results. They also found quasi-periodic response close to resonance.

Iu and Chia (1988) used Donnell's nonlinear shallow-shell theory to study free vibrations and post-buckling of clamped and simply supported, unsymmetrically laminated, cross-ply circular cylindrical shells. A multimode expansion was used without considering the companion mode, so that only free vibrations were investigated. Radial geometric imperfections were taken into account. The homogeneous solution of the stress function was retained, but the dependence on the axial coordinate was neglected. The equations of motion were obtained by using the Galerkin method and were studied by harmonic balance. Three asymmetric and three axisymmetric modes were used in the numerical calculations. In a later paper, Fu and Chia (1993) included in their model nonuniform boundary conditions around the edges. Softening-type or hardening-type nonlinearity was found, depending on the radius-to-thickness ratio. Only undamped free vibrations and buckling were investigated in this series of studies.

Large-amplitude vibrations of two vertical clamped circular cylindrical shells, partially filled with water to different levels, were experimentally studied by Chiba (1993). In this case, the responses displayed a general softening nonlinearity. The shells tested showed a larger nonlinearity when partially filled, compared with the empty and completely filled cases. Large-amplitude vibrations of four axially loaded, clamped circular cylindrical shells made of aluminum were experimentally studied by Gunawan (1998).

Ganapathi and Varadan (1996) used the finite-element method to study large-amplitude vibrations of doubly curved composite shells. Only free vibrations were investigated in the paper, using Donnell's nonlinear kinematics for shell deformation. A four-node finite element was developed with 5 dofs for each node. Ganapathi and Varadan (1996) also pointed out problems in the finite-element analysis of closed shells that are not present in open shells. The same approach was used to study numerically laminated composite circular cylindrical shells (Ganapathi and Varadan 1995).

Circular cylindrical shells with different boundary conditions were thoroughly investigated by Amabili (2003).

8.2 Theory

The following boundary conditions are imposed at the shell ends:

$$u = v = w = w_0 = 0, \quad \text{at}\, x = 0, L, \tag{8.1a–d}$$

$$M_x = -k(\partial w/\partial x) \quad \text{at}\, x = 0, L, \tag{8.1e}$$

and

$$\partial^2 w_0/\partial x^2 = 0 \quad \text{at}\, x = 0, L, \tag{8.1f}$$

where M_x is the bending moment per unit length and k is the stiffness per unit length of the elastic, distributed rotational springs at $x = 0, L$. Moreover, u, v and w must be continuous in θ. The boundary conditions (8.1a–c) restrain all the shell displacements at both edges. Equation (8.1e) represents an elastic rotational constraint at the shell edges. It gives any rotational constraint from zero moment ($M_x = 0$, unconstrained rotation) to perfectly clamped shell ($\partial w/\partial x = 0$, obtained as limit for $k \to \infty$), according to the value of k.

A base of shell displacements is used to discretize the system; the displacements u, v and w can be expanded by using the following expressions, which identically satisfy boundary conditions (8.1a–c)

$$u(x,\theta,t) = \sum_{m=1}^{M_1}\sum_{j=1}^{N}[u_{m,j,c}(t)\ \cos(j\theta) + u_{m,j,s}(t)\ \sin(j\theta)]\ \sin(\lambda_m x) + \sum_{m=1}^{M_2} u_{m,0}(t)\ \sin(\lambda_m x), \tag{8.2a}$$

$$v(x,\theta,t) = \sum_{m=1}^{M_1}\sum_{j=1}^{N}[v_{m,j,c}(t)\ \sin(j\theta) + v_{m,j,s}(t)\ \cos(j\theta)]\ \sin(\lambda_m x) + \sum_{m=1}^{M_2} v_{m,0}(t)\ \sin(\lambda_m x), \tag{8.2b}$$

$$w(x, \theta, t) = \sum_{m=1}^{M_1} \sum_{j=1}^{N} [w_{m,j,c}(t)\ \cos(j\theta) + w_{m,j,s}(t)\ \sin(j\theta)]\ \sin(\lambda_m x)$$
$$+ \sum_{m=1}^{M_2} w_{m,0}(t)\ \sin(\lambda_m x), \qquad (8.2c)$$

where j is the number of circumferential waves, m is the number of longitudinal half-waves, $\lambda_m = m\pi/L$ and t is the time; $u_{m,j}(t)$, $v_{m,j}(t)$ and $w_{m,j}(t)$ are the generalized coordinates, which are unknown functions of t; the additional subscript c or s indicates if the generalized coordinate is associated with the cosine or sine function in θ, except for v, for which the notation is reversed (no additional subscript is used for axisymmetric terms). The integers N, M_1 and M_2 must be selected with care in order to obtain the required accuracy and acceptable dimension of the nonlinear problem. Imperfection is assumed to be given by equation (5.52).

Excitation in the spectral neighborhood of the resonance of the mode with $m = 1$ longitudinal half-wave and n circumferential waves (mode with predominant radial oscillations), indicated as mode (m, n), is considered. The number of degrees of freedom that predicts the nonlinear response with good accuracy is 54. This result is based on a convergence study similar to the one reported in Chapter 7 for simply supported shells. In particular, only modes with an odd m value of longitudinal half-waves can be considered for reasons of symmetry in the expansions of v and w (if geometric imperfections with an even m value are not introduced); only even terms are necessary for u. Asymmetric modes having up to 12 longitudinal half-waves ($M_1 = M_2 = 12$) have been considered in the numerical calculations to achieve good accuracy. In fact, linear and nonlinear interaction among terms with different numbers of axial half-waves exists; these terms are not the linear modes of the shell with boundary conditions given by equations (8.1a–d). More terms are necessary for in-plane than for radial displacements. Torsional axisymmetric terms are not necessary.

The expansion used is

$$u(x, \theta, t) = \sum_{m=1}^{6} [u_{2m,n,c}(t)\ \cos(n\theta) + u_{2m,n,s}(t)\ \sin(n\theta)]\ \sin(\lambda_{2m} x)$$
$$+ \sum_{m=1}^{6} u_{2m,0}(t)\ \sin(\lambda_{2m} x)$$
$$+ [u_{2,2n,c}(t)\ \cos(2n\theta) + u_{2,2n,s}(t)\ \sin(2n\theta)]\ \sin(\lambda_2 x), \qquad (8.3a)$$

$$v(x, \theta, t) = \sum_{m=1}^{6} [v_{2m-1,n,c}(t)\ \sin(n\theta) + v_{2m-1,n,s}(t)\ \cos(n\theta)]\ \sin(\lambda_{2m-1} x)$$
$$+ [v_{1,2n,c}(t)\ \sin(2n\theta) + v_{1,2n,s}(t)\ \cos(2n\theta)]\sin(\lambda_1 x)$$
$$+ [v_{3,2n,c}(t)\ \sin(2n\theta) + v_{3,2n,s}(t)\ \cos(2n\theta)]\sin(\lambda_3 x), \qquad (8.3b)$$

$$w(x, \theta, t) = \sum_{m=1}^{6} [w_{2m-1,n,c}(t)\ \cos(n\theta) + w_{2m-1,n,s}(t)\ \sin(n\theta)]\ \sin(\lambda_{2m-1} x)$$
$$+ \sum_{m=1}^{6} w_{2m-1,0}(t)\ \sin(\lambda_{2m-1} x). \qquad (8.3c)$$

This expansion has 54 generalized coordinates (degrees of freedom). The dimension of the nonlinear dynamics is much smaller than the 54 dofs, and a reduced-order model can be built by using the techniques given in Chapter 6. However, this base is very intuitive and allows calculations without loosing the physical significance of each term.

Expansions (8.3a–c) are an extension of those developed for simply supported shells. In particular, more terms are necessary in order to have a good evaluation of the natural (linear) frequency because the functions used are different from the mode shape of the shell with constrained axial displacement ($u = 0$) at the shell ends. The addition of some extra term such, as $u_{4,2n,c}(t)$ and $u_{4,2n,s}(t)$, has been checked numerically and does not give any significant change in the shell response.

Equations (8.1) give the boundary conditions. In particular, equations (8.1a–d,f) are identically satisfied by the expansions of u, v, w and w_0. Moreover, the continuity in θ of all the displacements is also satisfied. On the other hand, equation (8.1e) can be rewritten in the following form:

$$M_x = \frac{Eh^3}{12(1-\nu^2)}(k_x + \nu k_\theta) = k\partial w/\partial x, \quad \text{at } x = 0, L, \tag{8.4}$$

In the case of zero stiffness of the distributed rotational springs, $k = 0$, equation (8.4) is identically satisfied for the assumed expansion, according to any shell theories with linear expressions of k_x and k_θ. In the case of k different from zero, an additional potential energy stored by the elastic rotational springs at the shell ends must be added. This potential energy U_R is given by

$$U_R = \frac{1}{2}\int_0^{2\pi} k\left\{\left[\left(\frac{\partial w}{\partial x}\right)_{x=0}\right]^2 + \left[\left(\frac{\partial w}{\partial x}\right)_{x=L}\right]^2\right\} \mathrm{d}\theta. \tag{8.5}$$

In equation (8.5), a nonuniform stiffness k (function of θ, simulating a nonuniform constraint) can be assumed.

In order to simulate clamped edges, corresponding to

$$\frac{\partial w}{\partial x} = 0, \quad \text{at } x = 0, L, \tag{8.6}$$

a very high value of the stiffness k must be assumed. In this case, equation (8.6) is satisfied by applying a condition on M_x. This approach is usually referred to as the artificial spring method, which can be regarded as a variant of the classical penalty method. The values of the spring stiffness simulating a clamped shell can be obtained (i) by studying the asymptotic convergence of natural frequencies by increasing the value of k, or (ii) by evaluating the edge stiffness of the shell. In fact, the natural frequencies of the system converge asymptotically from below to those of a clamped shell when k becomes very large.

8.3 Numerical Results

Calculations have been performed for a shell having the following dimensions and material properties: $L = 520$ mm, $R = 149.4$ mm, $h = 0.519$ mm, $E = 198$ GPa, $\rho = 7800$ kg/m^3 and $\nu = 0.3$. This shell has the fundamental mode with six circumferential waves and one longitudinal half-wave ($n = 6$, $m = 1$) for the boundary conditions given by equations (8.1) for any value of the stiffness k. The same shell, with simply supported edges (i.e. with $k = 0$ and condition [8.1a] replaced by $N_x = 0$, where N_x is the axial force per unit length) has the fundamental mode ($n = 5$, $m = 1$).

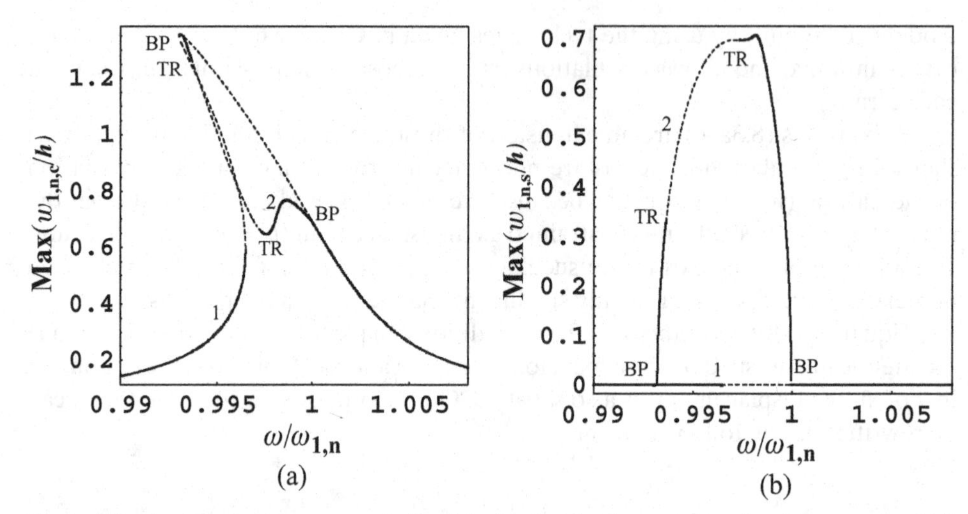

Figure 8.1. Amplitude-frequency relationship for the fundamental mode of the empty shell with free-edge rotation; $\zeta_{1,n} = 0.001$; $\tilde{f} = 3\,\text{N}$; Donnell's theory. (a) Amplitude of $w_{1,n,c}(t)$, driven mode. (b) Amplitude of $w_{1,n,s}(t)$, companion mode. 1, branch 1; 2, branch 2; BP, pitchfork bifurcation; TR, Neimark-Sacker bifurcation. ——, stable response; – –, unstable periodic response.

The nonlinear response of the empty shell in the spectral neighborhood of the fundamental frequency is shown in Figure 8.1 for the shell with free rotation ($M_x = 0$) and $u = 0$ at $x = 0, L$. The assumed modal damping is $\zeta_{1,n} = 0.001$ ($\zeta_{1,n} = \zeta_{1,n,c \text{ or } s}$). The fundamental mode ($n = 6$, $m = 1$) has a natural frequency of 313.7 Hz according to the Flügge theory of shells with an expansion at convergence of the solution (60 longitudinal modes). By using the expansion (8.3a–c), the natural frequency is 316.9 Hz for Donnell's theory and 309.9 Hz for Novozhilov's theory. For the fundamental mode, the radial displacement w is largely predominant and it is the only one shown. Only the two most important generalized coordinates, associated with driven and companion modes, are presented in Figure 8.1. However, the generalized coordinates associated with terms with more longitudinal half-waves also give an important contribution to the shell response. Qualitatively, the behavior is the same as observed for simply supported shells. The generalized coordinate $w_{1,n,c}$, which is directly excited by the a harmonic force $\tilde{f} = 3\,\text{N}$, presents a branch 1 displaying a softening-type nonlinearity; branch 1 corresponds to zero companion mode participation ($w_{1,n,s} = 0$). Branch 1 loses stability through a pitchfork bifurcation around the peak of the response. Through this bifurcation, branch 2 arises, which leads to companion mode participation. Branch 2 presents two Neimark-Sacker bifurcations; in between, the response is a stable quasi-periodic solution. Figure 8.1 has been obtained by Donnell's nonlinear theory (retaining in-plane inertia).

The same shell, but with clamped boundary conditions, has the fundamental mode ($n = 6$, $m = 1$) with a natural frequency of 315.1 Hz, according to the Flügge theory of shells with an expansion at convergence of the solution (60 longitudinal modes). By using the expansion (8.3a–c) and $k = 10^{10}$ N/rad, the natural frequency is 326.7 Hz for Donnell's theory. It can be easily observed that the natural frequency of the fundamental mode is very slightly increased by the rotational constraint with respect to the previous case. In fact, the shell is very thin and long enough to be little affected by this constraint, except in the two small regions close to the edges.

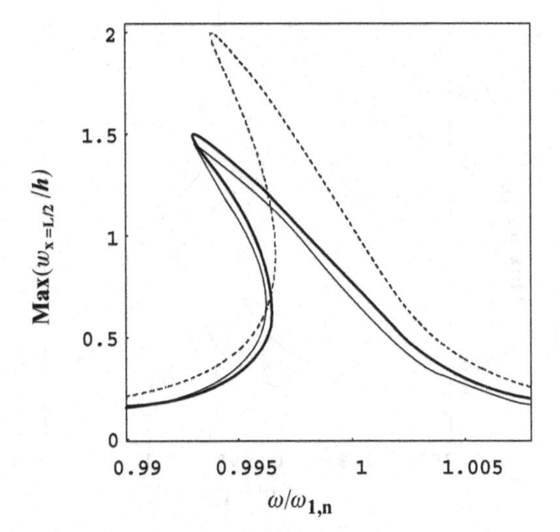

Figure 8.2. Amplitude-frequency relationship for the fundamental mode of the empty shell with different boundary conditions; branch 1 only; $\zeta_{1,n} = 0.001$; Donnell's theory. The shell displacement at the excitation point is reported. – –, simply supported shell ($n=5, m=1$), $\tilde{f}=2$ N; ——, clamped shell ($n=6$, $m=1$), $\tilde{f}=3$ N; ——, shell with $u=0$ and free rotation ($n=6$, $m=1$), $\tilde{f}=3$ N.

A comparison of branch 1 of the responses for (i) the shell with free rotation ($M_x = 0$) and $u = 0$ at $x = 0$, L, (ii) clamped shell and (iii) simply supported shell ($N_x = 0$, $M_x = 0$), for which the fundamental mode is ($n=5, m=1$) instead of ($n = 6$, $m = 1$), is provided in Figure 8.2. The model of the simply supported shell is the one obtained in Chapter 7. It is clearly shown that the axial constraint $u = 0$ largely increases the softening-type nonlinearity of the shell with respect to the constraint $N_x = 0$ (however, the nonlinearity also increases with the number of circumferential waves n). On the other end, for the studied thin shell, the effect of the rotational constraint is very small.

8.3.1 Comparison with Numerical and Experimental Results Available for Empty Shells

A first comparison has been performed for a clamped polyester shell experimentally tested by Chiba (1993) that has the following dimensions and material properties: $L = 480$ mm, $R = 240$ mm, $h = 0.254$ mm, $E = 4.65$ GPa, $\rho = 1400$ kg/m^3 and $\nu = 0.38$. The

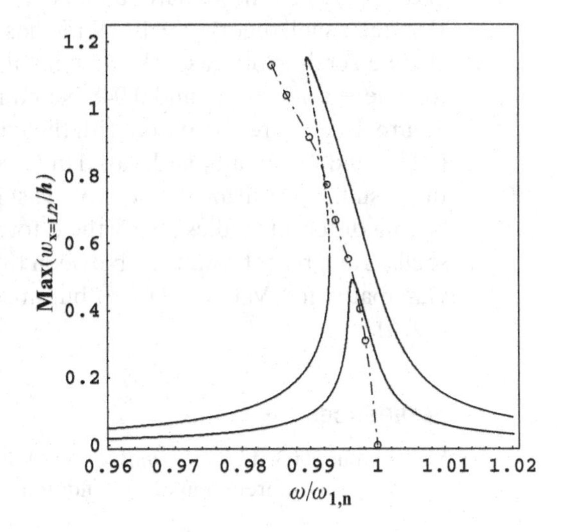

Figure 8.3. Amplitude-frequency relationship for mode ($n=15$, $m=1$) of the empty shell experimentally tested by Chiba (1993) for two force levels: $\tilde{f} = 0.008$ N and $\tilde{f} = 0.02$ N; branch 1 only; $\zeta_{1,n} = 0.0015$; Donnell's theory. The shell displacement at the excitation point $x = L/2$ is reported. ——, theoretical stable response; – –, theoretical unstable response; ○, experimental result from Chiba (1993); –·–, backbone curve fitting the experimental results.

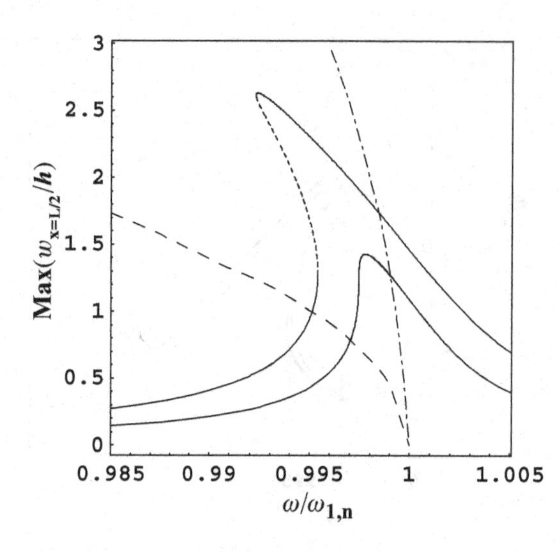

Figure 8.4. Amplitude-frequency relationship for mode (n = 8, m = 1) of the empty shell studied by Matsuzaki and Kobayashi (1969) and Ganapathi and Varadan (1996) for two force levels: $\tilde{f}$ = 0.025 N and $\tilde{f}$ = 0.048 N; branch 1 only; $\zeta_{1,n}$ = 0.0015; Donnell's theory. The shell displacement at the excitation point $x = L/2$ is reported. ——, theoretical stable response; – –, theoretical unstable response; – –, backbone curve from Matsuzaki and Kobayashi (1969); –·–, backbone curve from Ganapathi and Varadan (1996).

circular cylindrical shell had a longitudinal lap-joint seam, so that the axial symmetry of the shell was broken and the measured response of the vibration mode (n = 15, m = 1) did not present a traveling wave. For this reason, the companion modes were eliminated by the expansion of the shell displacements in the numerical calculation of this case. The natural frequency of mode (n = 15, m = 1) is 95.7 Hz, according to Donnell's shell theory; $k = 10^8$ N/rad has been used, which is large enough to simulate clamped ends in this case. The numerical responses, evaluated for two force levels $\tilde{f}$ = 0.008, and 0.02 N and modal damping $\zeta_{1,n}$ = 0.0015, are plotted in Figure 8.3 and are compared to the backbone curve (giving the free-vibration resonance; i.e. practically the maximum of the response versus the vibration amplitude) obtained by Chiba (1993). Figure 8.3 shows the same trend of softening-type nonlinearity computed by using the present approach and the experimental results obtained by Chiba (1993), but some quantitative difference is present for the largest excitation.

A second comparison has been performed with the numerical simulations of Matsuzaki and Kobayashi (1969) and Ganapathi and Varadan (1996), both of them for the fundamental mode (n = 8, m = 1) of the following clamped shell made of super-invar: L = 110 mm, R = 55 mm, h = 0.052 mm, E = 125.6 GPa, ρ = 7975.5 kg/m^3 and ν = 0.25. The natural frequency of the fundamental mode is 888.2 Hz, according to Donnell's shell theory; $k = 10^6$ N/rad has been used, which is large enough to simulate clamped ends in this case. The numerical responses (branch 1 only), evaluated for two force levels $\tilde{f}$ = 0.025 and 0.048 N and modal damping $\zeta_{1,n}$ = 0.0015, are plotted in Figure 8.4 and are compared with the backbone curves of Matsuzaki and Kobayashi (1969) and Ganapathi and Varadan (1996). A large difference is observed between the results of Matsuzaki and Kobayashi and Ganapathi and Varadan. Differences among different studies justify the effort for new research on nonlinear dynamics of shells. The present results lie between those of Matsuzaki and Kobayashi (1969) and Ganapathi and Varadan (1996) but are closer to those of Ganapathi and Varadan (1996).

REFERENCES

M. Amabili 2003 *AIAA Journal* **41**, 1119–1130. Nonlinear vibrations of circular cylindrical shells with different boundary conditions.

M. Chiba 1993 *ASME Journal of Pressure Vessel Technology* **115**, 381–388. Experimental studies on a nonlinear hydroelastic vibration of a clamped cylindrical tank partially filled with liquid.

Y. M. Fu and C. Y. Chia 1993 *International Journal of Non-Linear Mechanics* **28**, 313–327. Nonlinear vibration and postbuckling of generally laminated circular cylindrical thick shells with non-uniform boundary conditions.

M. Ganapathi and T. K. Varadan 1995 *Composite Structures* **30**, 33–49. Nonlinear free flexural vibrations of laminated circular cylindrical shells.

M. Ganapathi and T. K. Varadan 1996 *Journal of Sound and Vibration* **192**, 1–14. Large-amplitude vibrations of circular cylindrical shells.

L. Gunawan 1998 *Ph.D. Thesis*, Faculty of Aerospace Engineering, Technische Universiteit Delft, The Netherlands. Experimental study of nonlinear vibrations of thin-walled cylindrical shells.

V. P. Iu and C. Y. Chia 1988 *International Journal of Solid Structures* **24**, 195–210. Non-linear vibration and postbuckling of unsymmetric cross-ply circular cylindrical shells.

Y. Matsuzaki and S. Kobayashi 1969 *Transactions of the Japan Society for Aeronautical and Space Sciences* **12**, 55–62. A theoretical and experimental study of the nonlinear flexural vibration of thin circular cylindrical shells with clamped ends.

9 Vibrations of Circular Cylindrical Panels with Different Boundary Conditions

9.1 Introduction

This chapter addresses the linear (small amplitude) and nonlinear (large amplitude) vibrations of circular cylindrical panels. Simply and doubly curved panels are structural elements largely used in engineering applications, such as aeronautics, aerospace, cars, boats, buildings, trains and many others.

The linear vibrations of simply supported circular cylindrical panels are studied by using the Flügge-Lur'e-Byrne theory.

Nonlinear vibrations of circular cylindrical panels with different boundary conditions (including flexible rotational constraints) under radial harmonic excitations are studied by using Donnell's shell theory with in-plane inertia, as this gives practically the same results as for other refined classical theories for very thin isotropic shells (Amabili 2005). The solution is obtained by using the Lagrange equations of motion and up to 39 degrees of freedom (dofs). Numerical results show that a simply supported panel for mode (1,1) presents a significant softening nonlinearity, which turns to the hardening type for vibration amplitude larger than the shell thickness; a similar behavior is observed for the panel with fixed edges and free rotations; on the other hand, the same panel with free in-plane edges or clamped edges presents hardening nonlinearity.

A peculiar aspect of nonlinear vibrations of curved panels is the asymmetric oscillation with respect to the initial undeformed middle surface. In fact, the oscillation amplitude inward (i.e. in the direction of the center of curvature) is significantly larger than the amplitude outward.

Complex nonlinear vibrations are observed for large excitations, including period-doubling bifurcation, subharmonic, quasi-periodic, chaotic response and hyperchaos; they have been studied by using bifurcation diagrams of Poincaré maps, Lyapunov exponents and Lyapunov dimension.

The effect of geometric imperfections is investigated. Numerical and experimental results are presented and very satisfactorily compared.

9.1.1 Literature Review

The linear vibrations of curved panels are analyzed in the books of Leissa (1973) and Soedel (1993).

The first studies on geometrically nonlinear vibrations of classical simply supported, circular cylindrical shallow-shells are due to Reissner (1955), Grigolyuk (1955) and Cummings (1964). They used Donnell's nonlinear shallow-shell theory with a single-degree-of-freedom model. Leissa and Kadi (1971) studied linear and nonlinear free vibrations of doubly curved shallow panels, classical simply supported at the four edges. Donnell's nonlinear shallow-shell theory was used in a slightly modified version to take into account the meridional curvature. A single-mode expansion of the transverse displacement was used. Amabili (2005) studied the large-amplitude, forced vibrations of circular cylindrical panels with a rectangular base, classical simply supported at the four edges and subjected to radial harmonic excitation. The Donnell and the Novozhilov shell theories were used. In-plane inertia and geometric imperfections were taken into account. Convergence of the solution was shown and differences between the Donnell and the Novozhilov nonlinear shell theories were fully negligible. A very satisfactory comparison with the results obtained by Kobayashi and Leissa (1995) was given. Interaction of modes having integer ratios among their natural frequencies, giving rise to internal resonances, are also discussed.

Raouf (1993) studied the nonlinear free vibrations of simply supported, curved orthotropic panels by using Donnell's nonlinear shallow-shell theory. A single-mode analysis was carried out. Results show that thin circular cylindrical panels display softening nonlinearity when the ratio between the radius and length (R/L) of the panel is smaller than 1.25 or 1.5 for the orthotropic composite material used, but for R/L not too close to zero. Otherwise they display a hardening nonlinearity. Elishakoff et al. (1987) conducted a nonlinear analysis of small vibrations of an imperfect cylindrical panel around a static equilibrium position.

Not many studies deal with nonlinear vibrations of circular cylindrical panels with boundary conditions different from classical simply supported edges. In particular, Chia (1987a,b) studied laminated shallow cylindrical panels with mixed boundary conditions resting on an elastic foundation. Zero transverse displacement is assumed at the edges where rotational springs are introduced. In-plane boundary conditions are zero tangential force and given force orthogonal to the edges. A single-mode analysis is carried out by using Donnell's nonlinear shallow-shell theory. Fu and Chia (1989) extended this study by considering a multimode solution.

Tsai and Palazotto (1991) investigated the dynamic responses of circular cylindrical panels by developing a 36 dofs, curved, quadrilateral, thin-shell finite element, incorporating large displacement/rotation and parabolic transverse shear strains through the thickness.

Yamaguchi and Nagai (1997) studied vibrations of a shallow cylindrical panel with a rectangular boundary, with zero transverse deflection and free rotation at the boundaries and with in-plane elastic constraints orthogonal to the edges and free in-plane edges in the direction tangential to the edges. Donnell's nonlinear shallow-shell theory was used. The response of the panel was of the softening type over the whole range of the possible stiffness values of the in-plane springs (elastic support), becoming hardening for a vibration amplitude of the order of the shell thickness. In-plane constraints reduce the softening nonlinearity, which turns to hardening for smaller vibration amplitudes. The objective of the study was to investigate regions of chaotic motion; these regions were identified by means of Poincaré maps and Lyapunov exponents. It was found that, when approaching the static instability point (due to the constant acceleration load), chaotic shell behavior might be observed.

A finite-element approach to nonlinear dynamics of shells was developed by Sansour et al. (1997; 2002). They implemented a finite shell element based on a specifically developed nonlinear shell theory. In this shell theory, a linear distribution of the transverse normal strains was assumed, giving rise to a quadratic distribution of the displacement field over the shell thickness. They developed a time integration scheme for large numbers of degrees of freedom. In fact, problems can arise in finite-element formulations of nonlinear problems with a large number of degrees of freedom due to the instability of integrators. Chaotic behavior was found for a circular cylindrical panel simply supported on the straight edges, free on the curved edges and loaded by a point excitation having a constant value plus a harmonic component. The constant value of the load was assumed to give three equilibrium points (one unstable) in the static case.

Free nonlinear vibrations of doubly curved, laminated, clamped shallow panels, including circular cylindrical panels, were studied by Abe et al. (2000). Both the first-order shear deformation theory and the Donnell theory were used. Results obtained neglecting in-plane and rotary inertia are very close to those obtained retaining these effects. Two modes were considered to interact in the nonlinear analysis of the second mode (2,1) (two circumferential and one longitudinal half-waves), but a single mode was used for the first mode (1,1).

Comparison of theoretical and experimental results are given by Amabili (2006) for modes (1,1) and (2,1). Finally the effect of boundary conditions on the trend of nonlinearity is studied by Amabili (2007), together with large excitation giving rise to complex and chaotic responses.

9.2 Linear Vibrations

A thin circular cylindrical panel made of isotropic, homogeneous and linearly elastic material is considered, so that classic theories of shells are applicable. A cylindrical coordinate system $(O; x, r, \theta)$, having the origin O at the center of one end of the panel, is considered. The displacements of an arbitrary point of coordinates (x, θ) on the middle surface of the panel are denoted by u, v and w, in the axial, circumferential and radial directions, respectively; w is taken positive inward and the Flügge-Lur'e-Byrne theory of shells is used.

The panel is assumed to have simply supported edges, which give the boundary conditions

$$v = w = N_x = M_x = 0, \quad \text{at } x = 0, L, \tag{9.1a}$$

$$u = w = N_\theta = M_\theta = 0, \quad \text{at } \theta = 0, \alpha, \tag{9.1b}$$

where N is the normal force per unit length, M is the bending moment per unit length, L is the length of the panel in longitudinal direction and α is the angular extension of the panel in circumferential direction. The equations of motion are given in equations (5.12) and (5.13) and are the same as those of a circular cylindrical shell complete around the circumference. The solution of equations

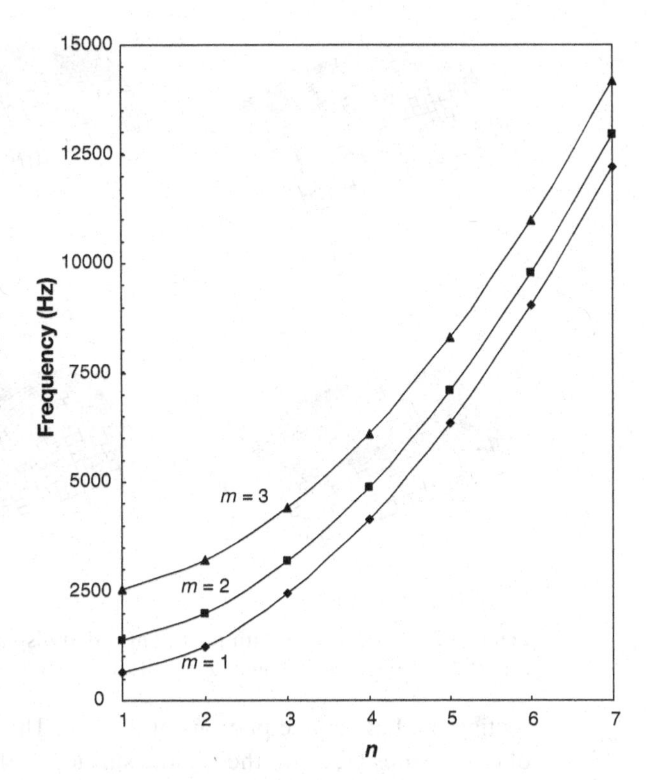

Figure 9.1. Natural frequencies of a simply supported circular cylindrical panel versus the number n of circumferential half-waves, for different number m of longitudinal half-waves. ♦, $m = 1$; ■, $m = 2$; ▲, $m = 3$.

(5.12) and (5.13), satisfying exactly all the boundary conditions (9.1), is given in the form:

$$u = C_1 \cos(\lambda_m x) \sin(\alpha_n \theta) \cos(\omega t), \tag{9.1c}$$

$$v = C_2 \sin(\lambda_m x) \cos(\alpha_n \theta) \cos(\omega t), \tag{9.1d}$$

$$w = C_3 \sin(\lambda_m x) \sin(\alpha_n \theta) \cos(\omega t), \tag{9.1e}$$

where $\lambda_m = m\pi/L$, $\alpha_n = n\pi/\alpha$, $m = 1, 2, \ldots$ and $n = 1, 2, \ldots$ are the number of longitudinal and circumferential half-waves, respectively, ω is the circular frequency of natural vibration and t is time. The eigenvalue problem that gives the natural frequencies and mode shapes of the simply supported circular cylindrical panel is given by equation (5.14), that is, the same as the complete circular cylindrical shell, with the substitution of n with α_n.

A simply supported circular cylindrical panel with $L = 0.1$ m, $R = 1$ m, $h = 0.001$ m, angular width between supports $\alpha = 0.1$ rad (i.e. the panel has a length equal to the circumferential width), Young's modulus $E = 206$ GPa, mass density $\rho = 7800$ kg/m^3 and Poisson's ratio $\nu = 0.3$ is considered. The natural frequencies of radial modes, which are those with the lowest natural frequencies, are given in Figure 9.1. Mode shapes are given in Figure 9.2.

9.3 Nonlinear Vibrations

A circular cylindrical panel with the cylindrical coordinate system $(O; x, r, \theta)$, having the origin O at the curvature center of one end of the panel, is considered, as shown in Figure 9.3. The Donnell nonlinear shell theory is used with an energy approach by

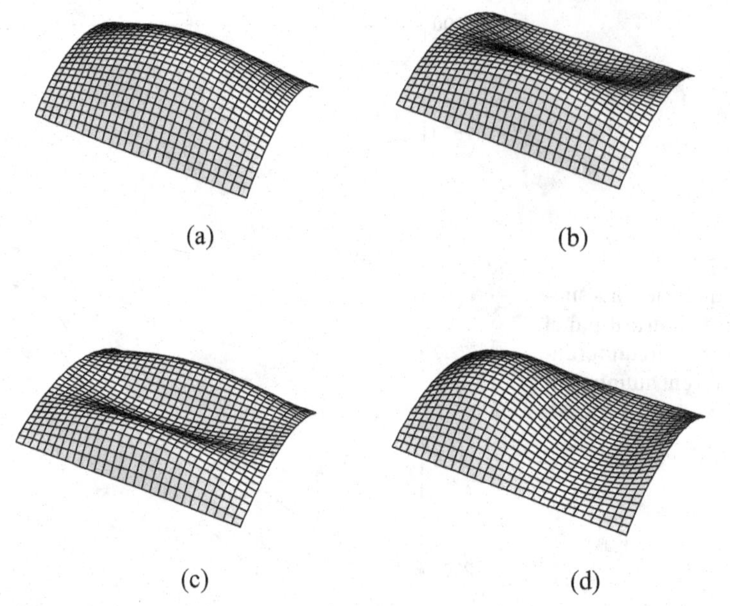

Figure 9.2. Mode shapes with prevalent radial displacement identified by the wavenumbers (n, m). (a) $n=1, m=1$; (b) $n=2, m=1$; (c) $n=3, m=1$; (d) $n=1, m=2$.

writing the Lagrange equations of motion. The displacements of an arbitrary point of coordinates (x, θ) on the middle surface of the panel are denoted by u, v and w, in the axial, circumferential and radial directions, respectively; w is taken positive outward. Initial imperfections of the circular cylindrical panel associated with zero initial tension are denoted by radial displacement w_0, also positive outward. Only radial initial imperfections are considered.

The following boundary conditions are introduced:

Model A

$$w = w_0 = N_x = M_x = \partial^2 w_0/\partial x^2 = 0, \quad N_{x,y} = -kv, \quad \text{at } x = 0, a, \tag{9.2a–f}$$

$$w = w_0 = N_y = M_y = \partial^2 w_0/\partial y^2 = 0, \quad N_{y,x} = -ku, \quad \text{at } y = 0, b, \tag{9.3a–f}$$

Model B

$$v = w = w_0 = N_x = M_x = \partial^2 w_0/\partial x^2 = 0, \quad \text{at } x = 0, a, \tag{9.4a–f}$$

$$u = w = w_0 = N_y = M_y = \partial^2 w_0/\partial y^2 = 0, \quad \text{at } y = 0, b, \tag{9.5a–f}$$

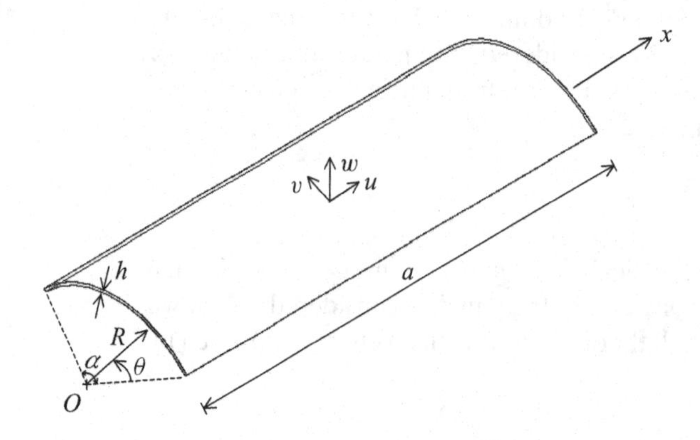

Figure 9.3. Geometry of the panel, coordinate system and symbols.

Model C

$$u = v = w = w_0 = 0, \quad M_x = \frac{Eh^3}{12(1-\nu^2)}(k_x + \nu k_y) = c\partial w/\partial x, \quad at\ x = 0, a, \tag{9.6a–e}$$

$$u = v = w = w_0 = 0, \quad M_y = \frac{Eh^3}{12(1-\nu^2)}(k_y + \nu k_x) = c\partial w/\partial y, \quad at\ y = 0, b, \tag{9.7a–e}$$

where $y = R\theta$, $b = R\alpha$, k is the distributed in-plane spring stiffness (N/m^2) where springs are distributed along the panel edges in the edge direction, M_x and M_y are the bending moments per unit length on the edges orthogonal to x and y, respectively, N_x and N_y are the normal forces per unit length and $N_{x,y}$ is the shear force per unit length. c is the stiffness per unit length of the elastic, distributed rotational springs placed at the four edges, $x = 0, a$ and $y = 0, b$. w is restrained at the four panel edges for all three models.

In model A, equations (9.2a–f) and (9.3a–f) give fully free in-plane boundary conditions for $k = 0$ and classical simply supported conditions in the limit case $k \to \infty$. This model was developed by Amabili (2006).

In model B, equations (9.4a–f) and (9.5a–f) give the classical simply supported boundary conditions. The problem was studied by Amabili (2005).

Model C gives fixed edges in-plane with free rotation for $c = 0$ and a perfectly clamped panel ($\partial w/\partial x = 0$ and $\partial w/\partial y = 0$) obtained as the limit for $c \to \infty$. The problem has been studied by Amabili (2007).

9.3.1 Mode Expansion

In order to reduce the system to finite dimensions, the middle-surface displacements u, v and w are expanded by using the following approximate functions, which satisfy identically the geometric boundary conditions:

Model A

$$w(x, y, t) = \sum_{m=1}^{M_1}\sum_{n=1}^{N_1} w_{m,n}(t)\sin(m\pi x/a)\sin(n\pi y/b), \tag{9.8a}$$

$$u(x, y, t) = \sum_{m=1}^{M_2}\sum_{n=0}^{N_2} u_{m,n}(t)\cos(m\pi x/a)\cos(n\pi y/b), \tag{9.8b}$$

$$v(x, y, t) = \sum_{m=0}^{M_3}\sum_{n=1}^{N_3} v_{m,n}(t)\cos(m\pi x/a)\cos(n\pi y/b); \tag{9.8c}$$

Model B

$$w(x, \theta, t) = \sum_{m=1}^{M_1}\sum_{n=1}^{N_1} w_{m,n}(t)\sin(m\pi x/a)\sin(n\pi y/b), \tag{9.9a}$$

$$u(x, \theta, t) = \sum_{m=1}^{M_2}\sum_{n=1}^{N_2} u_{m,n}(t)\cos(m\pi x/a)\sin(n\pi y/b), \tag{9.9b}$$

$$v(x, \theta, t) = \sum_{m=1}^{M_3}\sum_{n=1}^{N_3} v_{m,n}(t)\sin(m\pi x/a)\cos(n\pi y/b); \tag{9.9c}$$

Model C

$$w(x, y, t) = \sum_{m=1}^{M_1} \sum_{n=1}^{N_1} w_{m,n}(t) \sin(m\pi x/a) \sin(n\pi y/b), \tag{9.10a}$$

$$u(x, y, t) = \sum_{m=1}^{M_2} \sum_{n=1}^{N_2} u_{m,n}(t) \sin(m\pi x/a) \sin(n\pi y/b), \tag{9.10b}$$

$$v(x, y, t) = \sum_{m=1}^{M_3} \sum_{n=1}^{N_3} v_{m,n}(t) \sin(m\pi x/a) \sin(n\pi y/b), \tag{9.10c}$$

where m and n are the numbers of half-waves in x and y directions, respectively, and t is the time; $u_{m,n}(t)$, $v_{m,n}(t)$ and $w_{m,n}(t)$ are the generalized coordinates that are unknown functions of t. $M_\#$ and $N_\#$ indicate the terms necessary in the expansion of the displacements, where $\# = 1, 2, 3$.

Only out-of-plane initial geometric imperfections of the panel are assumed; they are associated with zero initial stress. The imperfection w_0 is expanded in the same form of w, that is, in a double Fourier sine series satisfying the boundary conditions (9.2b,e) and (9.3b,e) at the panel edges

$$w_0(x, y) = \sum_{m=1}^{\tilde{M}} \sum_{n=1}^{\tilde{N}} A_{m,n} \sin(m\pi x/a) \sin(n\pi y/b), \tag{9.11}$$

where $A_{m,n}$ are the modal amplitudes of imperfections and $\tilde{N}$ and $\tilde{M}$ are integers indicating the number of terms in the expansion.

9.3.2 Satisfaction of Boundary Conditions

9.3.2.1 Model A

The geometric boundary conditions, equations (9.2a,b,e) and (9.3a,b,e), are exactly satisfied by the expansions of u, v, w and w_0. On the other hand, equations (9.2d) and (9.3d) can be rewritten in the following form:

$$M_x = \frac{Eh^3}{12(1-\nu^2)} (k_x + \nu k_\theta) = 0, \quad at\ x = 0, a, \tag{9.12a}$$

$$M_y = \frac{Eh^3}{12(1-\nu^2)} (k_\theta + \nu k_x) = 0, \quad at\ y = 0, b. \tag{9.12b}$$

Equations (9.12a,b) are identically satisfied for the expressions of k_x and k_θ of the Donnell nonlinear shell theory. Moreover, the following constraints, equations (9.2c) and (9.3c), must be satisfied:

$$N_x = \frac{Eh}{1-\nu^2} (\varepsilon_{x,0} + \nu \varepsilon_{\theta,0}) = 0, \quad at\ x = 0, a, \tag{9.13a}$$

$$N_y = \frac{Eh}{1-\nu^2} (\varepsilon_{\theta,0} + \nu \varepsilon_{x,0}) = 0, \quad at\ y = 0, b. \tag{9.13b}$$

Equations (9.13a,b) are not identically satisfied. Eliminating zero terms at the panel edges, equations (9.13a,b) can be rewritten as:

$$\frac{\partial \hat{u}}{\partial x} + \frac{1}{2}\left(\frac{\partial w}{\partial x}\right)^2 + \frac{\partial w}{\partial x}\frac{\partial w_0}{\partial x} + \nu \frac{\partial (v + \hat{v})}{\partial y} = 0, \quad \text{at } x = 0, a, \tag{9.14a}$$

$$\frac{\partial \hat{v}}{\partial y} + \frac{1}{2}\left(\frac{\partial w}{\partial y}\right)^2 + \frac{\partial w}{\partial y}\frac{\partial w_0}{\partial y} + \nu \frac{\partial (u + \hat{u})}{\partial x} = 0, \quad \text{at } y = 0, b, \tag{9.14b}$$

where $\hat{u}$ and $\hat{v}$ are terms added to the expansion of u and v, given in equation (9.8b,c), in order to satisfy exactly the boundary conditions $N_x = 0$ and $N_y = 0$. The term v in equation (9.14a) is eliminated because it gives a linear relationship between u and v, which is satisfied by using the minimization of energy in the process of building the Lagrange equations of motion. In fact, this is equivalent to the Rayleigh-Ritz method, and it therefore requires satisfying only the geometrical boundary conditions. Similarly u is eliminated in equation (9.14b). Therefore, $\hat{u}$ and $\hat{v}$ are reduced to second-order terms in the panel displacement (assuming geometric imperfections of the same order of $w_{m,n}$). Nontrivial calculations for Donnell's nonlinear theory give

$$\hat{u}(t) = -\sum_{n=1}^{N_1}\sum_{m=1}^{M_1} \frac{m\pi}{a}\left\{\frac{1}{2} w_{m,n}(t)\sin(n\pi y/b)\right.$$
$$\times \sum_{k=1}^{N_1}\sum_{s=1}^{M_1} \frac{s}{m+s} w_{s,k}(t)\sin(k\pi y/b)\sin\left[(m+s)\pi x/a\right] + w_{m,n}(t)\sin(n\pi y/b)$$
$$\left.\times \sum_{j=1}^{\tilde{N}}\sum_{i=1}^{\tilde{M}} \frac{i}{m+i} A_{i,j}\sin(j\pi y/b)\,\sin\left[(m+i)\pi x/a\right]\right\}, \tag{9.15a}$$

$$\hat{v}(t) = -\sum_{n=1}^{N_1}\sum_{m=1}^{M_1} \frac{n\pi}{b}\left\{\frac{1}{2} w_{m,n}(t)\sin(m\pi x/a)\right.$$
$$\times \sum_{k=1}^{N_1}\sum_{s=1}^{M_1} \frac{k}{n+k} w_{s,k}(t)\,\sin(s\pi x/a)\sin\left[(n+k)\pi y/b\right] + w_{m,n}(t)\sin(m\pi x/a)$$
$$\left.\times \sum_{j=1}^{\tilde{N}}\sum_{i=1}^{\tilde{M}} \frac{j}{n+j} A_{i,j}\sin(i\pi x/a)\sin\left[(n+j)\pi y/b\right]\right\}. \tag{9.15b}$$

Actually, equations (9.14a,b) can be satisfied by energy minimization by avoiding the introduction of equations (9.15a,b). But the choice of the expansions of u and v becomes very tricky; that is, all the terms involved in equations (9.15a,b) must be inserted in the expansion as additional degrees of freedom in order to predict the system behavior with accuracy. This has been verified numerically.

The axial displacement u will be given by $u + \hat{u}$, where u is given by equation (9.8b) and $\hat{u}$ is given by equation (9.15a); similarly, for v that will be given by $v + \hat{v}$.

Finally, boundary conditions (9.2f) and (9.3f) must also be satisfied. They give

$$N_{x,y} = \frac{Eh}{2(1+\nu)}\gamma_{x,y} = -kv, \quad \text{at } x = 0, a, \tag{9.16a}$$

$$N_{y,x} = \frac{Eh}{2(1+\nu)}\gamma_{y,x} = -ku, \quad \text{at } y = 0, b. \tag{9.16b}$$

Eliminating zero terms at the panel edges, equations (9.16a,b) can be rewritten as

$$\frac{Eh}{2(1+\nu)}\left[\frac{\partial u}{\partial y}+\frac{\partial v}{\partial x}\right]_{x=0,a} = -kv \quad \text{at } x=0,a, \tag{9.17a}$$

$$\frac{Eh}{2(1+\nu)}\left[\frac{\partial u}{\partial y}+\frac{\partial v}{\partial x}\right]_{y=0,b} = -ku \quad \text{at } y=0,b. \tag{9.17b}$$

In the case of zero stiffness of the distributed springs, $k=0$, equations (9.17a,b) give a linear condition, which is satisfied by using the minimization of energy in the Lagrange equations of motion. Therefore, no additional terms in the expansion are introduced. In the case of k different from zero, an additional potential energy stored by the elastic springs at the shell edges must be added. This potential energy U_K is given by

$$U_K = \frac{1}{2}\int_0^b k\{[(v)_{x=0}]^2+[(v)_{x=a}]^2\}\,\mathrm{d}y + \frac{1}{2}\int_0^a k\{[(u)_{y=0}]^2+[(u)_{y=a}]^2\}\,\mathrm{d}x. \tag{9.18}$$

In equation (9.18), a nonuniform stiffness k (simulating a nonuniform constraint) can be assumed. In order to simulate classical simply supported edges, corresponding to $v=0$ at $x=0,a$, and $u=0$ at $y=0,b$, a very high value of the stiffness k must be assumed.

9.3.2.2 Model B

Boundary conditions (9.4a,b,c,e,f) and (9.4a,b,c,e,f) are identically satisfied by the expansions (9.4a–c) and (9.11). Conditions (9.4d) and (9.5d) are not identically satisfied. According to Donnell's theory, and eliminating null terms at the panel edges, equations (9.4d) and (9.5d) can be rewritten as:

$$\frac{\partial \hat{u}}{\partial x}+\frac{1}{2}\left(\frac{\partial w}{\partial x}\right)^2+\frac{\partial w}{\partial x}\frac{\partial w_0}{\partial x}+\nu\frac{\partial \hat{v}}{\partial y}=0 \quad \text{at } x=0,a, \tag{9.19a}$$

$$\frac{\partial \hat{v}}{\partial y}+\frac{1}{2}\left(\frac{\partial w}{\partial y}\right)^2+\frac{\partial w}{\partial y}\frac{\partial w_0}{\partial y}+\nu\frac{\partial \hat{u}}{\partial x}=0 \quad \text{at } y=0,b, \tag{9.19b}$$

where $\hat{u}$ and $\hat{v}$ are terms that must be added to the expansion of u and v, given in equation (9.9b,c), in order to satisfy exactly the in-plane boundary conditions $N_x=0$ and $N_y=0$. Because $\hat{u}$ and $\hat{v}$ are second-order terms in the panel displacement as shown by equations (9.19a,b), they have not been inserted in the second-order terms that involve u and v in equations (9.19a,b).

Nontrivial calculations give

$$\begin{aligned}\hat{u}(t) = -\sum_{n=1}^{N_1}\sum_{m=1}^{M_1}\frac{m\pi}{a}\Bigg\{&\frac{1}{4}\sum_{k=1}^{N_1}\sum_{s=1}^{M_1} s\,w_{m,n}w_{s,k}[\cos[(n-k)\pi y/b]-\cos[(n+k)\pi y/b]]\\ &\times\sin[(m+s)\pi x/a]/(m+s)+w_{m,n}(t)\sin(n\pi y/b)\\ &\times\sum_{j=1}^{\tilde{N}}\sum_{i=1}^{\tilde{M}}\frac{i}{m+i}A_{i,j}\sin(j\pi y/b)\sin[(m+i)\pi x/a]\Bigg\},\end{aligned} \tag{9.20a}$$

$$\hat{v}(t) = -\sum_{n=1}^{N}\sum_{m=1}^{M}\frac{n\pi}{b}\left\{\frac{1}{4}\sum_{k=1}^{N}\sum_{s=1}^{M} k\,w_{m,n}w_{s,k}[\cos[(m-s)\pi x/a] - \cos[(m+s)\pi x/a]]\right.$$
$$\times \sin[(n+k)\pi y/b]/(n+k) + w_{m,n}(t)\sin(m\pi x/a)$$
$$\left.\times \sum_{j=1}^{\tilde{N}}\sum_{i=1}^{\tilde{M}}\frac{j}{n+j}A_{i,j}\sin(i\pi x/a)\sin[(n+j)\pi y/b]\right\}, \tag{9.20b}$$

where the time dependence of $w_{m,n}$ from t has been dropped for simplicity.

9.3.2.3 Model C

The boundary conditions (9.6e) and (9.7e) are not identically satisfied by the expansions (9.10a–c), but they give a linear condition. In the case of zero stiffness c of rotational distributed springs, equations (9.6e) and (9.7e) are identically satisfied. In the case of c different from zero, an additional potential energy stored by the elastic rotational springs at the panel edges must be added in the minimization of the system energy. This potential energy U_R is given by

$$U_R = \frac{1}{2}\int_0^b c\left\{\left[\left(\frac{\partial w}{\partial x}\right)_{x=0}\right]^2 + \left[\left(\frac{\partial w}{\partial x}\right)_{x=a}\right]^2\right\}\mathrm{d}y$$
$$+\frac{1}{2}\int_0^a c\left\{\left[\left(\frac{\partial w}{\partial y}\right)_{y=0}\right]^2 + \left[\left(\frac{\partial w}{\partial y}\right)_{y=b}\right]^2\right\}\mathrm{d}x. \tag{9.21}$$

In equation (9.21), a nonuniform stiffness c (function of x or y, simulating a nonuniform constraint) can be assumed. In order to simulate clamped edges in numerical calculations, a very high value of the stiffness c must be assumed. This approach is usually referred to as the artificial spring method, which can be regarded as a variant of the classical penalty method. The values of the spring stiffness simulating a clamped panel can be obtained by studying the convergence of the linearized solution by increasing the value of c. In fact, it has been found that the natural frequencies of the system converge asymptotically to those of a clamped panel when c becomes very large.

9.3.3 Solution

The Lagrange equations of motion are built exactly as in Chapter 7, taking into account that, in the present case, there is no energy due to fluid-structure interaction and that for models A and C it is necessary to add the energies U_K and U_R, respectively, to the potential energy U of the system. In particular, the elastic strain energy of the panel is given by equation (2.9), and the kinetic energy T_S of a circular cylindrical panel, by neglecting rotary inertia, is given by

$$T_S = \frac{1}{2}\rho h\int_0^a\int_0^b(\dot{u}^2 + \dot{v}^2 + \dot{w}^2)\,\mathrm{d}x\,\mathrm{d}y, \tag{9.22}$$

Table 9.1. *Natural frequency of mode (1, 1) for different boundary conditions*

Model	Boundary condition	Natural frequency (Hz)
A	$k=0$	549.3 (19 dofs) 539.4 (50 dofs)
A	$k=4\times10^9$ N/m^2	597.0 (19 dofs)
B	Simply supported	636.9 (9 dofs)
C	$c=0$	912.0 (39 dofs) 912.0 (79 dofs)
C	$c=5\times10^4$ N	1211.2 (39 dofs) 1168.3 (79 dofs)

where ρ is the mass density of the panel. In equation (9.22), the overdot denotes the time derivative.

9.4 Numerical Results

Numerical calculations have been performed for the harmonic response of a circular cylindrical panel having the following dimensions and material properties: length between supports $a = 0.1$ m, radius of curvature $R = 1$ m, thickness $h = 0.001$ m, angular width between supports $\alpha = 0.1$ rad (i.e. the panel has length equal to the circumferential width), Young's modulus $E = 206$ GPa, mass density $\rho = 7800$ kg/m^3 and Poisson's ratio $\nu = 0.3$. A panel with the same dimension ratios ($R/a = 10$, $h/a = 0.01$, $b/a = 1$, $\nu = 0.3$) was studied by Kobayashi and Leissa (1995) and Amabili (2005; 2007).

The following generalized coordinates have been used for model A, giving a 19 dofs model: $w_{1,1}, w_{1,3}, w_{3,1}, w_{3,3}, u_{1,0}, u_{1,2}, u_{1,4}, u_{3,0}, u_{3,2}, u_{3,4}, v_{0,1}, v_{2,1}, v_{4,1}, v_{0,3}, v_{2,3}, v_{4,3}, v_{0,5}, v_{2,5}, v_{4,5}$. Convergence of this model was shown in Amabili (2006).

The following generalized coordinates have been used for model B, giving a 9 dofs model: $w_{1,1}, u_{1,1}, u_{3,1}, u_{1,3}, u_{3,3}, v_{1,1}, v_{1,3}, v_{3,1}, v_{3,3}$. Convergence of this model was shown in Amabili (2005).

The following generalized coordinates have been used for model C, giving a 39 dofs model: $w_{1,1}, w_{3,1}, w_{5,1}, w_{7,1}, w_{1,3}, w_{3,3}, w_{5,3}, w_{7,3}, w_{1,5}, w_{3,5}, w_{5,5}, w_{1,7}, w_{3,7}, u_{2,1}, u_{4,1}, u_{6,1}, u_{8,1}, u_{2,3}, u_{4,3}, u_{6,3}, u_{8,3}, u_{2,5}, u_{4,5}, u_{6,5}, u_{2,7}, u_{4,7}, v_{1,2}, v_{3,2}, v_{5,2}, v_{7,2}, v_{1,4}, v_{3,4}, v_{5,4}, v_{7,4}, v_{1,6}, v_{3,6}, v_{5,6}, v_{1,8}, v_{3,8}$. Convergence was studied by Amabili (2007).

Geometric imperfections are not considered.

The natural frequency of mode ($m=1$, $n=1$) for different boundary conditions is given in Table 9.1. For model A, which has a slow convergence of linear frequency, a model with 50 dofs has also been used for comparison. Model C has slow convergence in the case of high stiffness of rotational springs (almost clamped edges); in this case, a model with 79 dofs has been used for comparison.

The effect of the stiffness k (N/m^2) of distributed in-plane springs parallel to the panel edges on natural frequencies of mode (1,1) is shown in Figure 9.4, obtained with model A with 19 dofs. Figure 9.4 shows that k of the order of 10^{11} N/m^2 is necessary to simulate the classical simply supported panel.

The effect of the stiffness c (N/rad) of distributed rotational springs at panel edges on natural frequencies of mode (1,1) is shown in Figure 9.5, obtained with model C with 39 dofs. Figure 9.5 shows that c equal to or larger than 2×10^4 N/rad is necessary to simulate the clamped panel.

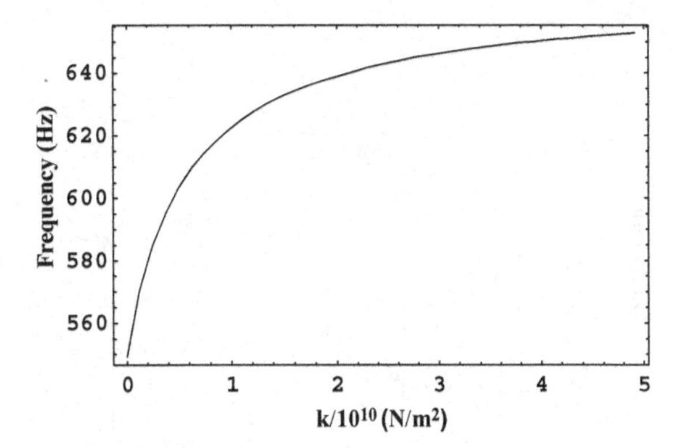

Figure 9.4. Natural frequency of mode (1,1) of supported panel computed with model A versus k; 19 dofs.

Large-amplitude forced vibrations are studied by using the software AUTO 97 (Doedel et al. 1998). The following nondimensional modal excitation on the generalized coordinate $w_{1,1}$ is introduced, and its amplitude is immediately related to the harmonic point force excitation $\tilde{f}$ at $(x = \tilde{x}, y = \tilde{y})$:

$$f = \frac{\tilde{f}\sin(\pi\tilde{x}/a)\sin(\pi\tilde{y}/b)}{h^2\rho\omega_{1,1}^2(a/2)(b/2)}.$$

Harmonic excitation of nondimensional amplitude $f = 0.021$ (corresponding to $\tilde{f} =$ 6.6 N for the simply supported panel) has been imposed at the center of the panel in the frequency range around resonance of the fundamental mode (1, 1). In all the numerical simulations, a modal damping $\zeta_{1,1} = 0.004$ has been assumed.

Figure 9.6 shows the effect of the boundary conditions on the nonlinear response (only the generalized coordinate $w_{1,1}$ is given, which is the most significant) of the panel considering models A and B. In fact, three different boundary conditions are compared: classical simply supported (model B) versus model A for $k = 0$ and $k = 4 \times 10^9$ N/m^2. The simply supported panel for mode (1,1) presents a significant softening nonlinearity, whereas model A with free in-plane edges ($k = 0$) presents hardening nonlinearity. The case for $k = 4 \times 10^9$ N/m^2 obviously lies in between. Comparison of Figures 9.6(a,b) also shows asymmetric behavior of the oscillation outward (maximum) and inward (minimum) with respect to the center of the curvature of the panel. In particular, this asymmetric behavior is enhanced for the simply supported panel.

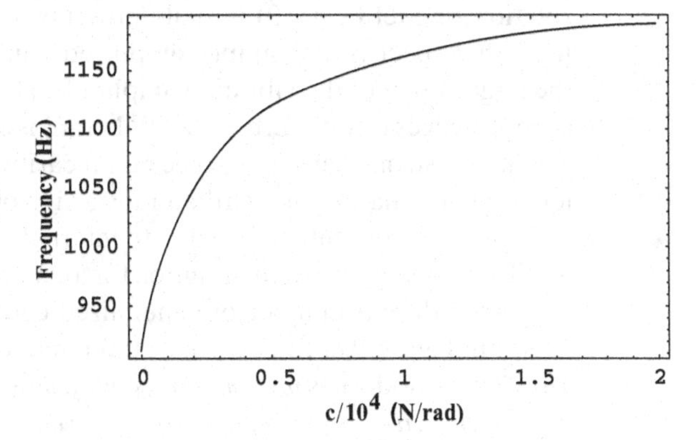

Figure 9.5. Natural frequency of mode (1,1) of panel with fixed edges computed with model C versus c; 39 dofs.

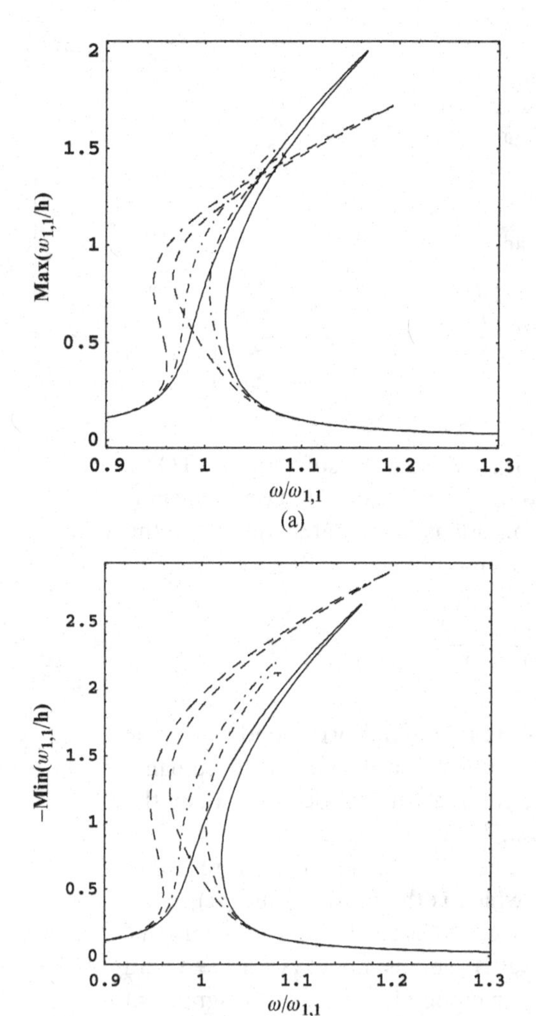

Figure 9.6. Nondimensional response of the panel for different boundary conditions versus nondimensional excitation frequency; mode (1,1), $f=0.021$, $\zeta_{1,1}=0.004$ – –, classical simply supported panel (model B), 9 dofs; –·–, model A with $k=4\times10^9$ N/m^2, 19 dofs; —, model A with $k=0$ (in-plane free edges), 19 dofs. (a) Maximum of generalized coordinate $w_{1,1}$. (b) Minimum of $w_{1,1}$.

Figure 9.7 shows the effect of the boundary conditions on the nonlinear response (in this case, the transverse displacement w at the center of the panel is shown, due to the significant contribution of several generalized coordinates for model C) of the panel considering models B and C. The panel with fixed in-plane edges and free rotation (model C, $c=0$) initially presents an enhanced softening-type nonlinearity with respect to the simply supported panel, turning to the hardening type for the larger value of the vibration amplitude. However, the behavior of the panel with clamped edges (model C, $c=5\times10^4$ N/rad) is completely different and always shows a relatively strong hardening-type nonlinearity. Also for model C, asymmetric behavior of the oscillation outward and inward is observed. Response for model C with $c=0$ presents a peculiar loop due to internal resonances.

The convergence of the solution for a clamped panel (model C, $c=5\times10^4$ N/rad) versus a different number of generalized coordinates retained in the expansion is shown in Figure 9.8. In particular, three models are compared: 24, 27, and 39 dofs. The 27 dofs model has $w_{1,1}, w_{3,1}, w_{5,1}, w_{1,3}, w_{3,3}, w_{5,3}, w_{1,5}, w_{3,5}, w_{5,5}, u_{2,1}, u_{4,1}, u_{6,1}, u_{2,3}, u_{4,3}, u_{6,3}, u_{2,5}, u_{4,5}, u_{6,5}, v_{1,2}, v_{3,2}, v_{5,2}, v_{1,4}, v_{3,4}, v_{5,4}, v_{1,6}, v_{3,6}, v_{5,6}$. The 24 dofs model has

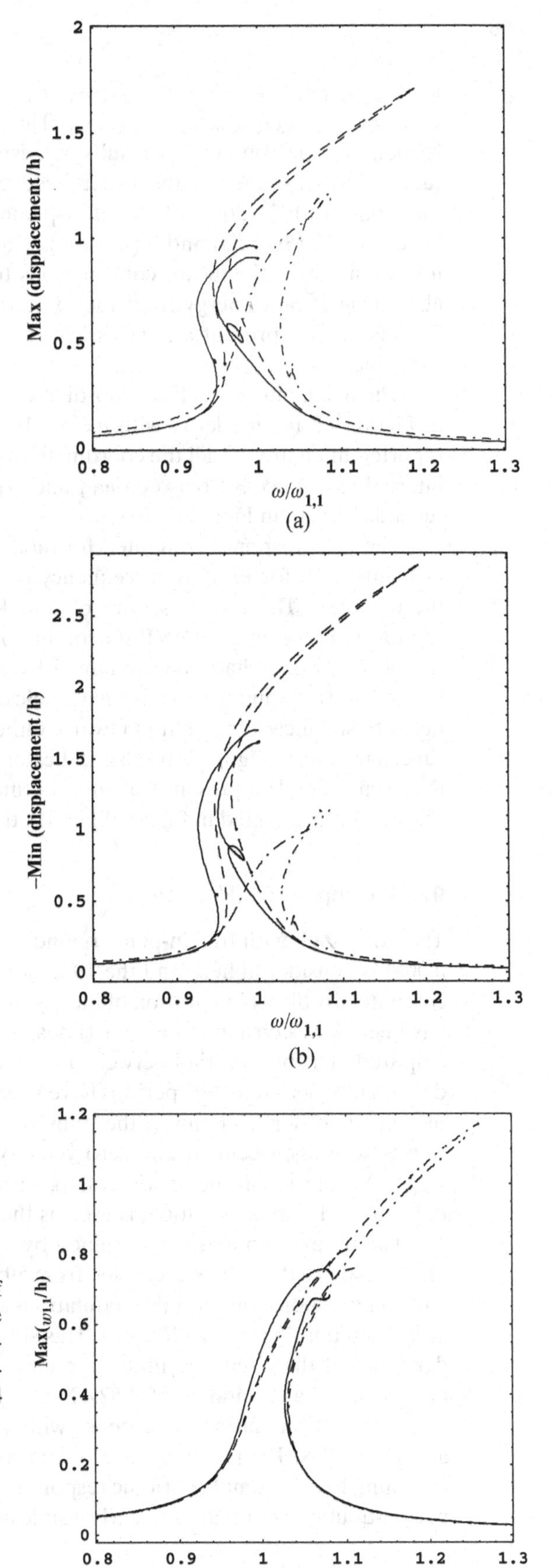

Figure 9.7. Nondimensional response (displacement measured at the panel center divided by h) of the panel for different boundary conditions versus nondimensional excitation frequency; mode (1,1), $f = 0.021$, $\zeta_{1,1} = 0.004$. – – classical simply supported panel (model B), 9 dofs; –·–, model C with $c = 5 \times 10^4$ N/rad (practically clamped), 39 dofs; —, model C with $c = 0$ (fixed edges), 39 dofs. (a) Maximum transverse displacement at the center of the panel. (b) Minimum transverse displacement at the center of the panel.

Figure 9.8. Convergence of model C ($c = 5 \times 10^4$ N/rad) for nonlinear forced vibrations of the clamped panel; nondimensional response of generalized coordinate $w_{1,1}$ versus nondimensional excitation frequency; fundamental mode (1,1), $f = 0.021$, $\zeta_{1,1} = 0.004$. –·–, 24 dofs; – –, 27 dofs; —, 39 dofs.

$w_{1,1}$, $w_{3,1}$, $w_{5,1}$, $w_{1,3}$, $w_{3,3}$, $w_{1,5}$, $u_{2,1}$, $u_{4,1}$, $u_{6,1}$, $u_{2,3}$, $u_{4,3}$, $u_{6,3}$, $u_{2,5}$, $u_{4,5}$, $u_{6,5}$, $v_{1,2}$, $v_{3,2}$, $v_{5,2}$, $v_{1,4}$, $v_{3,4}$, $v_{5,4}$, $v_{1,6}$, $v_{3,6}$, $v_{5,6}$. The three models give a very close trend of hardening-type nonlinearity results but, especially for the 24 dofs model, the amplitude of the response of the generalized coordinate $w_{1,1}$ is different. In fact, for the model with 24 dofs only 6 out-of-plane coordinates are included in the model, instead of 13 (39 dofs) and 9 (27 dofs). This result indicates that, even if $w_{5,3}$, $w_{3,5}$, $w_{5,5}$ do not give significant contributions to the trend of nonlinear response, they absorb significant energy from the excitation. It can be observed that the 27 and 39 dofs models present a nonclassical response with a strange tip, due to internal resonances.

The five main generalized coordinates associated with the panel response given in Figure 9.7 for model C with $c = 5 \times 10^4$ N/rad (practically clamped edges) are reported in Figure 9.9; all the coordinates have significant amplitude in this case. An internal resonance 3:1 between $w_{1,1}$ and $w_{3,1}$ is detected close to 1.04 $\omega_{1,1}$, giving a secondary peak in Figure 9.9(b).

The time response computed for model C with $c = 5 \times 10^4$ N has been plotted in Figure 9.10 for excitation frequency $\omega = 1.06\ \omega_{1,1}$, that is, close to the peak of the response. These results were obtained by direct integration of the equations of motion by using the DIVPAG routine of the Fortran library IMSL, whereas all the previous ones had been obtained by using AUTO 97. Results show that the generalized coordinate $w_{1,1}$ has a harmonic response almost without superharmonics, but with significant translation inward. Other generalized coordinates present large superharmonics. Figure 9.10 also indicates the phase relationship with respect to the excitation. The presence of superharmonics and zero-frequency (mean value) component is clarified in Figure 9.11 with the frequency spectra.

9.4.1 Nonperiodic Response

The same shell with free in-plane boundary conditions (model A with $k = 0$, $\zeta_{1,1} = 0.004$) is considered here and the 19 dofs model is used. Poincaré maps have been computed by direct integration of the equations of motion. The excitation frequency has been kept constant, $\omega = \omega_{1,1}$ (linear resonance condition), and the excitation amplitude has been varied between 0 and about 5952 N. The force range has been divided into 800 steps, 600 periods have been skipped each time the force is changed of a step in order to eliminate the transient motion. The initial condition at the first step is zero displacement and zero velocity for all the variables. At the following steps, the solution at the previous step, with the addition of a small perturbation in order to find a stable solution, is used as the initial condition.

The bifurcation diagram obtained by all these Poincaré maps is shown in Figure 9.12 where the load is decreased from 5952.4 N to 0. Simple periodic motion, subharmonic response, amplitude modulations and chaotic response have been detected, as indicated in Figures 9.12(a,c–e). This shows the very rich and complex nonlinear dynamics of the circular cylindrical panel subject to large harmonic excitation. In particular, for an excitation of 5952.4 N, a chaotic response is obtained, which is transformed into a subharmonic response with a period nine times the excitation period around 5200 N. Then the response becomes quasi-periodic (amplitude modulation), returning to the a simply periodic response around 4850 N. Around 4500 N there is a period-doubling bifurcation, clearly visible in Figure 9.12(c), after which the response

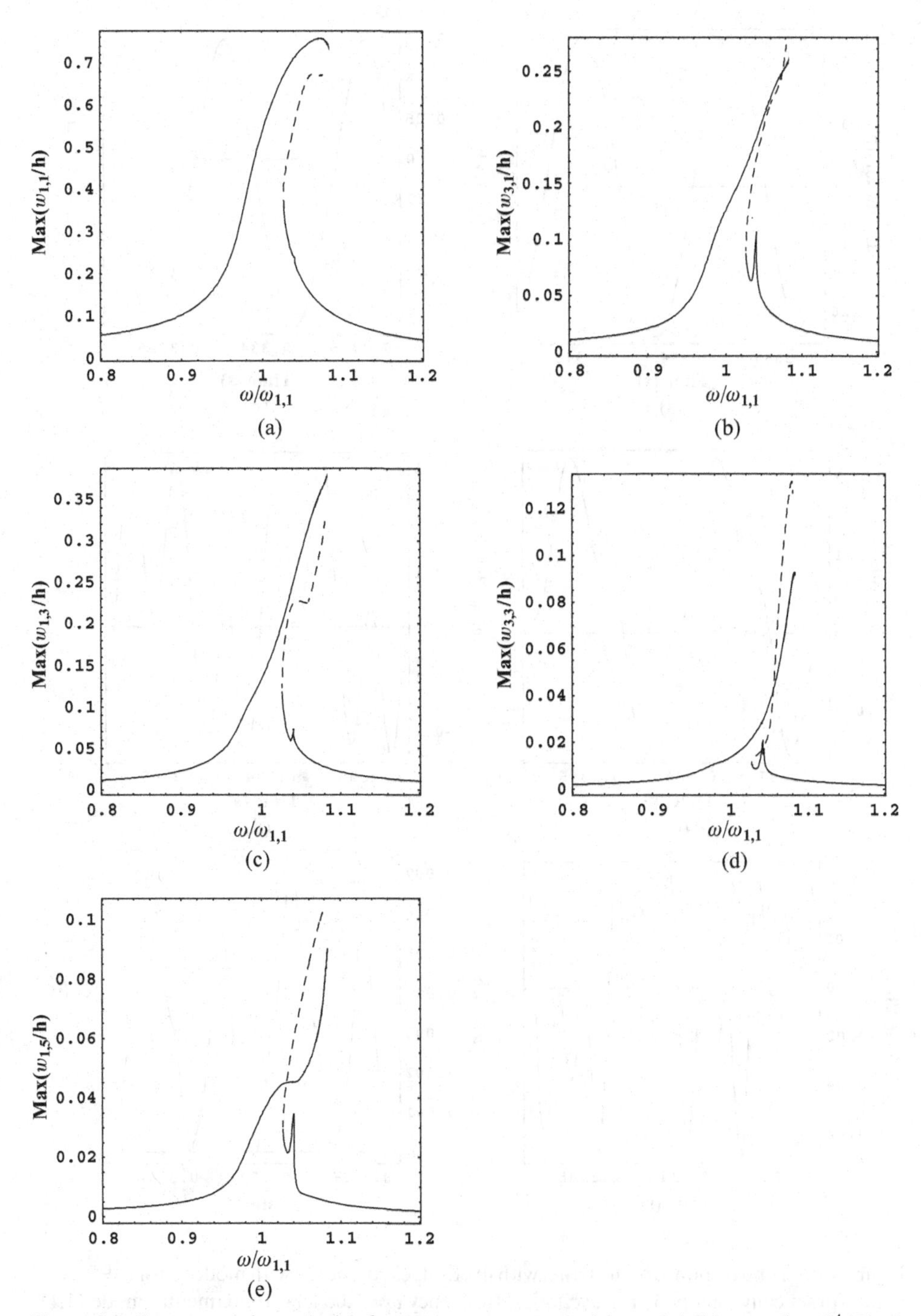

Figure 9.9. Response of panel with fixed edges computed with model C for $c = 5 \times 10^4$ N/rad (practically clamped); fundamental mode (1,1), $f = 0.021$, $\zeta_{1,1} = 0.004$, 39 dofs. —, stable periodic response; – –, unstable periodic response. (a) Maximum of the generalized coordinate $w_{1,1}$. (b) Maximum of the generalized coordinate $w_{3,1}$. (c) Maximum of the generalized coordinate $w_{1,3}$. (d) Maximum of the generalized coordinate $w_{3,3}$. (e) Maximum of the generalized coordinate $w_{1,5}$.

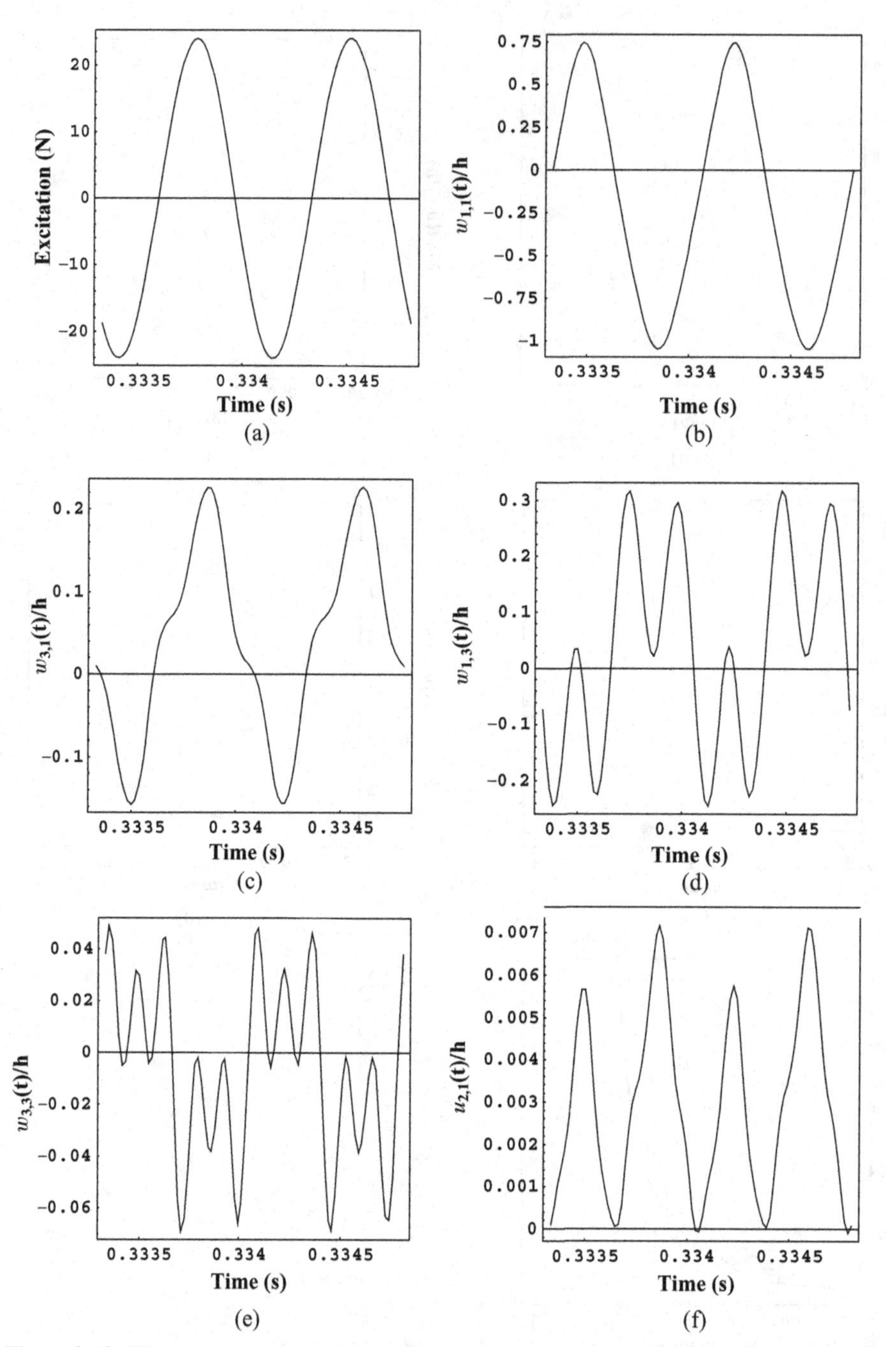

Figure 9.10. Time response of the panel with fixed edges computed with model C for $c = 5 \times 10^4$ N/rad (practically clamped) for excitation frequency $\omega = 1.06\ \omega_{1,1}$; fundamental mode (1,1), $f = 0.021$, $\zeta_{1,1} = 0.004$, 39 dofs. (a) Force excitation. (b) Generalized coordinate $w_{1,1}$. (c) Generalized coordinate $w_{3,1}$. (d) Generalized coordinate $w_{1,3}$. (e) Generalized coordinate $w_{3,3}$. (f) Generalized coordinate $u_{2,1}$.

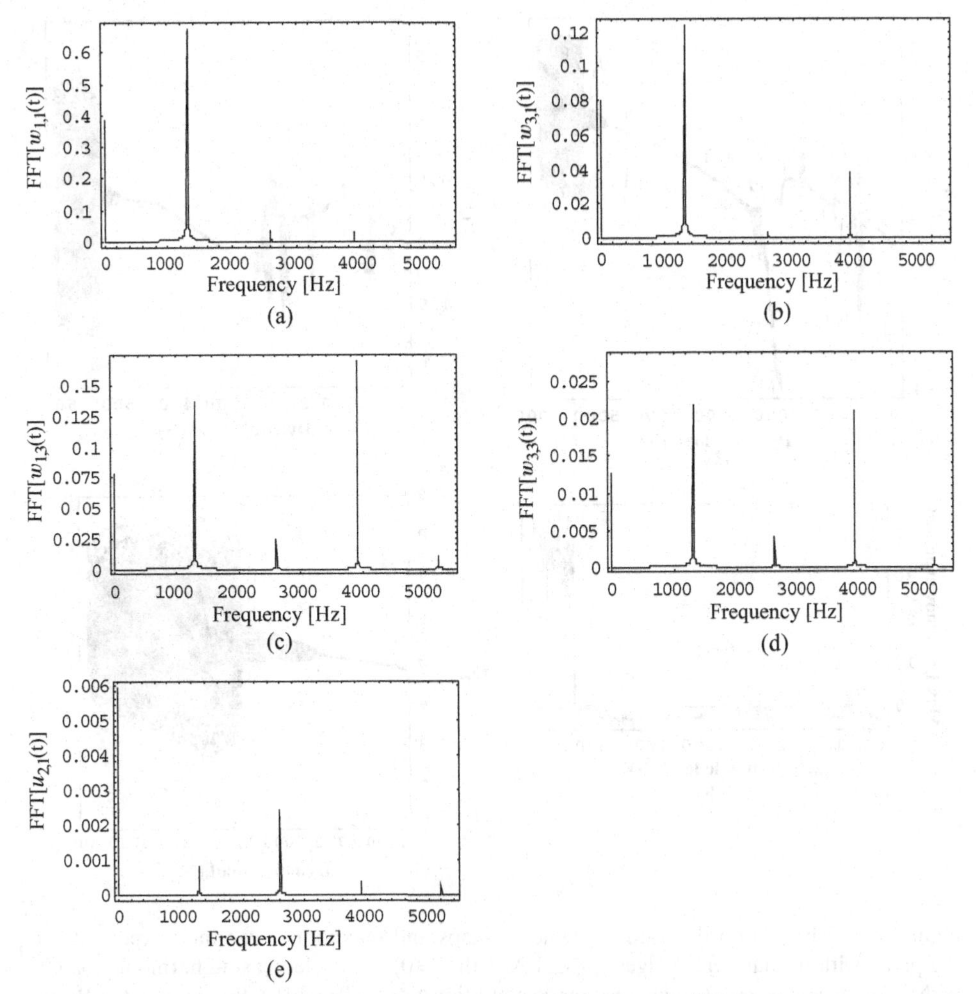

Figure 9.11. Frequency spectrum of the response of the panel with fixed edges computed with model C for $c = 5 \times 10^4$ N/rad (practically clamped) for excitation frequency $\omega = 1.06\ \omega_{1,1}$; fundamental mode (1,1), $f = 0.021$, $\zeta_{1,1} = 0.004$, 39 dofs. (a) Generalized coordinate $w_{1,1}$. (b) Generalized coordinate $w_{3,1}$. (c) Generalized coordinate $w_{1,3}$. (d) Generalized coordinate $w_{3,3}$. (e) Generalized coordinate $u_{2,1}$.

again becomes simply periodic. In the range between 2500 and 800 N, several regions of quasi-periodic response appear. A period-doubling bifurcation is detected around 670 N, which ends in a chaotic region at 580 N. After that, a simply periodic oscillation is detected, with a final jump to the undisturbed configuration at zero excitation.

Figure 9.12(b) gives the maximum Lyapunov exponent σ_1 associated with the bifurcation diagram. It can be easily observed that: (i) for periodic forced vibrations $\sigma_1 < 0$, (ii) for amplitude-modulated response (quasi-periodic) $\sigma_1 = 0$ and (iii) for chaotic response $\sigma_1 > 0$. Therefore, σ_1 can be conveniently used for identification of the system dynamics. A three-dimensional representation of the bifurcation diagram for the generalized coordinate $w_{1,1}$ in the displacement, velocity and load space is

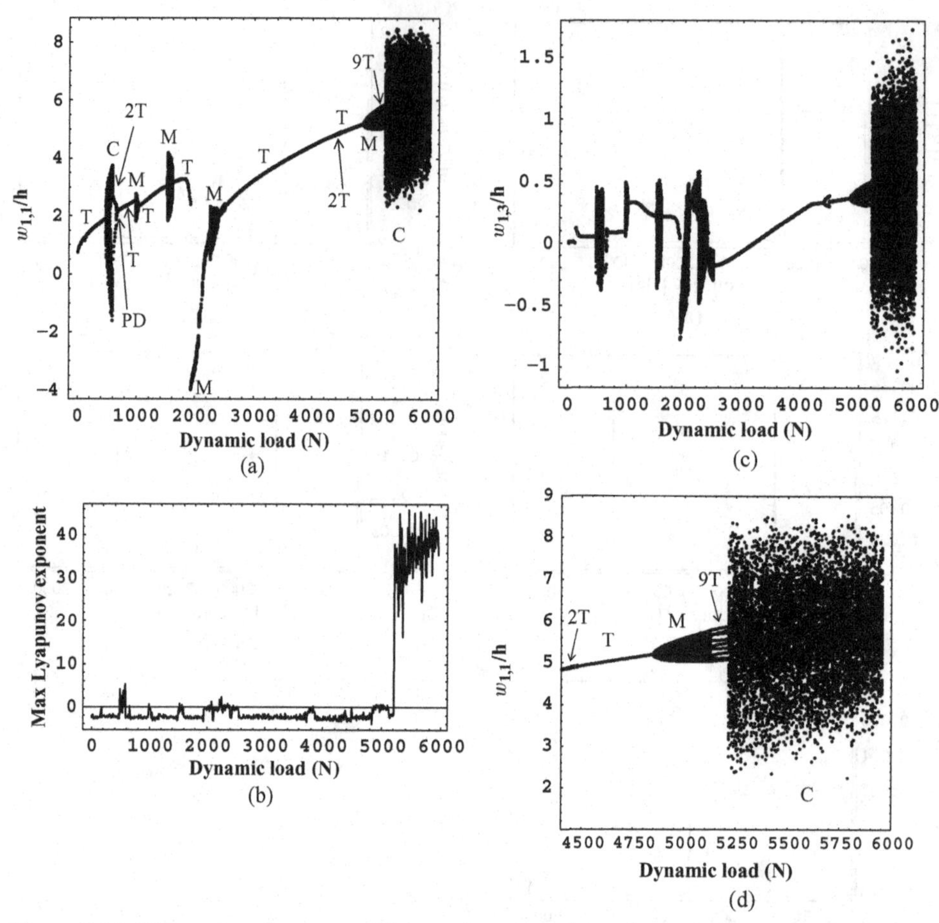

Figure 9.12. Bifurcation diagram of Poincaré maps and maximum Lyapunov exponent for the panel with in-plane free edges (model A with $k=0$) under decreasing harmonic load $\tilde{f}$ with frequency $\omega = \omega_{1,1}$ (linear resonance condition); $\zeta_{1,1} = 0.004$; 19 dofs model. (a) Bifurcation diagram: generalized coordinate $w_{1,1}$; T, response period equal to excitation period; 2T, periodic response with two times the excitation period; 9T, periodic response with nine times the excitation period; PD, period-doubling bifurcation; M, amplitude modulations; C, chaos. (b) Maximum Lyapunov exponent. (c) Bifurcation diagram: generalized coordinate $w_{1,3}$. (d) Bifurcation diagram: generalized coordinate $w_{1,1}$, enlarged scale. (e) Three-dimensional representation of the bifurcation diagram: generalized coordinate $w_{1,1}$.

shown in Figure 9.12(e), whereas in Figures 9.12(a,c,d) the bifurcation diagrams have been projected on a plane orthogonal to the velocity axis.

The study of the complete spectrum of the Lyapunov exponents is often reported only for simple systems. In the present case, all 38 Lyapunov exponents have been evaluated for the case with excitation $\tilde{f} = 5952.4$ N, corresponding to chaotic response, and are given in Figure 9.13. In this case, four positive Lyapunov exponents have been identified, allowing us to classify this response as hyperchaos. The Lyapunov dimension in this case is $d_L = 27.56$. The shape of the curve joining the exponents in Figure 9.13 is nearly antisymmetric. Moreover, most of the exponents have a very similar value (slightly negative), which can be related to damping; they form a characteristic, nearly horizontal segment.

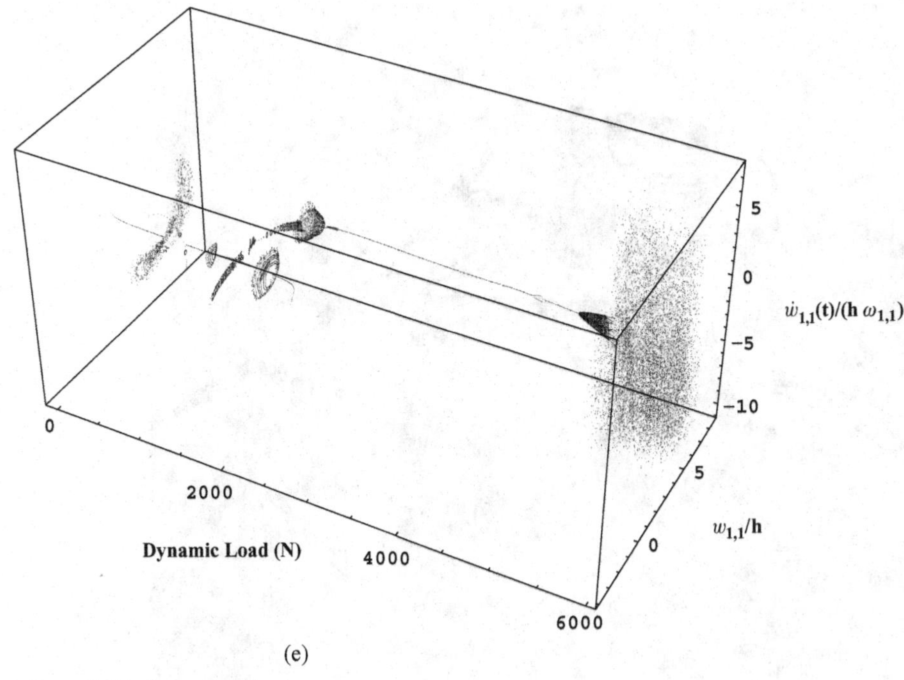

(e)

Figure 9.12. (*continued*)

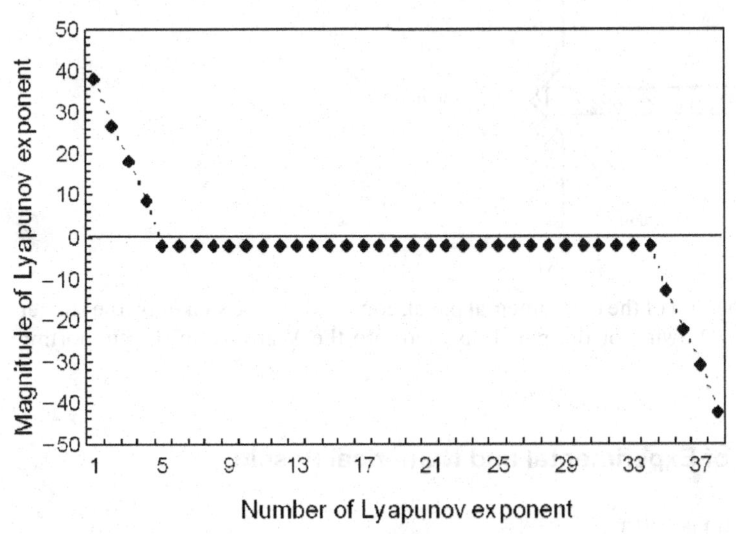

Figure 9.13. All 38 Lyapunov exponents for the panel with in-plane free edges (model A with $k = 0$); excitation frequency $\omega = \omega_{1,1}$ (linear resonance condition); $\tilde{f} = 5952.4$ N; $\zeta_{1,1} = 0.004$; 19 dofs model.

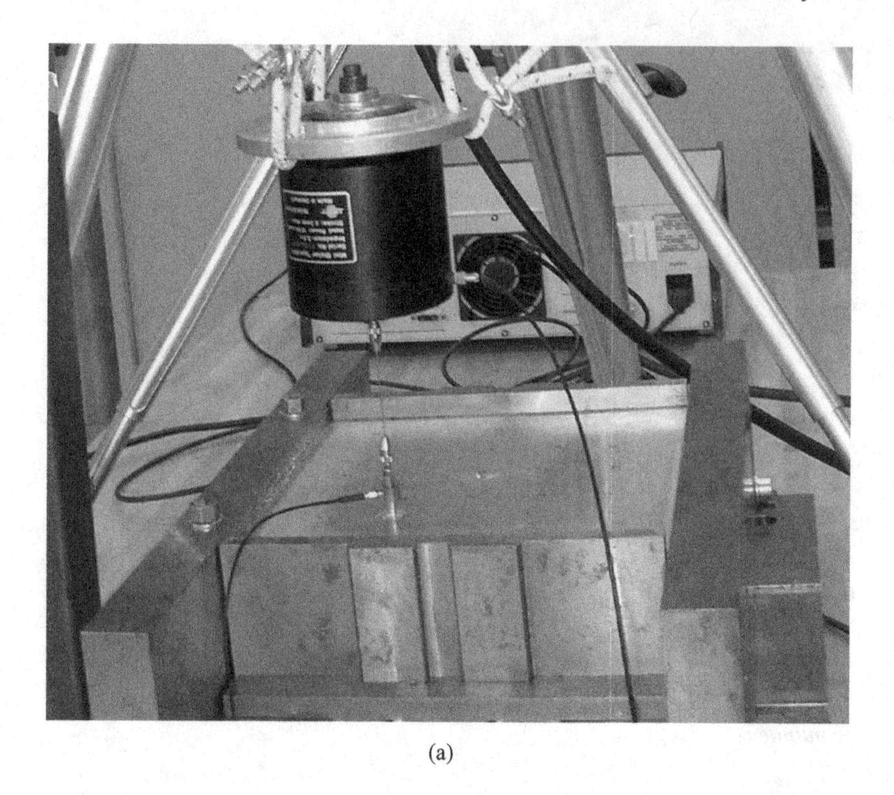

(a)

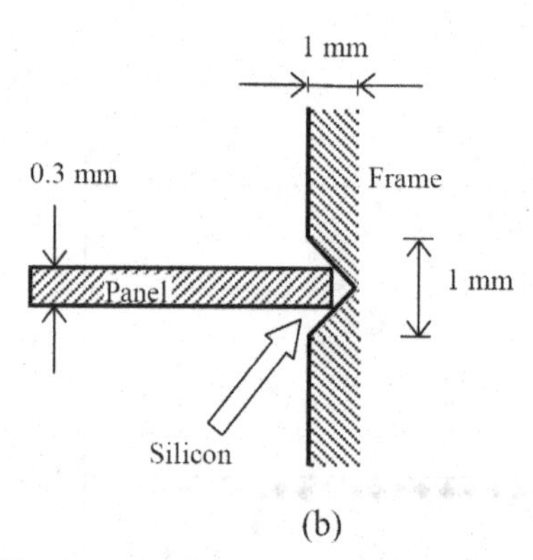

(b)

Figure 9.14. (a) Photograph of the experimental panel, connected to the shaker by the stinger and the load cell. (b) Drawing of the panel inserted into the V-groove in the supporting frame.

9.5 Comparison of Experimental and Numerical Results

9.5.1 Experimental Results

Tests have been conducted on a stainless steel panel with the following dimensions and material properties: $a = 0.199$ m, $R = 2$ m, $\alpha = 0.066$ rad, $h = 0.0003$ m, $E = 195$ GPa, $\rho = 7800\,\text{kg/m}^3$ and $\nu = 0.3$. The panel was inserted into a heavy, rectangular steel frame made of several thick parts (see Figure 9.14[a,b]), having V-grooves designed to hold the panel and to avoid transverse (radial) displacements at the edges.

Silicon was placed into the grooves to fill any gap between the panel and the grooves. Almost all the in-plane displacements normal to the edges were allowed because the constraint given by silicon on these displacements was very small. In-plane displacements parallel to the edges were elastically constrained by the silicon. Therefore, the experimental boundary conditions are close to those given by model A, with k assuming a relatively small value.

The panels have been subjected to (i) burst-random excitation to identify the natural frequencies and perform a modal analysis by measuring the panel response on a grid of points, (ii) harmonic excitation, increasing or decreasing by very small steps the excitation frequency in the spectral neighborhood of the lowest natural frequencies, to characterize nonlinear responses in the presence of large-amplitude vibrations (step-sine excitation). The excitation has been provided by an electrodynamical exciter (shaker), model B&K 4810. A piezoelectric miniature force transducer B&K 8203 weighing 3.2 g, glued to the panel and connected to the shaker with a stinger, measured the force transmitted. The panel response was measured by using a very accurate laser Doppler vibrometer Polytec (sensor head OFV-505 and controller OFV-5000) in order to have noncontact measurement without introduction of inertia. The time responses were measured by using the Difa Scadas II front-end, connected to a workstation, and the software CADA-X 3.5b of LMS for signal processing, data analysis, experimental modal analysis and excitation control. The same front-end was used to generate the excitation signal. The CADA-X closed-loop control was used to keep constant the value of the excitation force for any excitation frequency during the measurement of the nonlinear response.

Geometric imperfections of the panel have been detected by using a three-dimesional laser scanning system VI-910 Minolta to measure the actual panel surface. The contour plot indicating the deviation from the ideal panel surface is reported in Figure 9.15. Geometric imperfections are always present in actual panels. Actually, in the panels tested these imperfections are associated with initial stresses, which have been minimized with accurate positioning in the frame. These initial stresses are not measured and are not taken into account in the modeling.

Figure 9.16 shows the measured oscillation (displacement, directly measured by using the Polytec laser Doppler vibrometer with displacement decoder DD-200 in the OFV-5000 controller; measurement position at the center of the panel) around the fundamental frequency, that is, mode (1, 1), versus the excitation frequency for three different force levels: 0.01, 0.05, 0.1, 0.15 and 0.2 N. The excitation point was at $\tilde{x} = a/4$ and $\tilde{y} = b/3$. The level of 0.01 N gives a good evaluation of the natural (linear) frequency, identified at 96.2 Hz. The closed-loop control used in the experiments keeps constant the amplitude of the harmonic excitation force, after filtering the signal from the load cell in order to use only the harmonic component with the given excitation frequency. The measured oscillation reported in Figure 9.16 has been filtered in order to eliminate any frequency except the excitation frequency (first harmonic of the response). Experiments have been performed increasing and decreasing the excitation frequency (up and down). The frequency step used in this case is 0.025 Hz; 16 periods have been measured with 128 points per period and 200 periods have been waited before data acquisition every time that the frequency is changed. The hysteresis between the two curves (up = increasing frequency; down = decreasing frequency)

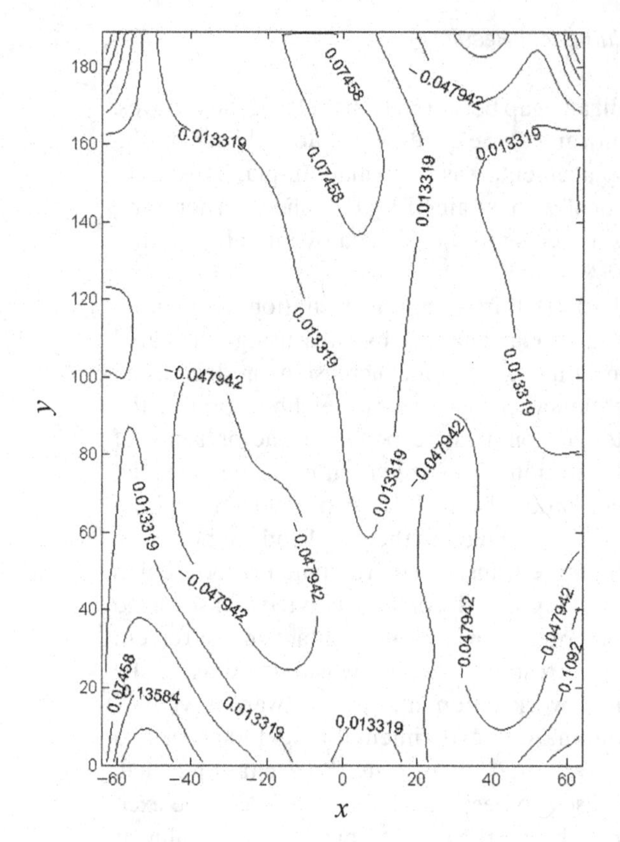

Figure 9.15. Contour plot indicating measured geometric imperfections as deviation from the ideal panel surface. Deviations are in millimeters.

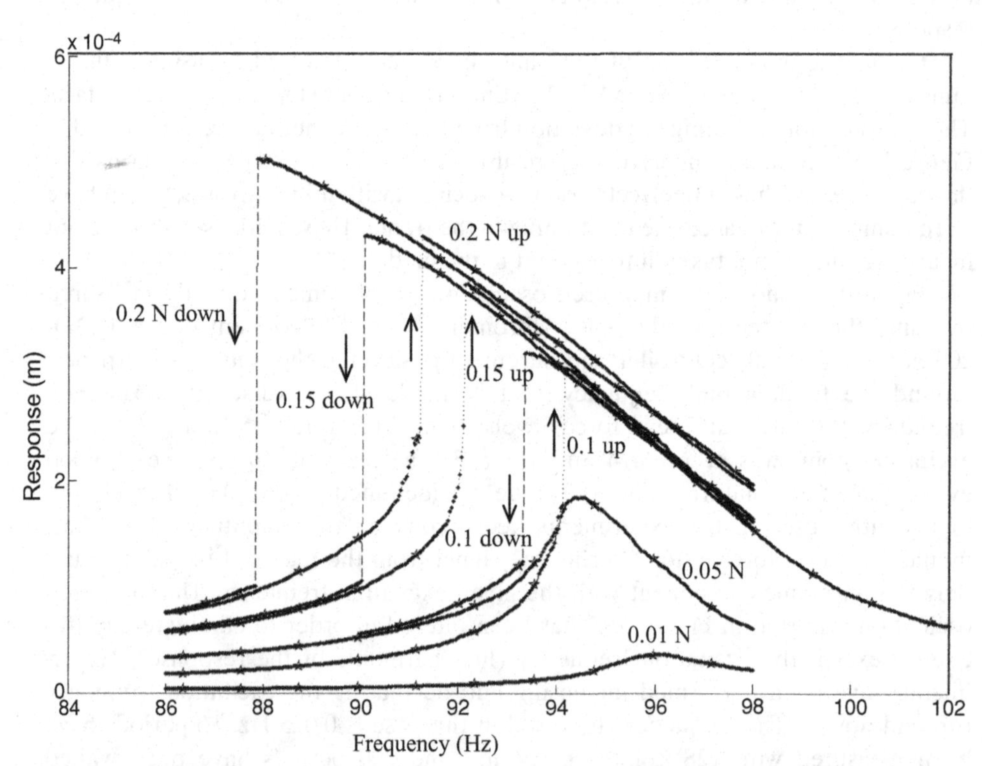

Figure 9.16. Experimental oscillatory displacement (first-harmonic) versus excitation frequency for different excitation levels measured at the center of the panel; fundamental mode (1,1). •, experimental point; – –, connecting line (down); . . . , connecting line (up); →, direction of movement along the line.

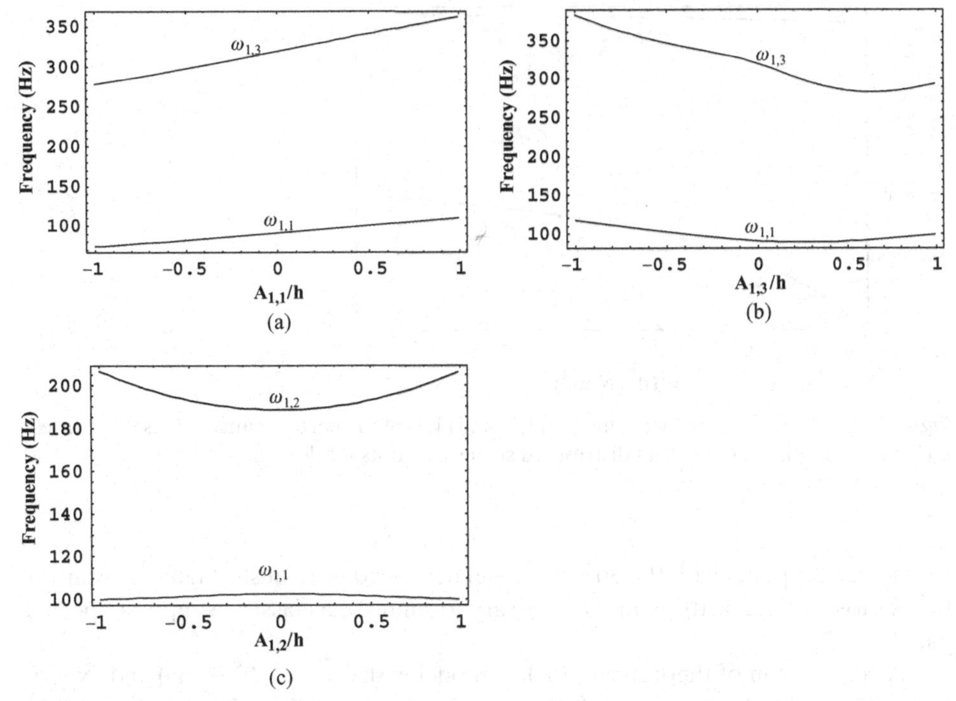

Figure 9.17. Natural frequency of the panel versus geometric imperfections; $k=0$. (a) Effect of $A_{1,1}$, 38 dofs model. (b) Effect of $A_{1,3}$, 38 dofs model. (c) Effect of $A_{1,2}$, 36 dofs model.

is clearly visible for the three larger excitation levels (0.1, 0.15 and 0.2 N). Sudden increments (jumps) of the vibration amplitude are observed when increasing and decreasing the excitation frequency; these indicate softening-type nonlinearity.

It must be observed that the force input around resonance was very distorted with respect to the imposed pure sinusoidal excitation; this is probably the reason for the lack of perfect superposition of part of the "up" and "down" responses.

9.5.2 Comparison of Numerical and Experimental Results

The effects of geometric imperfections $A_{1,1}$ and $A_{1,3}$ on the natural frequency of modes (1,1) and (1,3) are shown in Figures 9.17(a,b), respectively; results have been obtained for $k=0$ with a 38 dofs model including the following generalized coordinates: $w_{1,1}$, $w_{1,3}$, $w_{3,1}$, $w_{3,3}$, $w_{1,5}$, $w_{5,1}$, $u_{1,0}$, $u_{1,2}$, $u_{1,4}$, $u_{1,6}$, $u_{1,8}$, $u_{1,10}$, $u_{1,12}$, $u_{1,14}$, $u_{1,16}$, $u_{1,18}$, $u_{1,20}$, $u_{3,0}$, $u_{3,2}$, $u_{3,4}$, $u_{3,6}$, $u_{3,8}$, $v_{0,1}$, $v_{2,1}$, $v_{4,1}$, $v_{6,1}$, $v_{8,1}$, $v_{10,1}$, $v_{12,1}$, $v_{14,1}$, $v_{16,1}$, $v_{18,1}$, $v_{20,1}$, $v_{0,3}$, $v_{2,3}$, $v_{4,3}$, $v_{6,3}$, $v_{8,3}$. The effect of imperfection $A_{1,2}$ on the natural frequencies of modes (1,1) and (1,2) is presented in Figure 9.17(c), obtained for $k=0$ with a 36 dofs model including $w_{1,2}$, $w_{1,1}$, $w_{1,3}$, $w_{3,1}$, $w_{3,3}$, $w_{1,4}$, $u_{1,1}$, $u_{1,3}$, $u_{1,5}$, $u_{3,1}$, $u_{3,3}$, $u_{3,5}$, $u_{1,0}$, $u_{1,2}$, $u_{1,4}$, $u_{3,0}$, $u_{3,2}$, $u_{3,4}$, $u_{5,0}$, $u_{5,2}$, $u_{5,4}$, $v_{0,2}$, $v_{2,2}$, $v_{4,2}$, $v_{0,4}$, $v_{2,4}$, $v_{4,4}$, $v_{0,1}$, $v_{2,1}$, $v_{4,1}$, $v_{0,3}$, $v_{2,3}$, $v_{4,3}$, $v_{0,5}$, $v_{2,5}$, $v_{4,5}$.

The effect of the stiffness k (N/m^2) of distributed springs parallel to the panel edges on natural frequencies of modes (1,1) and (1,2) is shown in Figure 9.18, obtained

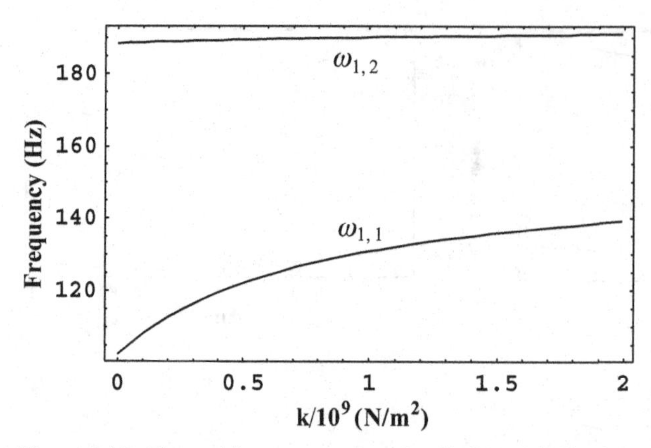

Figure 9.18. Natural frequency of modes (1,1) and (1,2) of the perfect panel versus the stiffness k (N/m^2) of in-plane, tangential distributed springs; 36 dofs model.

for the perfect panel with the 36 dofs model described previously. Figure 9.18 shows that k larger than 2×10^9 N/m^2 is necessary to simulate a classical simply supported panel.

A comparison of theoretical (36 dofs model with $k = 3 \times 10^8$ N/m^2) and experimental results for excitation $\tilde{f} = 0.15$N at ($\tilde{x} = a/4$, $\tilde{y} = b/3$) is shown in Figure 9.19 (damping $\zeta_{1,1} = 0.012$, assumed to be the same for all the generalized coordinates). Comparison of numerical and experimental results is excellent for both the first harmonic and the mean value. Calculations have been obtained introducing the

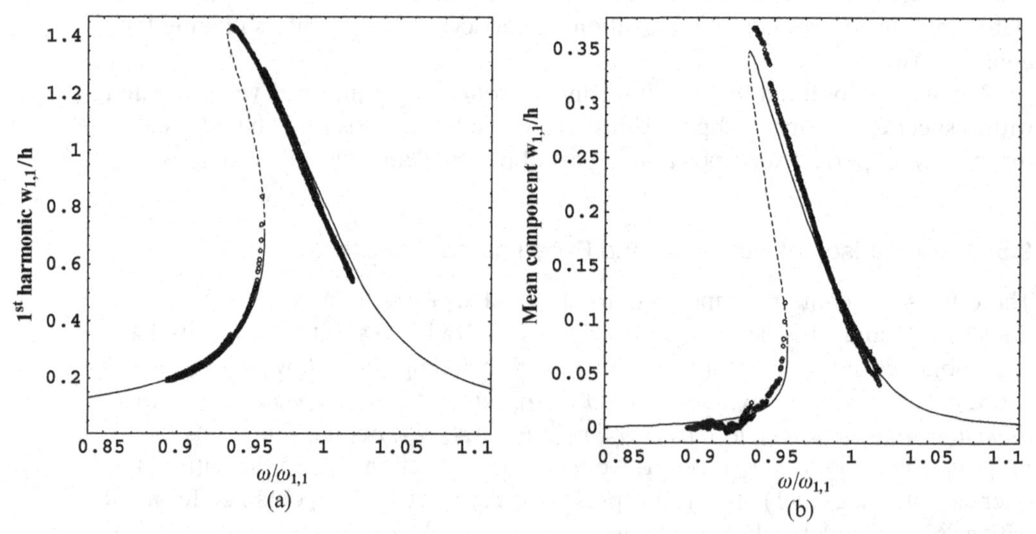

Figure 9.19. Comparison of numerical and experimental results for the nondimensional response of the panel versus nondimensional excitation frequency; mode (1,1), $\tilde{f} = 0.15$ N, $\zeta_{1,1} = 0.012$, $A_{1,1} = 0.35h$; $k = 2 \times 10^8$ N/m$_2$, 36 dofs. ○, experimental data; —, stable theoretical solutions; – – –, unstable theoretical solutions. (a) First harmonic. (b) Mean value.

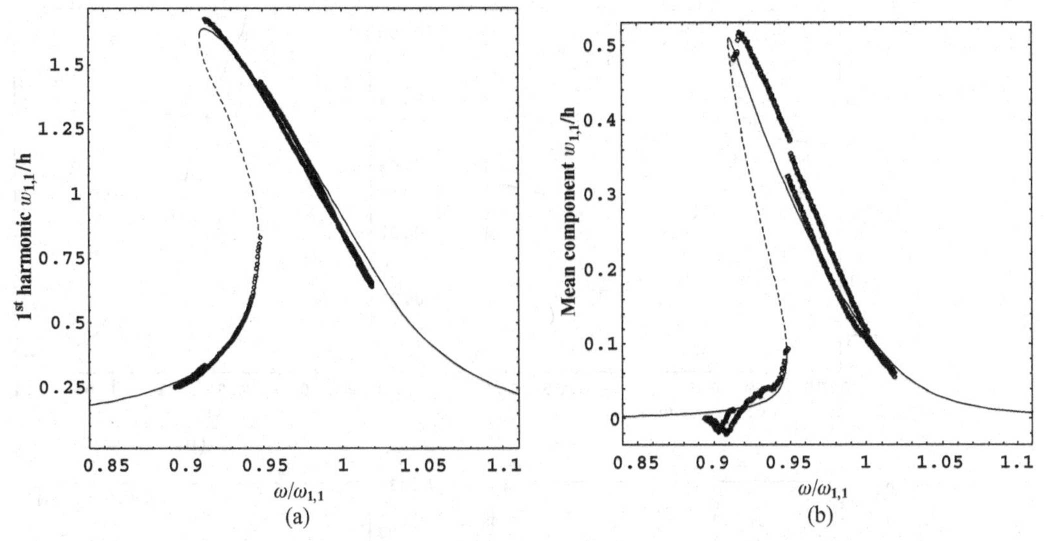

Figure 9.20. Comparison of numerical and experimental results for the nondimensional response of the panel versus nondimensional excitation frequency; mode (1,1), $\tilde{f} = 0.2$ N, $\zeta_{1,1} = 0.011$, $A_{1,1} = 0.35h$; $k = 2 \times 10^8$ N/m^2, 36 dofs. ○, experimental data; —, stable theoretical solutions; – –, unstable theoretical solutions. (a) First harmonic. (b) Mean value.

geometric imperfection $A_{1,1} = 0.35\ h$, having the form of mode (1,1), which is of the same order of magnitude as measured imperfections of the plate surface, as reported in Figure 9.15. Although the first harmonic of the response is the most significant one because it is directly excited, the mean value (constant value, at zero frequency) of the response indicates an asymmetric oscillation of the panel inward and outward. Additional harmonics of the response are much smaller. The assumed value of $k = 3 \times 10^8$ N/m^2 is compatible with the experimental boundary condition; the value of damping has been identified by using the nonlinear experimental response.

A second comparison of theoretical and experimental results is shown in Figure 9.20 for excitation $\tilde{f} = 0.2$ N (damping $\zeta_{1,1} = 0.011$). This comparison is also excellent for both the first harmonic and the mean value. In Figures 9.19 and 9.20 the indication of stability of the solution is also given; however, it is the classical one of a system with softening-type nonlinearity. The value of damping has been identified by using the nonlinear experimental response also in this case and shows a slightly increased damping with the increased excitation. The presence of nonlinear damping, in general increasing with the amplitude of oscillation, was observed previously and discussed in Chapter 5.

The five main generalized coordinates associated with the panel response given in Figure 9.19 are reported in Figure 9.21 for completeness. In particular, the asymmetry of the response of $w_{1,1}$ shows large difference between maximum (outward) and minimum (inward) oscillation, giving rise to the mean value of the response previously shown. This behavior is investigated with more accuracy in Figure 9.22, where the time response has been plotted for excitation frequency $\omega = 0.95\ \omega_{1,1}$, that is, close to the peak of the response. These results were obtained by direct integration of the equations of motion by using the DIVPAG routine of the Fortran library

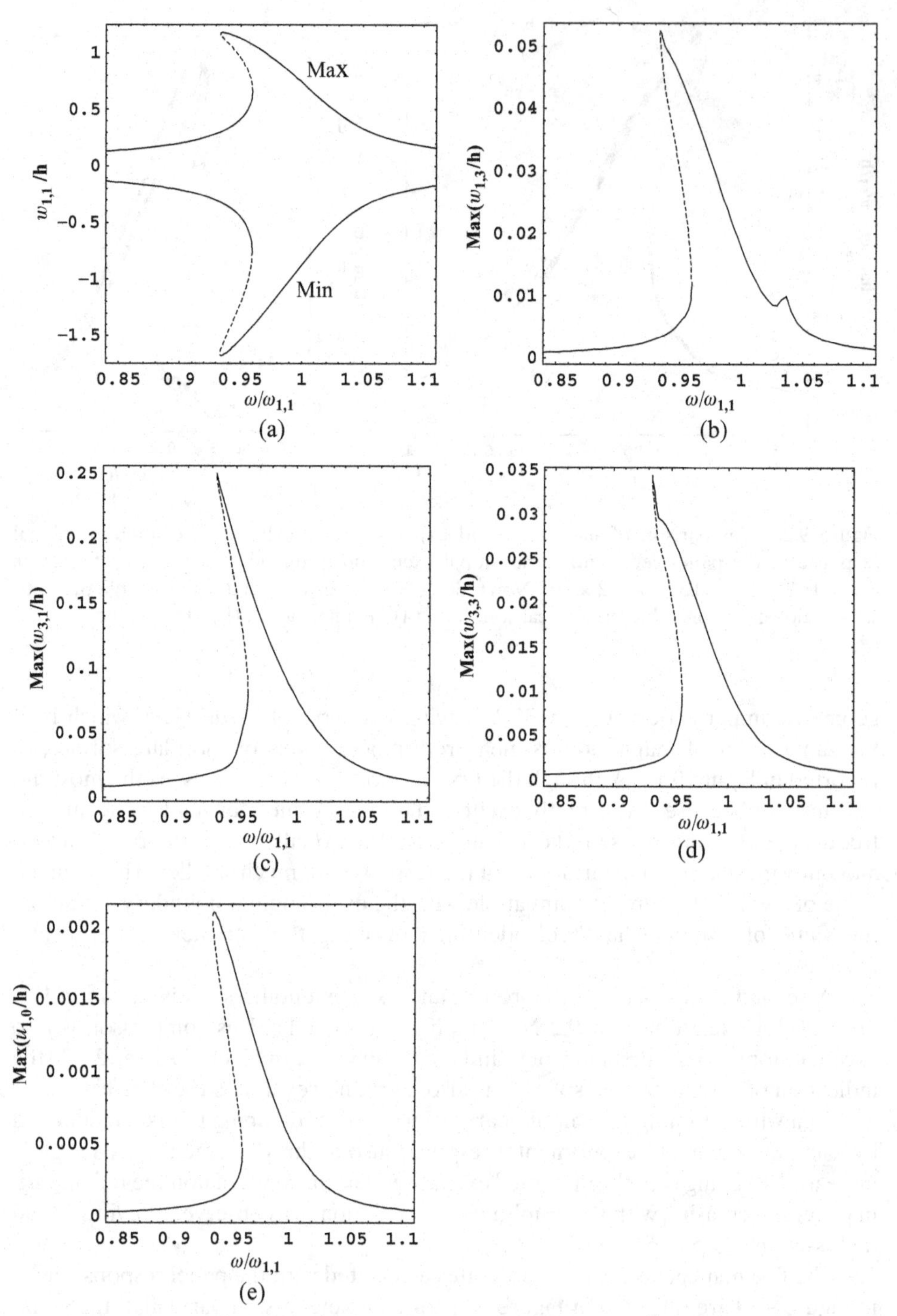

Figure 9.21. Response of the panel; fundamental mode (1,1), $\tilde{f} = 0.15$ N, $\zeta_{1,1} = 0.012$, $A_{1,1} = 0.35h$; $k = 2 \times 10^8$ N/m^2, 36 dofs. ——, stable periodic response – –, unstable periodic response. (a) Maximum and minimum of the generalized coordinate $w_{1,1}$. (b) Maximum of the generalized coordinate $w_{1,3}$. (c) Maximum of the generalized coordinate $w_{3,1}$. (d) Maximum of the generalized coordinate $w_{3,3}$. (e) Maximum of the generalized coordinate $u_{1,0}$.

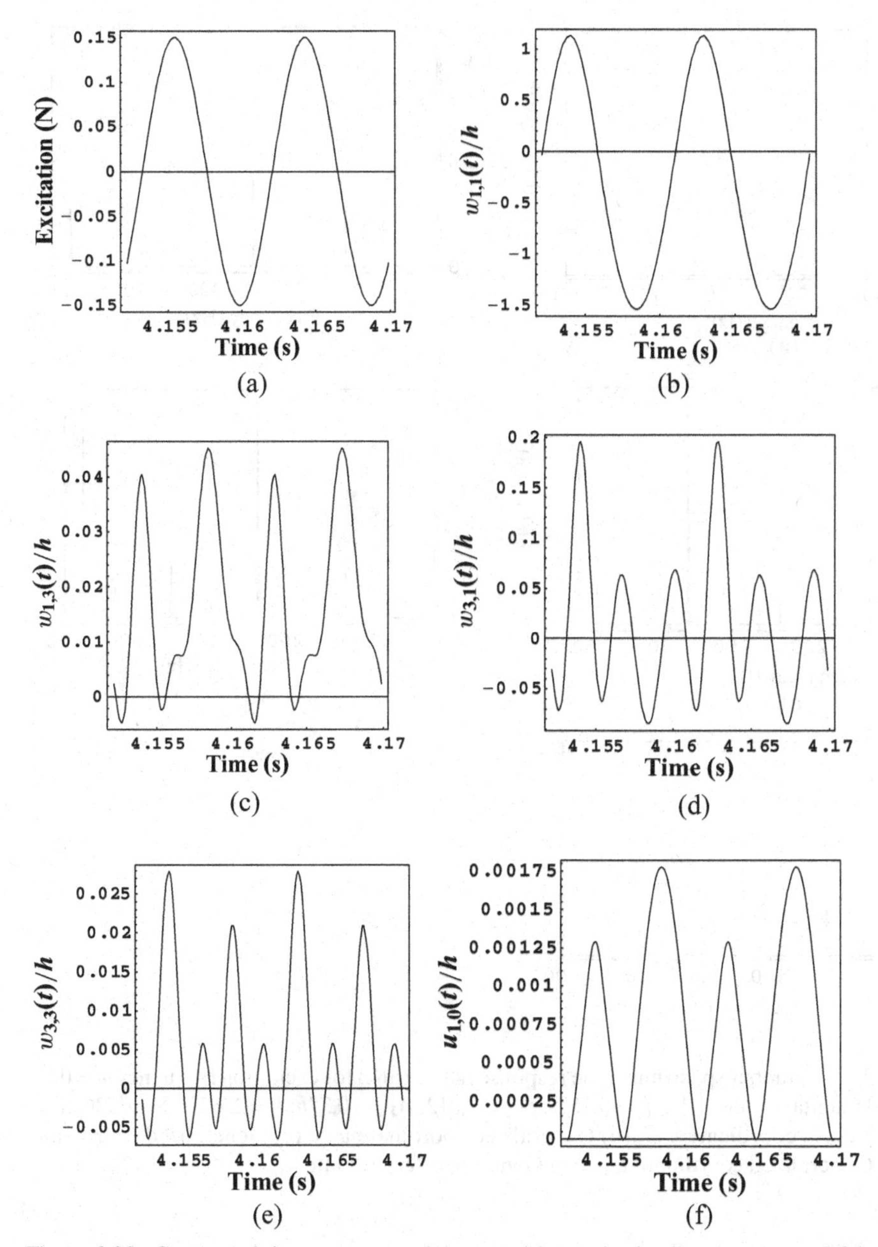

Figure 9.22. Computed time response of the panel for excitation frequency $\omega = 0.95\,\omega_{1,1}$; fundamental mode (1,1), $\tilde{f} = 0.15$ N, $\zeta_{1,1} = 0.012$, $A_{1,1} = 0.35h$; $k = 2 \times 10^8$ N/m^2, 36 dofs. (a) Force excitation. (b) Generalized coordinate $w_{1,1}$. (c) Generalized coordinate $w_{1,3}$. (d) Generalized coordinate $w_{3,1}$. (e) Generalized coordinate $w_{3,3}$. (f) Generalized coordinate $u_{1,0}$.

IMSL, whereas all the previous ones were obtained by using AUTO 97. Figure 9.22 also indicates the phase relationship with respect to the excitation. The presence of superharmonics and the zero-frequency (mean value) component is clarified in Figure 9.23 with the frequency spectra.

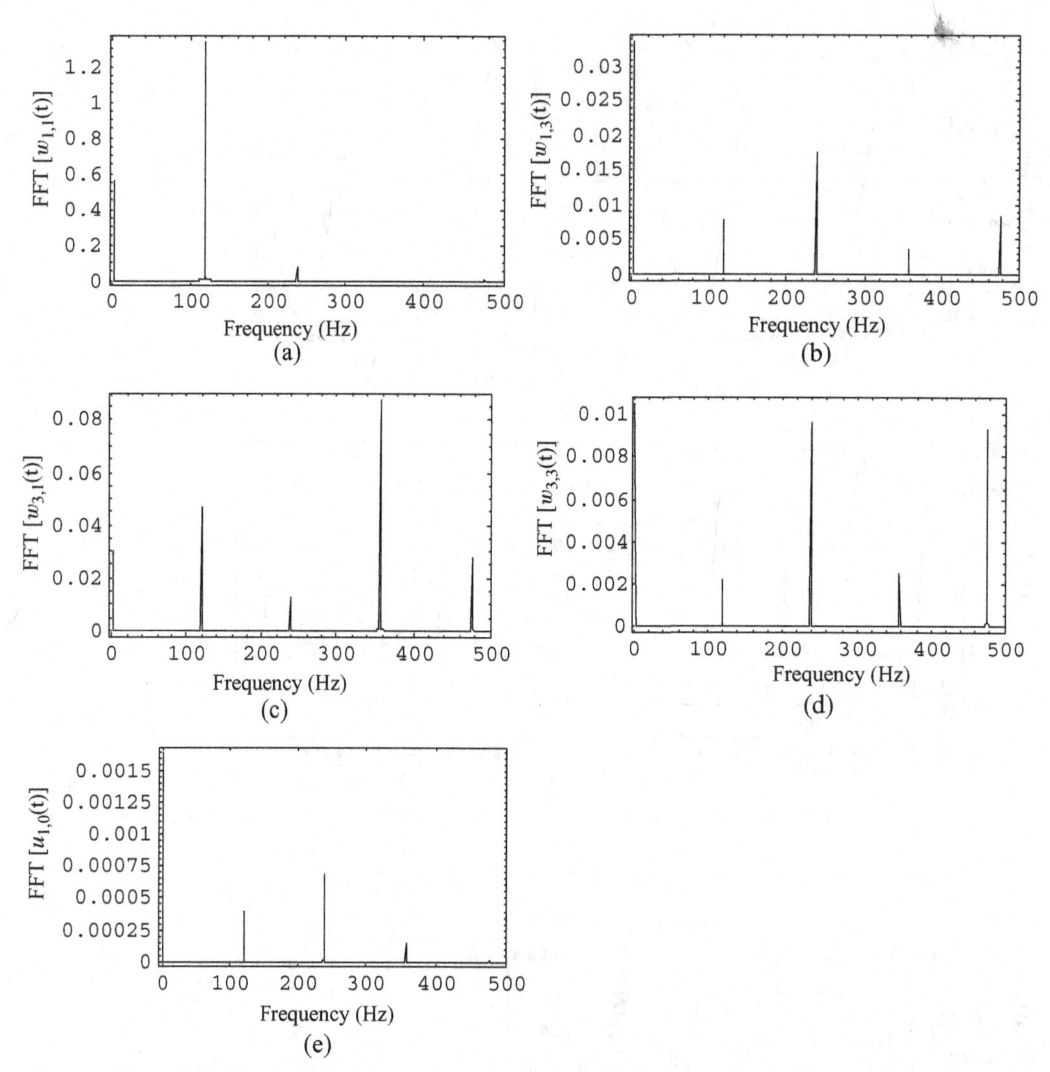

Figure 9.23. Frequency spectrum of the response of the panel for excitation frequency $\omega = 0.95\ \omega_{1,1}$; fundamental mode (1,1), $\tilde{f} = 0.15$ N, $\zeta_{1,1} = 0.012$, $A_{1,1} = 0.35h$; $k = 2 \times 10^8$ N/m^2, 36 dofs. (a) Generalized coordinate $w_{1,1}$. (b) Generalized coordinate $w_{1,3}$. (c) Generalized coordinate $w_{3,1}$. (d) Generalized coordinate $w_{3,3}$. (e) Generalized coordinate $u_{1,0}$.

REFERENCES

A. Abe, Y. Kobayashi and G. Yamada 2000 *Journal of Sound and Vibration* **234**, 405–426. Non-linear vibration characteristics of clamped laminated shallow shells.

M. Amabili 2005 *Journal of Sound and Vibration* **281**, 509–535. Nonlinear vibrations of circular cylindrical panels.

M. Amabili 2006 *Journal of Sound and Vibration* **298**, 43–72. Theory and experiments for large-amplitude vibrations of circular cylindrical panels with geometric imperfections.

M. Amabili 2007 *Journal of Applied Mechanics* **74**, 645–657. Effect of boundary conditions on nonlinear vibrations of circular cylindrical panels.

C. Y. Chia 1987a *International Journal of Solids and Structures* **23**, 1123–1132. Non-linear free vibration and postbuckling of symmetrically laminated orthotropic imperfect shallow cylindrical panels with two adjacent edges simply supported and the other edges clamped.

C. Y. Chia 1987b *International Journal of Engineering Sciences* **25**, 427–441. Nonlinear vibration and postbuckling of unsymmetrically laminated imperfect shallow cylindrical panels with mixed boundary conditions resting on elastic foundation.

B. E. Cummings 1964 *AIAA Journal* **2**, 709–716. Large-amplitude vibration and response of curved panels.

E. J. Doedel, A. R. Champneys, T. F. Fairgrieve, Y. A. Kuznetsov, B. Sandstede and X. Wang 1998 *AUTO 97: Continuation and Bifurcation Software for Ordinary Differential Equations (with HomCont)*. Concordia University, Montreal, Canada.

I. Elishakoff, V. Birman and J. Singer 1987 *Journal of Sound and Vibration* **114**, 57–63. Small vibrations of an imperfect panel in the vicinity of a non-linear static state.

Y. M. Fu and C. Y. Chia 1989 *International Journal of Non-Linear Mechanics* **24**, 365–381. Multi-mode non-linear vibration and postbuckling of anti-symmetric imperfect angle-ply cylindrical thick panels.

E. I. Grigolyuk 1955 *Prikladnaya Matematika i Mekhanika* **19**, 376–382. Vibrations of circular cylindrical panels subjected to finite deflections (in Russian).

Y. Kobayashi and A. W. Leissa 1995 *International Journal of Non-Linear Mechanics* **30**, 57–66. Large amplitude free vibration of thick shallow shells supported by shear diaphragms.

A. W. Leissa 1973 *Vibrations of Shells*. NASA SP-288, U.S. Government Printing Office, Washington, DC, USA (1993 reprinted by the Acoustical Society of America).

A. W. Leissa and A. S. Kadi 1971 *Journal of Sound and Vibration* **16**, 173–187. Curvature effects on shallow shell vibrations.

R. A. Raouf 1993 *Composites Engineering* **3**, 1101–1110. A qualitative analysis of the nonlinear dynamic characteristics of curved orthotropic panels.

E. Reissner 1955 Nonlinear effects in vibrations of cylindrical shells. Ramo-Wooldridge Corporation Report AM5-6, USA.

C. Sansour, W. Wagner, P. Wriggers and J. Sansour 2002 *International Journal of Non-Linear Mechanics* **37**, 951–966. An energy-momentum integration scheme and enhanced strain finite elements for the non-linear dynamics of shells.

C. Sansour, P. Wriggers and J. Sansour 1997 *Nonlinear Dynamics* **13**, 279–305. Nonlinear dynamics of shells: theory, finite element formulation, and integration schemes.

W. Soedel 1993 *Vibrations of Shells and Plates,* 2nd edition. Marcel Dekker, New York, USA.

C. T. Tsai and A. N. Palazotto 1991 *International Journal of Non-Linear Mechanics* **26**, 379–388. On the finite element analysis of non-linear vibration for cylindrical shells with high-order shear deformation theory.

T. Yamaguchi and K. Nagai 1997 *Nonlinear Dynamics* **13**, 259–277. Chaotic vibration of a cylindrical shell-panel with an in-plane elastic-support at boundary.

10 Nonlinear Vibrations and Stability of Doubly Curved Shallow-Shells: Isotropic and Laminated Materials

10.1 Introduction

Doubly curved shells are largely used in aeronautics and aerospace and are subjected to dynamic loads that can cause vibration amplitudes of the order of the shell thickness, giving rise to significant nonlinear phenomena. In order to reduce the weight, traditional materials are often substituted with laminated panels. This justifies the study of nonlinear vibrations of isotropic and laminated curved panels.

Nonlinear (large amplitude) forced vibrations of doubly curved shallow-shells are initially studied by using Donnell's theory retaining in-plane inertia and the Lagrange equations. The effect of the geometry and curvature are investigated for isotropic shells. Then, nonlinear free vibrations of laminated composite shells are studied by using both the Donnell and the first-order shear deformation theories in order to compare numerical results. It is observed that a shear deformation theory should be adopted for moderately thick laminated shells for which the ratio between the thickness and the largest of the in-plane curvilinear dimensions is equal or larger than 0.04.

The stability of a spherical shell under static normal load is discussed. Finally, the example of buckling analysis of the external tank of the NASA space shuttle, taking into account the effect of initial geometric imperfections, is performed following the study of Nemeth et al. (2002).

10.1.1 Literature Review

Leissa and Kadi (1971) studied linear and nonlinear free vibrations of doubly curved shallow-shells with rectangular boundaries, simply supported at the four edges and without in-plane constraints. Donnell's nonlinear shallow-shell theory was used. A single-mode expansion of the transverse displacement was used.

Large-amplitude vibrations of shallow-shells such as elliptic paraboloids, parabolic cylinders and hyperbolic paraboloids, with zero displacements and rotational springs at the four boundaries, were studied by Bhattacharya (1976). Sinharay and Banerjee (1985) studied nonlinear vibrations of shallow spherical and noncircular cylindrical shells. Furthermore, the same elliptic paraboloid, parabolic cylinder and hyperbolic paraboloid previously studied by Bhattacharya (1976) were analyzed and results were compared. In the case of cylindrical panels, the edges were immovable and elastically restrained against rotation. The authors used a simple mode expansion and an energy approach. Sinharay and Banerjee (1986) also studied shallow spherical and circular cylindrical shells of nonuniform thickness with a single-mode approach.

Chia (1988) investigated doubly curved panels with rectangular base by using Donnell's nonlinear shallow-shell theory and a single-mode expansion in all the numerical calculations, for both vibration shape and initial imperfection. Only the backbone curves, relative to free vibrations, were obtained.

Hui (1990) studied free nonlinear vibrations of rectangular plates and axisymmetric shallow spherical shells with geometric imperfections. Backbone curves, indicating an initial softening-type nonlinearity, turning to hardening for larger amplitudes, were obtained. A single-mode approach was used. Response to initial condition in the presence of viscous damping was also investigated.

Librescu and Chang (1993) investigated geometrically imperfect, doubly curved, undamped, laminated composite shallow-shells. The nonlinear third-order shear deformation theory was used. The nonlinearity was attributed to finite deformations of the panel due to in-plane loads and imperfections. Only small-amplitude free vibrations superimposed on this finite initial deformation were studied. A single mode was used to describe the free vibrations and the initial imperfections. Geometric imperfections of the panels with the same shape as the mode investigated significantly lowered the natural frequency. Librescu and Chang (1993) also studied the post-buckling stability. In fact, curved panels are characterized by an unstable post-buckling behavior, in the sense that they are subject to a snap-through instability.

Kobayashi and Leissa (1995) studied nonlinear free vibrations of doubly curved, thick shallow-shells. They used the nonlinear first-order shear deformation theory of shells in order to study thick shells. The rectangular boundaries of the panel were assumed to be simply supported at the four edges. A single-mode expansion was used for each of the three displacements and two rotations involved in the theory; in-plane and rotary inertia were neglected. The problem was reduced to a single degree of freedom describing the out-of-plane displacement. Numerical results were obtained for circular cylindrical, spherical and paraboloidal shallow-shells. Except for hyperbolic paraboloids, a softening behavior was found, becoming hardening for vibration amplitudes of the order of the shell thickness. However, increasing the radius of curvature, that is, approaching a flat plate, the behavior changed and became hardening. The effect of the shell thickness was also investigated.

Free vibrations of doubly curved, laminated, clamped shallow-shells with a rectangular boundary were studied by Abe et al. (2000). Both first-order shear deformation theory and classical shell theory (analogous to Donnell's theory) were used. Results obtained neglecting in-plane and rotary inertia are very close to those obtained retaining these effects. Two modes were considered to interact in the nonlinear analysis for higher modes, whereas a single mode was used for the mode with one circumferential and one longitudinal half-wave.

Soliman and Gonçalves (2003) studied large-amplitude, forced vibrations and stability of axisymmetric shallow spherical shells (spherical cups, i.e. with circular boundary). Transition to chaos for large harmonic loads, approaching the static stability limit, was investigated. Thomas et al. (2005) and Touzé and Thomas (2006) investigated nonlinear vibrations and 1:1:2 internal resonance of free-edge thin spherical caps by using a reduced-order model based on nonlinear normal modes.

Amabili (2005) studied large-amplitude vibrations of doubly curved shallow-shells with a rectangular base, simply supported at the four edges and subjected to harmonic excitation normal to the surface in the spectral neighborhood of the fundamental mode. Two different nonlinear strain-displacement relationships, from Donnell's and Novozhilov's shell theories, were used to calculate the elastic strain energy. In-plane inertia and geometric imperfections were taken into account. The solution was obtained by using the Lagrangian approach. Numerical results were compared with those available in the literature, and convergence of the solution was shown. Interaction of modes having integer ratio among their natural frequencies, giving rise to internal resonances, was discussed. Shell stability under static and dynamic loads was also investigated by using the continuation method, the bifurcation diagram from direct time integration and the calculation of the Lyapunov exponents and Lyapunov dimension.

10.2 Theoretical Approach for Simply Supported, Isotropic Shells

A doubly curved shallow (i.e. having a small rise compared with the smallest radius of curvature) shell with rectangular base is considered (as shown in Figure 2.1). A curvilinear coordinate system $(O; x, y, z)$, having the origin O at one edge of the panel is assumed; $x = \psi R_x$ and $y = \theta R_y$, where ψ and θ are the angular coordinates and R_x and R_y are the principal radii of curvature (constant); a and b are the curvilinear lengths of the edges and h is the shell thickness. The smallest radius of curvature at every point of the shell is larger than the greatest lengths measured along the middle surface of the shell. The displacements of an arbitrary point of coordinates (x, y) on the middle surface of the shell are denoted by u, v and w, in the x, y and z directions, respectively. w is taken positive outward from the center of the smallest radius of curvature. Initial imperfections of the shell associated with zero initial tension are denoted by out-of-plane displacement w_0, also positive outward. Only out-of-plane initial imperfections are considered.

Two different strain-displacement relationships for thin shells are used in the present study in order to compare results. These theories are: (i) Donnell's and (ii) Novozhilov's nonlinear shell theories.

The elastic strain energy U_S of the shell is given by equation (2.9). The kinetic energy T_S of the shell, by neglecting rotary inertia, is given by equation (9.22).

The virtual work W done by the external forces is written as

$$W = \int_0^a \int_0^b (q_x u + q_y v + q_z w)\, \mathrm{d}x\, \mathrm{d}y, \tag{10.1}$$

where q_x, q_y and q_z are the distributed forces per unit area acting in x, y and normal directions, respectively. Initially, only a single harmonic normal force is considered; therefore, $q_x = q_y = 0$. The external, normal, distributed load q_z applied to the shell, because of the radial concentrated force $\tilde{f}$, is given by

$$q_z = \tilde{f}\delta(x - \tilde{x})\delta(y - \tilde{y})\cos(\omega t), \tag{10.2}$$

where ω is the excitation frequency, t is the time, δ is the Dirac delta function, $\tilde{f}$ gives the normal force amplitude positive in w direction and $\tilde{x}$ and $\tilde{y}$ give the position of the point of application of the force. Here, the point excitation is located at $\tilde{x} = a/2$, $\tilde{y} = b/2$. Equation (10.2) can be rewritten in the following form:

$$W = \tilde{f} \cos(\omega t)\, (w)_{x=a/2,\ y=b/2} \cdot \tag{10.3}$$

The classical simply supported boundary conditions at the four edges of the shell are

$$v = w = w_0 = N_x = M_x = \partial^2 w_0/\partial x^2 = 0, \quad \text{at } x = 0, a, \tag{10.4a–f}$$

$$u = w = w_0 = N_y = M_y = \partial^2 w_0/\partial y^2 = 0, \quad \text{at } y = 0, b, \tag{10.5a–f}$$

where N is the normal force and M is the bending moment per unit length.

In order to reduce the system to finite dimensions, the middle surface displacements u, v and w are expanded by using approximate functions. A base of panel displacements is used to discretize the system. The displacements u, v and w can be expanded by using equations (9.9a–c), which satisfy identically the geometric boundary conditions (10.4a,b) and (10.5a,b). A sufficiently accurate model for the mode ($m = 1, n = 1$) in the absence of internal resonances has been shown to have 9 degrees of freedom (dofs). In particular, the following terms in equation (9.9) have been used: $m = 1, 3$ and $n = 1, 3$ in equations (9.9b,c) and $m = 1$ and $n = 1$ in equation (9.9a).

The normal imperfection w_0 is expanded by using equation (9.11).

10.2.1 Boundary Conditions

Equations (10.4) and (10.5) give the boundary conditions for a simply supported shell. Equations (10.4a,b,c,f) and (10.5a,b,c,f) are identically satisfied by the expansions of u, v, w and w_0. Equations (10.4e) and (10.5e) are satisfied by any classical shell theory without nonlinear terms in the changes of curvature k_x and k_y. On the other hand, equations (10.4d) and (10.5d) are not identically satisfied for all the nonlinear shell theories. According to Donnell's nonlinear theory and eliminating null terms at the panel edges, equations (10.4d) and (10.5d) can be rewritten as

$$\left[\frac{\partial \hat{u}}{R_x\,\partial\psi} + \frac{1}{2}\left(\frac{\partial w}{R_x\,\partial\psi}\right)^2 + \frac{\partial w}{R_x\,\partial\psi}\frac{\partial w_0}{R_x\,\partial\psi} + \nu\frac{\partial \hat{v}}{R_y\,\partial\theta}\right]_{\psi=0,\ a/R_x} = 0, \tag{10.6}$$

$$\left[\frac{\partial \hat{v}}{R_y\,\partial\theta} + \frac{1}{2}\left(\frac{\partial w}{R_y\,\partial\theta}\right)^2 + \frac{\partial w}{R_y\,\partial\theta}\frac{\partial w_0}{R_y\,\partial\theta} + \nu\frac{\partial \hat{u}}{R_x\,\partial\psi}\right]_{\theta=0,b/R_y} = 0, \tag{10.7}$$

where $\hat{u}$ and $\hat{v}$ are terms that must be added to the expansion of u and v, given in equations (9.9b,c), in order to satisfy exactly the in-plane boundary conditions $N_x = 0$ and $N_y = 0$. Because $\hat{u}$ and $\hat{v}$ are second-order terms in the panel displacement, they have not been inserted in the second-order terms that involve u and v in equations (10.6) and (10.7).

Nontrivial calculations give three contributions, related to the second, third and fourth term, respectively, in equation (10.6),

$$\hat{u}(t) = \hat{u}_1(t) + \hat{u}_2(t) + \hat{u}_3(t), \tag{10.8}$$

where

$$\hat{u}_1(t) = -\frac{1}{2}\int \left[\left(\frac{\partial w}{\partial x}\right)^2\right]_{x=0,a} \mathrm{d}x$$

$$= -\frac{1}{2}\sum_{n=1}^{N}\sum_{m=1}^{M}(m\pi/a)\left\{w_{m,n}(t)\sin(n\pi y/b)\sum_{k=1}^{N}\sum_{s=1}^{M}\frac{s}{m+s}w_{s,k}(t)\right.$$

$$\left.\times\sin(k\pi y/b)\sin\left[(m+s)\pi x/a\right]\right\}, \tag{10.9}$$

$$\hat{u}_2(t) = -\int \left[\frac{\partial w}{\partial x}\frac{\partial w_0}{\partial x}\right]_{x=0,a} \mathrm{d}x$$

$$= -\sum_{n=1}^{N}\sum_{m=1}^{M}(m\pi/a)\left\{w_{m,n}(t)\sin(n\pi y/b)\sum_{j=1}^{\tilde{N}}\sum_{i=1}^{\tilde{M}}\frac{i}{m+i}A_{i,j}\right.$$

$$\left.\times\ \sin(j\pi y/b)\sin\left[(m+i)\pi x/a\right]\right\}, \tag{10.10}$$

$$\hat{u}_3(t) = -\nu\int\left[\frac{\partial v}{\partial y}\right]_{x=0,a} \mathrm{d}x. \tag{10.11}$$

It can be observed that solution (10.9) is not unique, but the one proposed satisfies boundary conditions and deformation energy considerations. Then, in order to satisfy boundary conditions (10.4a), yields $[\partial v/\partial y]_{x=0,a} = 0$. Therefore, equation (10.11) gives

$$\hat{u}_3(t) = -\nu\int\left[\frac{\partial v}{\partial y}\right]_{x=0,a} \mathrm{d}x = 0. \tag{10.12}$$

Similar expressions are obtained by integration of equation (10.7). Finally, the following equations are obtained, which satisfy conditions (10.6) and (10.7):

$$\hat{u}(t) = -\sum_{n=1}^{N}\sum_{m=1}^{M}(m\pi/a)\left\{\frac{1}{2}w_{m,n}(t)\sin(n\pi y/b)\sum_{k=1}^{N}\sum_{s=1}^{M}\frac{s}{m+s}w_{s,k}(t)\sin(k\pi y/b)\right.$$

$$\times\sin\left[(m+s)\pi x/a\right] + w_{m,n}(t)\sin(n\pi y/b)$$

$$\left.\times\sum_{j=1}^{\tilde{N}}\sum_{i=1}^{\tilde{M}}\frac{i}{m+i}A_{i,j}\sin(j\pi y/b)\sin\left[(m+i)\pi x/a\right]\right\}, \tag{10.13}$$

$$\hat{v}(t) = -\sum_{n=1}^{N}\sum_{m=1}^{M}(n\pi/b)\left\{\frac{1}{2}w_{m,n}(t)\sin(m\pi x/a)\sum_{k=1}^{N}\sum_{s=1}^{M}\frac{k}{n+k}w_{s,k}(t)\sin(s\pi x/a)\right.$$

$$\times\sin\left[(n+k)\pi y/b\right] + w_{m,n}(t)\sin(m\pi x/a)$$

$$\left.\times\sum_{j=1}^{\tilde{N}}\sum_{i=1}^{\tilde{M}}\frac{j}{n+j}A_{i,j}\sin(i\pi x/a)\sin\left[(n+j)\pi y/b\right]\right\}. \tag{10.14}$$

A simplified version of equations (10.13) and (10.14) that can be used in the case of small interaction among mode (m, n) and other modes is

$$\hat{u}(t) = -\frac{1}{8}\left[a_{m,n}(t) + b_{m,n}(t)\cos(2n\pi y/b)\right]\sin(2m\pi x/a) - (m\pi/a)w_{m,n}(t)\sin(n\pi y/b)$$
$$\times \sum_{j=1}^{\tilde{N}}\sum_{i=1}^{\tilde{M}} \frac{i}{m+i} A_{i,j}\sin(j\pi y/b)\sin\left[(m+i)\pi x/a\right], \tag{10.15}$$

$$\hat{v}(t) = -\frac{1}{8}\left[c_{m,n}(t) + d_{m,n}(t)\cos(2m\pi x/a)\right]\sin(2n\pi y/b) - (n\pi/b)w_{m,n}(t)\sin(m\pi x/a)$$
$$\times \sum_{j=1}^{\tilde{N}}\sum_{i=1}^{\tilde{M}} \frac{j}{n+j} A_{i,j}\sin(i\pi x/a)\sin\left[(n+j)\pi y/b\right], \tag{10.16}$$

where

$$a_{m,n}(t) = (m\pi/a)w_{m,n}^2(t), \tag{10.17}$$
$$b_{m,n}(t) = -(m\pi/a)w_{m,n}^2(t) \tag{10.18}$$
$$c_{m,n}(t) = (n\pi/b)w_{m,n}^2(t), \tag{10.19}$$
$$d_{m,n}(t) = -(n\pi/b)w_{m,n}^2(t). \tag{10.20}$$

The simplified expressions (10.15) and (10.16) are obtained by assuming that all the generalized coordinates, except the three coordinates associated with the resonant mode (m, n), which are $u_{m,n}(t)$, $v_{m,n}(t)$, $w_{m,n}(t)$, are negligible because they are an infinitesimal of higher order.

For Novozhilov's nonlinear shell theory, the approximate equations (10.15) and (10.16) are still correct, but equations (10.17–10.20) are substituted by (boundary conditions are satisfied only approximately in this case)

$$a_{m,n}(t) = \frac{m\pi}{a}\left(w_{m,n}^2 + v_{m,n}^2\right) + \frac{a}{m\pi}\frac{u_{m,n}^2}{R_x^2} - \frac{2w_{m,n}u_{m,n}}{R_x} - \frac{\nu}{8}\frac{a}{m\pi}\left(\frac{n\pi}{b}\right)^2 u_{m,n}^2 \tag{10.21}$$

$$b_{m,n}(t) = -\frac{m\pi}{a}\left(w_{m,n}^2 - v_{m,n}^2\right) - \frac{a}{m\pi}\frac{u_{m,n}^2}{R_x^2} + \frac{2w_{m,n}u_{m,n}}{R_x}$$
$$+\frac{\nu}{8}\frac{a}{m\pi}\left(\frac{n\pi}{b}\right)^2 u_{m,n}^2 + \frac{1}{4}\frac{m\pi}{a}v_{m,n}^2, \tag{10.22}$$

$$c_{m,n}(t) = \frac{n\pi}{b}\left(w_{m,n}^2 + u_{m,n}^2\right) + \frac{b}{n\pi}\frac{v_{m,n}^2}{R_y^2} - \frac{2w_{m,n}v_{m,n}}{R_y} - \frac{\nu}{8}\frac{b}{n\pi}\left(\frac{m\pi}{a}\right)^2 v_{m,n}^2, \tag{10.23}$$

$$d_{m,n}(t) = -\frac{n\pi}{b}\left(w_{m,n}^2 - u_{m,n}^2\right) - \frac{b}{n\pi}\frac{v_{m,n}^2}{R_y^2} + \frac{2w_{m,n}v_{m,n}}{R_y}$$
$$+\frac{\nu}{8}\frac{b}{n\pi}\left(\frac{m\pi}{a}\right)^2 v_{m,n}^2 + \frac{1}{4}\frac{n\pi}{b}u_{m,n}^2. \tag{10.24}$$

10.2.2 Lagrange Equations of Motion

The nonconservative damping forces are assumed to be of the viscous type and are taken into account by using Rayleigh's dissipation function

$$F = \frac{1}{2}c\int_0^a\int_0^b \left(\dot{u}^2 + \dot{v}^2 + \dot{w}^2\right)\mathrm{d}x\,\mathrm{d}y, \tag{10.25}$$

where c has a different value for each term of the mode expansion. Simple calculations give

$$F = \frac{1}{2}\frac{ab}{4}\sum_{n=1}^{N}\sum_{m=1}^{M} c_{m,n}\left(\dot{u}_{m,n}^2 + \dot{v}_{m,n}^2 + \dot{w}_{m,n}^2\right). \tag{10.26}$$

The damping coefficient $c_{m,n}$ is related to the modal damping ratio $\zeta_{m,n}$ that can be evaluated from experiments by $\zeta_{m,n} = c_{m,n}/(2\mu_{m,n}\omega_{m,n})$, where $\omega_{m,n}$ is the natural circular frequency of mode (m, n) and $\mu_{m,n}$ is the modal mass of this mode, given by $\mu_{m,n} = \rho h(ab/4)$.

The virtual work done by the concentrated normal force $\tilde{f}$, given by equation (10.2) specialized for the expression of w given in equation (9.9a), is

$$W = \tilde{f}\cos(\omega t)\,(w)_{x=a/2,\ y=b/2} = \tilde{f}\cos(\omega t)\left[\sum_{i=1}^{\hat{M}}\sum_{j=1}^{\hat{N}}(-1)^{i+1}(-1)^{j+1}w_{2i-1,\,2j-1}(t)\right], \tag{10.27}$$

where $\hat{M} = \text{Integer Part}(1 + M/2)$ and $\hat{N} = \text{Integer Part}(1 + N/2)$.

In the presence of in-plane loads and pressure acting on the panel, additional virtual work is done by the external forces. Let us consider a time-dependent load $N(t)$ per unit length, applied at both the shell edges $x = 0, a$; $N(t)$ is positive in the x direction. In particular, $-N(t)$ is applied at $x = 0$ and $N(t)$ is applied at $x = a$. The distributed force q_x has the following expression:

$$q_x = N(t)\left[-\delta(x) + \delta(x - L)\right], \tag{10.28}$$

where δ is the Dirac delta function. The virtual work done by the load is

$$W = \int_0^a\!\!\int_0^b N(t)\left[-\delta(x) + \delta(x - a)\right]u\,\mathrm{d}x\,\mathrm{d}y = -\frac{4b}{\pi}N(t)\sum_{i=1}^{\hat{M}}\sum_{j=1}^{\hat{N}} u_{2i-1,\,2j-1}(t)/(2j-1). \tag{10.29}$$

In the case of uniform internal time-varying pressure $p_z(t)$, the normal distributed force q_z is obviously

$$q_z = p_z(t). \tag{10.30}$$

The virtual work done by pressure is

$$W = \int_0^a\!\!\int_0^b p_z(t)w\,\mathrm{d}x\,\mathrm{d}y = \frac{4ab}{\pi^2}p_z(t)\sum_{i=1}^{\hat{M}}\sum_{j=1}^{\hat{N}}\frac{w_{2i-1,\,2j-1}(t)}{(2i-1)(2j-1)}. \tag{10.31}$$

Equations (10.29) and (10.31) show that only modes with an odd number of half-waves in the x and y directions are directly excited by uniform in-plane loads and pressure.

Then, the Lagrange equations are obtained as in Chapter 4 with equations similar to (4.38–4.42).

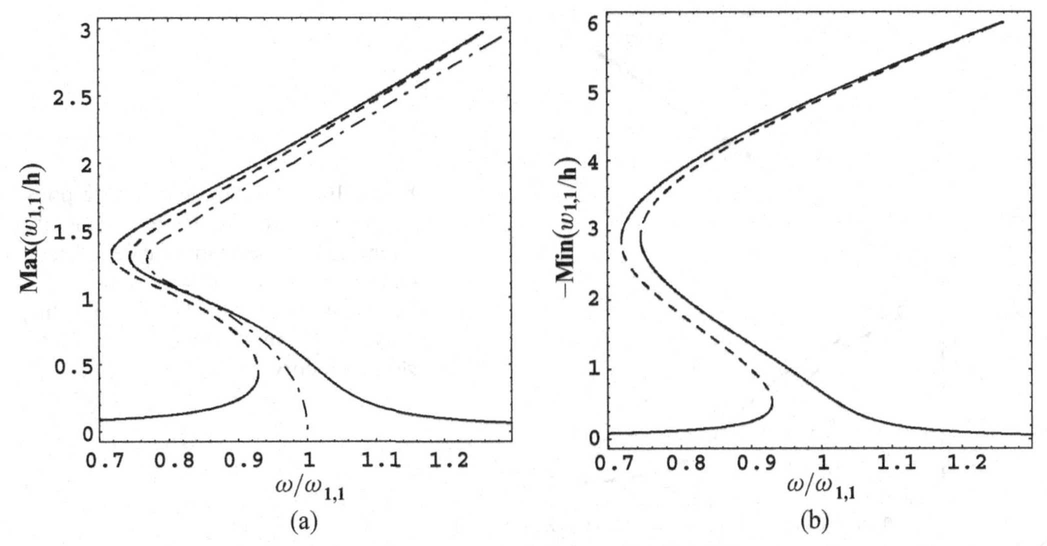

Figure 10.1. Amplitude of the response of the shell (generalized coordinate $w_{1,1}$) versus the excitation frequency; $R_x/R_y = 1$ (spherical shallow-shell); fundamental mode ($m = 1, n = 1$), $\tilde{f} = 31.2$ N and $\zeta_{1,1} = 0.004$; model with 9 dofs; Donnell's theory. —–, present stable results; – –, present unstable results; –·–, backbone curve from Kobayashi and Leissa (1995). (a) Maximum of the generalized coordinate $w_{1,1}$ (in a vibration period). (b) Minimum of the generalized coordinate $w_{1,1}$.

10.3 Numerical Results for Simply Supported, Isotropic Shells

Numerical calculations have been initially performed for doubly curved shallow-shells, simply supported at the four edges, having the following dimensions and material properties: curvilinear dimensions $a = b = 0.1$ m, radius of curvature $R_x = 1$ m, thickness $h = 0.001$ m, Young's modulus $E = 206 \times 10^9$ Pa, mass density $\rho = 7800$ kg/m^3 and Poisson's ratio $\nu = 0.3$. Shallow-shells with the same dimension ratios ($R_x/a = 10$, $h/a = 0.01$, $a/b = 1$, $\nu = 0.3$) were studied by Kobayashi and Leissa (1995) and Amabili (2005). In all the numerical simulations, a modal damping $\zeta_{1,1} = 0.004$ and harmonic force excitation at the center of the shell in z direction are assumed. If not explicitly specified, all calculations have been performed using Donnell's shell theory.

10.3.1 Case with $R_x/R_y = 1$, Spherical Shell

A spherical shallow-shell ($R_y = 1$ m) is considered. The frequency range around the fundamental frequency (mode [$m = 1$, $n = 1$] in this case, where m and n are the numbers of half-waves in x and y directions, respectively) is investigated. The fundamental frequency $\omega_{1,1}$ is 952.31 Hz according to Donnell's shell theory and 952.26 Hz according to Novozhilov's shell theory, that is, almost the same results for both theories. Other natural frequencies useful in the present study are (according to Donnell's theory): $\omega_{1,3} = \omega_{3,1} = 2575.9$ Hz, $\omega_{3,3} = 4472.3$ Hz. The amplitude of the harmonic force is $\tilde{f} = 31.2$ N.

Figure 10.1 shows the maximum (in the time period; this is positive, i.e. outward) and the minimum (negative, i.e. inward) of the shell response in z direction in the spectral neighborhood of the fundamental mode ($m = 1$, $n = 1$) versus the

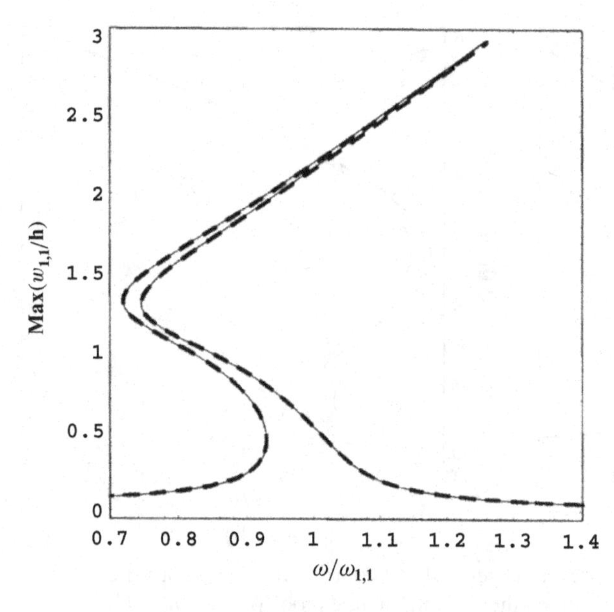

Figure 10.2. Amplitude of the response of the shell versus the excitation frequency; $R_x/R_y = 1$ (spherical shell); fundamental mode ($m = 1$, $n = 1$), $\tilde{f} = 31.2$ N and $\zeta_{1,1} = 0.004$; model with 9 dofs. —, Donnell's theory; – –, Novozhilov's theory.

excitation frequency. Calculations have been performed with the 9 dofs model. This model includes the following terms in equations (9.9a–c): $w_{1,1}, u_{1,1}, v_{1,1}, u_{3,1}, v_{3,1}, u_{1,3}, v_{1,3}, u_{3,3}, v_{3,3}$, that is, the same used for the simply supported rectangular plate (see Chapter 4) and the simply supported circular cylindrical panel in Chapter 9. Results are compared to those obtained by Kobayashi and Leissa (1995), where only the backbone curve is given. The present results with 9 dofs are quite close to those of Kobayashi and Leissa (considering that the backbone curve, indicating the maximum of the response for different force excitations, approximately passes through the middle of the forced response curve computed in the present calculation) and shows a softening-type nonlinearity, turning to hardening for vibration amplitudes about two times larger than the shell thickness. Comparing Figures 10.1(a) and 10.1(b) it is evident that displacement inward is about two times larger than displacement outward during a vibration period.

Comparison of results obtained by using Novozhilov's and Donnell's nonlinear shell theories is given in Figure 10.2. Results are practically coincident in this case ($m = 1, n = 1$) and have been computed by using 9 dofs. Results indicate that, at least for the thin shallow-shell studied here, there is no advantage in using the more refined Novozhilov shell theory, which is much more time consuming in numerical computations.

In order to check the convergence of the solution, the response has been calculated with a larger model, including 22 dofs. This model has the following additional generalized coordinates with respect to the 9 dofs model: $w_{1,3}, w_{3,1}, w_{3,3}, u_{1,5}, u_{3,5}, u_{5,1}, u_{5,3}, u_{5,5}, v_{1,5}, v_{3,5}, v_{5,1}, v_{5,3}, v_{5,5}$. Comparison of the response computed with the 22 dofs and the 9 dofs models is given in Figure 10.3, where the backbone curve of Kobayashi and Leissa (1995) is also shown. The results of the 22 dofs model are moved slightly to the left with respect to the smaller 9 dofs model and present a more complex curve, specially in the frequency region around $0.9\omega_{1,1}$. In fact, for excitation frequency $\omega = 0.9\,\omega_{1,1}$, there is a 3:1 internal resonance with modes ($m = 3, n = 1$) and ($m = 1$, $n = 3$), giving $3\omega = \omega_{3,1} = \omega_{1,3}$. A second relationship between natural frequencies that leads to internal resonances is for $\omega = 0.77\,\omega_{1,1}$, where $6\omega = \omega_{3,3}$. This more complex response, due to internal resonances, can be observed in Figure 10.4(a–c)

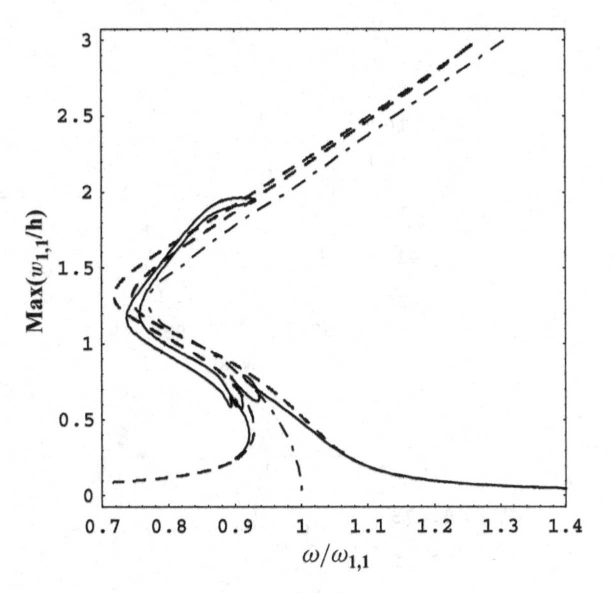

Figure 10.3. Amplitude of the response of the shell versus the excitation frequency; $R_x/R_y = 1$ (spherical shell); fundamental mode ($m = 1$, $n = 1$), $\tilde{f} = 31.2$ N and $\zeta_{1,1} = 0.004$; Donnell's theory. —–, 22 dofs model; – –, 9 dofs model; –·–, backbone curve from Kobayashi and Leissa (1995).

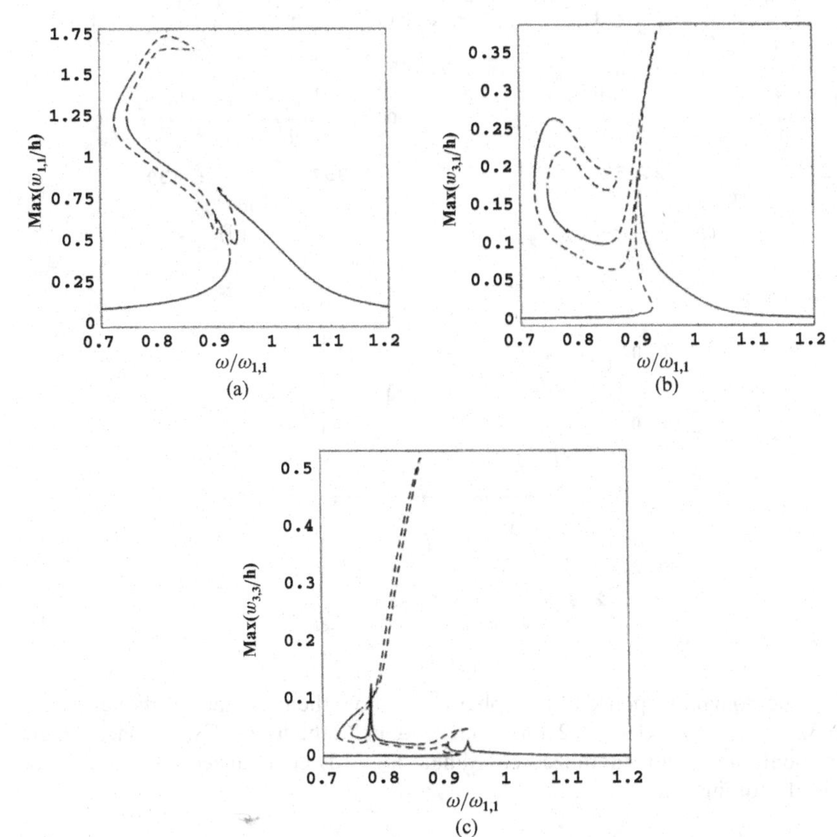

Figure 10.4. Amplitude of the response of the shell versus the excitation frequency; $R_x/R_y = 1$ (spherical shell); fundamental mode ($m = 1$, $n = 1$), $\tilde{f} = 31.2$ N and $\zeta_{1,1} = 0.004$; 22 dofs model; Donnell's theory. —–, stable solutions; – –, unstable solutions. (a) Maximum of $w_{1,1}$ (b) Maximum of $w_{3,1}$. (c) Maximum of $w_{3,3}$.

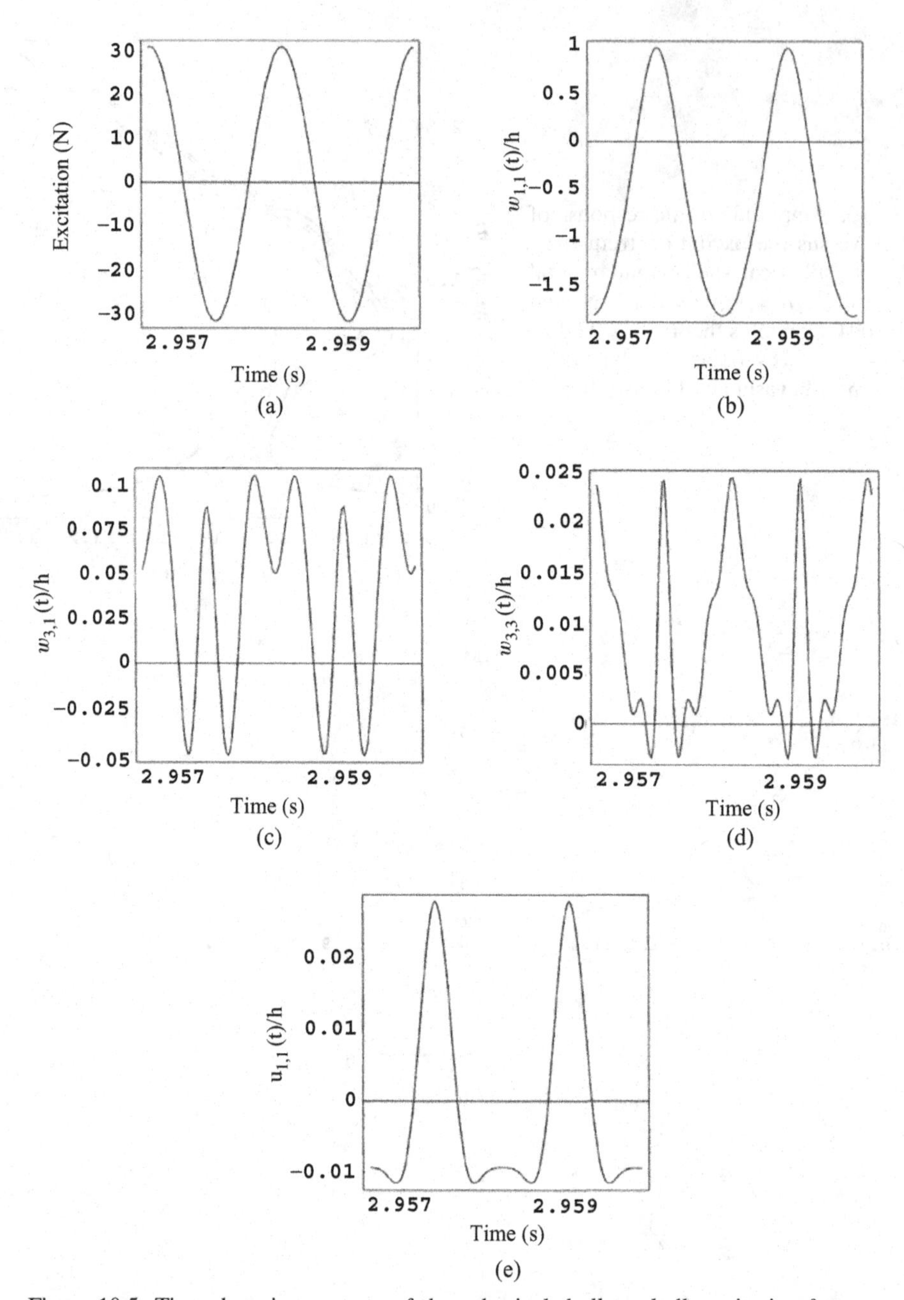

Figure 10.5. Time-domain response of the spherical shallow-shell; excitation frequency $\omega = 0.8\omega_{1,1}$, $\tilde{f} = 31.2$ N and $\zeta_{1,1} = 0.004$; 22 dofs model; Donnell's theory. (a) Excitation. (b) Generalized coordinate $w_{1,1}$. (c) Generalized coordinate $w_{3,1}$. (d) Generalized coordinate $w_{3,3}$. (e) Generalized coordinate $u_{1,1}$.

where the generalized coordinates involved are shown (for the symmetry of the spherical shell, $w_{3,1} = w_{1,3}$). In synthesis, the main difference between the 9 dofs model and the 22 dofs model is related the internal resonances that cannot be studied with the 9 dofs model because $w_{1,3}$, $w_{3,1}$, $w_{3,3}$ are not included. Otherwise, the trend of nonlinearity does not change much if investigated with the 9 or 22 dofs model.

It is of great importance to analyze the results in the time domain by using direct integration of the equations of motion. Figure 10.5 shows the most significant

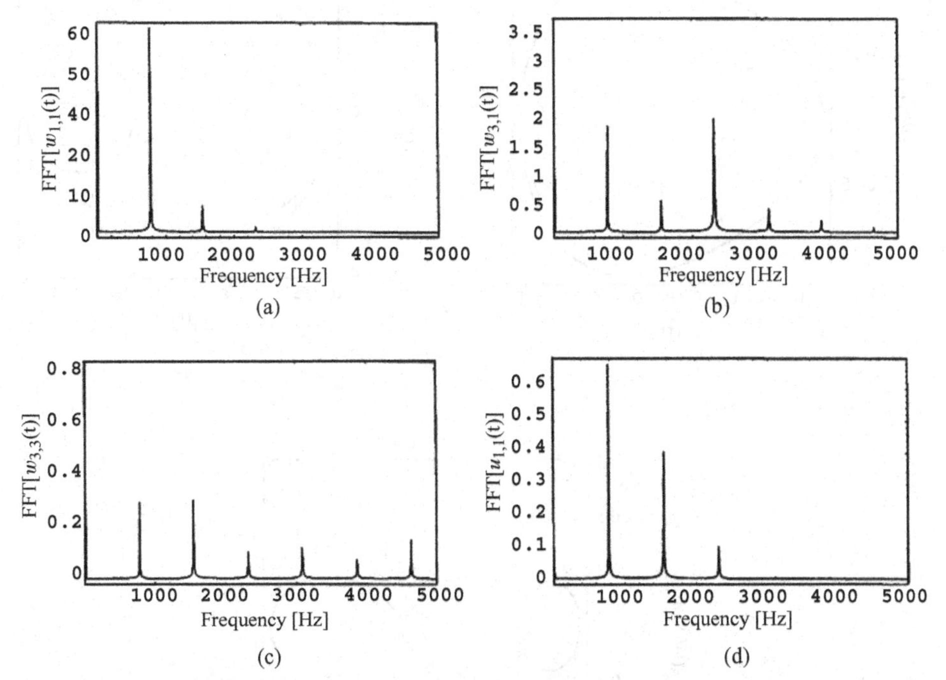

Figure 10.6. Spectrum of the response of the spherical shallow-shell; excitation frequency $\omega = 0.8\omega_{1,1}$, $\tilde{f} = 31.2$ N and $\zeta_{1,1} = 0.004$; 22 dofs. (a) Generalized coordinate $w_{1,1}$. (b) Generalized coordinate $w_{3,1}$. (c) Generalized coordinate $w_{3,3}$. (d) Generalized coordinate $u_{1,1}$.

(i.e. those with larger amplitude) generalized coordinates for excitation frequency $\omega = 0.8\omega_{1,1}$ (with the 22 dofs model and initial conditions to catch the higher-amplitude stable solution, which is in the softening-type region). The generalized coordinates related to normal displacement w clearly indicate that there is a strong asymmetry between vibration inward and outward. In fact, the vibration is much larger inward (negative) than outward (positive), as previously observed for circular cylindrical panels; this is related to the curvature of the shell. In particular, for the generalized coordinate $w_{1,1}$, which has the largest amplitude, the inward displacement is about 1.8 times bigger than the outward displacement. Moreover, there is a large contribution of higher harmonics, as clearly indicated by the frequency spectra in Figure 10.6. The generalized coordinate $w_{1,1}$ (associated with the resonant mode, with $u_{1,1}$ and $v_{1,1}$) is less affected by these higher harmonics. Phase plane diagrams in Figure 10.7 are also provided to better understand the asymmetry of inward and outward vibration; the loops are related to higher harmonics.

10.3.2 Case with $R_x/R_y = -1$, Hyperbolic Paraboloidal Shell

A shallow hyperbolic paraboloidal shell with $R_x/R_y = -1(R_y = -1$ m) is considered. The fundamental frequency $\omega_{1,1}$ is 488.1 Hz according to Donnell's shell theory. Other natural frequencies useful in the present study are (according to Donnell's theory): $\omega_{1,3} = \omega_{3,1} = 2528.7$ Hz, $\omega_{3,3} = 4396.6$ Hz. The amplitude of the harmonic force is $\tilde{f} = 4.37$ N.

Figure 10.8 shows the maximum of the shell response in z direction in the spectral neighborhood of the fundamental mode ($m = 1$, $n = 1$) versus the excitation frequency. Calculations have been obtained with the 9 and 22 dofs models. Results

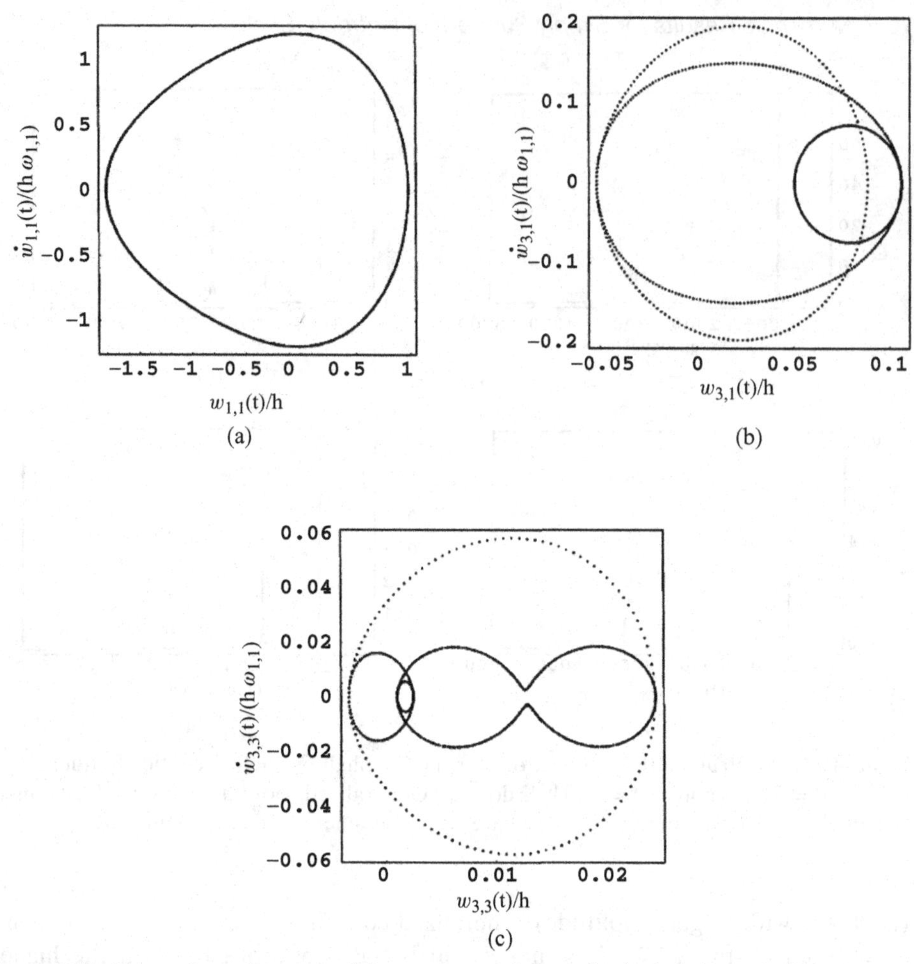

Figure 10.7. Phase plane diagram of the response of the spherical shallow-shell; excitation frequency $\omega = 0.8\omega_{1,1}$, $\tilde{f} = 31.2$ N and $\zeta_{1,1} = 0.004$; 22 dofs. (a) Generalized coordinate $w_{1,1}$. (b) Generalized coordinate $w_{3,1}$. (c) Generalized coordinate $w_{3,3}$.

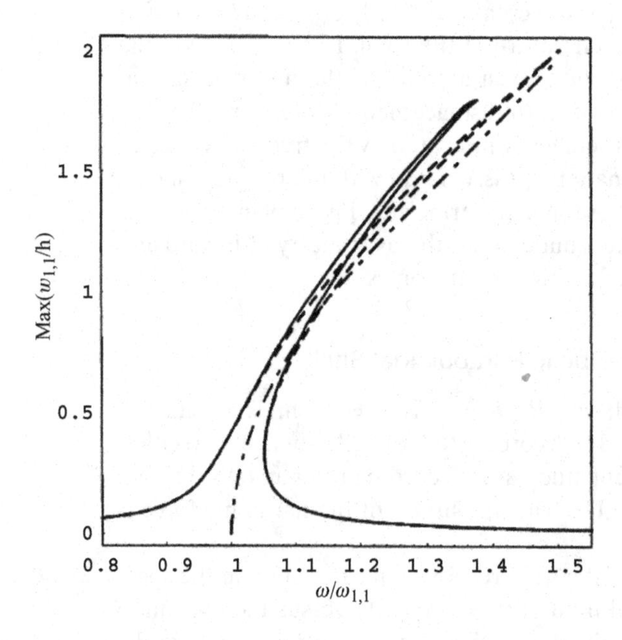

Figure 10.8. Amplitude of the response of the shell versus the excitation frequency; $R_x/R_y = -1$ (hyperbolic paraboloidal shallow-shell); fundamental mode ($m = 1$, $n = 1$), $\tilde{f} = 4.37$ N and $\zeta_{1,1} = 0.004$; Donnell's theory. ——, 22 dofs model; – –, 9 dofs model; –·–, backbone curve from Kobayashi and Leissa (1995).

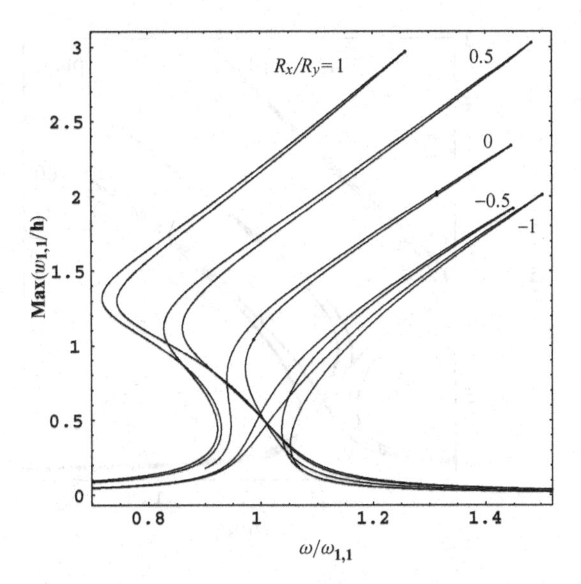

Figure 10.9. Effect of the curvature aspect ratio R_x/R_y on the shell response (maximum of the generalized coordinate $w_{1,1}$) versus the excitation frequency; fundamental mode ($m = 1$, $n = 1$); $\zeta_{1,1} = 0.004$; 9 dofs model; Donnell's theory.

are compared with those obtained by Kobayashi and Leissa (1995), where only the backbone curve is given. The present results with 9 dofs are very close to those obtained by Kobayashi and Leissa and show a hardening-type nonlinearity. It is observed that the displacement in z and $-z$ direction is perfectly symmetric during a vibration period. The results of the 22 dofs model are moved slightly to the left with respect to the smaller 9 dofs model, showing good accuracy of the 9 dofs model in this case, where internal resonances do not appear.

10.3.3 Effect of Different Curvature

Figure 10.9 synthesizes the maximum responses for the 9 dofs models for different shell curvature aspect ratios R_x/R_y. It is clearly shown that for $R_x/R_y = 1$ (spherical), 0.5 and 0 (circular cylindrical), the shallow-shells considered exhibit softening-type behavior turning to hardening type for vibration amplitude of the order of the shell thickness. The softening behavior becomes weaker with the decrement of the curvature aspect ratio R_x/R_y. In particular, the softening behavior of the spherical shallow-shell for vibration amplitude at about 1.3 times the shell thickness is very strong. On the other hand, for $R_x/R_y = -0.5, -1$ (hyperbolic paraboloid), the shell has a strong hardening-type behavior without the presence of the softening region.

The effect of the curvature on the maximum response of spherical shallow-shells is investigated in Figure 10.10, obtained with the 9 dofs model and Donnell's nonlinear shell theory. Here, the radius of curvature $R_x = R_y$ is varied with respect to the value 1 m used previously, whereas all the other geometric and material characteristics are kept constant. In particular, three responses for $R_x/a = 10$, 20, 100 are presented (solid line) and compared with the response of a flat square plate (dashed line) of the same dimension and material, computed with the same theory presented in Chapter 4 and the 9 dofs model. It must be observed that the flat plate is the limiting case for a spherical shallow-shell when $R_x/a \to \infty$. Figure 10.10 shows that, by increasing the curvature ratio R_x/a the shell response changes gradually from (i) strongly softening turning to hardening for a vibration amplitude of about two times the shell thickness, to (ii) fully hardening, which is the typical behavior of flat plates. It is curious that

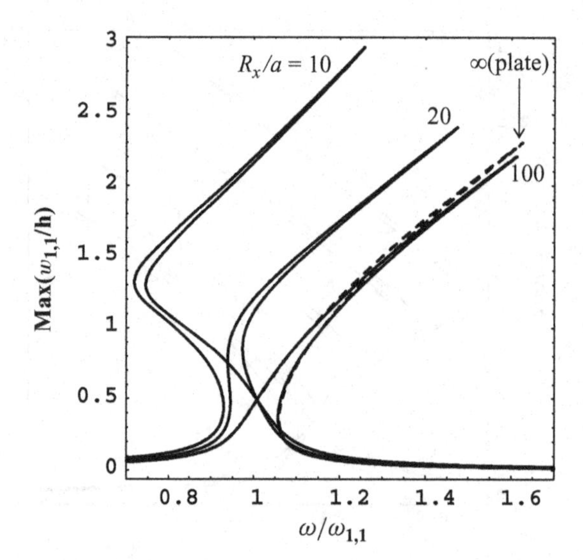

Figure 10.10. Effect of the curvature ratio R_x/a on the response (maximum of the generalized coordinate $w_{1,1}$) of spherical shallow-shells versus the excitation frequency; fundamental mode ($m = 1$, $n = 1$); $\zeta_{1,1} = 0.004$; 9 dofs model; Donnell's theory. For $R_x/a = 10$, $\omega_{1,1} = 952.3$ Hz and $\tilde{f} = 31.2$ N; for $R_x/a = 20$, $\omega_{1,1} = 637.1$ Hz and $\tilde{f} = 11.16$ N; for $R_x/a = 100$, $\omega_{1,1} = 495.4$ Hz and $\tilde{f} = 5.56$ N; for $R_x/a = \infty$ (square plate), $\omega_{1,1} = 488.6$ Hz and $\tilde{f} = 5.4$ N.

a flat plate ($R_x/a \to \infty$) presents a slightly weaker hardening-type behavior than the spherical shallow-shell with $R_x/a = 100$. However, the difference is practically negligible.

10.4 Buckling of Simply Supported Shells under Static Load

The same spherical shallow-shell studied in Section 10.3.1 is investigated here with the 22 dofs model. A static load is applied at the center of the shell in $-z$ direction and the shell deformations are reported in Figures 10.11(a–c). A buckling instability is clearly identified at about 350 N, where the load-displacement curve in Figures 10.11(a,d) presents an unstable region, and the shell deformation jumps from a relatively small value to a significantly larger one without any load increment.

This type of instability is usually referred as snap-through instability. The region of jumps from small to large deformations is delimited by two foldings (saddle-node bifurcations) of the equilibrium solution. In this region, three static equilibrium configurations of the shell are possible for any load: two of them are stable solutions and one is unstable. This kind of instability is characteristic of shells under normal loads. The shell deformations just before and after the snap-through are shown in Figures 10.11(e,f). Note the enlarged scale of the vertical axis.

In practical applications, the snap-through instability under static load is associated with a transient dynamic solution, as a consequence of the sudden jump from a configuration of small deformation, before the buckling, to a large post-buckling deformation. The buckling load of shells is highly sensitive to geometric imperfections.

10.5 Theoretical Approach for Clamped Laminated Shells

The theoretical approach used here is taken from Abe et al. (2000). The first-order shear deformation theory is used to study large-amplitude free vibrations of laminated shallow-shells. In particular, the displacements of a generic point of the shell are related to the middle-plane displacements by using equations (2.91a–c), where the

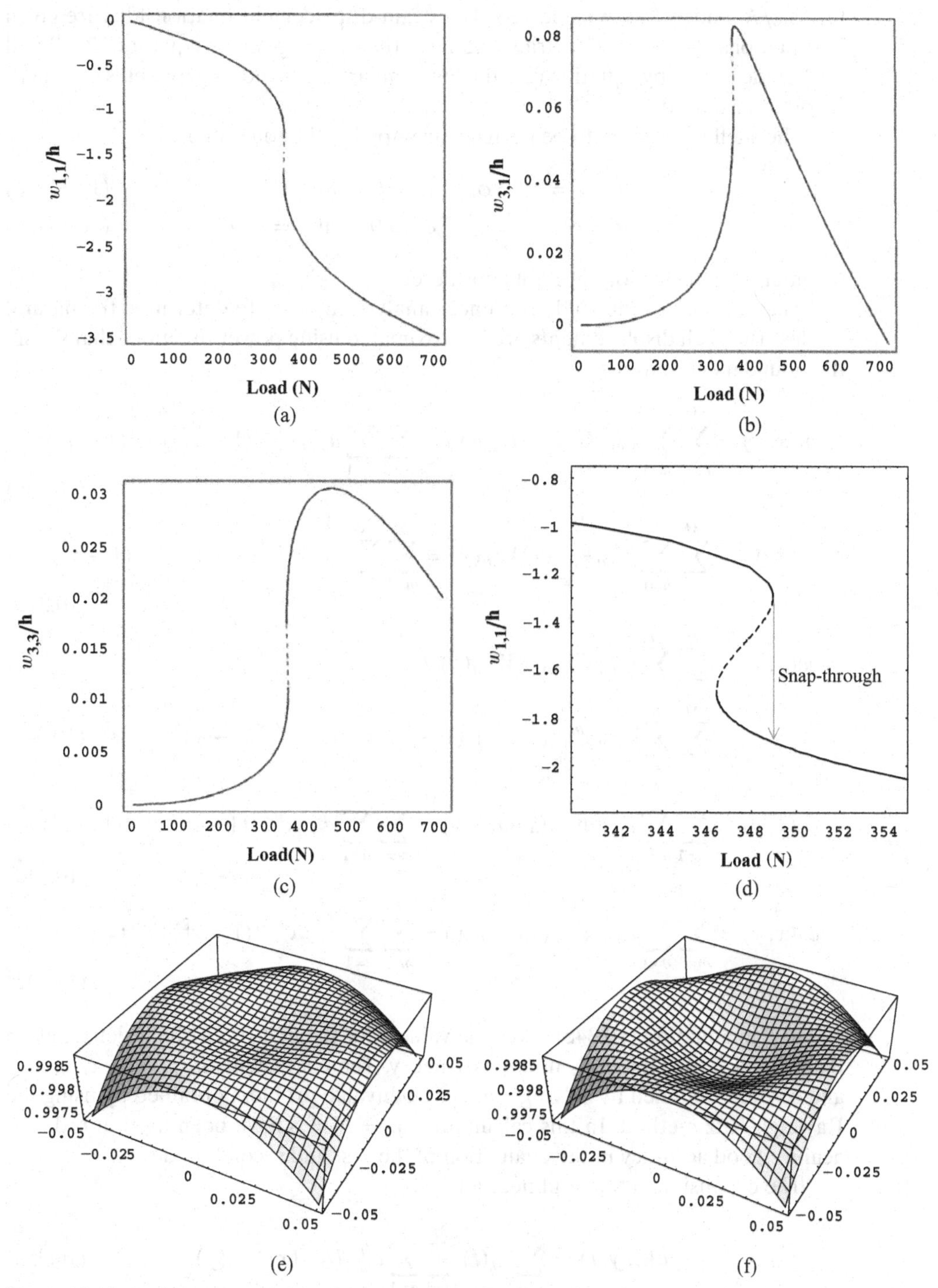

Figure 10.11. Deformation of the spherical shallow-shell under static concentrated load at the shell center in $-z$ direction; 22 dofs model. (a) Generalized coordinate $w_{1,1}$. (b) Generalized coordinate $w_{3,1}$. (c) Generalized coordinate $w_{3,3}$. (d) Zoom of (a) with indication of the jump in the deformation amplitude occurring at snap-through instability. (e) Deformed shell shape for load of 348.9 N (just before the snap-through instability). (f) Deformed shell shape for load of 350 N (just after the snap-through instability).

terms z/R_1 and z/R_2 are neglected. The strain-displacement relationships are given by equations (2.94a–h). The strain energy of the shell is given by equation (2.97) and the kinetic energy by equation (2.101). For comparison purposes, Donnell's nonlinear theory is also used.

The shell is assumed to be perfectly clamped at the four edges:

$$u = v = w = \phi_1 = \phi_2 = 0, \quad \text{at } x = 0, a, \tag{10.32a–e}$$

$$u = v = w = \phi_1 = \phi_2 = 0, \quad \text{at } y = 0, b. \tag{10.33a–e}$$

Geometric imperfections are not considered.

The first step of the study is a linear analysis in order to determine the natural modes. The shell displacements are approximated using power functions that satisfy the boundary conditions

$$u(x, y) = \sum_{m=1}^{M}\sum_{n=1}^{N} a_{m,n} U_{x,m}(x) U_{y,n}(y) = \sum_{m=1}^{M}\sum_{n=1}^{N} a_{m,n} x^{m-1}(1-x^2) y^{n-1}(1-y^2), \tag{10.34a}$$

$$v(x, y) = \sum_{m=1}^{M}\sum_{n=1}^{N} b_{m,n} V_{x,m}(x) V_{y,n}(y) = \sum_{m=1}^{M}\sum_{n=1}^{N} b_{m,n} x^{m-1}(1-x^2) y^{n-1}(1-y^2), \tag{10.34b}$$

$$w(x, y) = \sum_{m=1}^{M}\sum_{n=1}^{N} c_{m,n} W_{x,m}(x) W_{y,n}(y)$$
$$= \sum_{m=1}^{M}\sum_{n=1}^{N} c_{m,n} x^{m-1}(1+x)^{\beta}(1-x)^{\beta} y^{n-1}(1+y)^{\beta}(1-y)^{\beta}, \tag{10.34c}$$

$$\phi_1(x, y) = \sum_{m=1}^{M}\sum_{n=1}^{N} d_{m,n} \Phi_{x,m}(x) \Phi_{y,n}(y) = \sum_{m=1}^{M}\sum_{n=1}^{N} d_{m,n} x^{m-1}(1-x^2) y^{n-1}(1-y^2), \tag{10.34d}$$

$$\phi_2(x, y) = \sum_{m=1}^{M}\sum_{n=1}^{N} e_{m,n} \Psi_{x,m}(x) \Psi_{y,n}(y) = \sum_{m=1}^{M}\sum_{n=1}^{N} e_{m,n} x^{m-1}(1-x^2) y^{n-1}(1-y^2), \tag{10.34e}$$

where β in equation (10.34c) takes the value 1 for the first-order shear deformation theory and 2 for Donnell's nonlinear theory. The coefficients $a_{m,n}$, $b_{m,n}$, $c_{m,n}$, $d_{m,n}$ and $e_{m,n}$ are obtained by solving a linear eigenvalue problem obtained by using the Rayleigh-Ritz method. In this calculation, $M = N = 8$ have been used in order to achieve good accuracy in the evaluation of the first and second mode.

The expansion in the nonlinear analysis is

$$u(x, y, t) = \sum_{i=1}^{K} u_i(t) \sum_{m=1}^{M}\sum_{n=1}^{N} a_{m,n}^{(i)} U_{x,m}(x) U_{y,n}(y), \tag{10.35a}$$

$$v(x, y, t) = \sum_{i=1}^{K} v_i(t) \sum_{m=1}^{M}\sum_{n=1}^{N} b_{m,n}^{(i)} V_{x,m}(x) V_{y,n}(y), \tag{10.35b}$$

$$w(x, y, t) = \sum_{i=1}^{K} w_i(t) \sum_{m=1}^{M}\sum_{n=1}^{N} c_{m,n}^{(i)} W_{x,m}(x) W_{y,n}(y), \tag{10.35c}$$

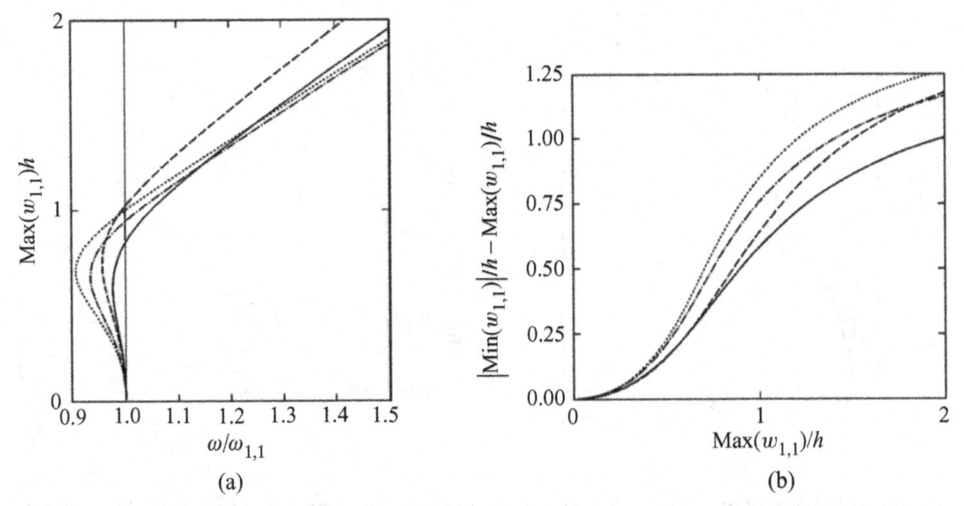

Figure 10.12. Effect of lamination sequence on nonlinear free vibration of mode (1, 1). (a) Frequency-response curve (backbone curve). (b) $|\text{Min}(w_{1,1})| - \text{Max}(w_{1,1})$ versus $\text{Max}(w_{1,1})$. —, $\theta = 45°/-45°/45°$; – –, $\theta = 0°/90°/0°$; –·–, $\theta = 45°/-45°$; …., $\theta = 0°/90°$. Adapted from Abe et al. (2000).

$$\phi_1(x, y, t) = \sum_{i=1}^{K} \phi_i(t) \sum_{m=1}^{M} \sum_{n=1}^{N} d_{m,n}^{(i)} \Phi_{x,m}(x) \Phi_{y,n}(y), \tag{10.35d}$$

$$\phi_2(x, y, t) = \sum_{i=1}^{K} \psi_i(t) \sum_{m=1}^{M} \sum_{n=1}^{N} e_{m,n}^{(i)} \Psi_{x,m}(x) \Psi_{y,n}(y), \tag{10.35e}$$

where $u_i(t)$, $v_i(t)$, $w_i(t)$, $\phi_i(t)$ and $\psi_i(t)$ are the generalized coordinates and the coefficients $a_{m,n}$, $b_{m,n}$, $c_{m,n}$, $d_{m,n}$ and $e_{m,n}$ have the superscript i, which indicates the i-th natural modes. In the calculation of the nonlinear free vibration of the first mode $(1, 1)$, $K = 1$ is used for simplicity; therefore, only a single mode is used. Moreover, in order to reduce the number of degrees of freedom to a single one, in-plane and rotary inertia are neglected so that $u_i(t)$, $v_i(t)$, $\phi_i(t)$ and $\psi_i(t)$ are expressed as functions of $w_i(t)$ by solving the equilibrium equations.

10.6 Numerical Results for Vibrations of Clamped Laminated Shells

Calculations have been performed for laminated shallow-shells with graphite-epoxy layers. Each layer has the same thickness and material properties: $E_1 = 138$ GPa, $E_2 = 8.96$ GPa, $G_{12} = 7.1$ GPa, $G_{13} = E_2/2$, $\nu_{12} = 0.3$. The shear correction factor $K_i^2 = 5/6$ has been used.

The effect of the lamination sequence on large-amplitude free vibrations of spherical shells is investigated in Figure 10.12 for mode (1, 1) by using the first-order shear deformation theory. All the shells have the following geometry: $a/b = 1$, $R_x = 10a$, $R_x = R_y$, $h = a/100$. In particular, Figure 10.12(a) shows the trend of the nonlinearity for the maximum of the generalized coordinate $w_{1,1}$, which is outward. Figure 10.12(b) gives the difference between the absolute value of the minimum (inward) and the maximum (outward) displacements. For all the lamination sequences studied, the shells initially present softening nonlinearity, which

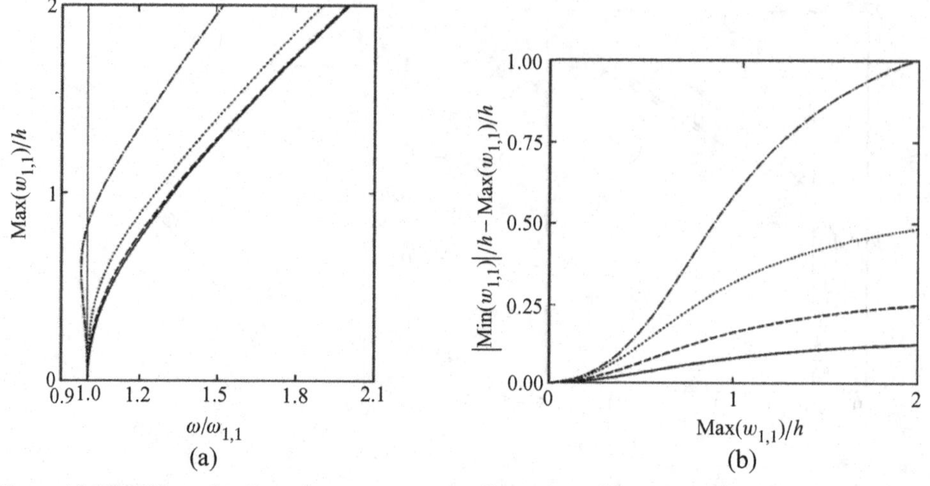

Figure 10.13. Effect of radius of curvature on nonlinear free vibration of mode (1, 1). Symmetric laminated ($\theta = 45°/-45°/45°$) spherical shells. (a) Frequency-response curve (backbone curve). (b) $|\text{Min}(w_{1,1})| - \text{Max}(w_{1,1})$ versus $\text{Max}(w_{1,1})$. —–, $R_x = 100a$; – –, $R_x = 50a$; …., $R_x = 25a$; –·–, $R_x = 10a$. Adapted from Abe et al. (2000).

turns to hardening for vibration amplitude at about the shell thickness. In particular, the nonlinearity is stronger for the lamination sequences that induce strong bending-stretching coupling ($\theta = 0°/90°, 45°/-45°$) than for symmetric lamination sequences that do not induce this effect ($\theta = 45°/-45°/45°$, $0°/90°/0°$); here, θ is the angle defined in Figure 2.9.

Symmetric angle-ply ($\theta = 45°/-45°/45°$) laminated spherical shells of different curvature are investigated in Figure 10.13. Mode (1, 1) is considered by using the first-order shear deformation theory. The shells have the following geometric parameters: $a/b = 1$, $R_x = R_y$, $h = a/100$. For $R_x = 10a$, the shell has an initial softening behavior, whereas by increasing the radius of curvature, the nonlinearity gradually becomes the strong hardening type.

The effects of the curvature ratio R_x/R_y and aspect ratio a/b of symmetric angle-ply laminated shallow-shells are studied in Figure 10.14. Mode (1, 1) is considered by using the first-order shear deformation theory. Among the cases studied, only the spherical shell with square base ($a/b = 1$) presents an initial softening behavior. As previously observed, for the hyperbolic paraboloid ($R_x/R_y = -1$), the vibration amplitude inward and outward is the same.

Finally, Figure 10.15 shows the nonlinear vibration characteristics of symmetric angle-ply laminated spherical shells with different thickness ratio (h/a). Mode (1, 1) is considered by using both the first-order shear deformation theory (thick line) and Donnell's nonlinear theory (thin line) in order to compare them. The natural frequency $\omega_{1,1}$, used to normalize the curves in Figure 10.15(a), has been obtained by using the first-order shear deformation theory; $\omega_{1,1}$ is overpredicted by using Donnell's theory for $h/a = 0.04$ and 0.1, as shown in Figure 10.15(a). Results show that the hardening nonlinearity becomes stronger by increasing the shell thickness; on the other hand, the asymmetry of the vibration amplitude inward and outward decreases with the thickness. Further, when increasing the thickness, results obtained by using the Donnell theory deviate from those obtained by using the first-order shear deformation theory. It can be observed that, in the present case, the first-order shear

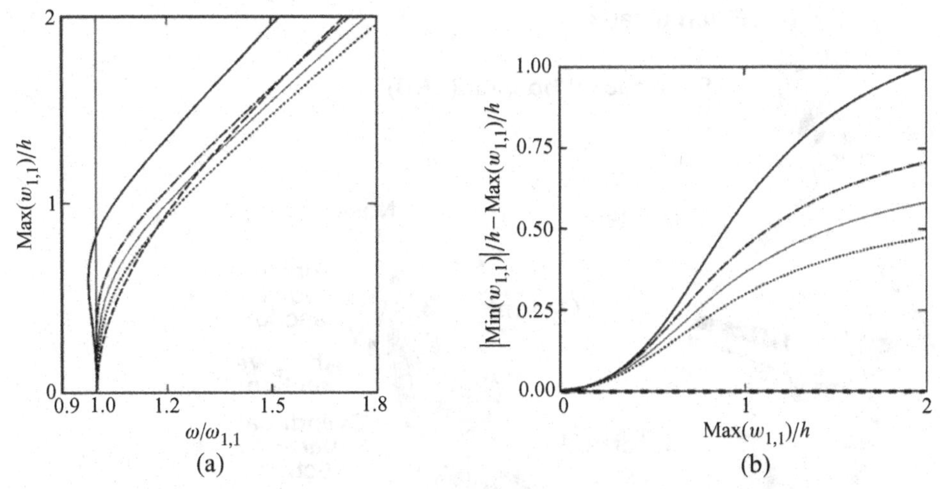

Figure 10.14. Effects of curvature and aspect ratios on nonlinear free vibration of mode (1, 1). Symmetric laminated ($\theta = 45°/-45°/45°$) spherical shells. (a) Frequency-response curve (backbone curve). (b) $|\text{Min}(w_{1,1})| - \text{Max}(w_{1,1})$ versus $\text{Max}(w_{1,1})$. —, $R_x/R_y = 1$, $a/b = 1$; —, $R_x/R_y = 0, a/b = 1$; – –, $R_x/R_y = -1, a/b = 1$; –·–, $R_x/R_y = 1, a/b = 1.5$; ...., $R_x/R_y = 1, a/b = 2$. Adapted from Abe et al. (2000).

deformation theory should be used for thickness ratio $h/a \geq 0.04$ in order to achieve good accuracy.

10.7 Buckling of the Space Shuttle Liquid-Oxygen Tank

In this section, the buckling analysis of the NASA space shuttle's liquid-oxygen tank is discussed by following the study of Nemeth et al. (1996; 2002). Figure 10.16 shows the

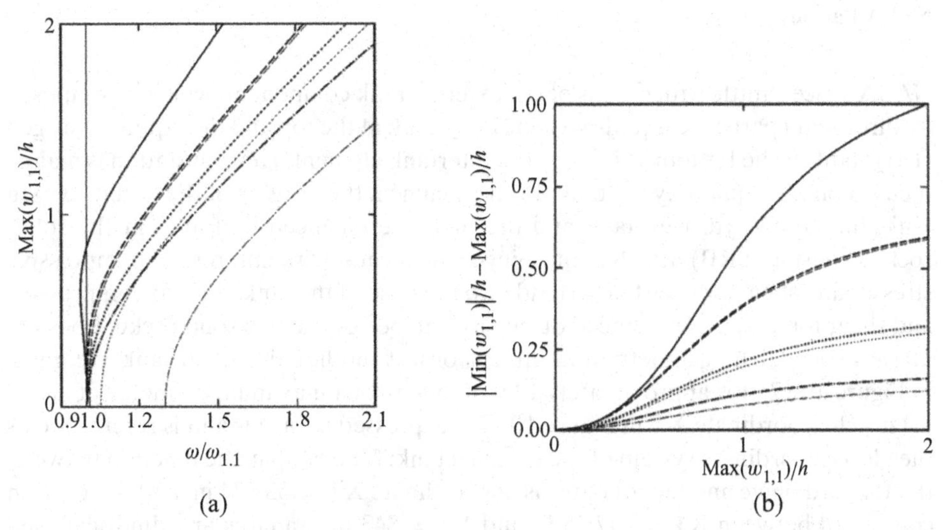

Figure 10.15. Effect of thickness radius on nonlinear free vibration of mode (1, 1). Symmetric laminated ($\theta = 45°/-45°/45°$) spherical shells. (a) Frequency-response curve (backbone curve). (b) $|\text{Min}(w_{1,1})| - \text{Max}(w_{1,1})$ versus $\text{Max}(w_{1,1})$. —, $h/a = 0.01$; – –, $h/a = 0.02$; ...., $h/a = 0.04$; –·–, $h/a = 0.1$. Thick line, first-order shear deformation theory; thin line, Donnell's nonlinear theory. Adapted from Abe et al. (2000).

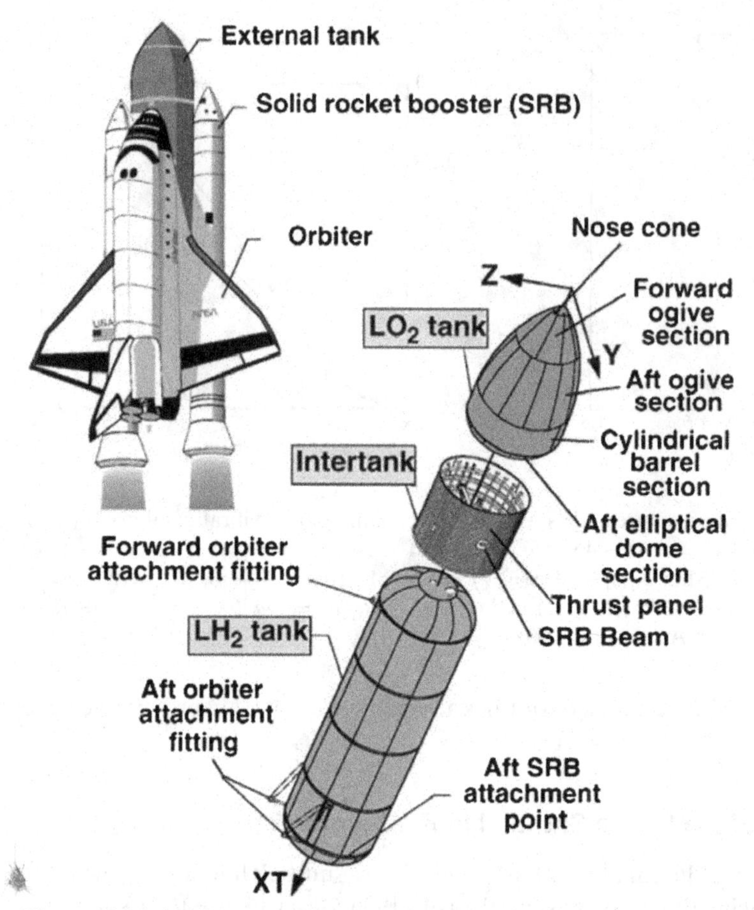

Figure 10.16. NASA space shuttle with detail of the external tank components. LO_2, liquid oxygen; LH_2, liquid hydrogen. From Nemeth et al. (2002). Courtesy of Dr. M. P. Nemeth of NASA Langley.

NASA space shuttle with details of the external tank component, which is composed by three main parts: the liquid-oxygen (LO_2) tank at the top and the liquid-hydrogen (LH_2) tank at the bottom, joined by the intertank element. Here, attention is mainly focused on the liquid-oxygen tank. Prior to launch, the weights of the liquid-oxygen tank, the liquid-hydrogen tank and the fuels are balanced by forces at the solid-rocket-booster (SRB) attachment points, which cause circumferential compressive stresses and shear stresses that extend into the nose of the tank. Similar compressive and shear forces are experienced during ascent, before the two solid rocket boosters are jettisoned. The geometry and dimensions of the liquid-oxygen tank are given in Figure 10.17; it is approximately 14.9 m long with a maximum diameter of about 8.4 m. The coordinate XT in Figure 10.17 is expressed in inches and is referred to as the global coordinate system of the external tank; for example, the junction between the forward ogive and the aft ogive is at coordinate XT = 536.74 in, and the portion comprised between XT = 747.35 in and XT = 843 in is a circular cylindrical shell (barrel in Figure 10.17). The liquid-oxygen tank has a forward T-ring and an aft Y-ring that support a slosh baffle in order to prevent fuel sloshing during ascent. The tank is made of gore and barrel panels that are stretch formed, chemically milled and then welded together. The panels have tapered thickness to reduce structural mass.

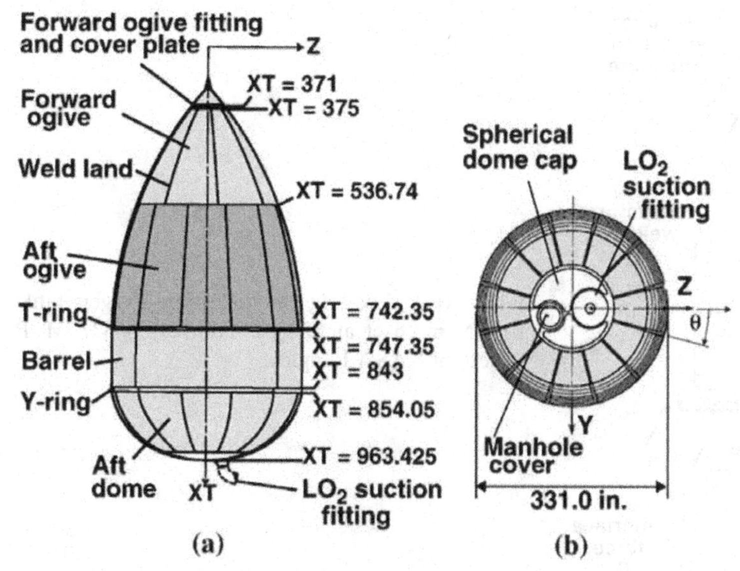

Figure 10.17. Liquid-oxygen (LO_2) tank components; values of the XT coordinates are given in inches. (a) Side view. (b) Aft view. From Nemeth et al. (2002). Courtesy of Dr. M. P. Nemeth of NASA Langley.

The external tank was originally made of 2219 aluminum alloy and was coated with a thermal protection system given by a sprayed-on foam insulation; this is referred to as the standard-weight external tank and has a structural mass of 30,000 kg. Two full-scale experimental tests were conducted on the standard-weight liquid-oxygen tank at the end of the 1970s at the NASA Marshall Spaceflight Center during the design development program of the space shuttle. Both test specimens exhibited unexpected buckling. Local and global buckling can cause the thermal protection system to separate from the tank, with a consequent catastrophic failure of the vehicle. Therefore, great care must be taken in design to prevent buckling. Calculations show that buckling is highly sensitive to initial geometric imperfections. For this reason, imperfections were measured on the test specimens, but not everywhere. In particular, relatively large welding-induced imperfections were detected next to the circumferential welds in the barrel section. The maximum outward amplitude of the initial imperfection was 6.8 mm, and the maximum inward amplitude was 5.2 mm; these amplitudes correspond to approximately twice the minimum skin thickness in the region of the barrel.

A new super lightweight external tank was introduced in order to save 3600 kg of weight, increasing the payload. This tank is made of 2195 aluminum-lithium alloy and is thinner than the previous one and has a less-uniform thickness distribution in order to optimize the material. The new tank flew for the first time in 1998. New buckling experiments have not been performed and design was based on finite-element calculation based on the STAGS nonlinear structural analysis code for general shells.

Four critical loads configuration were individuated: two prelaunch conditions (full liquid-hydrogen tank and empty liquid-oxygen tank, and both full tanks without ullage pressure) and two flight conditions (early booster ascent and a $3g$ axial acceleration at end-of-flight condition). The loading characteristics are given in Figure 10.18. Loads consist of wind or aerodynamic pressure, structural weight or inertia,

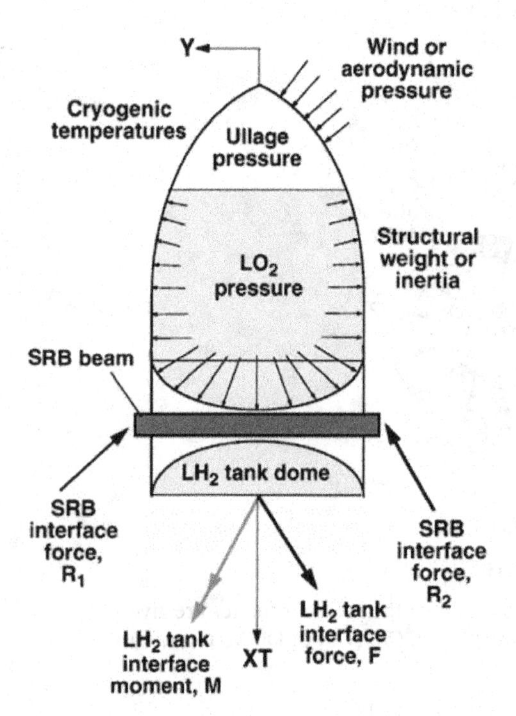

Figure 10.18. Loading on the liquid-oxygen tank. From Nemeth et al. (2002). Courtesy of Dr. M. P. Nemeth of NASA Langley.

pressure exerted on the shell wall by the liquid oxygen (including fluid weight or inertia), ullage pressure inside the tank, interface forces exerted by each SRB, interface force and moment between the intertank and the liquid-hydrogen tank and thermally induced loads due to the cryogenic fuel temperatures and the aerodynamic heating.

The following results are relative to the standard-weight liquid-oxygen tank experimentally tested at the NASA Marshall Spaceflight Center to simulate the flight conditions corresponding to early booster ascent. The liquid-oxygen tank and the intertank were mounted vertically to a liquid-hydrogen tank simulator giving a self-equilibrated line load applied to the bottom of the intertank and to two rigid vertical posts at the SRB attachment points. A uniformly distributed circumferential line load of 6.2×10^6 N was applied to the tank at XT = 852.8 in. The tank was filled with water at room temperature to XT = 455 in (corresponding to a depth of about 12.9 m) maintaining an ullage pressure in the tank. Decreasing the ullage pressure at 3927 Pa, the tank unexpectedly buckled in the forward ogive between XT = 455 and 475 in and between $\theta = 253°$ and $277°$. Under these conditions, the reactions at the SRB attachments were 5.76×10^6 N. The loading conditions during the test are synthesized in Figure 10.19.

The forward ogive has a minimum thickness of 2 mm. The finite-element mesh used in the STAGS nonlinear structural analysis code is shown in Figure 10.20(a) and corresponds to 159,993 dofs. It is capable of modeling short-wavelength buckling modes. The shape of the buckling mode in the present case is shown in Figure 10.20(b).

Numerical results are presented in Figure 10.21, where the effect of geometric imperfections on the buckling load of the forward ogive is shown. Geometric imperfections of amplitude A and the same mode shape shown in Figure 10.20(b) are introduced; they are normalized with respect to the minimum thickness of the shell $t = 2$ mm. In Figure 10.21, the experimental buckling load is at $p_a = 1$, where p_a is the load factor that multiplies all the loads acting on the shell. Linear calculations find

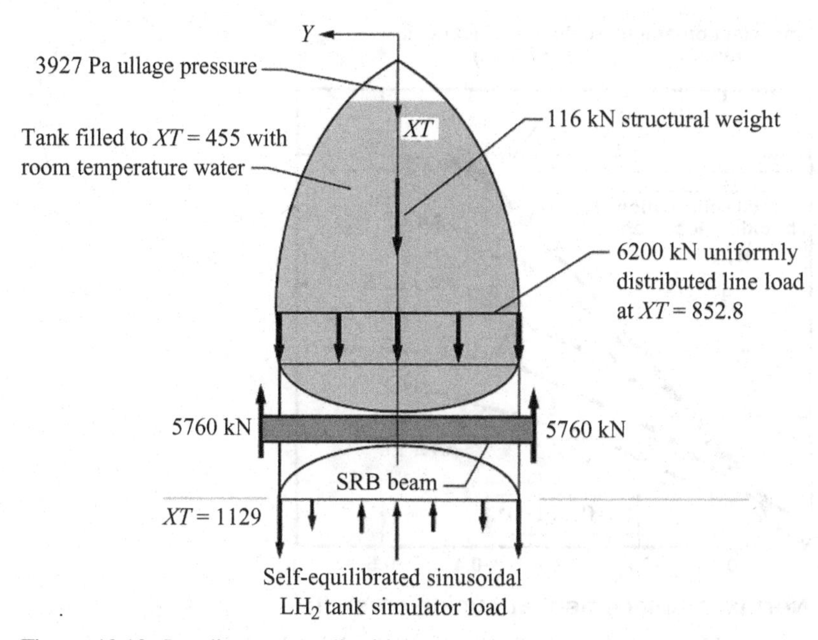

Figure 10.19. Loading on the liquid-oxygen tank during experimental test. Adapted from Nemeth et al. (1996). Courtesy of Dr. M. P. Nemeth of NASA Langley.

buckling at $p_a = 1.14$ for the perfect shell. Nonlinear calculations indicate buckling load at $p_a = 1.18$ for the perfect shell; this value is largely decreased by introducing geometric imperfections, giving $p_a = 0.53$ for geometric imperfection $A/t = 1$. All curves in Figure 10.21, giving the load factor versus the maximum normal displacement at the buckle crest, approach a horizontal tangent as the load factor increases, which is characteristic of an incipient snap-through instability. In fact, with the finite-element program utilized, the unstable part of the equilibrium solution cannot be followed (see Figure 10.11[d] where the snap-through instability is shown) and the

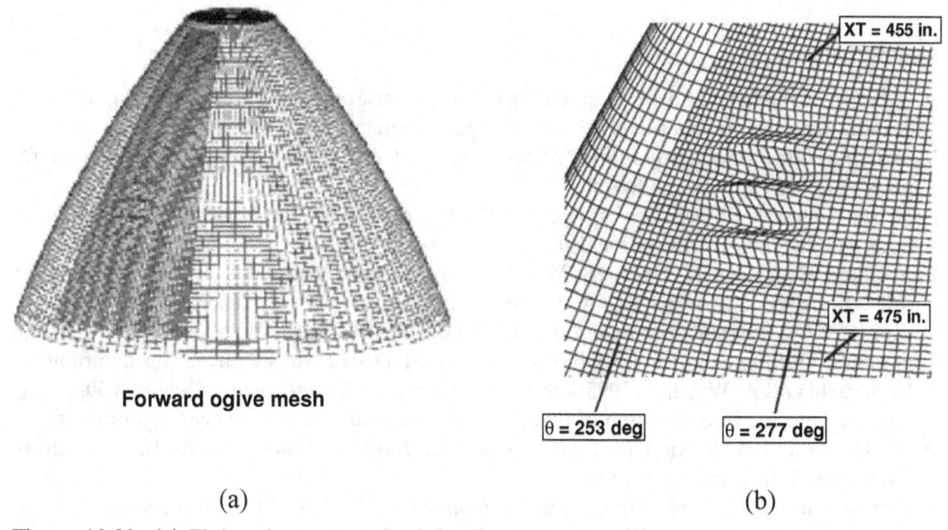

Figure 10.20. (a) Finite-element mesh of the forward ogive. (b) Buckling mode shape. From Nemeth et al. (2002). Courtesy of Dr. M. P. Nemeth of NASA Langley.

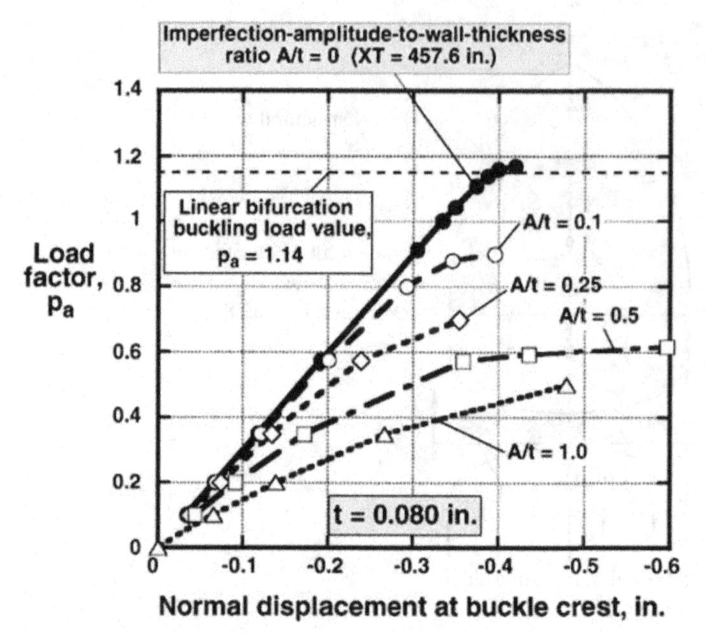

Figure 10.21. Effect of imperfection amplitude on the buckling load of the forward ogive. A, imperfection amplitude; t, minimum thickness. Linear buckling $p_a = 1.14$; experimental buckling $p_a = 1$. From Nemeth et al. (2002). Courtesy of Dr. M. P. Nemeth of NASA Langley.

calculations had to be stopped because of numerical difficulties. For the forward ogive, a small geometric imperfection $A/t = 0.064$ gives exactly the buckling load measured during experiments. Unfortunately, geometric imperfections on this area of the forward ogive were not measured.

As a general rule, it is convenient to introduce geometric imperfections of the order of the shell thickness in the numerical model if actual geometric imperfections are not measured. This gives a reasonable guarantee of a safe design.

REFERENCES

A. Abe, Y. Kobayashi and G. Yamada 2000 *Journal of Sound and Vibration* **234**, 405–426. Non-linear vibration characteristics of clamped laminated shallow shells.

M. Amabili 2005 *International Journal of Non-Linear Mechanics* **40**, 683–710. Non-linear vibrations of doubly curved shallow shells.

A. P. Bhattacharya 1976 *Journal of Applied Mechanics* **43**, 180–181. Nonlinear flexural vibrations of thin shallow translational shells.

C. Y. Chia 1988 *Ingenieur-Archiv* **58**, 252–264. Nonlinear analysis of doubly curved symmetrically laminated shallow shells with rectangular platform.

D. Hui 1990 *Journal of Vibration and Acoustics* **112**, 304–311. Accurate backbone curves for a modified-Duffing equation for vibrations of imperfect structures with viscous damping.

Y. Kobayashi and A. W. Leissa 1995 *International Journal of Non-Linear Mechanics* **30**, 57–66. Large amplitude free vibration of thick shallow shells supported by shear diaphragms.

A. W. Leissa and A. S. Kadi 1971 *Journal of Sound and Vibration* **16**, 173–187. Curvature effects on shallow shell vibrations.

L. Librescu and M.-Y. Chang 1993 *Acta Mechanica* **96**, 203–224. Effects of geometric imperfections on vibration of compressed shear deformable laminated composite curved panels.

M. P. Nemeth, V. O. Britt, T. J. Collins and J. H. Starnes Jr. 1996 *NASA Technical Paper 3616*. Nonlinear analysis of the Space Shuttle superlightweight external fuel tank.

M. P. Nemeth, R. D. Young, T. J. Collins and J. H. Starnes Jr. 2002 *International Journal of Non-Linear Mechanics* **37**, 723–744. Effects of initial imperfections on the non-linear response of the space shuttle superlightweight liquid-oxygen tank.

G. C. Sinharay and B. Banerjee 1985 *International Journal of Non-Linear Mechanics* **20**, 69–78. Large-amplitude free vibrations of shallow spherical shell and cylindrical shell – a new approach.

G. C. Sinharay and B. Banerjee 1986 *AIAA Journal* **24**, 998–1004. Large-amplitude free vibrations of shells of variable thickness – a new approach.

M. S. Soliman and P. B. Gonçalves 2003 *Journal of Sound and Vibration* **259**, 497–512. Chaotic behaviour resulting in transient and steady state instabilities of pressure-loaded shallow spherical shells.

O. Thomas, C. Touzé and A. Chaigne 2005 *International Journal of Solids and Structures* **42**, 3339–3373. Non-linear vibrations of free-edge thin spherical caps: modal interaction rules and 1:1:2 internal resonance.

C. Touzé and O. Thomas 2006 *International Journal of Non-linear Mechanics* **41**, 678–692. Non-linear behaviour of free-edge shallow spherical shells: effect of the geometry.

11 Meshless Discretization of Plates and Shells of Complex Shape by Using the R-Functions

11.1 Introduction

In order to solve the system of nonlinear partial differential equations governing the nonlinear dynamics of shells and plates, it is necessary to discretize the problem, for example, by using admissible functions. The construction of the basis functions, satisfying the boundary conditions, is a very difficult task in the case of domains with complex shape. In fact, a very small number of studies address the nonlinear dynamics of structural elements of complicated geometry. The R-functions theory, developed by the Ukrainian mathematician V. L. Rvachev, is one of the possible approaches to solving the problem. It allows us to build the basis functions in analytical form for an arbitrary domain and different boundary conditions, including mixed boundary conditions. However, the R-functions are still little known in the Western scientific community.

The R-functions are a powerful tool to obtain meshless discretization of bidimensional and three-dimensional domains of complex shape, and they can be applied to moving boundaries. This approach is particularly convenient to study nonlinear vibration compared with the finite-element method. In fact, it allows us to discretize shells and plates of complex shape with a very small number of degrees of freedom. The method has a great potential to develop commercial computer codes.

In this chapter, the R-functions are introduced and then applied to obtain series expansions of displacements that satisfy exactly the boundary conditions in the case of complex shape of shells and plates.

11.1.1 Literature Review

The Ukrainian mathematician Rvachev (1963; 1982) proposed the R-functions theory and applied it to solve many engineering problems (Rvachev and Sheiko 1995; Rvachev et al. 2000). Rvachev and his school in Kharkov developed a specific computer program to discretize partial differential equations in complicated domains by using the R-functions method (Rvachev and Shevchenko 1988). Many applications to statics and dynamics of plates have been addressed (Rvachev and Kurpa 1987, 1998; Kurpa et al. 2003). Nonlinear vibrations of plates of complex geometry have been studied by Kurpa et al. (2005b). Vibrations of shallow-shells have been investigated by Kurpa et al. (2005a) with a linear theory, and by Kurpa et al. (2007) with a geometrically nonlinear theory.

11.2 The R-Functions Method

The R-functions method (RFM) allows us to search for a solution of boundary value problems in a form that satisfies exactly all the boundary conditions and contains functions to be determined in order to satisfy the differential equations governing the problem in approximate way.

11.2.1 Boundary Value Problems with Homogeneous Dirichlet Boundary Conditions

Kantorovich (Kantorovich and Krylov 1958) introduced a solution u to homogeneous Dirichlet conditions:

$$u|_{\partial\Omega} = 0, \tag{11.1}$$

where $\partial\Omega$ is the boundary of the domain. This solution has the form:

$$u = \omega P, \tag{11.2a}$$

where the function ω vanishes at the boundary ($\omega|_{\partial\Omega} = 0$) and is positive in the interior of Ω, and P is an unknown function that allows us to satisfy the differential equations governing the problem. Because ω is identically zero at the boundary $\partial\Omega$, then no matter what the indefinite component P (differentiable up to the required order) is, the function u given by equation (11.2a) will satisfy the boundary condition (11.1) exactly. The function P is determined in order to satisfy the differential equations; therefore, in general, it can be expressed in approximate way by using a series expansion of basis functions ψ_i

$$P = \sum_{i=1}^{N} C_i \psi_i, \tag{11.2b}$$

where C_i are scalar coefficients. Thus, the solution takes the form:

$$u = \omega \sum_{i=1}^{N} C_i \psi_i. \tag{11.2c}$$

The undetermined coefficients C_i can be found numerically, for example, by using variational (e.g. Lagrange equations) or Galerkin methods.

Kantorovich's idea did not have particular success because (i) no available technique existed to construct such real functions ω for domains of complex shape, and (ii) the same solution was not applicable to other types of boundary value problems.

Rvachev found the way to overcome both of the obstacles by creating the R-functions theory. R-functions are elementary functions f whose sign is completely defined by the sign of their arguments; that is, for any R-function f exists a Boolean function $F(X_1, X_2, \ldots, X_n)$ of Boolean variables X_n which satisfies the equality

$$F(S_2(x_1), S_2(x_2), \ldots, S_2(x_n)) = S_2(f(x_1, x_2, \ldots, x_n)).$$

The two-valued predicate $S_2(x)$, which returns 0 or 1 according to the sign of the argument, is defined by

$$S_2(x) = \begin{cases} 0 & \forall \quad x < 0, \\ 1 & \forall \quad x \geq 0. \end{cases} \tag{11.3}$$

Functions that satisfy these properties are, for example, xyz, $x + y + \sqrt{xy + x^2 + y^2}$ and $xy + z + |z - xy|$. R-functions behave as continuous analogs of logical Boolean functions. Every Boolean function has infinity analog R-functions. The best-known system of R-functions is given by the R-conjunction $(x \wedge_\alpha y)$ and the R-disjunction $(x \vee_\alpha y)$, which are defined as

$$x \wedge_\alpha y = \frac{1}{1+\alpha}\left(x + y - \sqrt{x^2 + y^2 - 2\alpha xy}\right), \tag{11.4}$$

$$x \vee_\alpha y = \frac{1}{1+\alpha}\left(x + y + \sqrt{x^2 + y^2 - 2\alpha xy}\right), \tag{11.5}$$

where α is a continuous function satisfying the condition $-1 < \alpha(x, y) \leq 1$; the denial function, simply given by a minus, must be added to complete this system of R-functions. The R-conjunction (11.4) is an R-function whose companion Boolean function is logical "and" ($\wedge$), whereas equation (11.5) has companion Boolean function logical "or" ($\vee$). Note that the precise value of α is not important in many applications, and it is often set to a constant. If $\alpha = 1$ is taken, equations (11.4) and (11.5) become the functions $\mathrm{Min}(x, y)$ and $\mathrm{Max}(x, y)$, respectively. Setting $\alpha = 0$ in equations (11.4) and (11.5), the following simpler functions are obtained:

$$x \wedge_0 y = x + y - \sqrt{x^2 + y^2}, \tag{11.6}$$

$$x \vee_0 y = x + y + \sqrt{x^2 + y^2}. \tag{11.7}$$

R-functions are closed under composition. Therefore, the function ω can be obtained for domains of complicate shape (eventually time-varying), which can be represented by using primitive geometric regions, for example, defined by systems of inequalities. Using R-operations, such as R-disjunctions $(x \vee_\alpha y)$, which has analog Boolean union $\cup$, R-conjunction $(x \wedge_\alpha y)$, with analog Boolean intersection $\cap$, and R-denial $(-x)$, with analog Boolean absolute complement $^{-}$ (indicated with an overline), it is possible to construct the analytical expression of ω for almost any domain Ω.

The first step in order to build ω for the domain $\Omega = \Omega(Q)$, where Q is a point in the plane (x, y) or in the three-dimensional space, it is necessary to construct the so-called characteristic function (also named two-valued predicate) of domain Ω, which is defined as

$$\Omega = \Omega(Q) = \begin{cases} 0, & \forall Q \notin \Omega, \\ 1, & \forall Q \in \Omega. \end{cases} \tag{11.8}$$

The domain and its characteristic function are usually denoted with the same symbol.

In general, the characteristic function of domain Ω is obtained by applying simple operations $\wedge$, $\vee$, $^{-}$, that correspond to the Boolean operations intersection $\cap$, union $\cup$, and absolute complement $^{-}$, respectively, to subdomains Ω_i. A characteristic function Ω_i can be defined for each subdomain Ω_i:

$$\Omega_i(Q) = \begin{cases} 0, & \forall Q \notin \Omega_i, \\ 1, & \forall Q \in \Omega_i. \end{cases} \tag{11.9}$$

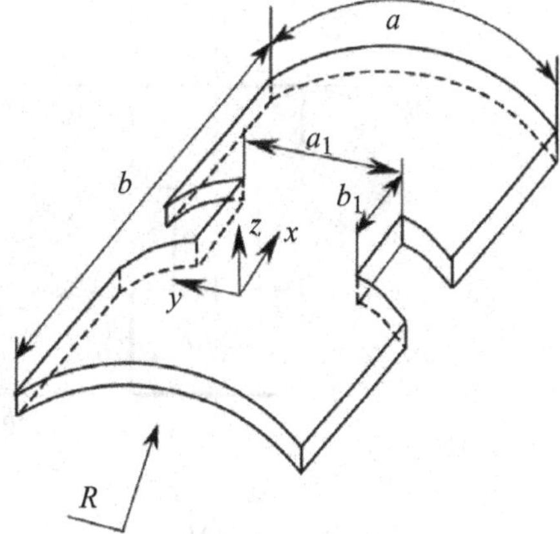

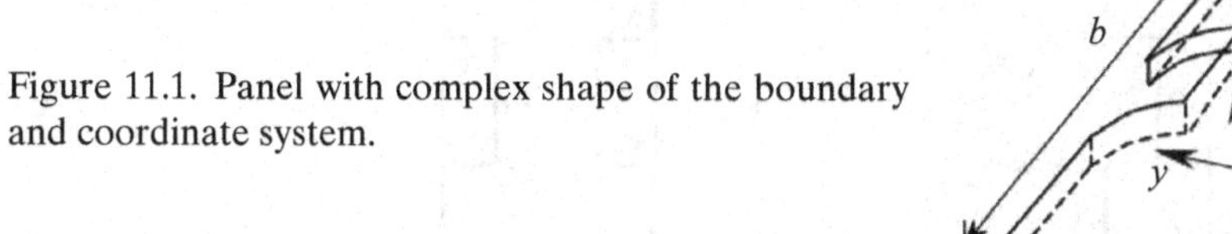

Figure 11.1. Panel with complex shape of the boundary and coordinate system.

Because $\Omega_i \in B_2\{0, 1\}$, that is, to the space of Boolean functions, then Ω_i may be used as an argument of the Boolean function F

$$\Omega = F(\Omega_1, \Omega_2, \ldots, \Omega_n), \tag{11.10}$$

which is the characteristic function of domain Ω. It is obvious that domain Ω is determined not only by the shape of subdomains Ω_i but also by the type of Boolean functions involved in F (union, intersection and absolute complement).

As next step in the determination of ω, the function η_i is introduced as the continuous analog of the characteristic function Ω_i and is defined as

$$\begin{aligned} \eta_i(Q) &> 0 \quad \forall Q \in \Omega_i, \\ \eta_i(Q) &< 0 \quad \forall Q \notin \Omega_i, \\ \eta_i(Q) &= 0 \quad \forall Q \in \partial\Omega_i. \end{aligned} \tag{11.11}$$

Let us assume that domain Ω is defined by the characteristic function (two-valued predicate) represented in equation (11.10), where $F(\Omega_1, \ldots, \Omega_n)$ is a known Boolean function. Then, the inequality $f(\eta_1, \ldots, \eta_n) \geq 0$ describes domain Ω, where $\omega = f(\eta_1, \ldots, \eta_n)$ is an R-function that corresponds to the Boolean function $F(\Omega_1, \ldots, \Omega_n)$. To construct the function $\omega(x, y)$, it is sufficient to perform a formal substitution of the characteristic function Ω_i with the continuous function η_i and the Boolean operations $\Omega_i \cap \Omega_j$, $\Omega_i \cup \Omega_j$, $\bar{\Omega}_i$ (intersection, union, absolute complement) with the corresponding symbols of R-operations, $\eta_i \wedge_\alpha \eta_j$, $\eta_i \vee_\alpha \eta_j$, $-\eta_i$. Following this substitution, the continuous function ω of the domain Ω is given by

$$\omega = f(\eta_1, \eta_2, \ldots, \eta_n). \tag{11.12}$$

11.2.1.1 Example: Shell with Complex Shape

In order to better understand the method, the function ω is built for the shallow-shell shown in Figure 11.1. The domain Ω representing the shell surface is given by the following Boolean operations on subdomains Ω_i:

$$\Omega = \Omega_1 \wedge \Omega_2 \wedge (\Omega_3 \vee \Omega_4). \tag{11.13}$$

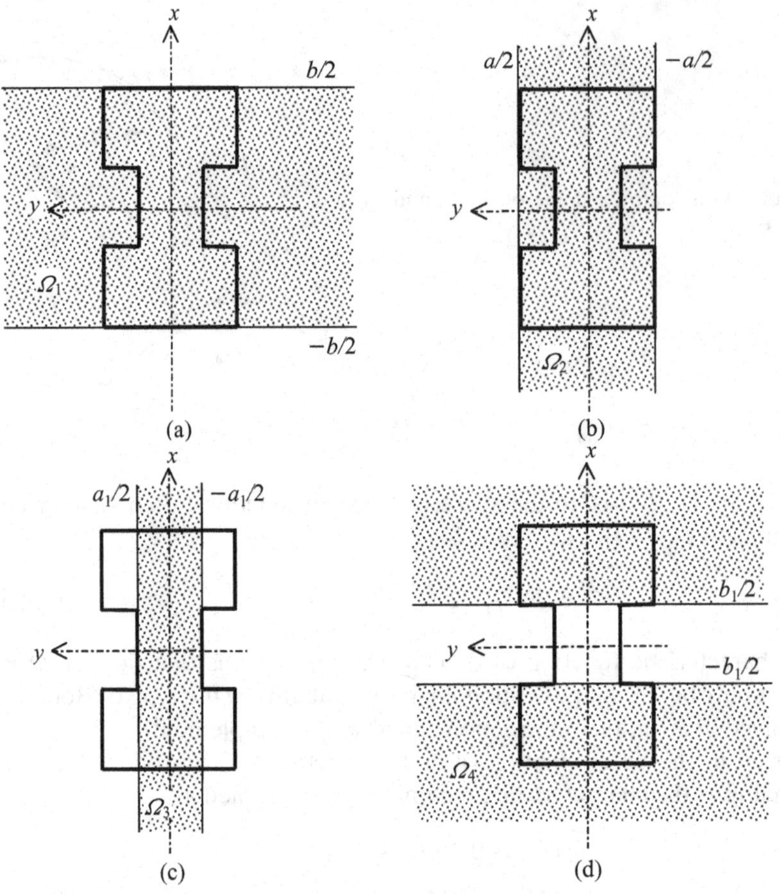

Figure 11.2. Subdomains used to build the function ω. (a) Subdomain Ω_1. (b) Subdomain Ω_2. (c) Subdomain Ω_3. (d) Subdomain Ω_4.

By using (11.13), the equation of the domain for the shell in Figure 11.1 may be written as

$$\omega(x, y) = \eta_1 \wedge_0 \eta_2 \wedge_0 (\eta_3 \vee_0 \eta_4), \tag{11.14}$$

where the functions $\eta_i, i = 1, \ldots 4$, are given by

$$\eta_1 = \left(\left(\frac{b}{2}\right)^2 - x^2\right) \Big/ b \geq 0 \quad \text{in } \Omega_1, \quad \eta_2 = \left(\left(\frac{a}{2}\right)^2 - y^2\right) \Big/ a \geq 0 \quad \text{in } \Omega_2, \tag{11.15a,b}$$

which are associated with horizontal (Ω_1) and vertical (Ω_2) strips in the plane confined between the straight lines $x = \pm b/2$ and $y = \pm a/2$, respectively (as shown in Figure 11.2), and

$$\eta_3 = \left(\left(\frac{a_1}{2}\right)^2 - y^2\right) \Big/ a_1 \geq 0 \quad \text{in } \Omega_3, \quad \eta_4 = \left(x^2 - \left(\frac{b_1}{2}\right)^2\right) \Big/ b_1 \geq 0 \quad \text{in } \Omega_4. \tag{11.15c,d}$$

In particular, Ω_3 is a vertical strip between $y = \pm a_1/2$ and Ω_4 is the region of the plane exterior to the horizontal strip delimited by $x = \pm b_1/2$, as shown in Figure 11.2.

Equation (11.14) is normalized up to the first order, that is, it satisfies the following conditions:

$$\omega(x, y) = 0, \quad \forall (x, y) \in \partial\Omega, \tag{11.16a}$$

$$\frac{\partial \omega}{\partial n} = -1, \quad \forall (x, y) \in \partial\Omega, \tag{11.16b}$$

where n is the normal to the boundary $\partial\Omega$ contained in the plane tangent to the panel at the boundary and directed outward the domain Ω. Inside the domain Ω, the inequality $\omega(x, y) > 0$ is verified.

11.2.2 Boundary Value Problems with Inhomogeneous Dirichlet Boundary Conditions

It is possible to extend the solution procedure of equation (11.2) to the case of inhomogeneous Dirichlet conditions given by

$$u|_{\partial\Omega} = \varphi_0, \tag{11.17}$$

where φ_0 is a function defined at the boundary $\partial\Omega$ and φ is its extension defined in all the domain Ω. The solution structure can be written in the following form:

$$u = \omega P + \varphi. \tag{11.18}$$

The general case of inhomogeneous Dirichlet conditions, in which the function φ_0 in equation (11.17) is specified as a piecewise function at the boundary $\partial\Omega$, can be solved by using the generalized Lagrange formula obtained by Rvachev (1982). The function φ_0 on section $\partial\Omega_i$ of the boundary $\partial\Omega$ of domain Ω is indicated as

$$\varphi_0(x, y)|_{\partial\Omega_i} = \varphi_i. \tag{11.19}$$

Then, the solution may be represented by using the generalized Lagrange formula:

$$u = \omega^k P + \frac{\sum_{i=1}^{n} \varphi_i \omega_i^{-k_i}}{\sum_{i=1}^{n} \omega_i^{-k_i}}, \quad \text{with } k_i \geq 1 \quad \text{and} \quad k = \max(k_i). \tag{11.20}$$

Solution (11.20) is given by a set of functions taking the prescribed value φ_i on section $\partial\Omega_i$ of the boundary $\partial\Omega$. A possible choice is $k = k_i = 1$, which makes the solution defined everywhere except at corner points. If $k_i > 1$ is chosen, the partial derivatives of the function u coincide on section $\partial\Omega_i$ with the corresponding partial derivatives of the function φ_i up to the $k_i - 1$ order. Equation (11.20) may be considered as the general solution structure for inhomogeneous Dirichlet problems.

11.2.3 Boundary Value Problems with Neumann and Mixed Boundary Conditions

Differential operators have been introduced by Rvachev to take into account boundary conditions of differential type. For example, for two-dimensional problems, these operators have the following form:

$$D_k = \sum_{i=0}^{k} C_k^i \left(\frac{\partial \omega}{\partial x}\right)^{k-i} \left(\frac{\partial \omega}{\partial y}\right)^{i} \frac{\partial^k}{\partial x^{k-i}\, \partial y^i}, \tag{11.21a}$$

$$T_k = \sum_{i=0}^{k} (-1)^{k-i}\, C_k^i \left(\frac{\partial \omega}{\partial x}\right)^{i} \left(\frac{\partial \omega}{\partial y}\right)^{k-i} \frac{\partial^k}{\partial x^{k-i}\, \partial y^i}, \tag{11.21b}$$

where $C_k^i = k(k-1)\ldots(k-i+1)/i!$ and $k \geq 1$; $\omega(x, y) = 0$ is, as usual, the normalized equation of $\partial\Omega$ (or of section $\partial\Omega_i$). Any function $u(x, y) \in C^n(\Omega)$ satisfies the following expressions:

$$D_k(u) = \frac{\partial^k u}{\partial n^k}, \quad T_k(u) = \frac{\partial^k u}{\partial \tau^k}, \quad \forall (x, y) \in \partial\Omega, \tag{11.21c,d}$$

$$(T_{k-m}D_m)(u) = (D_m T_{k-m})(u) = \frac{\partial^k u}{\partial n^m \, \partial \tau^{k-m}}, \quad \forall (x, y) \in \partial\Omega, \tag{11.21e}$$

where n and τ are the normal and the tangent to the boundary, respectively; both of them are contained in the plane tangent to the panel at the boundary, with the normal directed outward the shell domain.

In the case of Neumann boundary conditions in the form

$$\left.\frac{\partial u}{\partial n}\right|_{\partial\Omega} = \varphi_0, \tag{11.22}$$

the solution structure can be written in the following form:

$$u = P + \omega\varphi - \omega D_1(P) + \omega^2 P, \tag{11.23a}$$

where P and φ have been previously defined and D_1 is given by (11.21) for $k = 1$. Equation (11.23a) is based on the generalized Taylor series (Rvachev and Sheiko 1995). Similar procedures can be used for almost any boundary condition, including mixed boundary conditions. In particular, for boundary conditions of the type $u|_{\partial\Omega} = \partial u/\partial n|_{\partial\Omega} = 0$, which are those with clamped edges, the solution structure can be written in the following simple form:

$$u = \omega^2 P. \tag{11.23b}$$

11.2.4 Admissible Functions for Shells and Plates with Different Boundary Conditions

According to classical shell and plate theories, four boundary conditions are given at each edge. The boundary conditions considered in this section are

Clamped edge

$$u_n = 0, \quad v_n = 0, \quad w = \frac{\partial w}{\partial n} = 0, \tag{11.24}$$

where $u_n = ul + vm$, $v_n = -um + vl$ and l, m are the direction cosines of the edge with respect to the in-plane coordinates u and v (in the case of a curved edge, l and m are functions of the position).

In-plane immovable simply supported edge

$$u_n = 0, \quad v_n = 0, \quad w = 0, \quad M_n = -D\left(\frac{\partial^2 w}{\partial n^2} + \nu\frac{\partial^2 w}{\partial \tau^2}\right) = 0, \tag{11.25}$$

where M_n is the bending moment.

Classical simply supported edge

$$v_n = 0, \quad w = 0, \quad N_n = 0, \quad M_n = 0. \tag{11.26}$$

In-plane free simply supported edge

$$N_n = 0, \quad N_{n\tau} = 0, \quad w = 0, \quad M_n = 0, \tag{11.27}$$

where $N_n = N_x l^2 + N_y m^2 + 2N_{xy} lm$, $N_{n\tau} = N_{xy}(l^2 - m^2) + (N_y - N_x)lm$.

In-plane free simply supported edge with elastic distributed springs tangent to the edge

$$N_n = 0, \quad N_{n\tau} = -k_s v_n, \quad w = 0, \quad M_n = 0, \tag{11.28}$$

where k_s is the spring stiffness per unit length.

For nonlinear vibrations of shells and plates (classical theories), the displacements can written as

$$u(x, y, t) = \sum_{i=1}^{N_1} u_i(t)\, U_i(x, y) + \hat{u}, \tag{11.29a}$$

$$v(x, y, t) = \sum_{i=1}^{N_2} v_i(t)\, V_i(x, y) + \hat{v}, \tag{11.29b}$$

$$w(x, y, t) = \sum_{i=1}^{N_3} w_i(t)\, W_i(x, y), \tag{11.29c}$$

where $u_i(t)$, $v_i(t)$ and $w_i(t)$ are the generalized coordinates; $U_i(x,y)$, $V_i(x,y)$ and $W_i(x,y)$ are the admissible functions and $\hat{u}$ and $\hat{v}$ are nonlinear terms in $w_i(t)$ that must be added to the expansions of u and v in order to satisfy exactly the boundary conditions (e.g. see Section 10.2.1).

The structure of admissible functions corresponding to clamped edges, equations (11.24), and in-plane immovable simply supported edges, equations (11.25), is

$$U_i(x, y) = \omega P_1, \quad V_i(x, y) = \omega P_2, \quad W_i(x, y) = \omega^k P_3, \tag{11.30a–c}$$

where $\omega = 0$ is the normalized equation of the domain boundary. The parameter k in equations (11.30) depends on the type of boundary condition: $k = 1$ for in-plane immovable simply supported edges and $k = 2$ for clamped edges.

The following structure of the solution satisfies the classical simply supported edges, equations (11.26):

$$U_i(x, y) = \frac{\partial \omega}{\partial x} P_2 + \omega P_3, \quad V_i(x, y) = \frac{\partial \omega}{\partial y} P_2 + \omega P_4, \quad W_i(x, y) = \omega P_1. \tag{11.31a–c}$$

If the edges of the shell base are parallel to the coordinates axes, equations (11.31a–c) can be rewritten in a simpler form. As a consequence that $\partial\omega/\partial x = 0$ on the sides parallel to the axis x, and that $\partial\omega/\partial y = 0$ on the sides parallel to y, equations (11.31a–c) take the simplified form:

$$U_i(x, y) = \omega_1 P_3, \quad V_i(x, y) = \omega_2 P_4, \quad W_i(x, y) = \omega P_1, \tag{11.32a–c}$$

where $\omega_i = 0$, for $i = 1, 2$, are the equations of the domain edges on which the boundary conditions $u = 0$ or $v = 0$, respectively, should be satisfied.

If the panel has in-plane free simply supported edges, equations (11.27), the structure of the solution satisfying the kinematic boundary conditions is

$$U_i(x,y) = P_1, \quad V_i(x,y) = P_2, \quad W_i(x,y) = \omega P_3. \tag{11.33a–c}$$

In the case of boundary conditions represented by equation (11.28), the structure of the solution satisfying all the boundary conditions is

$$U_i(x,y) = P_1(1 - m^2 C_{spr}\,\omega) - \omega[D_1 P_1 - lm(1+\nu)T_1 P_1 + (\nu l^2 - m^2)T_1 P_2 - lmC_{spr} P_2], \tag{11.34a}$$

$$V_i(x,y) = P_2(1 - l^2 C_{spr}\,\omega) - \omega[D_1 P_2 + lm(1+\nu)T_1 P_2 + (l^2 - \nu m^2)T_1 P_1 - lmC_{spr} P_1], \tag{11.34b}$$

$$W_i(x,y) = P_3\left(\omega + \frac{1}{3}\tilde{\omega}\right) + \omega\left(h_1\frac{\partial P_3}{\partial x} + h_2\frac{\partial P_3}{\partial y}\right), \tag{11.34c}$$

where D_1 and T_1 are obtained by equations (11.21), $C_{spr} = 2(1+\nu)k_s/E$, $\omega = \omega(x,y)$ as usual, $\tilde{\omega} = \omega(x+h_1, y+h_2)$ and

$$h_1 = -\omega\left[x + \frac{1}{2}\omega(x,y), y\right] + \omega\left[x - \frac{1}{2}\omega(x,y), y\right],$$

$$h_2 = -\omega\left[x, y + \frac{1}{2}\omega(x,y)\right] + \omega\left[x, y - \frac{1}{2}\omega(x,y)\right].$$

In equations (11.30–11.34) the functions P_i, $i = 1, \ldots, 4$, are the unknown functions that are represented by

$$P_1 = \sum_{i=1}^{M_1} a_i\varphi_i(x,y), \quad P_2 = \sum_{i=M_1+1}^{M_2} b_i\eta_i(x,y), \tag{11.35a,b}$$

$$P_3 = \sum_{i=M_2+1}^{M_3} c_i\psi_i(x,y), \quad P_4 = \sum_{i=M_3+1}^{M_4} d_i\sigma_i(x,y). \tag{11.35c,d}$$

In equations (11.35) a_i, b_i, c_i and d_i are the coefficients defined in equation (11.2b) as C_i, to be determined by solving the corresponding eigenvalue (linear) problem, and $\varphi_i(x,y)$, $\eta_i(x,y)$, $\psi_i(x,y)$, $\sigma_i(x,y)$ are elements of a complete function system $\{\varphi_i(x,y)\}$, $\{\eta_i(x,y)\}$, $\{\psi_i(x,y)\}$, $\{\sigma_i(x,y)\}$; for example, power polynomials, Chebyshev polynomials, Legendre polynomials, trigonometric polynomials and functions or finite functions (e.g. splines) can be used in equations (11.35).

11.3 Numerical Results for a Shallow-Shell with Complex Shape

A circular cylindrical panel with a complex base, shown in Figure 11.1, is investigated. The dimension and material properties of the panel are: overall length $b = 0.199$ m, curvilinear width $a = 0.132$ m, length of the cut $b_1 = 0.041$ m, curvilinear width at the cut $a_1 = 0.092$ m, radius of curvature $R_x = 2$ m, thickness $h = 0.00028$ m, Young's modulus $E = 195$ GPa, mass density $\rho = 7800$ kg/m^3 and Poisson's ratio $\nu = 0.3$.

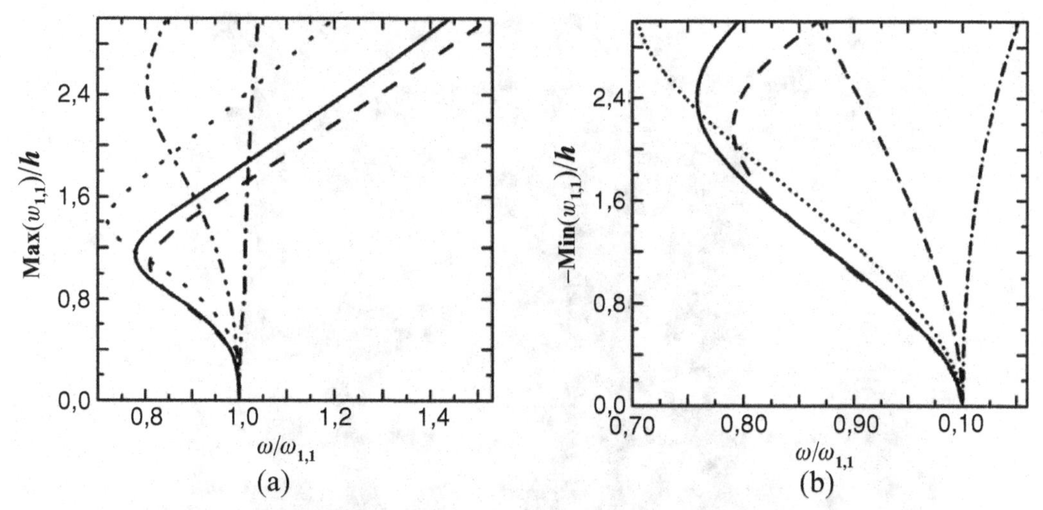

Figure 11.3. Effect of boundary conditions on amplitude-frequency curves for the panel with complex shape; RFM solution; —, clamped edges; – –, in-plane immovable simply supported edges; · · · ·, classical simply supported panel; – · – ·, in-plane free simply supported edges; – · · – · ·, in-plane free simply supported edges with additional distributed elastic tangential spring of stiffness per unit length $k_s = 75 \times 10^{11}$ N/m². (a) Maximum of the coordinate w in a vibration period. (b) Minus minimum of coordinate w.

In the present study, the power polynomials have been used in equations (11.35) to obtain numerical results. For the panel in Figure 11.1, assuming simply supported edges, because of the symmetry of the problem in x and y, the following polynomials have been used:

$$\{\varphi_i\} : 1, x^2, y^2, x^4, x^2y^2, y^4 \ldots, \tag{11.36a}$$

$$\{\eta_i\} : x, y, xy, x^3y, xy^3 \ldots, \tag{11.36b}$$

$$\{\psi_i\} : y, x^2y, y^3, x^4y, x^2y^3, y^5 \ldots, \tag{11.36c}$$

$$\{\sigma_i\} : x, x^3, xy^2, x^5, x^3y^2, xy^4 \ldots. \tag{11.36d}$$

In numerical calculations, globally 63 coefficients a_i, b_i, c_i and d_i have been used in the linear analysis. This number of terms corresponds to the eleventh power in the polynomials approximating u and v, and to the tenth power in the polynomial approximating w in equations (11.36a–d) and give convergence of the linear eigenfunctions and eigenvectors. In the nonlinear analysis, only mode (1, 1) has been considered and the model has been reduced to a single degree of freedom neglecting in-plane inertia; u and v are obtained as functions of w from the in-plane equilibrium equations. Obviously, this is a rough model, but it can be easily improved by using the same approach.

Figure 11.3 shows the backbone curve of the panel calculated in the present study by using the RFM for each of the five types of boundary conditions given in equations (11.24–11.28); the maximum, Figure 11.3(a), and the minimum, Figure 11.3(b), of the response at the center of the panel, in radial direction, are shown. It is interesting to note that, in the case of in-plane free simply supported edges of the panel, results indicate a very weak hardening-type nonlinearity. The backbone curves for the other types of boundary conditions indicate a softening-type nonlinearity.

(a)

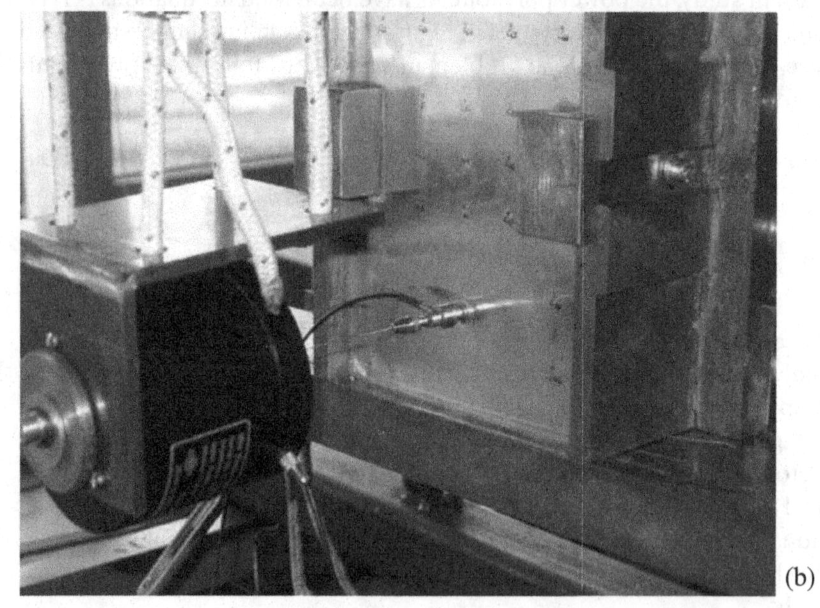
(b)

Figure 11.4. Experimental setup. (a) Curved panel with complex base. (b) Panel excited by the shaker at $\tilde{x} = 40$ mm and $\tilde{y} = a/2$.

11.4 Experimental Results and Comparison

Tests have been conducted on an aluminium circular cylindrical panel with the same geometry (but different material) used for calculation in Section 11.3: $b = 0.199$ m, $a = 0.132$ m, $b_1 = 0.041$ m, $a_1 = 0.092$ m, $R_x = 2$ m, $h = 0.00028$ m, $E = 70$ GPa,

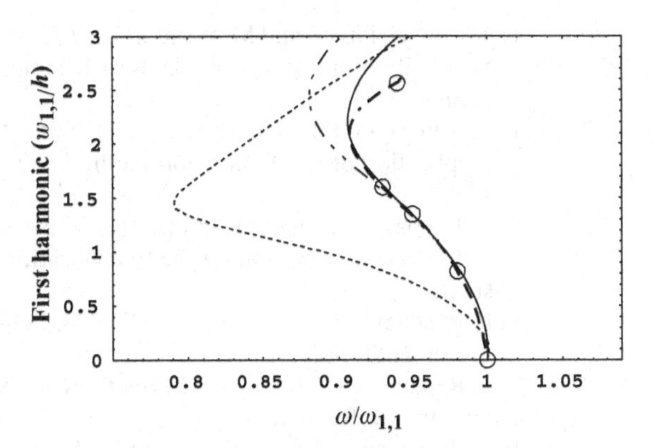

Figure 11.5. Comparison of numerical (for different stiffness values k_s) and experimental amplitude-frequency curves (first harmonic) of the panel with complex shape; fundamental mode (1,1). ○, Experimental value; – · –, line interpolating the experimental data; – –, simply supported panel ($k_s \to \infty$); –·–, $k_s = 197 \times 10^{11}$ N/m^2; —, $k_s = 52.6 \times 10^{11}$ N/m^2.

$\rho = 2700$ kg/m^3, $\nu = 0.33$. The panel was inserted into a heavy rectangular steel frame (see Figure 11.4), having V-grooves designed to hold the panel and to avoid transverse (radial) displacements at the edges; silicon was placed into the grooves to fill any gap between the panel and the grooves. In-plane displacements normal to the edges were allowed on the convex part of the domain because the constraint given by silicon on these displacements was very small. This may be not true for the concave part of the domain; in-plane displacements parallel to the edges were elastically constrained by the silicon. Therefore, the experimental boundary conditions are close to those given for in-plane free, simply supported boundary conditions with distributed springs parallel to the edges, as expressed by equations (11.28), at least for the convex part of the panel. For the two concave parts, the same boundary conditions are imposed on the model, but the actual constraints in the tested plate are probably more complex.

The experimental backbone curve is plotted in Figure 11.5 versus the theoretical calculations for aluminium panels with simply supported edges and in-plane free simply supported edges with additional elastic tangential spring of stiffness k_s. The theoretical first-harmonic component is plotted, which is approximately the average value between the oscillation amplitude outside and inside the shell. In particular, the theoretical curve for $k_s = 52.6 \times 10^{11}$ N/m^2 is extremely close to the experimental results, giving both a qualitative and quantitative validation of the theoretical approach. However, the natural frequency computed for $k_s = 52.6 \times 10^{11}$ N/m^2 is significantly larger than the measured one (156 Hz). But for $k_s = 0.63 \times 10^{11}$ N/m^2, a computed natural frequency of 160 Hz is obtained, which is extremely close to the experimental value. The indetermination of the experimental boundary conditions as well as a significant effect of geometric imperfections on both natural frequency and the trend of nonlinearity are the probable reason for differences between numerical and experimental results; geometric imperfections are not taken into account in the present study.

REFERENCES

L. V. Kantorovich and V. I. Krylov 1958 *Approximate Methods of Higher Analysis*. Interscience Publisher, New York, USA.

L. V. Kurpa, K. I. Lyubitska and A. V. Shmatko 2005a *Journal of Sound and Vibration* **279**, 1071–1084. Solution of vibration problems for shallow shells of arbitrary form by the R-function method.

L. Kurpa, G. Pilgun and M. Amabili 2007 *Journal of Sound and Vibration* **306**, 580–600. Nonlinear vibrations of shallow shells with complex boundary: R-functions method and experiments.

L. V. Kurpa, G. Pilgun and E. Ventsel 2005b *Journal of Sound and Vibration* **284**, 379–392. Application of the R-function method to nonlinear vibrations of thin plates of arbitrary shape.

L. V. Kurpa, V. L. Rvachev and E. Ventsel 2003 *Journal of Sound and Vibration* **261**, 109–122. The R-function method for the free vibration analysis of thin orthotropic plates of arbitrary shape.

V. L. Rvachev 1963 *Doklady AS USSR* **153**, 765–768. Analytical description of some geometric objects (in Russian).

V. L. Rvachev 1982 *Theory of R-Functions and Some of Its Applications*. Nauka Dumka, Kiev, Ukraine (in Russian).

V. L. Rvachev and L. V. Kurpa 1987 *The R-functions in Problems of Plate Theory*. Nauka Dumka, Kiev, Ukraine (in Russian).

V. L. Rvachev and L. V. Kurpa 1998 *Problems of Machine-Building* **1**, 33–53. The application of the R-functions theory for the investigation of plates and shells of arbitrary shape (in Russian).

V. L. Rvachev and T. I. Sheiko 1995 *Applied Mechanics Reviews* **48**, 151–188. R-functions in boundary value problems in mechanics.

V. L. Rvachev, T. I. Sheiko, V. Shapiro and I. Tsukanov 2000 *Computational Mechanics* **25**, 305–316. On completeness of RFM solution structures.

V. L. Rvachev and A. N. Shevchenko 1988 *Problem-Oriented Languages and Systems for Engineering Approaches*. Tekhnika, Kiev, Ukraine (in Russian).

12 Vibrations of Circular Plates and Rotating Disks

12.1 Introduction

Rotating disks are the principal components of many engineering systems, such as hard-disk drives for computers (see Figure 12.1), optical memory disks (CD and DVD), circular saws and turbines. They can be conveniently modeled as circular and annular plates by using the nonlinear equations developed in Section 1.4 of this book. When a spinning disk is subjected to a transverse load, as in the case of a reading/writing head flying over a memory disk, the disk can experience a critical speed resonance with catastrophically large vibration amplitudes whenever the rotating speed is close to a certain multiple of a natural frequency of the disk. This explains the reason for the choice of operating disks at subcritical speeds. However, the necessity of faster data access rates in disk drives is increasing the speed of such devices; this justifies the nonlinear study developed here.

In this chapter, the linear vibrations of fixed circular and annular plates is addressed first. Then, the nonlinear vibrations of free-edge circular plates is studied by using the von Kármán equations of motion for circular plates. Finally, the problem of rotating clamped-free disks is solved.

12.1.1 Literature Review

Linear vibrations of circular and annular plates are discussed in the book by Leissa (1969).

Tobias (1957) and Williams and Tobias (1963) were among the first to study nonlinear undamped vibrations of imperfect and perfect circular plates. Wah (1963) studied large-amplitude vibrations of circular plates by using the Berger equations, which are based on a simplification of the von Kármán theory; this approximation can lead to inaccurate results for some boundary conditions. Nonlinear vibrations of imperfect circular plates were also investigated by Efstathiades (1971). An excellent agreement between theoretical and experimental results for nonlinear vibrations of a clamped circular plate were obtained by Kung and Pao (1972). The effect of elastic constraints was addressed by Ramachandran (1974). Perfectly axisymmetric nonlinear circular plates were studied by Sridhar et al. (1975; 1978). A very interesting theoretical and experimental study on nonlinear vibrations of clamped circular

Figure 12.1. Hard-disk drive for computer. Luis Alonso Ocana/age fotostock.

plates with geometric imperfections was developed by Yamaki et al. (1981a,b). Circular plates on elastic foundation were analyzed by Dumir (1986). Subharmonic forced excitation of nonlinear circular plates were studied by Nayfeh and Vakakis (1994).

Touzé et al. (2002) studied nonlinear vibrations of free-edge circular plates by using a perturbation approach; perfect and imperfect plates were considered. Experimental results for this case were obtained by Thomas et al. (2003).

Tobias and Arnold (1957) were the first to discuss the effect of imperfections on the vibrations of spinning circular plates; they also performed experiments. Von Kármán-type nonlinear plate equations for axisymmetric rotating circular plates were derived by Nowinski (1964). The nonlinear vibrations of spinning disks near critical speed were also studied by Dugdale (1979). The seminal studies on nonlinear vibrations of perfect and imperfect spinning disks near critical speed are due to Raman and Mote (1999; 2001a,b).

The case of a disk spinning in air, giving rise to nonlinear aeroelastic flutter, was thoroughly investigated by Jana and Raman (2005). Flutter is an instability phenomenon; it arises when the equilibrium solution presents an Hopf bifurcation. D'Angelo and Mote (1993) obtained direct evidence that the flutter of spinning disks in air is due to the instability of a single reflected traveling wave. At preflutter speed, they observed the characteristic behavior of spinning disks: a backward traveling-wave frequency vanished at the disk critical speed, and the backward traveling wave became a single reflected traveling wave with increasing frequency at supercritical speeds. The vibration amplitude remained small at preflutter speeds. However, beyond the flutter speed, the amplitude of the unstable single reflected traveling wave grew rapidly.

12.2 Linear Vibrations of Circular and Annular Plates

An isotropic annular plate of constant thickness h, inner radius a_1 and outer radius a_2 is considered. In the case of linear vibrations, the equations of motion (1.181) and (1.182) are reduced to

$$D\nabla^4 w + \rho h \frac{\partial^2 w}{\partial t^2} = 0, \tag{12.1}$$

where the biharmonic operator in polar coordinates is $\nabla^4 = [\partial^2/\partial r^2 + (1/r)(\partial/\partial r) + \partial^2/(r^2\,\partial\theta^2)]^2$. In the case of axisymmetric boundary conditions, by using the separation of variables, the solution takes the following form:

$$w(r,\theta,t) = \sum_{m=0}^{\infty}\sum_{n=0}^{\infty} W_{m,n}(r)\cos(n\theta)e^{i\,\omega_{m,n}t}, \tag{12.2}$$

where

$$W_{m,n}(r) = c_{m,n}\mathrm{J}_n\left(\frac{\lambda_{m,n}\,r}{a_2}\right) + d_{m,n}\mathrm{Y}_n\left(\frac{\lambda_{m,n}\,r}{a_2}\right) + e_{m,n}\mathrm{I}_n\left(\frac{\lambda_{m,n}\,r}{a_2}\right) + g_{m,n}\,\mathrm{K}_n\left(\frac{\lambda_{m,n}\,r}{a_2}\right), \tag{12.3}$$

in which n and m represent the number of nodal diameters and circles, respectively; $c_{m,n}$, $d_{m,n}$, $e_{m,n}$ and $g_{m,n}$ are the mode shape coefficients, which are determined by the boundary conditions; J_n and Y_n are the Bessel functions of first and second kind, I_n and K_n are the modified Bessel functions of first and second kind, respectively, and $\lambda_{m,n}$ is the frequency parameter, which is also determined by the boundary conditions. As in Chapter 5 for circular cylindrical shells, for any asymmetric mode shape (i.e. for $n > 0$), there is a second mode, similar to the previous one, but orthogonal to it, with an angular shift of $\pi/2n$ (i.e. exchanging cosine and sine functions in the angular coordinate θ). The frequency parameter $\lambda_{m,n}$ is related to the circular natural frequency $\omega_{m,n}$ by

$$\omega_{m,n} = \frac{\lambda_{m,n}^2}{a_2^2}\sqrt{\frac{D}{\rho h}}. \tag{12.4}$$

In the case of circular plates, there is no central hole and the coefficients $d_{m,n}$ and $g_{m,n}$ are zero.

Table 12.1 gives the frequency parameters for annular plates for different ratios $a = a_2/a_1$ in the case of the clamped inner edge

$$w = \partial w/\partial r = 0 \qquad \text{at } r = a_1, \tag{12.5}$$

and the free outer edge

$$M_r = -D\left(\frac{\partial^2 w}{\partial r^2} + \frac{\nu}{r}\frac{\partial w}{\partial r} + \frac{\nu}{r^2}\frac{\partial^2 w}{\partial \theta^2}\right) = 0 \quad \text{at } r = a_2, \tag{12.6a}$$

$$Q_r + \frac{1}{r}\frac{\partial M_{r,\theta}}{\partial \theta} = -D\left[\frac{\partial}{\partial r}\nabla^2 w + \frac{1-\nu}{r^2}\frac{\partial^2}{\partial \theta^2}\left(\frac{\partial w}{\partial r} - \frac{w}{r}\right)\right] = 0 \quad \text{at } r = a_2, \tag{12.6b}$$

where M_r and $M_{r\theta}$ are the bending and twisting moments and Q_r is the transverse shear force. In-plane boundary conditions are not applied and have no effect on

Table 12.1. *Frequency parameters λ_{mn} for annular plates clamped at the inner edge and free at the outer edge for $\nu = 0.3$*

n	m	$a = 0.1$	$a = 0.3$	$a = 0.5$	$a = 0.7$	$a = 0.9$
0	0	2.0583	2.5806	3.6088	6.0787	18.557
0	1	1.8648	2.5596	3.6454	6.1234	18.578
0	2	2.3711	2.8206	3.8344	6.2670	18.640
1	0	5.0260	6.5278	9.2212	15.486	46.784
1	1	5.2603	6.6805	9.3114	15.533	46.798
1	2	6.0777	7.1376	9.5779	15.671	46.839
2	0	8.5965	11.111	15.610	26.089	78.457
2	1	8.8026	11.217	15.666	26.116	78.464
2	2	9.4963	11.536	15.832	26.196	78.489

natural vibrations by using classical plate linear theory. Disks in the hard-disk drives of computers can be modeled as clamped-free annular plates. Figure 12.2 shows the mode shape ($m = 0, n = 4$) of a clamped-free annular plate.

12.3 Nonlinear Vibrations of Circular Plates

The equations of motion (1.181) and (1.182) must be satisfied with the appropriate boundary conditions. In particular, free-edge circular plates are considered here ($a_1 = 0$). The boundary conditions are given by equations (12.6a,b), by a non-divergence condition at the center and by the following in-plane boundary conditions, which play a fundamental role in the nonlinear theory:

$$N_r = \frac{1}{r}\frac{\partial F}{\partial r} + \frac{1}{r^2}\frac{\partial^2 F}{\partial \theta^2} = 0 \quad \text{at } r = a_2, \tag{12.7a}$$

$$N_{r,\theta} = \frac{1}{r^2}\frac{\partial F}{\partial \theta} - \frac{1}{r}\frac{\partial^2 F}{\partial r\, \partial \theta} = 0 \quad \text{at } r = a_2, \tag{12.7b}$$

where F is the in-plane potential stress function.

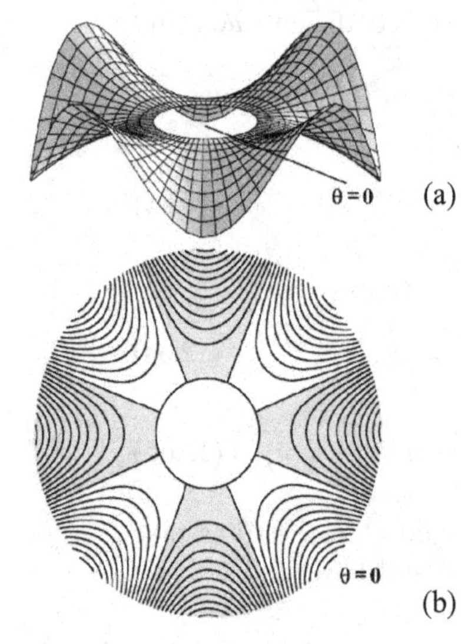

Figure 12.2. Mode shape ($m = 0$, $n = 4$) of a clamped-free annular plate; $a = 0.3$, $\nu = 0.27$. (a) Three-dimensional representation. (b) Contour plot. From Raman and Mote (2001a). Courtesy of Dr. A. Raman.

The displacement w is expanded by using the base of linear modes

$$w(r,\theta,t) = \sum_{m=0}^{M}\sum_{n=0}^{N}\left[A_{m,n}(t)\cos(n\theta) + B_{m,n}(t)\sin(n\theta)\right] W_{m,n}(r), \qquad (12.8)$$

where $A_{m,n}(t)$ and $B_{m,n}(t)$ are the generalized coordinates and $W_{m,n}(r)$ is defined in equation (12.3) with $d_{m,n} = g_{m,n} = 0$. Equation (12.8) can be written in the compact form:

$$w(r,\theta,t) = \sum_{i=1}^{\overline{N}} q_i(t)\, w_i(r,\theta), \qquad (12.9)$$

where $q_i(t)$ are the generalized coordinates and $\overline{N}$ is the number of degrees of freedom retained in the expansion (12.8). By introducing the functional

$$L(w,F) = \left[\frac{\partial^2 w}{\partial r^2}\left(\frac{\partial F}{r\,\partial r} + \frac{\partial^2 F}{r^2\,\partial\theta^2}\right) + \left(\frac{\partial w}{r\,\partial r} + \frac{\partial^2 w}{r^2\,\partial\theta^2}\right)\frac{\partial^2 F}{\partial r^2} - 2\left(\frac{\partial^2 w}{r\,\partial r\,\partial\theta} - \frac{\partial w}{r^2\,\partial\theta}\right)\left(\frac{\partial^2 F}{r\,\partial r\,\partial\theta} - \frac{\partial F}{r^2\,\partial\theta}\right)\right], \qquad (12.10)$$

equation (1.182) can be written as

$$\nabla^4 F = -\frac{Eh}{2} L(w,w). \qquad (12.11)$$

Substituting equation (12.9) into (12.11) leads to

$$\nabla^4 F = \sum_{i=1}^{\overline{N}}\sum_{j=1}^{\overline{N}} q_i(t) q_j(t)\, E_{i,j}(r,\theta), \qquad (12.12)$$

where

$$E_{i,j}(r,\theta) = -\frac{Eh}{2} L(w_i, w_j). \qquad (12.13)$$

The solution of equation (12.12) is given as $F = F_h + F_p$, where F_h is the homogeneous solution, which is zero in the present case for free-edge boundary conditions (in fact, zero in-plane forces are imposed at the edges), and F_p is the particular solution. Similarly to equation (12.9), the expansion of the particular solution can be written as (Touzé et al. 2002):

$$F_p(r,\theta,t) = \sum_{k=1}^{\overline{M}} \eta_k(t)\, \phi_k(r,\theta), \qquad (12.14)$$

so that

$$\nabla^4 F_p = \sum_{k=1}^{\overline{M}} \eta_k(t)\, \nabla^4 \phi_k(r,\theta), \qquad (12.15)$$

where $\eta_k(t)$ are unknown time functions and ϕ_k are chosen in order to satisfy the following equation

$$\nabla^4 \phi_k = \xi_k^4 \phi_k, \qquad (12.16)$$

where ξ_k is a real number. Boundary conditions (12.7a,b) are rewritten in terms of ϕ_k by using equation (12.14)

$$\frac{\partial \phi_k}{\partial r} + \frac{1}{r}\frac{\partial^2 \phi_k}{\partial \theta^2} = 0, \quad \frac{1}{r}\frac{\partial \phi_k}{\partial \theta} - \frac{\partial^2 \phi_k}{\partial r \partial \theta} = 0, \quad \text{at} \quad r = a_2. \tag{12.17}$$

For axisymmetric boundary conditions, it is possible to use the separation of the variables r and θ to represent ϕ_k in terms of $\varphi_i(r)\,[\cos(j\theta) + \sin(j\theta)]$. Substituting this expression in equations (12.17) leads to

$$\frac{\partial \varphi_i}{\partial r} - \frac{j^2}{r}\varphi_i = 0, \quad \frac{\partial \varphi_i}{\partial r} - \frac{1}{r}\varphi_i = 0, \quad \text{at}\, r = a_2, \tag{12.18a}$$

which, for $j \neq 1$ as in the present case, is equivalent to

$$\phi_k = 0, \quad \frac{\partial \phi_k}{\partial r} = 0, \quad \text{at}\, r = a_2. \tag{12.18b}$$

The problem given by equation (12.16) with boundary conditions (12.18b) is equivalent to equation (12.1) with clamped boundary conditions (12.5).

By substituting equation (12.15) into (12.12), considering equation (12.16), multiplying the result by ϕ_l and integrating over the plate surface S, the following result is obtained by taking into account the orthogonality of the eigenmodes ϕ_l:

$$\eta_k(t) = \sum_{i=1}^{\overline{N}} \sum_{j=1}^{\overline{N}} G_{i,j,k} q_i(t)\, q_j(t), \tag{12.19}$$

where

$$G_{i,j,k} = \frac{\int_S E_{i,j} \phi_k \,\mathrm{d}S}{\xi_k^4 \int_S \phi_k^2 \,\mathrm{d}S}. \tag{12.20}$$

Finally, the particular solution is given by

$$F_p(r, \theta, t) = \sum_{k=1}^{\overline{M}} \left(\sum_{i=1}^{\overline{N}} \sum_{j=1}^{\overline{N}} G_{i,j,k} q_i(t)\, q_j(t) \right) \phi_k(r, \theta). \tag{12.21}$$

It can be observed that the non-zero terms in equation (12.21) correspond to axisymmetric and asymmetric shapes with $2n$ nodal diameters. Substituting equation (12.21) into (1.181) and applying the Galerkin method (i.e. multiplying by w_l and integrating over the plate surface S), the discretized equations of motion take the following form:

$$\ddot{q}_l(t) + 2\zeta_l\, \omega_l\, \dot{q}_l(t) + \omega_l^2 q_l(t) = \tilde{f}_l + \sum_{i=1}^{\overline{N}} \sum_{j=1}^{\overline{N}} \sum_{p=1}^{\overline{N}} \Gamma_{i,j,l,p}\, q_i(t)\, q_j(t)\, q_p(t), \quad \text{for}\ l = 1, \ldots, \overline{N}, \tag{12.22}$$

where

$$\omega_l^2 = \frac{D \int_S \nabla^4 w_l w_l \,\mathrm{d}S}{\rho h \int_S w_l^2 \,\mathrm{d}S}, \tag{12.23}$$

$$\tilde{f}_l = \frac{\int_S (f + p) w_l \,\mathrm{d}S}{\rho h \int_S w_l^2 \,\mathrm{d}S}, \tag{12.24}$$

$$\Gamma_{i,j,l,p} = -\frac{E}{2\rho}\sum_{k=1}^{\overline{M}} \frac{\int_S L(w_i, w_j)\phi_k \, \mathrm{d}S \int_S L(w_p, \phi_k) w_l \, \mathrm{d}S}{\xi_k^4 \int_S \phi_k^2 \, \mathrm{d}S \int_S w_l^2 \, \mathrm{d}S}. \tag{12.25}$$

Here, it is assumed to have zero aerodynamic pressure $p = 0$ and external harmonic point force excitation $f = \tilde{f}\,\delta(R\theta - R\tilde{\theta})\,\delta(r - \tilde{r})\cos(\omega t)$ applied at $(\tilde{r}, \tilde{\theta})$.

12.3.1 Numerical Results

A simplified analysis is developed by using only 2 degrees of freedom (dofs) in equation (12.8), $A_{1,n}(t)$ and $B_{1,n}(t)$, for a given mode with n nodal diameters. The equations of motion take the following nondimensional form (Touzé et al. 2002):

$$\ddot{q}_1 + \Omega_1^2 q_1 = \left[12(1-\nu^2)h^2/a_2^2\right]\left[\Gamma_1 q_1^3 + C_1 q_1 q_2^2 - 2\mu\,\dot{q}_1 + Q_1\cos(\Omega_n t_n)\right], \tag{12.26a}$$

$$\ddot{q}_2 + \Omega_2^2 q_2 = \left[12(1-\nu^2)h^2/a_2^2\right]\left[\Gamma_2 q_2^3 + C_2 q_2 q_1^2 - 2\mu\,\dot{q}_2\right], \tag{12.26b}$$

where $q_1(t) = A_{1,n}(t)a_2/h^2$, $q_2(t) = B_{1,n}(t)a_2/h^2$, $t_n = t/(a_2^2\sqrt{\rho h/D})$ is the nondimensional time and the dot denotes the derivative with respect to nondimensional time; $\Gamma_1, \Gamma_2, C_1, C_2, Q_1$ and μ are appropriate coefficients; Ω_1 and Ω_2 are the dimensionless natural frequencies and Ω_n is the dimensionless excitation frequency. In the case of a plate without geometric imperfections, $\Gamma_1 = \Gamma_2 = C_1 = C_2$ and $\Omega_1 = \Omega_2$.

The numerical results presented here were obtained by Touzé et al. (2002) by using a perturbation approach instead of a numerical solution of the equations of motion, as done in previous chapters.

Figure 12.3 shows the response to harmonic excitation of a free-edge circular plate with small imperfection modeled only as difference between Ω_1 and Ω_2; in particular, $\Omega_2 = \Omega_1 + 0.05$ with $\Omega_1 = 2\pi$. The amplitude of vibration, normalized with respect to h^2/a_2, is represented versus the detuning parameter σ which is related to the nondimensional excitation frequency by $\Omega_n = \Omega_1 + \varepsilon\,\sigma$, where $\varepsilon = 12(1-\nu^2)h^2/a_2^2$ is the small perturbation parameter, which is taken equal to 0.1 in Figure 12.3. The detuning parameter measures the closeness of the excitation frequency to the linear frequency of cosine mode (m, n). Both the vibration amplitudes A_1 and A_2 of the first harmonic of q_1 and q_2 (i.e. cosine and sine modes) are plotted in Figure 12.3(a). It is shown that branch 1 corresponds to a single-mode vibration, which shows a hardening-type nonlinearity that was previously observed in Chapter 4 for rectangular plates. However, because of the one-to-one internal resonance between cosine and sine modes, even in the presence of a small imperfection that breaks the perfect axisymmetry of the problem, branch 1 solution bifurcates, and branch 2 with a participation of both cosine and sine modes appears. The phases γ_1 and γ_2 of the first harmonic of q_1 and q_2 are plotted in Figure 12.3(b). Coupled-mode solution, branch 2, exhibits a phase difference between γ_1 and γ_2 nearly equal to $\pi/2$, giving rise to a wave traveling around the plate.

12.4 Nonlinear Vibrations of Disks Spinning Near a Critical Speed

The nonlinear model used here is for large-amplitude transverse vibrations of clamped-free disks under a space-fixed point force, rotating near a critical speed. The model falls in the category of the nongyroscopic systems because the equations of motion are nongyroscopic in the corotating coordinates (that is, reference system rotating with the disk).

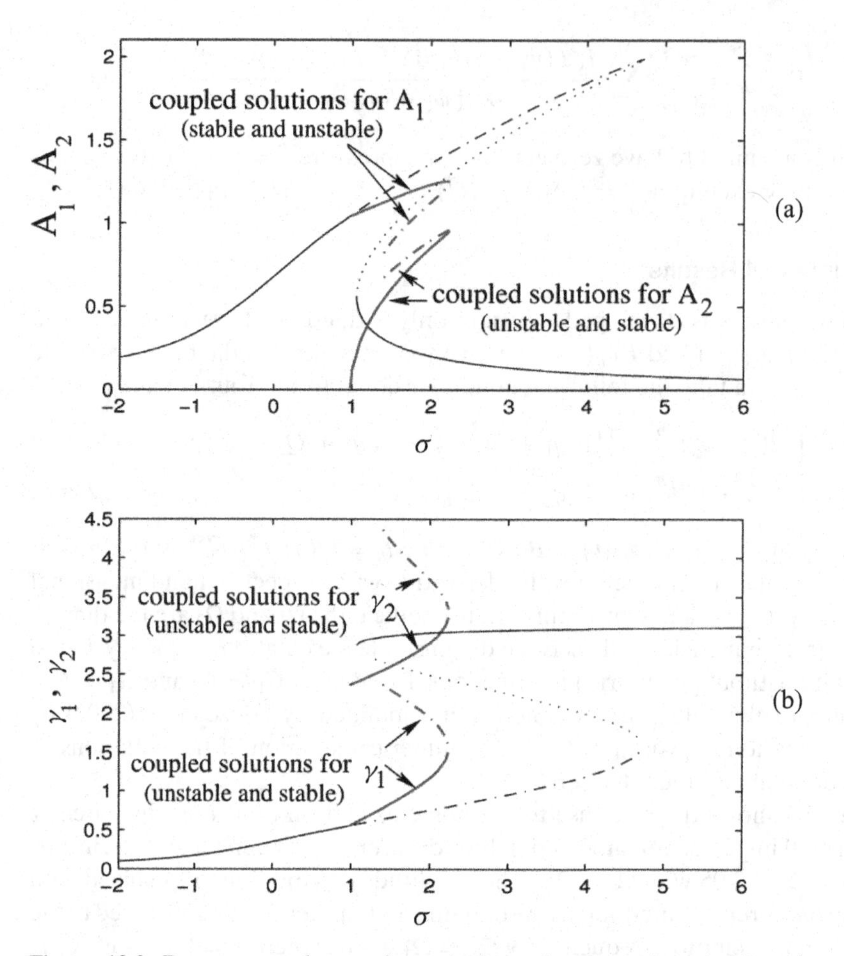

Figure 12.3. Response to harmonic excitation of a free-edge circular plate with small imperfection; $\Omega_1 = 2\pi$, $\Omega_2 = \Omega_1 + 0.05$, $\varepsilon = 0.1$, $\mu = 2$ and $Q_1 = 10$; $\Gamma_1 = \Gamma_2 = C_1 = C_2 = -5$. (a) Amplitude of the first harmonic of the response versus the detuning parameter. (b) Phase of the first harmonic of the response versus the detuning parameter. From Touzé et al. (2002). Courtesy of Dr. C. Touzé.

It must be observed that the flexural natural frequencies $\omega_{m,n}$ of the disk increase with the rotation speed Ω, so that they are indicated as $\omega_{m,n}(\Omega)$; mode shapes also change with Ω. The critical speed of mode (m, n) is the rotation speed Ω that gives $\omega_{m,n}(\Omega) = n\Omega$. At this speed, a lateral excitation f given to the disk causes unbounded response in the absence of damping; this is the critical speed resonance.

By introducing the in-plane stress function F, the von Kármán equation of motion for spinning circular and annular plates at speed Ω in the corotating reference system are

$$D\nabla^4 w + ch\dot{w} + \rho h\ddot{w} = f + p + L(w, F) - \rho h\Omega^2 r\left(\frac{1}{2} r\nabla^2 w + \frac{\partial w}{\partial r}\right), \quad (12.27)$$

which makes a pair with the compatibility equation

$$\nabla^4 F = -\frac{1}{2} Eh\, L(w, w) - 2\rho h(1 - \nu)\Omega^2. \quad (12.28)$$

Equations (12.27) and (12.28) reduce to equations (1.181) and (1.182) for $\Omega = 0$. The function F is obtained as $F = F_h + F_p + F_\Omega$, where F_h and F_p are obtained for $\Omega = 0$ and F_Ω is the particular solution of the equation

$$\nabla^4 F_\Omega = -2\rho h(1-\nu)\Omega^2. \tag{12.29}$$

The boundary conditions for a clamped-free disk are given by equations (12.5–12.7) and by

$$u_r = \frac{1}{Eh}\left[\frac{\partial^2 F}{\partial r^2} - \nu\left(\frac{1}{r}\frac{\partial F}{\partial r} + \frac{1}{r^2}\frac{\partial^2 F}{\partial \theta^2}\right)\right] = 0, \quad \text{at } r = a_1, \tag{12.30a}$$

$$u_\theta = \frac{1}{Eh}\left[\frac{\partial^3 F}{\partial r^3} + \frac{1}{r}\frac{\partial^2 F}{\partial r^2} - \frac{1}{r^2}\frac{\partial F}{\partial r} + \frac{2+\nu}{r^2}\frac{\partial^2 F}{\partial r\,\partial \theta} - \frac{3+\nu}{r^3}\frac{\partial^2 F}{\partial \theta^2}\right] = 0, \quad \text{at } r = a_1, \tag{12.30b}$$

where u_r and u_θ are the in-plane radial and circumferential displacements, respectively.

The external transverse load of constant amplitude, which is fixed in a fixed reference system, is a rotating point force in the corotating reference system. It is given by

$$f = \tilde{f}\,\delta(R\theta - R\Omega t)\,\delta(r - \tilde{r}), \tag{12.31}$$

where $\tilde{r}$ is the radius of force application; $a_1 \le \tilde{r} \le a_2$.

The displacement w is expanded by using equation (12.8). Following Raman and Mote (2001a), it is assumed that at the critical speed resonance of mode (m, n) the contribution of modes with different radial and circumferential wavenumber is negligible, therefore excluding internal resonances different form the one-to-one between sine and cosine modes. With the further assumption of harmonic oscillations,

$$A_{m,n}(t) = A_1 \sin(n\Omega t + \phi_1), \tag{12.32a}$$

$$B_{m,n}(t) = A_1 \sin(n\Omega t + \phi_2), \tag{12.32b}$$

equation (12.8) can be simplified into

$$w(r,\theta,t) = W(r)\left[A_1 \sin(n\Omega t + \phi_1)\cos(n\theta) + A_2 \sin(n\Omega t + \phi_2)\sin(n\theta)\right]. \tag{12.33a}$$

By using a trigonometric transformation, equation (12.33a) can be simplified into

$$w(r,\theta,t) = W(r)\left[B_1 \sin(n\Omega t + n\theta + \psi_1) + B_2 \sin(n\Omega t - n\theta + \psi_2)\right], \tag{12.33b}$$

where B_1, ψ_1 and B_2, ψ_2 are the amplitudes and phases of the forward and backward traveling waves, respectively. In particular,

$$B_1 = \sqrt{A_1^2 + A_2^2 - 2A_1A_2\sin(\phi_1 - \phi_2)}/2, \tag{12.34a}$$

$$B_2 = \sqrt{A_1^2 + A_2^2 + 2A_1A_2\sin(\phi_1 - \phi_2)}/2, \tag{12.34b}$$

$$\tan\psi_1 = \frac{A_1 \sin\phi_1 - A_2 \sin\phi_2}{A_1 \cos\phi_1 + A_2 \sin\phi_2}, \tag{12.35a}$$

$$\tan\psi_2 = \frac{A_1 \sin\phi_1 + A_2 \sin\phi_2}{A_1 \cos\phi_1 - A_2 \sin\phi_2}. \tag{12.35b}$$

The solution (12.33b) of the discretized equations of motion gives traveling waves in the same sense as the rotation of the disk, referred to as forward-traveling waves (FTW), and in the opposite sense, named backward-traveling waves (BTW). As a consequence of the corotating reference system, the point force is rotating; a forward-traveling wave is rotating in the sense opposite to the point force; a backward-traveling wave is rotating in the same sense as the point force.

The amplitudes B_1 and B_2 of the forward-traveling wave (FTW) and backward-traveling wave (BTW), respectively, are expressed in the corotating reference system in equations (12.33b) and (12.34). The transformation to the ground-fixed reference system can be easily obtained. A ground-fixed angular coordinate φ is introduced, so that

$$\varphi = \theta - \Omega t, \tag{12.36}$$

noting that $\varphi = 0$ is the location of the space-fixed force given by equation (12.31).

By using the transformation (12.36), equation (12.33b) can be rewritten as

$$\hat{w}(r, \varphi, t) = W(r)\,[B_1 \sin(2n\Omega t + n\varphi + \psi_1) + B_2 \sin(-n\varphi + \psi_2)], \tag{12.37}$$

where $\hat{w}$ is the flexural displacement in the fixed reference. Equation (12.37) presents a BTW response at zero rad/s (standing deformation) of amplitude B_2 and a FTW response at $2n\Omega$ rad/s of amplitude B_1.

12.4.1 Numerical Results

A linear analysis, performed eliminating all nonlinear terms from the equations of motion, concludes that the only steady-state solution is a backward-traveling wave with equal amplitude of sine and cosine modes, whose amplitude is unbounded in the absence of damping exactly at the critical speed. Viewed from a fixed reference, this backward traveling wave appears as a standing deformation at critical speed (Raman and Mote 1999).

The numerical results presented here were obtained by Raman and Mote (1999). They are relative to the nonlinear case and were obtained by using a perturbation approach. The detuning parameter σ is defined in this case as $n\Omega/\omega_{m,n}(\Omega) = 1 + \varepsilon\sigma$, where $n\Omega$ is the excitation frequency and $\varepsilon = 12(1 - \nu^2)h^2/a_2^2$ as in the previous section.

Figure 12.4 shows the steady-state solutions for backward- and forward-traveling waves. In addition, unstable mode-localized backward-traveling waves (MBTW) are also given; these are amplitude-modulated waves, with sometimes strongly unequal amplitudes of sine and cosine modes. Figure 12.4(a) gives a three-dimensional representation of the amplitudes A_1 and A_2 in equation (12.33a) versus the detuning parameter σ; BTW is contained in the plane $A_1 = A_2$. The amplitude A_1 versus σ is shown in Figure 12.4(b) and clearly displays an hardening behavior for the BTW. The FTW, which is only obtained by nonlinear analysis, appears to a fixed observer as a wave rotating at an angular speed $n\Omega + \omega_{m,n}(\Omega)$.

Figure 12.5 shows the BTW, which is the only steady-state solutions in presence of damping. It displays a typical hardening behavior and can be easily compared with the solid-line curve in Figure 12.4(a).

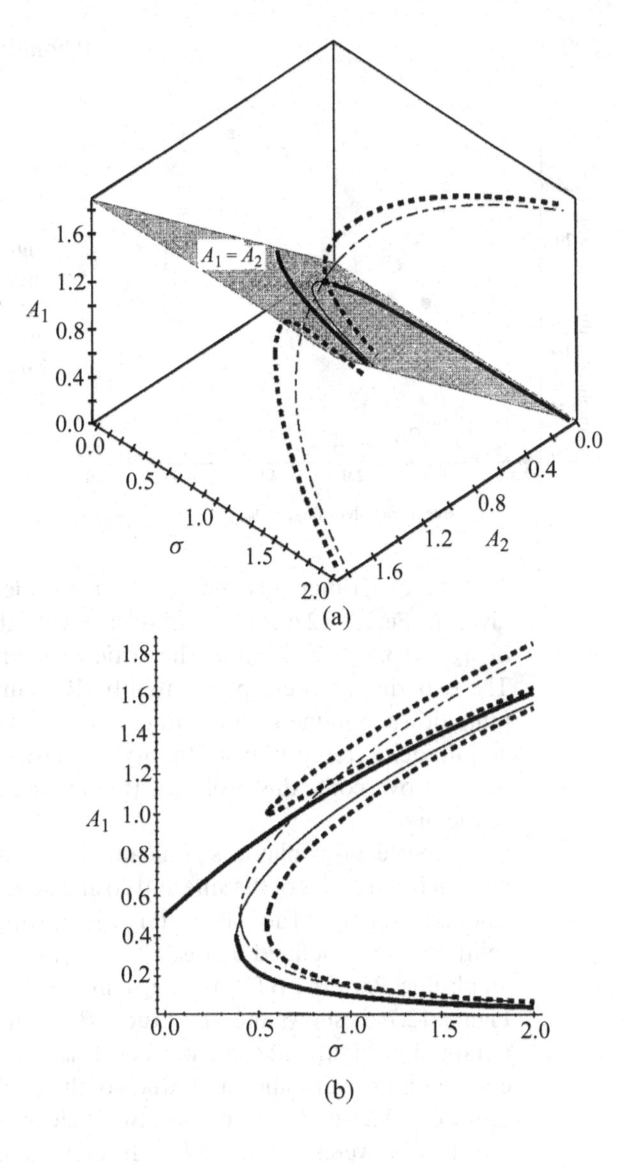

Figure 12.4. Steady-state solutions for backward, forward and mode-localized traveling waves of a spinning disk subjected to lateral point force. Bold curves are stable solutions and the rest are unstable solutions. Solid lines, BTW; dashed-dotted, MBTW; dotted, FTW. (a) Three-dimensional representation of the amplitudes A_1 and A_2 of the first harmonic of cosine and sine modes versus the detuning parameter σ. (b) Amplitude A_1 versus σ. From Raman and Mote (1999). Courtesy of Dr. A. Raman.

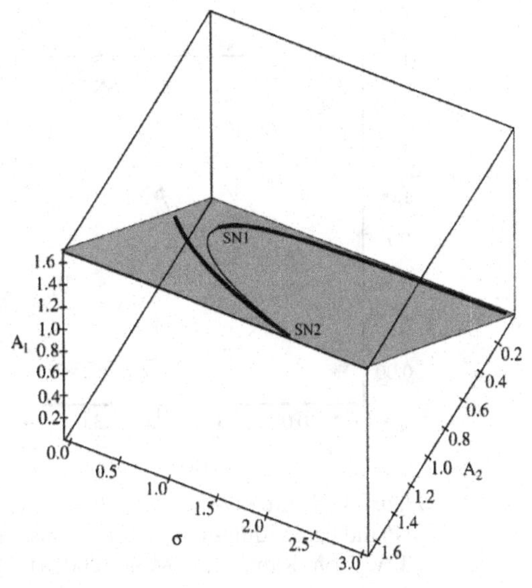

Figure 12.5. Steady-state solution for backward-traveling wave (BTW) in presence of non-dimensional damping $c = 0.06$. Bold solid line is stable solution and thin solid line is unstable solution. From Raman and Mote (1999). Courtesy of Dr. A. Raman.

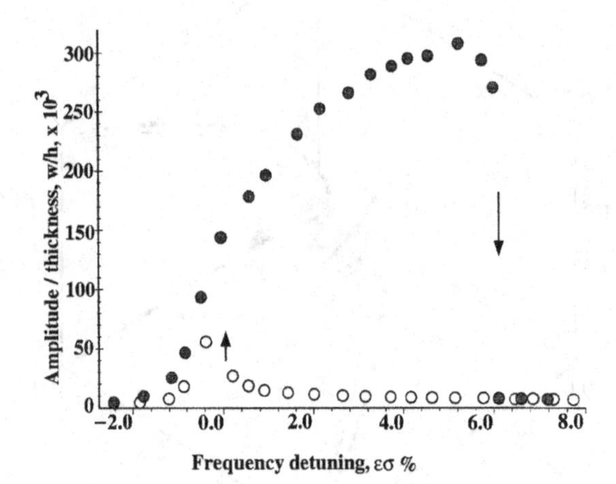

Figure 12.6. Experimentally measured amplitude of backward-traveling wave (BTW); mode ($m = 0$, $n = 4$). Solid circle: increasing rpm; empty circle: decreasing rpm. From Raman and Mote (1999). Courtesy of Dr. A. Raman.

The experimentally measured amplitude of BTW for mode ($m = 0$, $n = 4$) is given in Figure 12.6 for a steel disk having the following geometry: $a_2 = 356$ mm, $a_1/a_2 = 0.3$, $h = 0.775$ mm. The critical speed for mode ($m = 0$, $n = 4$) is 2580 rpm. The experiments were performed by Raman and Mote (1999). In Figure 12.6 the jump down, obtained at detuning $\varepsilon\sigma \approx 6\%$, is preceded by a slight reduction of the amplitude, differently from theory. This was attributed to aerodynamic effects. In order to overcome this problem, Raman and Mote (2001b) performed experiments in vacuum.

Imperfections play a significant role in actual spinning disk, as they split the natural frequencies of the sine and cosine modes that become no more in perfect one-to-one resonance. The parameter β is introduced to measure the linear frequency splitting and is defined as $(\omega_{m,n})_{\sin}/(\omega_{m,n})_{\cos} = 1 - \varepsilon\beta/2$. Figure 12.7(a) shows the amplitude B_2 of the BTW response in case of small imperfection $\beta = -0.128$, while Figure 12.7(b) shows the amplitude B_1 of the FTW response. The BTW response obtained in Figure 12.6 for perfect disk is still present on the first branch (with a corresponding very small amplitude of the FTW response), but a new branch appears (due to saddle-node bifurcation SN3) close to $\sigma = 0.5$ and it becomes stable near $\sigma = 1.5$ (between saddle-node bifurcations SN5 and SN6). The amplitudes of the

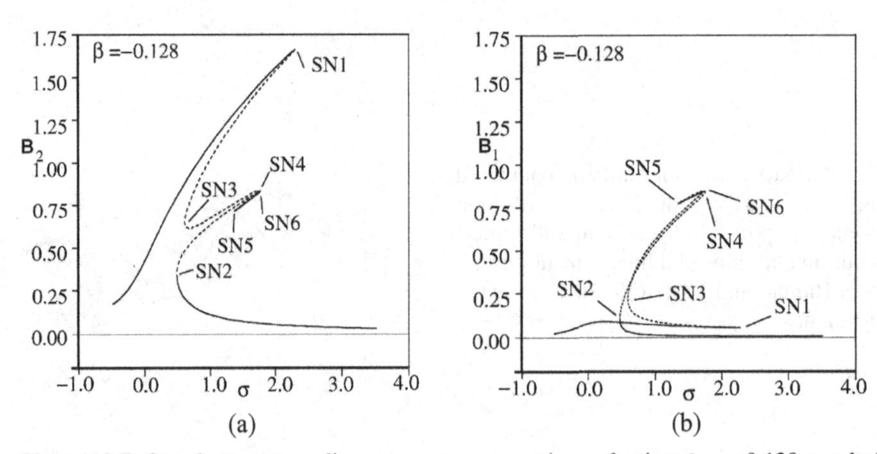

Figure 12.7. Steady-state traveling-wave response at imperfection $\beta = -0.128$; symbols SN1-SN6 indicate saddle-node bifurcations. (a) Amplitude B_2 of the BTW; (b) amplitude B_1 of the FTW. From Raman and Mote (2001a). Courtesy of Dr. A. Raman.

BTW and FTW responses are nearly equal in this new branch. It can be observed that, in this case, the phase $\phi_1 \approx \phi_2$. Along this branch of the solution, the disk vibrates nearly in a sine (or cosine) mode as viewed in the corotating system. Therefore, this solution is referred as pseudo-standing wave. Actually, the solution appears as a very slowly and discontinuously propagating wave in the corotating system.

REFERENCES

C. D'Angelo III and C. D. Mote Jr. 1993 *Journal of Sound and Vibration* **168**, 15–30. Aerodinamically excited vibration and flutter of a thin disk rotating at supercritical speed.

D. S. Dugdale 1979 *International Journal of Engineering Science* **17**, 745–756. Non-linear vibration of a centrally clamped rotating disc.

P. C. Dumir 1986 *Journal of Sound and Vibration* **107**, 253–263. Non-linear vibration and postbuckling of isotropic thin circular plates on elastic foundations.

G. J. Efstathiades 1971 *Journal of Sound and Vibration* **16**, 231–253. A new approach to the large-deflection vibrations of imperfect circular disks using Galerkin's procedure.

A. Jana and A. Raman 2005 *Journal of Fluids and Structures* **20**, 993–1006. Nonlinear aeroelastic flutter phenomena of a flexible disk rotating in an unbounded fluid.

G. C. Kung and Y.-H. Pao 1972 *Journal of Applied Mechanics* **39**, 1050–1054. Nonlinear flexural vibrations of a clamped circular plate.

A. W. Leissa 1969 *Vibration of Plates*. NASA SP-160, U.S. Government Printing Office, Washington, DC, USA (1993 reprinted by the Acoustical Society of America).

T. A. Nayfeh and A. F. Vakakis 1994 *International Journal of Non-Linear Mechanics* **29**, 233–245. Subharmonic travelling waves in a geometrically non-linear circular plate.

J. Nowinski 1964 *Journal of Applied Mechanics* **31**, 72–78. Non-linear transverse vibrations of a spinning disk.

J. Ramachandran 1974 *Journal of the Acoustical Society of America* **55**, 880–882. Large-amplitude vibrations of elastically restrained circular plates.

A. Raman and C. D. Mote Jr. 1999 *International Journal of Non-Linear Mechanics* **34**, 139–157. Non-linear oscillations of circular plates near a critical speed resonance.

A. Raman and C. D. Mote Jr. 2001a *International Journal of Non-Linear Mechanics* **36**, 261–289. Effects of imperfection on the non-linear oscillations of circular plates spinning near critical speed.

A. Raman and C. D. Mote Jr. 2001b *International Journal of Non-Linear Mechanics* **36**, 291–305. Experimental studies on the non-linear oscillations of imperfect circular disks spinning near critical speed.

S. Sridhar, D. T. Mook and A. H. Nayfeh 1975 *Journal of Sound and Vibration* **41**, 359–373. Nonlinear resonances in the forced responses of plates. Part I: symmetric responses of circular plates.

S. Sridhar, D. T. Mook and A. H. Nayfeh 1978 *Journal of Sound and Vibration* **59**, 159–170. Nonlinear resonances in the forced responses of plates. Part II: asymmetric responses of circular plates.

O. Thomas, C. Touzé and A. Chaigne 2003 *Journal of Sound and Vibration* **265**, 1075–1101. Asymmetric non-linear forced vibrations of free-edge circular plates. Part II: experiments.

S. A. Tobias 1957 *Proceedings of the Institution of Mechanical Engineers* **171**, 691–701. Free undamped non-linear vibrations of imperfect circular disks.

S. A. Tobias and R. N. Arnold 1957 *Proceedings of the Institution of Mechanical Engineers* **171**, 669–690. The influence of dynamical imperfection on the vibration of rotating disks.

C. Touzé, O. Thomas and A. Chaigne 2002 *Journal of Sound and Vibration* **258**, 649–676. Asymmetric non-linear forced vibrations of free-edge circular plates. Part I: theory.

T. Wah 1963 *Journal of the Engineering Mechanics Division, Proceedings of the ASCE* **89**, 1–15. Vibration of circular plates at large amplitudes.

C. J. H. Williams and S. A. Tobias 1963 *Journal of Mechanical Engineering Science* **5**, 325–335. Forced undamped non-linear vibrations of imperfect circular discs.

N. Yamaki, K. Otomo and M. Chiba 1981a *Journal of Sound and Vibration* **79**, 23–42. Non-linear vibrations of a clamped circular plate with initial deflection and initial edge displacement. Part I: theory.

N. Yamaki, K. Otomo and M. Chiba 1981b *Journal of Sound and Vibration* **79**, 43–59. Non-linear vibrations of a clamped circular plate with initial deflection and initial edge displacement. Part II: experiment.

13 Nonlinear Stability of Circular Cylindrical Shells under Static and Dynamic Axial Loads

13.1 Introduction

Circular cylindrical shells are very stiff structural elements with optimal use of the material. Therefore, they are also very light. This is one of the reasons why they are used for rockets (see Figure 13.1). Similarly to other thin-walled structures, the main strength analysis of circular cylindrical shells is a stability analysis; in fact, these structural elements buckle much before the failure stress of the material is reached.

In this chapter, the stability and the postcritical behavior of circular cylindrical shells under the action of axial static and periodic loads are investigated. Because of the strongly subcritical nature of the pitchfork bifurcation associated with buckling, even if the static compression load is much smaller than the critical load, shells collapse under small perturbation with a jump from the trivial equilibrium configuration to the stable bifurcated solution. Moreover, circular cylindrical shells subjected to axial loads are highly sensitive to geometric imperfections.

Periodic axial loads generate large-amplitude asymmetric vibrations due to period-doubling bifurcation of axisymmetric small-amplitude vibration. Period-doubling bifurcation arises for frequency of axial load close to twice the natural frequency of an asymmetric mode; this is usually referred as parametric instability. In fact, most of the studies are based on Donnell's nonlinear shallow-shell theory, so that axial loads do not appear directly in the equations of motion obtained with this shell theory, where only radial loads are directly inserted. They appear through boundary conditions, giving the so-called parametric excitation in the equation of motion. For the other shell theories introduced in Chapters 1 and 2, axial loads give direct excitation to the generalized coordinates used in the expansion of the axial displacement u.

Both the Sanders-Koiter and Donnell's nonlinear shallow-shell theories are used and their results are compared, following the study of Pellicano and Amabili (2006). Lagrange equations (for the Sanders-Koiter theory) and the Galerkin method (for Donnell's nonlinear shallow-shell theory) are applied in order to discretize the problem. Convergence tests are carried out in order to ensure good accuracy of the results.

13.1.1 Literature Review

Babcock (1983), in his review on shell stability, observed the huge number of papers published on the subject. He focused his attention on the most important topics on

Figure 13.1. NASA *Apollo 11* spacecraft with the *Saturn V* rocket at launch.

this field: post-buckling, imperfection sensitivity, dynamic buckling, plastic buckling and experiments. It is now clear that the most important types of imperfections are the geometrical ones; unfortunately, it is not simple to relate them to the knockdown factor.

Calladine (1995) claimed that "there are strong grounds for supposing that locked-in initial stresses on account of imperfect initial geometry and the static indeterminacy of boundary conditions of real shells have a pronounced effect on the buckling performance." He showed several interesting approximate formulas for actual buckling prediction; he discussed the problem of coincident buckling modes and concluded that great attention should be addressed to initial geometric imperfections as well as to locked-in stresses, caused, for example, by fixed boundaries. Another review on shell buckling is due to Teng (1996).

A fundamental study on the static stability of thin circular cylindrical shells under axial load is due to von Kármán and Tsien (1941). They developed a nonlinear theory in order to explain big discrepancies between linear theories and experiments and introduced a form of the radial deflection that was able to reproduce the diamond-shaped pattern observed in buckling experiments.

The effect of geometric imperfections on the buckling of thin-walled circular cylindrical shells has been deeply studied theoretically, numerically and experimentally by Yamaki (1984). Shells with variable thickness were studied by Elishakoff et al. (2001).

Even though the stability of shells subjected to periodic axial loads received less attention with respect to the static buckling, the literature is rich in interesting papers;

many of them are from Eastern Europe or the former Soviet Union, as shown in the extensive review by Amabili and Païdoussis (2003) on nonlinear vibrations and dynamics of shells, and in the review by Kubenko and Koval'chuk (1998), where interesting results were reported about parametric vibrations.

One of the first studies on parametric instability of circular cylindrical shells is due to Yao (1963), who reduced parametric oscillations of circular cylindrical shells to the well-known Mathieu equation and studied the stability bounds. Yao (1965) extended his previous study by using Donnell's nonlinear shallow-shell theory. Vijayaragham and Evan-Iwanowski (1967) studied theoretically and experimentally the dynamic instability of clamped-free, seismically excited circular cylindrical shells. Vol'mir and Ponomarev (1973) studied the nonlinear response of orthotropic circular cylindrical shells subjected to (i) a time-varying axial load (constant load plus harmonic component) plus a static external pressure and (ii) a harmonic external pressure and static axial load. They used Donnell's nonlinear shallow-shell theory and a mode expansion involving three terms: one asymmetric and two axisymmetric. In particular, the first axisymmetric term is analogous to the term used by Evensen (1967) and the second one is just a constant radial displacement of the shell. Continuity of the circumferential displacement was satisfied. The system was reduced to single degree of freedom.

The dynamic stability and the longitudinal resonance of simply supported cylindrical shells under axial load were analyzed by Koval (1974). Hsu (1974) and Nagai and Yamaki (1978) made interesting considerations about the parametric instability and developed models suitable for describing the shell dynamics. In particular, Nagai and Yamaki (1978) studied circular cylindrical shells subjected to compressive axial periodic forces (constant force plus harmonic load) by means of Donnell's nonlinear shallow-shell theory. They observed that the classical membrane approach is inaccurate when the vibration contribution of the axisymmetric modes is not negligible.

An interesting experimental study on parametric resonances of clamped-free circular cylindrical shells is due to Bondarenko and Galaka (1977), who observed a particularly violent instability. This instability takes place suddenly when a transition between stable to unstable regions occurs (or between two different instability regions involving different modes).

Tamura and Babcock (1975) studied finite circular cylindrical shells with imperfections subjected to step axial loads. A modified Donnell's nonlinear shallow-shell theory was used to account for the axial inertia in an approximate manner. Koval'chuk and Krasnopol'skaya (1980) introduced the effect of geometric imperfections, using Donnell's nonlinear shallow-shell theory and a simple three-mode expansion: driven, involving $\cos(n\theta)$, and companion, involving $\sin(n\theta)$, and axisymmetric modes were considered.

Popov et al. (1998) analyzed the parametric instability of an infinitely long circular cylindrical shell introducing an expansion with 3 degrees of freedom (dofs). The effect of internal resonances between the two asymmetric modes was pointed out. Popov (2003) used a continuation technique to study nonlinear oscillations and parametric instabilities of an infinitely long cylindrical shell.

Pellicano and Amabili (2003) studied nonlinear vibrations and dynamical instability of simply supported cylindrical shells, under the action of longitudinal dynamic force, including fluid-structure interaction. Donnell's nonlinear shallow-shell theory was used and a multimode approach was developed. Catellani et al. (2004) developed a multimode approach to analyze the correlation of parametric instability with shell collapse, considering geometric imperfections. Pellicano and Amabili (2006) extended their previous study by using the Sanders-Koiter nonlinear shell theory and deeply analyzing the effect of geometric imperfections. Chaotic oscillations were detected and investigated by using Lyapunov exponents.

Gonçalves and Del Prado (2002) analyzed the dynamic buckling of a perfect circular cylindrical shell under axial static and dynamic loads. Donnell's nonlinear shallow-shell theory was used and the membrane theory was considered to evaluate in-plane stresses. The partial differential operator was discretized by the Galerkin method, using a relatively large modal expansion. However, no companion mode participation was considered and the boundary conditions were dropped by assuming an infinitely long shell. Escape from potential well (i.e., attractive solution) was analyzed in detail, and a correlation of this phenomenon with the parametric resonance was given.

An analytical approach was developed by Jansen (2005) in order to simulate dynamic step and periodic axial loads acting on isotropic and anisotropic shells, showing that simple periodic responses can be simulated through low-dimensional models. The effect of in-plane inertia was included in the Donnell nonlinear shell theory. It was found that, neglecting in-plane inertia, a moderate underestimation of the instability region is obtained.

13.2 Theoretical Approach

The Lagrange equations of motion for the Sanders-Koiter nonlinear shell theory are obtained in the same way as in Chapter 7. In particular, the displacements u, v and w are expanded by using equations (7.4a–c) and geometric imperfections by using equation (5.52). The only difference with respect to Chapter 7 is that the axial boundary condition in the present case is

$$N_x = \frac{\tilde{P}(t)}{2\pi R}, \tag{13.1}$$

where $\tilde{P}(t)$ is the time-dependent axial load applied at both shell ends. $\tilde{P}(t)$ is positive in the x direction, that is, giving traction to the shell. In particular, $-\tilde{P}(t)$ is applied at $x = 0$ and $\tilde{P}(t)$ is applied at $x = L$.

In the presence of axial loads acting on the shell, additional virtual work is done by the external forces. Let us consider a time-dependent axial load $\tilde{P}(t) = -P + P_D \cos \omega t$ (expressed in newtons) uniformly distributed on the shell ends. The axial distributed force q_x has the following expression:

$$q_x = \frac{\tilde{P}(t)}{2\pi R}\left[-\delta(x) + \delta(x - L)\right], \tag{13.2}$$

where δ is the Dirac function. The virtual work done by the axial load is

$$W = \int_0^{2\pi}\int_0^{L} \frac{\tilde{P}(t)}{2\pi R}\left[-\delta(x) + \delta(x - L)\right] u \,\mathrm{d}x\, R\,\mathrm{d}\theta = -2\tilde{P}(t)\sum_{m=1}^{M_2} u_{m,0}(t). \tag{13.3}$$

For Donnell's nonlinear shallow-shell theory, the following expansion of the radial displacement is used:

$$
\begin{aligned}
w = {} & w_{1,n,c}(t)\sin(\eta)\cos(n\theta) + w_{1,n,s}(t)\sin(\eta)\sin(n\theta) + w_{3,n,c}(t)\sin(3\eta)\cos(n\theta)\\
& + w_{3,n,s}(t)\sin(3\eta)\sin(n\theta) + w_{1,2n,c}(t)\sin(\eta)\cos(2n\theta) + w_{1,2n,s}(t)\sin(\eta)\sin(2n\theta)\\
& + w_{1,0}(t)\sin(\eta) + w_{3,0}(t)\sin(3\eta) + w_{5,0}(t)\sin(5\eta) + w_{7,0}(t)\sin(7\eta)\\
& + w_{9,0}(t)\sin(9\eta),
\end{aligned}
\tag{13.4}
$$

where $\eta = \pi x/L$ and $w_{m,n,c}(t)$, $w_{m,n,s}(t)$ and $w_{m,0}(t)$ are the generalized coordinates.

Radial geometric imperfections having the same shape as the modes included in the expansion of the radial displacement are introduced:

$$
\begin{aligned}
w_0 = {} & \tilde{A}_{1,n}\sin(\eta)\sin(n\theta) + \tilde{A}_{3,n}\sin(3\eta)\sin(n\theta)\\
& + \tilde{A}_{1,2n}\sin(\eta)\sin(2n\theta) + \tilde{A}_{1,0}\sin(\eta) + \tilde{A}_{3,0}\sin(3\eta),
\end{aligned}
\tag{13.5}
$$

where $\tilde{A}_{m,n}$ and $\tilde{A}_{m,0}$ are the modal imperfection amplitudes. The equations of motion for Donnell's nonlinear shallow-shell theory are obtained as in Chapter 5.

13.3 Numerical Results

A benchmark problem, studied by Popov et al. (1998), Gonçalves and Del Prado (2002) and Pellicano and Amabili (2003; 2006) has been considered. This benchmark shell has the following characteristics: $h = 2 \times 10^{-3}$ m, $R = 0.2$ m, $L = 0.4$ m, $E = 2.1 \times 10^{11}$ N/m^2, $\nu = 0.3$, $\rho = 7850$ kg/m^3. For this shell, the classical buckling theory (Yamaki 1984) predicts the buckling load $P_{cl} = 2\pi Eh^2/\sqrt{3(1-\nu^2)} = 2\pi R \times 2.54 \times 10^6$ N.

Although a single model has been built by using Donnell's nonlinear shallow-shell theory (see equation [13.4]), several models have been built by using the Sanders-Koiter nonlinear theory of shells in order to perform a convergence analysis:

- Model A, 25 dofs, does not include sine modes in equations (7.4a–c); that is, only $u_{m,j,c}$, $v_{m,j,c}$ and $w_{m,j,c}$ are included. In particular, the following terms are considered: asymmetric modes $(1, n)$, $(1, 2n)$, $(3, n)$, $(3, 2n)$, $(1, 3n)$ for u, v and w; axisymmetric modes (1, 0), (3, 0), (5, 0), (7, 0), (9, 0) for u and w.
- Model B, 22 dofs, does not include sine modes. In particular, the following modes are considered: asymmetric modes $(1, n)$, $(1, 2n)$, $(3, n)$ for u and w and $(1, n)$, $(1, 2n)$, $(3, n)$, $(3, 2n)$, $(9, n)$, $(1, 4n)$ for v; axisymmetric modes (1, 0), (3, 0), (5, 0), (7, 0), (9, 0) for u and w.
- Model C, 30 dofs, includes sine modes. In particular, the following modes are considered: asymmetric modes $(1, n)$, $(1, 2n)$, $(3, n)$ for u and w and $(1, n)$, $(1, 2n)$, $(3, n)$, $(3, 2n)$ for v; axisymmetric modes (1, 0), (3, 0), (5, 0), (7, 0), (9, 0) for u and w.
- Model D, 20 dofs, the same as model C, but sine modes are not considered.

Where not specified, results have been obtained with the Sanders-Koiter shell theory.

13.3.1 Static Bifurcations

When the perfect shell is compressed, it presents to a small-amplitude axisymmetric deformation (see Figure 13.2), which has been calculated by using linearized

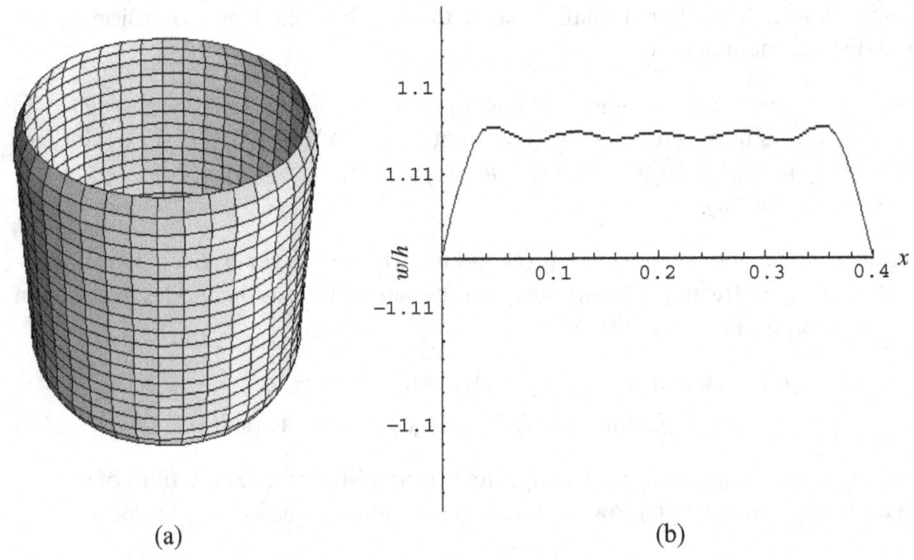

Figure 13.2. Pre-buckling deformation on a perfect shell under axial static load $P/P_{cl} = 0.4$. (a) Three-dimensional representation. (b) Longitudinal section. Normalized amplitude with respect to the shell thickness h; Donnell's shallow-shell theory.

Donnell's shallow-shell equations for $P/P_{cl} = 0.4$. The pre-buckling deformation of a perfect circular cylindrical shell is axisymmetric and the amplitude of asymmetric modes is zero. The edge effect is clearly visible in Figure 13.2.

The buckling mode considered in the present analysis is $n = 5$. Increasing the axial load, the shell loses stability. By using model B, one finds that for $P/P_{cl} = 0.947$ the shell loses stability through a pitchfork bifurcation; this is the actual critical load, which is slightly reduced with respect to the classical value. Results presented in Figure 13.3 have been obtained by means of the continuation software AUTO 97. At the bifurcation point, $P/P_{cl} = 0.947$, two new solutions, bifurcated branches, are found; these branches are initially unstable and subcritical. Asymmetric and axisymmetric modes participate to the shell deformation. In order to follow these solutions, the axial load must be reduced to the folding at $P/P_{cl} = 0.311$. It should be noted that the upper branch presents an amplitude smaller than the lower branch; indeed, it is well known that circular cylindrical shells are stiffer outward than inward. In Figure 13.3, the thick dashed line represents the solution obtained by Gonçalves and Del Prado (2002). Their bifurcation is at $P/P_{cl} = 1$ because they neglected the pre-buckling effect; the postcritical behavior is in good agreement, except for large amplitudes, where the difference between the more accurate Sanders-Koiter and Donnell's nonlinear shallow-shell theories becomes significant. Pellicano and Amabili (2003), by using Donnell's nonlinear shallow-shell theory, obtained the buckling load at $P/P_{cl} = 0.95$ and folding at $P/P_{cl} = 0.2$. Using model A, which has an increased number of degrees of freedom with respect to model B, we obtain the critical load $P/P_{cl} = 0.949$ and the folding at $P/P_{cl} = 0.316$, that is, almost the same results as obtained with model B.

Results show that, because of the strongly subcritical pitchfork bifurcation, even if the compression load is much smaller than the critical load, the shell collapses under small perturbation with a jump from the trivial equilibrium configuration to the stable bifurcated solution. Indeed, for $0.3 < P/P_{cl} < 0.95$, there are three stable

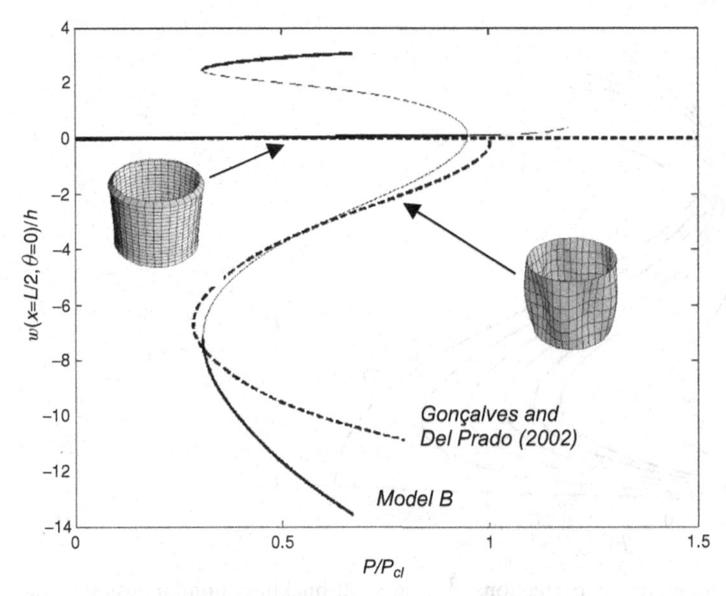

Figure 13.3. Shell buckling: nondimensional displacement at $x = L/2$ and $\theta = 0$, versus nondimensional static load. **—**, stable solutions, model B; —, unstable solutions, model B; **– –**, Gonçalves and Del Prado (2002).

equilibrium positions and two unstable ones. The unperturbed equilibrium position is represented by a small-amplitude axisymmetric deformation; its basin of attraction is reduced as P approaches P_{cl} and a perturbation can cause a jump to one of the buckled positions, with collapse of the shell; this is the snap-through instability.

A convergence test is performed with model A (25 dofs); then, several modes are eliminated from model A. Results of the convergence test can be summarized: mode $(1, 3n)$, with $n = 5$, can be eliminated without loss of accuracy; mode $(3, n)$ cannot be eliminated; mode $(3, 2n)$ can be eliminated only from the expansion of u and w. The result of this convergence analysis is that models C and D can be used without significant loss of accuracy.

Geometric imperfections are considered in the static buckling behavior presented in Figure 13.4, where the bifurcation path is presented for several imperfection amplitudes. Imperfection $\tilde{A}_{1,5}$ is given on asymmetric mode $(1, 5)$ with positive value; the pitchfork bifurcation is immediately destroyed by the imperfection and is replaced by a folding (saddle-node bifurcation). Indeed, in Figure 13.4 one can see that the imperfect shell exhibits a continuous deformation path. However, after the first folding, which is the one with the smaller deformation, the equilibrium solution becomes unstable and regains stability at the second folding point. Therefore, the shell presents a snap-through instability, analogous to the one studied in Section 10.4 for a shallow spherical shell; the shell deformation jumps from a relatively small deformation to a significantly larger one without any load increment.

The effect of imperfections consists in a large reduction of the critical load, which is now identified by the first folding point, which gives the pre-buckling deformation of the shell. The second folding is only slightly moved to the left-hand side by geometric imperfections $\tilde{A}_{1,5}$.

The effect of asymmetric and axisymmetric imperfections is summarized in Table 13.1.

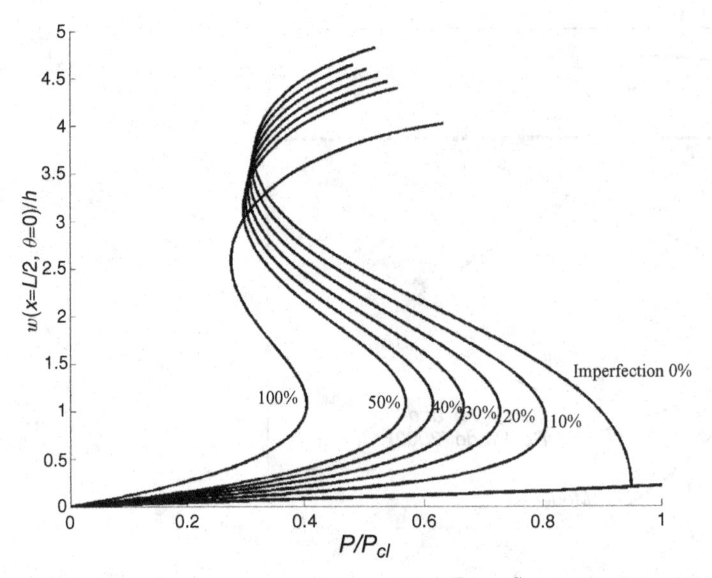

Figure 13.4. Effect of geometric imperfections $\tilde{A}_{1,5}$ on shell buckling; nondimensional displacement at $x = L/2$ and $\theta = 0$, versus nondimensional static load; model A. Imperfection amplitude: 100% = h.

13.3.2 Dynamic Loads

A periodic time-varying axial load is considered. This gives rise to a direct excitation of the axisymmetric generalized coordinate $u_{m,0}$ according to the Sanders-Koiter theory. The main interest is in the large-amplitude vibrations due to the period-doubling instability of asymmetric modes and not in the small-amplitude axisymmetric vibrations excited by dynamic axial loads. In fact, when the excitation frequency is close to twice the linear frequency of a shell asymmetric mode, a dynamic Mathieu-type instability takes place. The shell vibration is subharmonic and the vibration amplitude can be quite large.

Table 13.1. *Static buckling: effect of imperfections on modes (1, 5) and (1, 0). Results obtained with Sanders-Koiter theory if not differently specified.* P_{cr}, *critical load corresponding to the first folding;* P_{fold}, *folding load corresponding to the second folding;* P_{cl}, *classical buckling load*

$\tilde{A}_{1,5}/h$	$\tilde{A}_{1,0}/h$	P_{cr}/P_{cl}	P_{fold}/P_{cl}
0	0	0.95 (Donnell)	0.2 (Donnell)
0	0	0.949	0.316
0.1	0	0.805	0.312
0.2	0	0.727	0.308
0.3	0	0.666	0.304
0.4	0	0.615	0.300
0.5	0	0.56	0.297
1	0	0.405	0.276
0	0.2	1.038	0.326
0	0.4	1.135	0.338

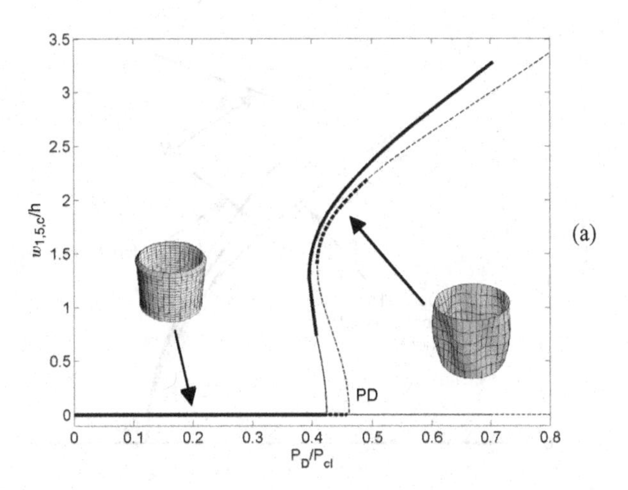

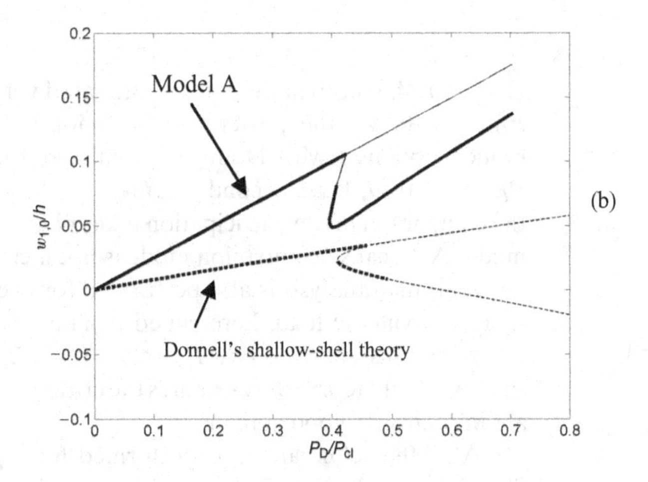

Figure 13.5. Dynamic instability; $P = 0$, $\omega/\omega_{1,5(0)} = 1.9$, $\zeta = 0.089$. —, Sanders-Koiter nonlinear theory, model A; – –, Donnell's nonlinear shallow-shell theory. Thick line, stable solution; thin line, unstable solution. (a) Amplitude of asymmetric mode (1, 5). (b) Axisymmetric mode (1, 0); PD, period-doubling bifurcation.

A first analysis is performed by using model A on a perfect shell considering a purely harmonic axial load of increasing magnitude, that is, no static preload is present; the excitation frequency is fixed to $\omega/\omega_{1,5(0)} = 1.9$ ($\omega_{1,5(0)} = 2\pi$ 484.22 rad/s is the fundamental frequency of the shell without initial compression by using the Sanders-Koiter theory and $\omega_{1,5(0)} = 2\pi$ 503.7 rad/s by using Donnell's shallow-shell theory), $P = 0$ and modal damping ratio $\zeta = 0.089$ for all the degrees of freedom. The dynamic axial load P_D is increased from zero to the onset of instability. Periodic solutions, bifurcations and stability are studied with AUTO 97. When the excitation amplitude is small, the shell vibrates axisymmetrically with small amplitude and the response is periodic (see Figure 13.5). When $P_D/P_{cl} = 0.424$, a period-doubling (PD) bifurcation appears; by increasing the dynamic load, the axisymmetric vibration becomes unstable. From the bifurcation point, a new branch of solution arises; it is slightly subcritical and initially unstable. This response is not only axisymmetric, and both asymmetric and axisymmetric modes are excited. Note that an amplitude h of oscillation corresponds to an acceleration $a = \omega^2 h$ ($\omega = 2\pi 484.22$ rad/s), which is about 1900 g!

Figure 13.5 shows a comparison among solutions obtained by using Donnell's nonlinear shallow-shell theory and model A (Sanders-Koiter). Donnell's nonlinear shallow-shell theory underestimates the axisymmetric vibration (see Figure 13.5[b]).

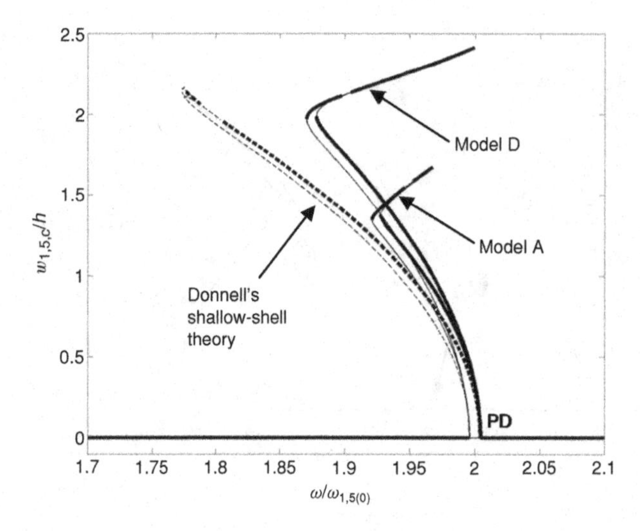

Figure 13.6. Dynamic instability; $P = 0$, $P_D/P_{cl} = 0.01$, $\zeta = 0.0008$. —, Sanders-Koiter nonlinear theory; – –, Donnell's nonlinear shallow-shell theory. Thick line, stable solution; thin line, unstable solution; PD, period-doubling bifurcation.

However, the bifurcation point is estimated very close to the Sanders-Koiter theory: $P_D/P_{cl} = 0.448$; the postcritical behavior is in good agreement. The bifurcated branch obtained with Donnell's nonlinear shallow-shell theory loses stability at $P_D/P_{cl} = 0.49$, for a second bifurcation; from this point, a new branch including the companion mode participation takes place. This bifurcation is not found by using model A, because companion mode is not included.

A similar analysis is also performed for excitation frequency $\omega/\omega_{1,5(0)} = 2$. The smallest dynamic load, here called dynamic critical load P_{Dcr}, which gives rise to parametric instability, is now $P_{Dcr}/P_{cl} = 0.387$. Also in this case Donnell's nonlinear shallow-shell theory overestimates the critical load ($P_{Dcr}/P_{cl} = 0.416$), but the results are in relatively good agreement.

A further comparison is performed for a perfect shell by considering a small damping ratio $\zeta = 0.0008$ and a constant dynamic excitation amplitude $P_D/P_{cl} = 0.01$; the excitation frequency is now varied. In Figure 13.6, a comparison between Donnell's nonlinear shallow-shell theory and the Sanders-Koiter theory (models A and D) is shown. When $\omega/\omega_{1,5}$ is close to 2, two period-doubling bifurcations are found for all the models, and a very good agreement of results is observed. Two subharmonic branches (response having a period twice the excitation period) appear from the bifurcation points and the postcritical behavior is now strongly subcritical (softening). For large-vibration amplitudes, the Sanders-Koiter theory predicts a change in the response behavior, which becomes supercritical (hardening). This behavior is not predicted by the less accurate Donnell's nonlinear shallow-shell theory. It should be noted that model D does not agree properly with model A for large-vibration amplitudes; this disagreement is due to the series truncation.

The effect of geometric imperfections is now considered: $\tilde{A}_{1,5}/h = 0.1$, $\tilde{A}_{1,0}/h = 0.1$; the fundamental frequency is 484.22 Hz for the perfect shell and 479.82 Hz for the shell with these geometric imperfections. The principal parametric instability is found for $\omega/\omega_{1,5(0)} = 1.98181$ ($\omega_{1,5(0)}$ is the natural frequency of the perfect shell without preload), which means that $\omega/\omega_{1,5} = 2$.

Simulations are performed by using model A with imperfections and $\omega/\omega_{1,5(0)} = 1.98181$, model A and Donnell's nonlinear shallow-shell theory without imperfections and $\omega/\omega_{1,5(0)} = 2$ (in this case, $\omega_{1,5(0)} = \omega_{1,5}$ because no imperfections nor static

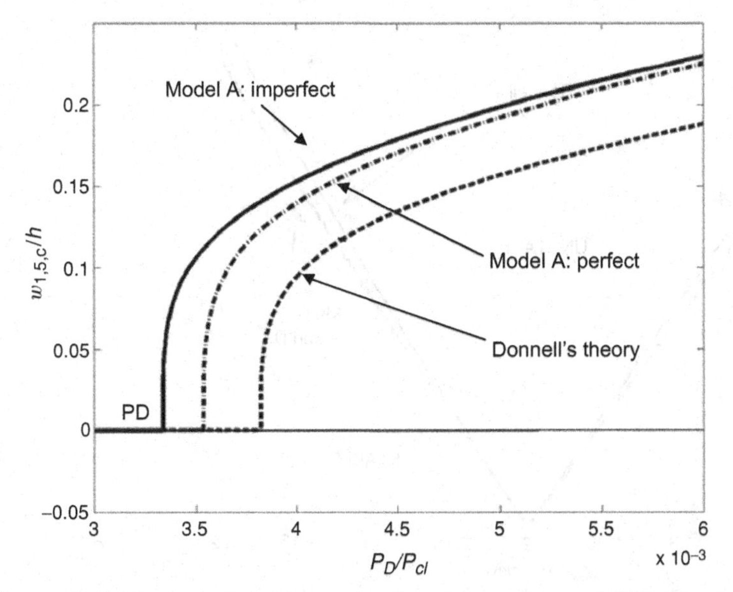

Figure 13.7. Dynamic instability; $P = 0$, $\omega/\omega_{1,5} = 2$, $\zeta = 0.0008$; model A. –·–, perfect shell, the Sanders-Koiter theory; —, shell with geometric imperfections $\tilde{A}_{1,5}/h = \tilde{A}_{1,0}/h = \tilde{A}_{3,0}/h = 0.1$, the Sanders-Koiter theory; – –, perfect shell, Donnell's nonlinear shallow-shell theory; PD, period-doubling bifurcation; thick line, stable solution; thin line, unstable solution.

preload are present). The damping ratio is $\zeta = 0.0008$. In Figure 13.7 the amplitude of the first asymmetric generalized coordinate is shown versus the excitation amplitude. Before the bifurcation, the amplitude is zero, but the shell is vibrating axisymmetrically, similarly to the case shown in Figure 13.5. Here, for the sake of brevity, the behavior of axisymmetric modes is not reported. Donnell's nonlinear shallow-shell theory predicts instability for $P_D/P_{cl} = 0.0038$; model A (perfect shell) finds $P_D/P_{cl} = 0.0035$ and model A with imperfections finds $P_D/P_{cl} = 0.0033$. Note that, with respect to Figure 13.5, the ratio P_D/P_{cl} is reduced about 100 times, because of the corresponding reduction in damping; moreover, the bifurcation is now supercritical. Results show a good agreement between the two theories; Donnell's nonlinear shallow-shell theory is sufficiently accurate, even though the in-plane inertia is neglected. Simulations show that the instability onset is not very sensitive to small geometric imperfections.

Results are now obtained by means of a two-parameter continuation by using the software AUTO 97, which allows for finding the minimum dynamic load versus the excitation frequency. In Figure 13.8, the principal instability region is presented for $\zeta = 0.0008$ and $P/P_{cl} = 0$: regions obtained with Donnell's nonlinear shallow-shell and Sanders-Koiter theories are very close. Geometric imperfection ($\tilde{A}_{1,5}/h = 0.15$) gives a translation of the instability boundary, without changing the minimum value of P_{Dcr}.

A general conclusion is that small geometric imperfections that have strong effects on the static buckling are not effective on the parametric instability onset (Figure 13.8). Similarly, the postcritical dynamic behavior is not sensitive to small geometric imperfections (Figure 13.7). However, a strong modification of the static post-buckling behavior should modify the dynamic behavior and the escape from the potential well associated with the stable equilibrium position.

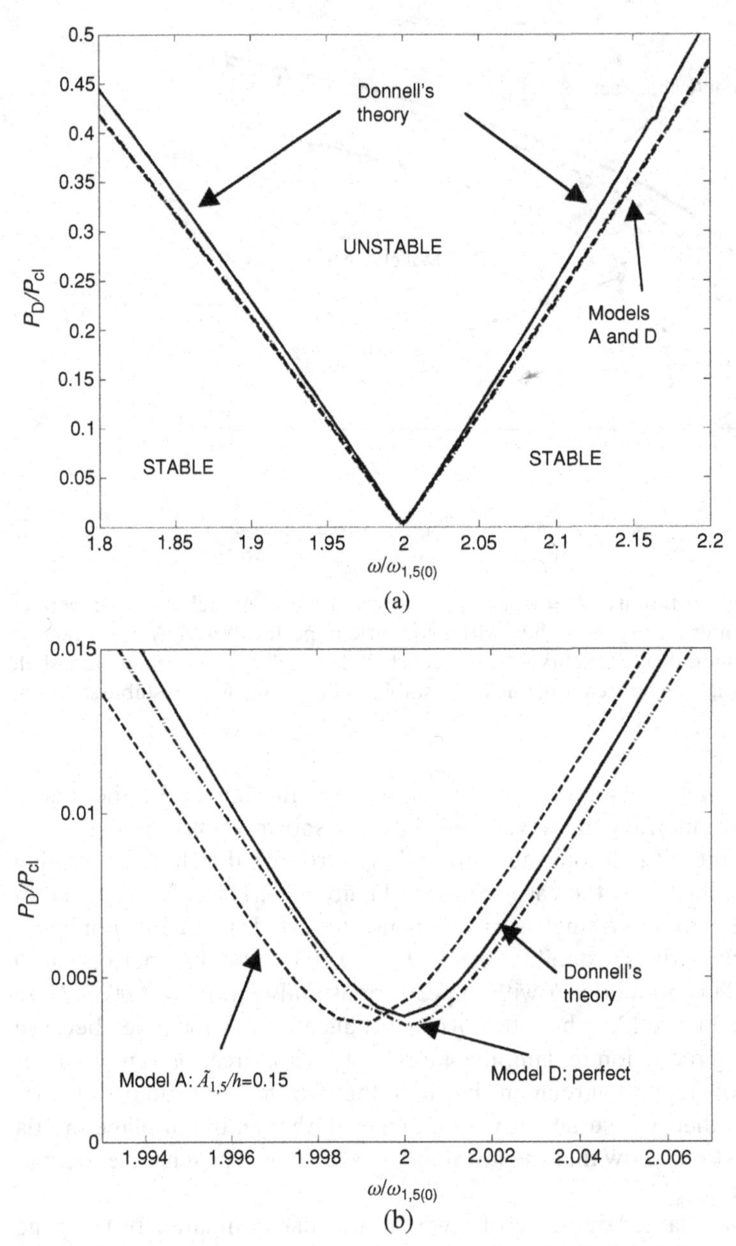

Figure 13.8. Principal instability region: comparison of shell theories and effect of imperfections; $P = 0$, $\zeta = 0.0008$. (a) Complete scenario. (b) Zoom of (a).

REFERENCES

M. Amabili and M. P. Païdoussis 2003 *Applied Mechanics Reviews* **56**, 349–381. Review of studies on geometrically nonlinear vibrations and dynamics of circular cylindrical shells and panels, with and without fluid-structure interaction.

C. D. Babcock 1983 *Journal of Applied Mechanics* **50**, 935–940. Shell stability.

A. A. Bondarenko and P. I. Galaka 1977 Parametric instability of glass-plastic cylindrical shells. *Soviet Applied Mechanics* **13**, 411–414.

C. R. Calladine 1995 *Thin-Walled Structures* **23**, 215–235. Understanding imperfection-sensitivity in the buckling of thin-walled shells.

G. Catellani, F. Pellicano, D. Dall'Asta and M. Amabili 2004 *Computers and Structures* **82**, 2635–2645. Parametric instability of a circular cylindrical shell with geometric imperfections.

I. Elishakoff, Y. Li and J. H. Starnes Jr. 2001 *Non-Classical Problems in the Theory of Elastic Stability*. Cambridge University Press, Cambridge, UK.

D. A. Evensen 1967 Nonlinear flexural vibrations of thin-walled circular cylinders. NASA TN D-4090.

P. B. Gonçalves and Z. J. G. N. Del Prado 2002 *Meccanica* **37**, 569–597. Nonlinear oscillations and stability of parametrically excited cylindrical shells.

C. S. Hsu 1974 On parametric excitation and snap-through stability problems of shells. In *Thin-Shell Structures: Theory, Experiment and Design*, pp. 103–131 (editors Y. C. Fung & E. E. Sechler). Prentice-Hall, Englewood Cliffs, NJ, USA.

E. Jansen 2005 *Nonlinear Dynamics* **39**, 349–367. Dynamic stability problems of anisotropic cylindrical shells via a simplified analysis.

L. R. Koval 1974 *Journal of Acoustical Society of America* **55**, 91–97. Effect of longitudinal resonance on the parametric stability of an axially excited cylindrical shell.

P. S. Koval'chuk and T. S. Krasnopol'skaya 1980 *Soviet Applied Mechanics* **15**, 867–872. Resonance phenomena in nonlinear vibrations of cylindrical shells with initial imperfections.

V. D. Kubenko and P. S. Koval'chuk 1998 *International Applied Mechanics* **34**, 703–728. Nonlinear problems of the vibration of thin shells (review).

K. Nagai and N. Yamaki 1978 *Journal of Sound and Vibration* **59**, 425–441. Dynamic stability of circular cylindrical shells under periodic compressive forces.

F. Pellicano and M. Amabili 2003 *International Journal of Solids and Structures* **40**, 3229–3251. Stability and vibration of empty and fluid-filled circular cylindrical shells under static and periodic axial loads.

F. Pellicano and M. Amabili 2006 *Journal of Sound and Vibration* **293**, 227–252. Dynamic instability and chaos of empty and fluid-filled circular cylindrical shells under periodic axial load.

A. A. Popov 2003 *Engineering Structures* **25**, 789–799. Parametric resonance in cylindrical shells: a case study in the nonlinear vibration of structural shells.

A. A. Popov, J. M. T. Thompson and F. A. McRobie 1998 *Journal of Sound and Vibration* **209**, 163–186. Low dimensional models of shell vibrations. Parametrically excited vibrations of cylindrical shells.

Y. S. Tamura and C. D. Babcock 1975 *Journal of Applied Mechanics* **42**, 190–194. Dynamic stability of cylindrical shells under step loading.

J. G. Teng 1996 *Applied Mechanics Reviews* **49**, 263–274. Buckling of thin shells: recent advances and trends.

A. Vijayaraghavan and R. M. Evan-Iwanowski 1967 *Journal of Applied Mechanics* **34**, 985–990. Parametric instability of circular cylindrical shells.

A. S. Vol'mir and A. T. Ponomarev 1973 *Mekhanika Polimerov* **3**, 531–539. Nonlinear parametric vibrations of cylindrical shells made of composite materials (in Russian).

T. von Kármán and H. S. Tsien 1941 *Journal of the Aeronautical Sciences* **8**, 303–312. The buckling of thin cylindrical shells under axial compression.

N. Yamaki 1984 *Elastic Stability of Circular Cylindrical Shells*. North-Holland, Amsterdam.

J. C. Yao 1963 *AIAA Journal* **1**, 457–468. Dynamic stability of cylindrical shells under static and periodic axial and radial load.

J. C. Yao 1965 *ASME Journal of Applied Mechanics* **32**, 109–115. Nonlinear elastic buckling and parametric excitation of a cylinder under axial load.

14 Nonlinear Stability and Vibration of Circular Shells Conveying Fluid

14.1 Introduction

Thin-walled circular cylindrical shell structures containing or immersed in flowing fluid may be found in many engineering and biomechanical systems. There are many applications of great interest in which shells are subjected to incompressible or subsonic flows. For example, thin cylindrical shells are used as thermal shields in nuclear reactors and heat shields in aircraft engines; as shell structures in jet pumps, heat exchangers and storage tanks; as thin-walled piping for aerospace vehicles. Furthermore, in biomechanics, veins, pulmonary passages and urinary systems can be modeled as shells conveying fluid.

Circular cylindrical shells conveying subsonic flow are addressed in this chapter; they lose stability by divergence (which is a static pitchfork bifurcation of the equilibrium, exactly as buckling) when the flow speed reaches a critical value. The divergence is strongly subcritical, becoming supercritical for larger amplitudes. It is very interesting to observe that the system has two or more stable solutions, related to divergence in the first or a combination of the first and second longitudinal modes, much before the pitchfork bifurcation occurs. This means that the shell, if perturbed from the initial configuration, has severe deformations causing failure much before the critical velocity predicted by the linear threshold. In particular, for the case studied in this chapter, the system can diverge for flow velocity three times smaller than the velocity predicted by linear theory, indicating, in this case, the necessity of using a nonlinear shell theory for engineering design.

In this chapter, a nonlinear shell theory is used because displacements larger than the shell thickness occur. However, a liner theory, that is, the potential flow theory, is used for the fluid. In fact, the deformation of the fluid domain is the same as for the shell, but it must be compared with the radius of the fluid domain and not with the shell thickness in order to verify if a linear model can be used. Because the shell radius is much larger than the shell thickness, a linear potential-flow model can be used in many applications.

14.1.1 Literature Review

The literature concerning the dynamic stability of circular cylindrical shells in the presence of internal or external axial flows is quite wide. The effects of an internal flow have been studied, for example, by Païdoussis and Denise (1972), Weaver and Unny (1973), Païdoussis et al. (1984; 1985), Païdoussis (2003) and Amabili and Garziera (2002a,b). As shown by Païdoussis et al. (1985) and Amabili and Garziera (2002b), the effect of viscous forces is not particularly large for short shells subjected to internal flow. In the previously mentioned studies, linear shell theories and potential flow theory are used.

The studies developed in the past for the stability of circular cylindrical shells in axial flow do not agree sufficiently well with experimental results, as pointed out by Horn et al. (1974). In particular, for subsonic Mach numbers, highly divergent and catastrophic instabilities have been measured experimentally for clamped-clamped copper shells excited by a fully developed turbulent flow (Horn et al. 1974).

The problem was solved for the first time by Amabili, Pellicano and Païdoussis (1999a,b; 2000a,b), who discovered the strongly subcritical postdivergence behavior. They used Donnell's nonlinear shallow-shell theory and a base of seven natural modes to deeply study the nonlinear vibrations and stability of circular shells conveying or immersed in subsonic flow. Amabili, Pellicano and Païdoussis (2002) extended this study by using refined nonlinear models up to 18 dofs. Karagiozis et al. (2008) applied the same method to clamped shells and successfully compared it with experimental data. Experiments, confirming the calculations, were performed by Karagiozis (2005) and Karagiozis et al. (2005).

Amabili, Pellicano and Païdoussis (2001) studied the nonlinear stability of shells immersed in axial incompressible flow.

Recently, computer programs based on the finite-element method (FEM) have been used to model artery and capillary (modeled with shell elements) conveying blood flow (see, e.g. Bathe and Kamm [2001]). However, FEM simulations of nonlinear dynamics and the stability of similar problems are still far from being reliable.

14.2 Fluid-Structure Interaction for Flowing Fluid

14.2.1 Fluid Model

A cylindrical coordinate system x, r, θ is introduced with the origin at one shell end. The displacements of the shell middle surface are indicated as usual with u, v, w in axial, circumferential and radial directions, respectively. The shell has radius R, length L and thickness h. The fluid-structure interaction is described by the linear potential flow theory. The shell is considered conveying incompressible flow. The fluid is assumed to be inviscid, and the flow to be isentropic and irrotational. The irrotationality property is the condition for the existence of a scalar potential function Ψ from which the velocity may be written as:

$$\mathbf{v} = -\nabla\Psi. \tag{14.1}$$

The potential Ψ consists of two components: one due to the mean flow associated with the undisturbed flow velocity U in the axial direction, and the unsteady perturbation potential Φ associated with the shell motion. Thus,

$$\Psi = -\,Ux + \Phi. \tag{14.2}$$

The potential of the perturbation velocity satisfies the Laplace equation

$$\nabla^2 \Phi = \frac{\partial^2 \Phi}{\partial x^2} + \frac{\partial^2 \Phi}{\partial r^2} + \frac{1}{r}\frac{\partial \Phi}{\partial r} + \frac{1}{r^2}\frac{\partial^2 \Phi}{\partial \theta^2} = 0. \tag{14.3}$$

The perturbed pressure P may be related to the velocity potential by Bernoulli's equation for unsteady fluid flow,

$$-\frac{\partial \Phi}{\partial t} + \frac{1}{2}V^2 + \frac{P}{\rho_F} = \frac{P_S}{\rho_F}, \tag{14.4}$$

where $V^2 = \nabla\Psi \cdot \nabla\Psi$, P_S is the stagnation pressure, ρ_F is the fluid mass density and t is time. The pressure P in the fluid domain can be written as

$$P = \overline{P} + p, \tag{14.5}$$

where $\overline{P}$ is the mean pressure and p is the perturbation pressure, assumed positive outward the shell. For small perturbations, $V^2 \cong U^2 - 2U(\partial\Phi/\partial x)$, and equations (14.4) and (14.5) give the stagnation pressure $P_S = \overline{P} + \frac{1}{2}\rho_F U^2$, so that it is fixed for an assumed mean flow velocity. Then, equation (14.4) gives the following expression for the perturbation pressure:

$$p = \rho_F\left(\frac{\partial \Phi}{\partial t} + U\frac{\partial \Phi}{\partial x}\right). \tag{14.6}$$

14.2.2 Shell Expansion

Donnell's nonlinear shallow-shell theory, equations (1.160) and (1.161), is used. The radial displacement w is assumed to be positive inward the shell. For simply supported shells, the following expansion of the radial displacement is used:

$$w(x, \theta, t) = \sum_{m=1}^{M} [A_{m,n}(t)\cos(n\theta) + B_{m,n}(t)\sin(n\theta)]\sin(\lambda_m x) + \sum_{m=1}^{M_0} A_{(2m-1),0}(t)\sin(\lambda_{(2m-1)} x), \tag{14.7}$$

where $\lambda_m = m\pi/L$. In equation (14.7) $M = M_0 = 6$ has been used for the refined model (18 dofs) and $M = 2$ and $M_0 = 3$ has been used for the reduced model (7 dofs).

Results show a uniform convergence of the solution when adding from two to six longitudinal modes. Modes with $2n$ and $3n$ circumferential waves can be added to the solution, but they do not play an important role.

For clamped shells, the following expansion can be used (Karagiozis 2005; Karagiozis et al. 2008):

$$w(x, \theta, t) = \sum_{m=1}^{M}\sum_{n=1}^{N} [A_{m,n}(t)\cos(n\theta) + B_{m,n}(t)\sin(n\theta)]\varphi_m(x) + \sum_{m=1}^{M_0} A_{2m-1,0}(t)\varphi_{2m-1}(x), \tag{14.8}$$

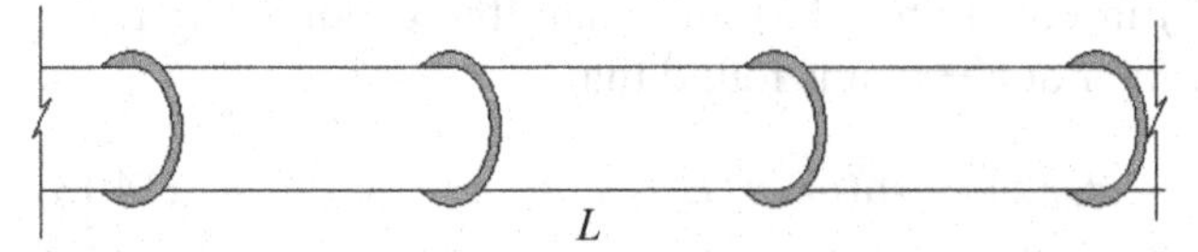

Figure 14.1. Infinite circular cylindrical shell periodically supported at distance L (Karagiozis et al. 2008). Courtesy of Dr. Karagiozis.

where φ_m is the eigenfunctions for a clamped-clamped beam defined by

$$\varphi_m(x) = \cosh(\lambda_m x/L) - \cos(\lambda_m x/L) - a_m \left[\sinh(\lambda_m x/L) - \sin(\lambda_m x/L)\right], \quad (14.9)$$

where λ_m is the corresponding dimensionless eigenvalues obtained from solving the characteristic equation $\cosh\lambda_m \cos\lambda_m = 1$; m is the axial wavenumber (equal to the number of half-waves along the shell), n the circumferential wavenumber and $a_m = [\cosh\lambda_m - \cos\lambda_m]/[\sinh\lambda_m - \sin\lambda_m]$.

The functions φ_m can be expanded into Fourier sine series. For obtaining the equations describing the fluid-structure interaction, the $\sin(n\theta)$ modes can be momentarily eliminated because they give exactly the same equations obtained for $\cos(n\theta)$ modes. Therefore, for the present scope, both equations (14.7) and (14.8) can be written in the generic form:

$$w(x,\theta,t) = \sum_{m=1}^{M}\sum_{n=0}^{N} A_{m,n}(t)\sin(m\pi x/L)\cos(n\theta) = \sum_{m=1}^{M}\sum_{n=0}^{N} w_{m,n}(x,\theta,t). \quad (14.10)$$

14.2.3 Fluid-Structure Interaction

When a fluid flow is considered, its behavior is not only related to what happens along the shell but also to how the flow is constrained before and after the shell. For this reason, it is necessary to assume boundary conditions also outside the shell. Since any system is different, and since different inflow and outflow boundary conditions do not affect the results as greatly if they are taken sufficiently away from the shell, simplified models can be conveniently introduced.

In this study, the fluid domain is assumed to be a cylinder of infinite extent, inside a periodically supported shell of infinite length (see Figure 14.1), so that it is possible to employ the method of separation of variables to obtain the velocity potential. The distance between periodical supports is L. This means that the shell radial displacement w is assumed to be a periodic function with main period $2L$, and the same is verified for the velocity potential and the perturbation pressure.

If no cavitation occurs at the fluid-shell interface, the boundary condition expressing the contact between the shell wall and the flow is

$$\left(\frac{\partial\Phi}{\partial r}\right)_{r=R} = \left(\frac{\partial w}{\partial t} + U\frac{\partial w}{\partial x}\right). \quad (14.11)$$

By using the method of separation of variables, Φ has the following form:

$$\Phi(x,r,\theta,t) = \sum_{m=1}^{M}\sum_{n=0}^{N} \phi_m(x)\,\psi_{m,n}(r)\,\cos(n\theta)\,f_{m,n}(t). \quad (14.12)$$

Substituting equation (14.12) in equation (14.3) and using the condition that the velocity potential must be regular at $r = 0$, it is found that

$$\phi_m(x) = \sin(m\pi x/L), \tag{14.13}$$

$$\psi_{m,n}(r) = \mathrm{I}_n(m\pi r/L), \tag{14.14}$$

where I_n is the modified Bessel functions of the order n of the first kind. Equation (14.11) is then satisfied by taking

$$\Phi = \sum_{m=1}^{M}\sum_{n=0}^{N} \frac{L}{m\pi}\frac{\mathrm{I}_n(m\pi r/L)}{\mathrm{I}'_n(m\pi R/L)}\left(\frac{\partial w_{m,n}}{\partial t} + U\frac{\partial w_{m,n}}{\partial x}\right), \tag{14.15}$$

where I'_n is the derivative of I_n with respect to the argument. The perturbation pressure at the shell wall, by using equations (14.6) and (14.15), is given by

$$p = \rho_F \sum_{m=1}^{M}\sum_{n=0}^{N} \frac{L}{m\pi}\frac{\mathrm{I}_n(m\pi R/L)}{\mathrm{I}'_n(m\pi R/L)}\left(\frac{\partial}{\partial t} + U\frac{\partial}{\partial x}\right)^2 w_{m,n}. \tag{14.16}$$

A minus sign appears on the right-hand side of equations (14.11), (14.15) and (14.16) in the case that w is assumed to be positive outward.

14.2.4 Nonlinear Equations of Motion with the Galerkin Method

By introducing equation (14.7) or (14.8) into Donnell's nonlinear shallow-shell theory, equations (1.160) and (1.161), and introducing the expression of the perturbation pressure p given by equation (14.16), the nonlinear equations of motion can be obtained by using the Galerkin method with a procedure analogous to that used in Chapter 5 for circular cylindrical shells without flow.

14.2.5 Energy Associated with Flow and Lagrange Equations

This section presents the alternative solution obtained by using the Lagrange equations of motion instead of the Galerkin method. Calculations, which are presented for radial displacement w assumed positive inward, give results that are unchanged in the case of w assumed positive outward.

The mean flow potential $-Ux$ does not give any time-varying contribution to the energy of the flowing fluid. In fact, by using the Green's theorem, the total energy associated with the flow is given by

$$E_{TF} = \frac{1}{2}\rho_F \iiint_{\Gamma} \mathbf{v}\cdot\mathbf{v}\,d\Gamma = \frac{1}{2}\rho_F \iiint_{\Gamma} \nabla\Psi\cdot\nabla\Psi\,d\Gamma$$

$$= \frac{1}{2}\rho_F \iint_{\Omega} \left(\Psi\frac{\partial\Psi}{\partial n}\right)\bigg|_{\Omega} d\Omega, \tag{14.17}$$

where Γ and Ω are the cylindrical fluid volume inside the shell (delimited by the length L) and the boundary surface of this volume, respectively, and n is the coordinate along the normal to the boundary, taken positive outward. Equation (14.17) shows that the fluid energy is a kinetic energy, even if it can be divided in three qualitatively different terms that will be identified in a few passages.

Equation (14.17) integrated over the shell internal surface and on the circular surfaces at $x = 0, L$ gives

$$E_{TF} = \frac{1}{2}\rho_F \int_0^{2\pi}\int_0^{L} (\Psi\,\partial\Psi/\partial r)\Big|_{r=R} \mathrm{d}x\, R\mathrm{d}\theta + \frac{1}{2}\rho_F \int_0^{2\pi}\int_0^{R} (\Psi\,\partial\Psi/\partial x)\Big|_{x=L} r\,\mathrm{d}\theta\,\mathrm{d}r$$
$$-\frac{1}{2}\rho_F \int_0^{2\pi}\int_0^{R} (\Psi\,\partial\Psi/\partial x)\Big|_{x=0} r\,\mathrm{d}\theta\,\mathrm{d}r. \quad (14.18)$$

In equation (14.18), $\partial\Psi/\partial r = \partial\Phi/\partial r$, $\partial\Psi/\partial x = -U + \partial\Phi/\partial x$, $\Psi|_{x=L} = -UL + \Phi$ and $\Psi|_{x=0} = \Phi$. Equation (14.18) is therefore reduced to

$$E_{TF} = E_F + \frac{1}{2}\rho_F \int_0^{2\pi}\int_0^{L} (-U x\,\partial\Phi/\partial r)\Big|_{r=R} \mathrm{d}x\, R\mathrm{d}\theta$$
$$+\frac{1}{2}\rho_F \int_0^{2\pi}\int_0^{R} -UL\,(\partial\Phi/\partial x)\Big|_{x=L} r\,\mathrm{d}\theta\,\mathrm{d}r + \frac{1}{2}\rho_F U^2\pi R^2 L, \quad (14.19)$$

where E_F is the energy associated with the perturbation potential, which is given by

$$E_F = \frac{1}{2}\rho_F \int_0^{2\pi}\int_0^{L} \left(\Phi\frac{\partial\Phi}{\partial r}\right)_{r=R} \mathrm{d}x\, R\mathrm{d}\theta, \quad (14.20)$$

The two integrals in equation (14.19) give zero for $n \neq 0$ because the result of the integration between 0 and 2π of $\cos(n\theta)$ is zero (see equation [14.15] for the expression of Φ). For $n = 0$, it is necessary to substitute L with $2L$, which is the longitudinal period of motion, and the two integrals vanish. In fact, in Section 14.2.3, w was assumed to be a periodic function with main period $2L$ in order to obtain the variable separation for Φ.

As already anticipated, the mean flow potential $-Ux$ gives no contribution to the evaluation of the flow energy, excluding a constant term, which does not affect the shell dynamics. Therefore, only E_F must be evaluated. By using equation (14.11), equation (14.20) is transformed into the following expression:

$$E_F = \frac{1}{2}\rho_F \int_0^{2\pi}\int_0^{L} (\Phi)_{r=R}\left(\frac{\partial w}{\partial t} + U\frac{\partial w}{\partial x}\right)\mathrm{d}x\, R\mathrm{d}\theta$$
$$= \frac{1}{2}\rho_F \int_0^{2\pi}\int_0^{L} \sum_{m=1}^{M}\sum_{l=1}^{M}\sum_{n=0}^{N}\sum_{k=0}^{N} \frac{L}{m\pi}\frac{\mathrm{I}_n(m\pi R/L)}{\mathrm{I}'_n(m\pi R/L)}$$
$$\times\left(\dot{w}_{m,n} + U\frac{\partial w_{m,n}}{\partial x}\right)\left(\dot{w}_{l,k} + U\frac{\partial w_{l,k}}{\partial x}\right)\mathrm{d}x\, R\mathrm{d}\theta$$
$$= \frac{1}{2}\rho_F \int_0^{2\pi}\int_0^{L} \sum_{m=1}^{M}\sum_{l=1}^{M}\sum_{n=0}^{N}\sum_{k=0}^{N} \frac{L}{m\pi}\frac{\mathrm{I}_n(m\pi R/L)}{\mathrm{I}'_n(m\pi R/L)}$$
$$\times\left[\dot{w}_{m,n}\dot{w}_{l,k} + U\left(\dot{w}_{m,n}\frac{\partial w_{l,k}}{\partial x} + \dot{w}_{l,k}\frac{\partial w_{m,n}}{\partial x}\right) + U^2\frac{\partial w_{m,n}}{\partial x}\frac{\partial w_{l,k}}{\partial x}\right]\mathrm{d}x\, R\mathrm{d}\theta. \quad (14.21)$$

Equation (14.21) shows that the energy E_F can be conveniently divided into three terms having different contributions of time functions and their derivatives:

$$E_F = T_F + E_G - V_F. \tag{14.22}$$

The first and second of the three terms on the right-hand side can be identified as the kinetic and gyroscopic energies, respectively; an opposite sign is introduced for the potential energy V_F for convenience.

The kinetic energy T_F of the fluid associated with the perturbation potential, by using the orthogonality of $\sin(m\pi x/L)$ in $(0, L)$ and of $\cos(n\theta)$ in $(0, 2\pi)$, is given by

$$T_F = \frac{1}{2}\rho_F \sum_{m=1}^{M}\sum_{n=0}^{N} \int_0^{2\pi}\int_0^{L} \frac{L}{m\pi}\frac{\mathrm{I}_n(m\pi R/L)}{\mathrm{I}'_n(m\pi R/L)}\dot{w}_{m,n}^2\, \mathrm{d}x\, R\mathrm{d}\theta. \tag{14.23}$$

The potential energy V_F, by using the orthogonality of $\cos(m\pi x/L)$ in $(0, L)$ and of $\cos(n\theta)$ in $(0, 2\pi)$, is

$$V_F = -\frac{1}{2}\rho_F \sum_{m=1}^{M}\sum_{n=0}^{N} \int_0^{2\pi}\int_0^{L} \frac{L}{m\pi}\frac{\mathrm{I}_n(m\pi R/L)}{\mathrm{I}'_n(m\pi R/L)} U^2 \left(\frac{\partial w_{m,n}}{\partial x}\right)^2 \mathrm{d}x\, R\mathrm{d}\theta. \tag{14.24}$$

Equation (14.24) shows that V_F is negative, that is, the stiffness of the system is decreasing with U. This explains the shell instability at high-flow velocities U.

The gyroscopic energy E_G associated with the perturbation potential is

$$E_G = \frac{1}{2}\rho_F \sum_{m=1}^{M}\sum_{l=1}^{M}\sum_{n=0}^{N}\sum_{k=0}^{N} \int_0^{2\pi}\int_0^{L} \frac{UL}{m\pi}\frac{\mathrm{I}_n(m\pi R/L)}{\mathrm{I}'_n(m\pi R/L)} \left(\dot{w}_{m,n}\frac{\partial w_{l,k}}{\partial x} + \dot{w}_{l,k}\frac{\partial w_{m,n}}{\partial x}\right) \mathrm{d}x\, R\mathrm{d}\theta. \tag{14.25}$$

One can easily verify that E_G is globally zero in the case of harmonic vibrations. This proves that the system is conservative and no energy is dissipated; in fact, the fluid is assumed to be inviscid. Note that equation (14.25) expresses a particular coupling between modes through damping that is characteristic of gyroscopic systems, which are systems with mass transport.

In the present case, the Lagrange equations of motion given by equation (3.98) are rewritten as

$$\frac{\mathrm{d}}{\mathrm{d}t}\left[\frac{\partial(T_S + T_F + E_G)}{\partial \dot{q}_j}\right] - \frac{\partial(T_S + T_F + E_G)}{\partial q_j} + \frac{\partial(U_S + V_F)}{\partial q_j} = Q_j,\ j = 1, \ldots, N, \tag{14.26}$$

where $\partial T_S/\partial q_j = 0$ and $\partial T_F/\partial q_j = 0$.

The gyroscopic energy E_G can be written in the following vectorial notation:

$$E_G = \frac{1}{2}\mathbf{q}^{\mathrm{T}}\mathbf{G}\,\dot{\mathbf{q}}, \tag{14.27}$$

where $\mathbf{G}$ is the gyroscopic matrix, which the following antisymmetric form with zeros on the diagonal:

$$\mathbf{G} = \begin{pmatrix} 0 & & \mathbf{B} \\ & \ddots & \\ -\mathbf{B}^{\mathrm{T}} & & 0 \end{pmatrix}, \tag{14.28}$$

where **B** is a triangular submatrix. Equation (14.28) gives immediately the relationship

$$\mathbf{G}^{\mathrm{T}} = -\mathbf{G}. \tag{14.29}$$

As a consequence of

$$\frac{\partial}{\partial \dot{q}_j}\dot{\mathbf{q}} = \begin{pmatrix} 0 \\ \vdots \\ 1 \\ \vdots \\ 0 \end{pmatrix} = \mathbf{J}, \tag{14.30}$$

where **J** is the unit vector that has only the j-th unitary term different from zero, it is possible to write

$$\frac{\mathrm{d}}{\mathrm{d}t}\left(\frac{\partial E_G}{\partial \dot{q}_j}\right) = \frac{\mathrm{d}}{\mathrm{d}t}\left[\frac{1}{2}\mathbf{q}^{\mathrm{T}}\mathbf{G}\,\mathbf{J}\right] = \frac{1}{2}\dot{\mathbf{q}}^{\mathrm{T}}\mathbf{G}\,\mathbf{J}. \tag{14.31}$$

Similarly, the following expression is obtained:

$$-\frac{\partial E_G}{\partial q_j} = -\frac{1}{2}\mathbf{J}^{\mathrm{T}}\mathbf{G}\,\dot{\mathbf{q}} = -\frac{1}{2}\dot{\mathbf{q}}^{\mathrm{T}}\mathbf{G}^{\mathrm{T}}\mathbf{J} = \frac{1}{2}\dot{\mathbf{q}}^{\mathrm{T}}\mathbf{G}\,\mathbf{J}. \tag{14.32}$$

By comparing the right-hand terms in equations (14.31) and (14.32), the following relationship is obtained:

$$\frac{\mathrm{d}}{\mathrm{d}t}\left(\frac{\partial E_G}{\partial \dot{q}_j}\right) = -\frac{\partial E_G}{\partial q_j}. \tag{14.33}$$

By substituting equation (14.33) into (14.26), the final expression of Lagrange equations of motion is obtained:

$$\frac{\mathrm{d}}{\mathrm{d}t}\left[\frac{\partial\,(T_S + T_F)}{\partial \dot{q}_j}\right] - 2\frac{\partial\, E_G}{\partial q_j} + \frac{\partial\,(U_S + V_F)}{\partial q_j} = Q_j, \quad j = 1, \ldots, N. \tag{14.34}$$

14.2.6 Solution of the Associated Eigenvalue Problem

The equations of motion for linear (canceling all the nonlinear terms from the equations of motion [14.34], which are only obtained from U_S) vibrations of circular cylindrical shells conveying flowing fluid can be written in the following form:

$$\mathbf{M}\ddot{\mathbf{u}} + \mathbf{C}\dot{\mathbf{u}} + \mathbf{K}\mathbf{u} = \mathbf{F}, \tag{14.35}$$

where **u** is the displacement vector of the middle surface of the shell, **M** is the mass matrix of the fluid-filled shell, **C** is the damping matrix, which in the present case is given by the sum of the gyroscopic matrix **G** due to the flow and by the structural damping matrix, **K** is the stiffness matrix of the shell with flow and **F** is the vector of external forcing. The gyroscopic matrix **G** is not proportional to **M** and **K**; therefore, equation (14.35) gives complex modes, which, differently from natural modes, have a shape evolving with time.

Assume that the solution of equation (14.35) has the form

$$\mathbf{u} = \mathbf{U}e^{\lambda t}, \tag{14.36}$$

where **U** is a complex vector indicating displacements and not depending on time. By introducing the new vector **W**, indicating velocity,

$$\mathbf{W} = \lambda\,\mathbf{U}, \tag{14.37}$$

and by using equation (14.36), it is possible to rewrite equation (14.35) as

$$\lambda\,\mathbf{MW} + \mathbf{CW} + \mathbf{KU} = \mathbf{F}. \tag{14.38}$$

Equation (14.38) can be transformed into

$$\lambda\,\mathbf{W} = -\mathbf{M}^{-1}\mathbf{KU} - \mathbf{M}^{-1}\mathbf{CW} + \mathbf{M}^{-1}\mathbf{F}. \tag{14.39}$$

Equations (14.37) and (14.39) can be written in the following state-space representation:

$$\lambda \begin{Bmatrix} \mathbf{U} \\ \mathbf{W} \end{Bmatrix} = \begin{bmatrix} \mathbf{0} & \mathbf{I} \\ -\mathbf{M}^{-1}\mathbf{K} & -\mathbf{M}^{-1}\mathbf{C} \end{bmatrix} \begin{Bmatrix} \mathbf{U} \\ \mathbf{W} \end{Bmatrix} + \begin{Bmatrix} \mathbf{0} \\ \mathbf{M}^{-1}\mathbf{F} \end{Bmatrix}. \tag{14.40}$$

Introducing the state-space variable $\mathbf{Z} = \{\mathbf{U}, \mathbf{W}\}^{\mathrm{T}}$ and the matrix **A**,

$$\mathbf{A} = \begin{bmatrix} \mathbf{0} & \mathbf{I} \\ -\mathbf{M}^{-1}\mathbf{K} & -\mathbf{M}^{-1}\mathbf{C} \end{bmatrix}, \tag{14.41}$$

equation (14.40), after elimination of the forcing term in order to study free vibrations, takes the following standard eigenvalue form:

$$\mathbf{AZ} = \lambda\mathbf{Z}. \tag{14.42}$$

The eigenvalues λ are complex conjugates and can be written as

$$\lambda = -\alpha \pm i\omega, \tag{14.43}$$

and solution (14.36) takes the form

$$\mathbf{u} = \mathbf{U}e^{-\alpha t}e^{i\omega t}, \tag{14.44}$$

where ω is the circular frequency of vibration of the complex mode and α is immediately related to the damping ratio ζ by $\alpha = \zeta\,\omega$ if α is positive. Instability occurs in case of negative value of α.

14.3 Numerical Results for Stability

A circular cylindrical shell, periodically simply supported and containing flowing water, is assumed to have the following characteristics: $L/R = 2$, $h/R = 0.01$, $E = 206 \times 10^9$ Pa, $\rho = 7850$ kg/m^3, $\rho_F = 1000$ kg/m^3 and $\nu = 0.3$. A nondimensional fluid velocity V is introduced for convenience, defined as in Weaver and Unny (1973) by $V = U/\{(\pi^2/L)[D/(\rho h)]^{1/2}\}$, with $D = Eh^3/[12(1-\nu^2)]$. Similarly, a nondimensional, generally complex, eigenvalue Ω is defined as $\Omega = \lambda/\{(\pi^2/L^2)[D/(\rho h)]^{1/2}\}$, λ being the corresponding complex eigenvalue in equation (14.40).

Figure 14.2, obtained by canceling the nonlinear terms from the equations of motion, shows the imaginary and real parts of the nondimensional eigenvalues Ω versus the nondimensional flow velocity for modes having five circumferential waves ($n = 5$), including the fundamental mode ($n = 5$, $m = 1$ for $V = 0$). Results have been obtained by using nine longitudinal modes in the expansion, although only the first three longitudinal modes are reported in Figure 14.2.

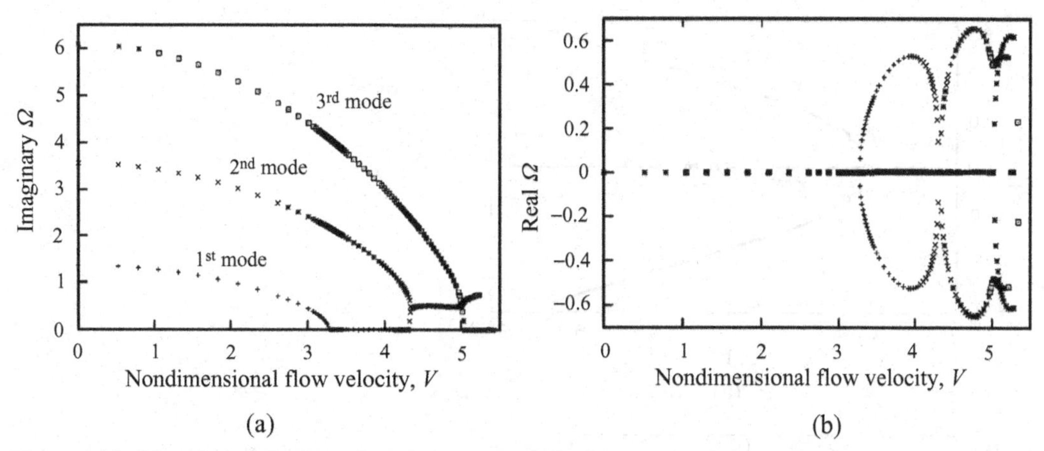

Figure 14.2. First three nondimensional eigenvalues Ω with $n = 5$ circumferential waves for the simply supported shell conveying water versus the nondimensional flow velocity. (a) Imaginary part of Ω (frequency). (b) Real part of Ω (damping).

In Figure 14.2(a), the curves give the nondimensional frequency (imaginary part of Ω) of the shell versus the nondimensional fluid velocity V. Eigenvalues, in general, decrease with V until reaching the value zero; at this point, the real part of Ω becomes positive, giving instability; this instability is associated to a pitchfork bifurcation at $V = 3.33$ and is referred as divergence. The lowest curve corresponds to the mode with one longitudinal wave ($m = 1$) for $V = 0$; the second curve corresponds to $m = 2$, and the third one to $m = 3$, for $V = 0$. The mode shape is changed by the flow, becoming complex, and the first mode can present two longitudinal waves at some instants for large flow velocities.

Calculations show that buckled equilibrium solutions arise from the pitchfork bifurcation point at $V = 3.33$ found by linear analysis. In particular, two different types of equilibrium solutions are possible after divergence: (i) in-phase modes, with all asymmetric modes (i.e. those with $n > 0$) in equation (14.7) having only sine (or only cosine) terms along the angular coordinate θ; and (ii) orthogonal modes, with asymmetric modes in equation (14.7) having alternatively non-zero sine and cosine terms in θ depending on whether the longitudinal wavenumber m is odd or even.

Figure 14.3, which relates to configuration (i) (i.e. in-phase modes), shows the nondimensional generalized coordinates versus V, starting from an initially undisturbed configuration at $V = 0$ calculated with the 18 dofs model. The trivial solution presents a pitchfork bifurcation at $V = 3.33$ from which an axisymmetric bifurcating surface in the $(A_{1,n}, B_{1,n}, V)$ space comes out. In fact, the shell does not possess a preferential direction of buckling because of its axisymmetry. Figure 14.3(a) shows the intersection of this surface with the plane $B_{1,n} = 0$, which gives branch 1. The bifurcated solution (postdivergence, branch 1) related to the first longitudinal mode becomes stable after a folding at $V = 1.15$. The divergence is strongly subcritical for small deformations, becoming supercritical for larger amplitudes. At $V = 4.33$, a second solution (divergence, branch 2) appears, associated with the second longitudinal mode. It is interesting to observe that the axisymmetric deformation associated with shell divergence is inward (see Figure 14.3[c]). The nondimensional velocity $V = 1.15$ is the threshold of stability under perturbation; for larger velocities, catastrophic divergence, associated with very large shell deformation, is possible given enough perturbation to the shell (e.g. small bubbles or flow perturbations). Therefore,

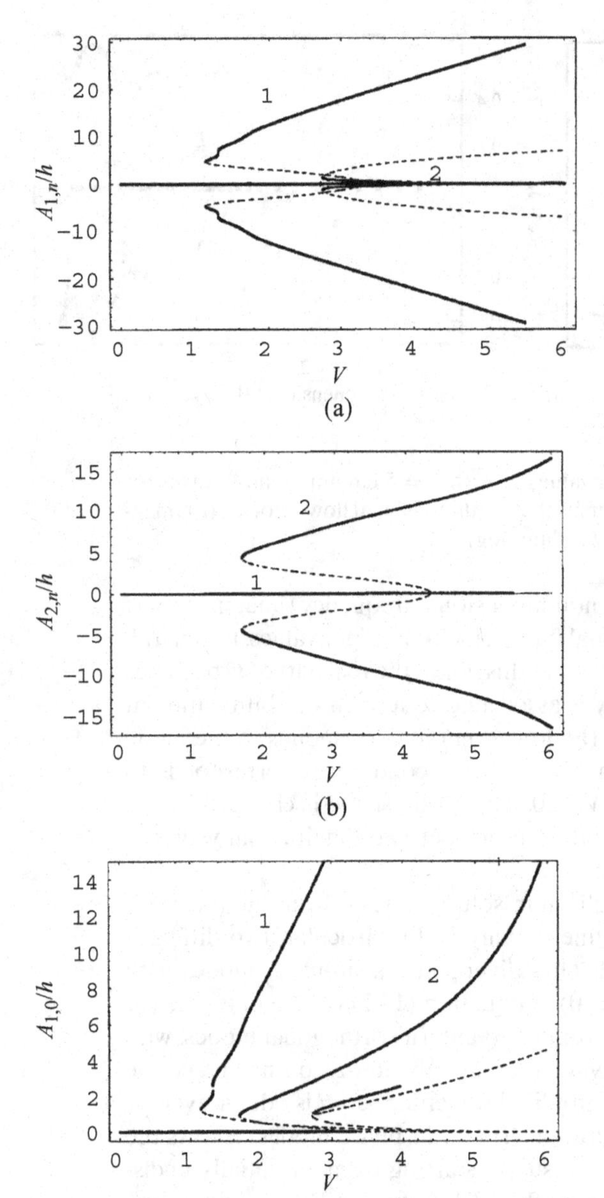

Figure 14.3. Bifurcation diagram: amplitude of static solutions (divergence) versus the nondimensional flow velocity V; in-phase modes; 18 dofs model. —–, Stable branches; – –, unstable branches. (a) Amplitude of the first longitudinal mode $A_{1,n}/h$. (b) Amplitude of the second longitudinal mode $A_{2,n}/h$. (c) Amplitude of the first axisymmetric mode $A_{1,0}/h$.

instability can occur for flow velocity that is almost three times less then the instability point predicted by linear theory. This justifies using a time-consuming nonlinear analysis to predict the stability threshold.

14.4 Comparison of Numerical and Experimental Stability Results

Experiments on stability of circular cylindrical shells conveying water flow were performed in the water tunnel of McGill University by Karagiozis (2005) under the supervision of Professor Païdoussis (Karagiozis et al. 2005; 2008). The water-flow apparatus involves a modification of the vertical test section of a water tunnel such that the flow from the upper part of a 203-mm-diameter test section is channeled into the test shell, which has a considerably smaller diameter, as shown in Figure 14.4. The test shell is

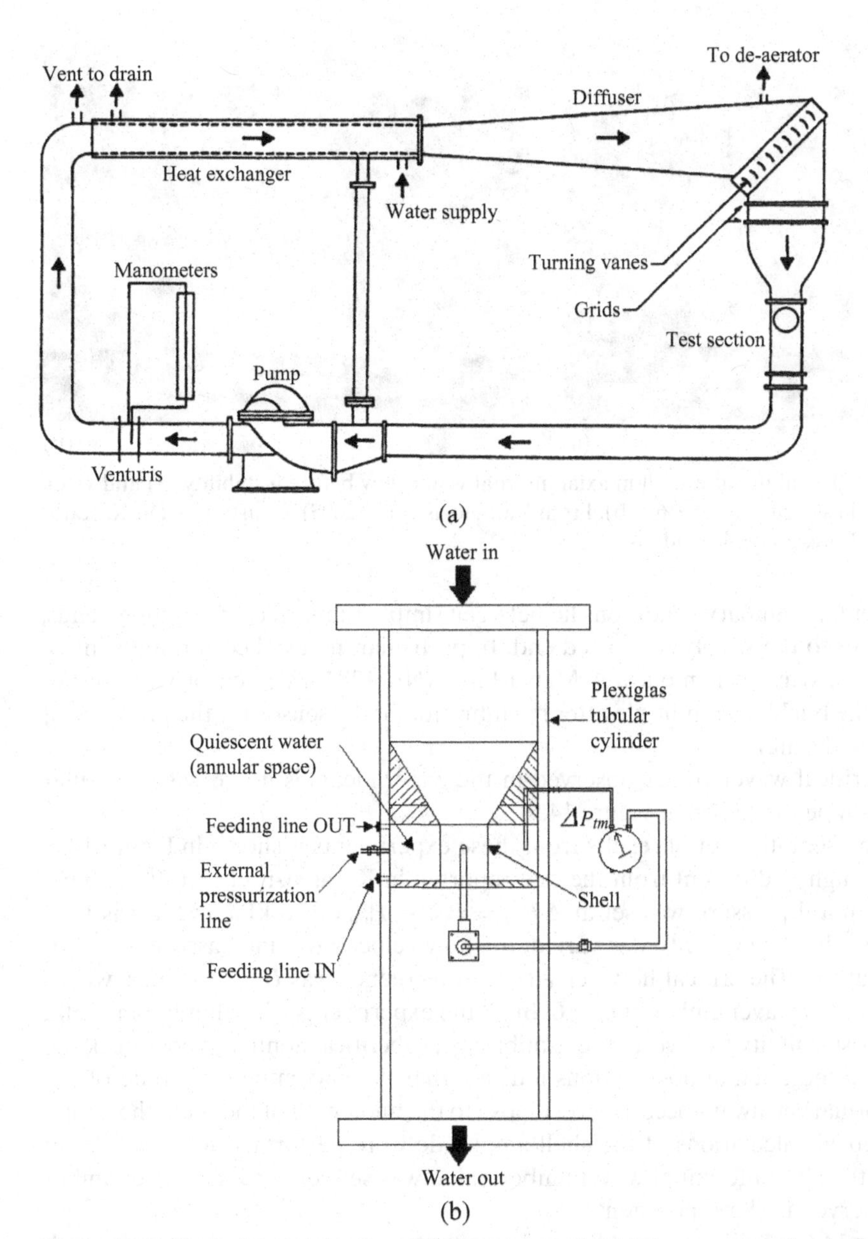

Figure 14.4. View of the test section of the water tunnel, with the apparatus for shells subjected to internal water flow mounted in it. (a) Schematic of the water tunnel (Païdoussis 2003). (b) Schematic of the test section (Karagiozis et al. 2005). Courtesy of Dr. K. Karagiozis and Professor M. P. Païdoussis.

surrounded by quiescent water; the pressure in that fluid region was made higher than the pressure in the internally flowing fluid at mid-length of the shell, that is, $P_{\text{ann}} > P_{\text{inn}}$ at $x = L/2$, as shown in the figure. Thus, there was a net inward-acting transmural pressure. This was necessary to achieve flow-induced instability for some of the shells because of the limited maximum flow velocity available in the water tunnel.

Experiments were conducted on aluminium shells glued to copper rings. The tested shells have the following dimensions and material properties: $L = 0.1225$ m; $R = 0.041125$ m; $h = 0.000137$ m; $\rho = 2720$ kg/m^3; $\nu = 0.38$ and $E = 70 \times 10^9$ Pa. The rings sat on plastic supports in the test section and were sealed with silicon rubber.

(a) (b)

Figure 14.5. The aluminum shell in axial internal water flow before instability (a) and when divergence has occurred ($n = 6$). (b). From Karagiozis et al. (2005). Courtesy of Dr. K. Karagiozis and Professor M. P. Païdoussis.

Experimental boundary conditions lie between simply supported and clamped ends, being closer to the simply supported end. In particular, no axial constraint is introduced for axisymmetric modes. A Matsushita ANR 1282 laser sensor was used to measure the buckling amplitude after recalibration of the sensor for the presence of Plexiglas and water.

The critical wavenumber observed in the experiments is $n = 6$ with a regular buckling shape (as shown in Figure 14.5).

A representative set of results from these experiments is shown in Figure 14.6; this set is slightly different from the one reported in Karagiozis et al. (2005; 2008). The transmural pressure was set at $\Delta P_{tm} = P_{\text{ann}} - P_{\text{inn}} = 5.8$ kPa and it was kept constant as the flow velocity was varied; the flow velocity was increased until instability occurred. The critical flow velocity for divergence was $U_c = 16.0$ m/s, with a circumferential wavenumber of $n = 6$. In all the experiments with aluminium shells, the shell lost stability by divergence, exhibiting a subcritical nonlinear post-buckling behavior. Experimental observations indicate that the maximum amplitude of the shell deformation always occurred very close to the midlength of the shell; therefore, all theoretical calculations of the shell amplitude were performed for $x = L/2$. In all cases, the circumferential wavenumber value was set equal to the wavenumber $n = 6$ observed in the experiments.

Figure 14.6 shows a reasonably good agreement between numerical results with the 18 dofs model for simply supported edges and the experimental results. Also in this case, a very large difference (about three times) is observed between the instability point predicted by linear theory and the actual value under perturbations. Results for clamped shells obtained with a 7 dofs model are also shown for comparison.

14.5 Numerical Results for Nonlinear Forced Vibrations

14.5.1 Periodic Response

A harmonic modal excitation $f_{1,5} = 0.03\, h^2 \rho \omega_{1,5}^2$ with a frequency close to the natural frequency of the mode ($n = 5, m = 1$) is considered for the same simply supported shell as studied in Section 14.3, with the 7 dofs model having the following generalized

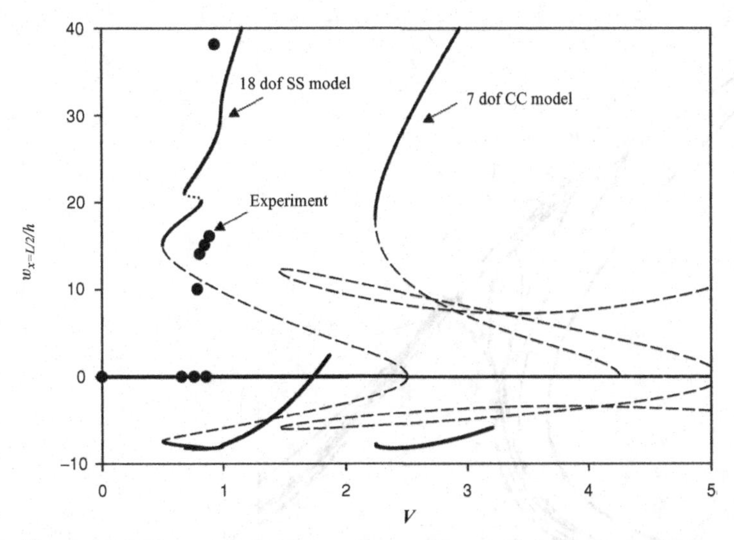

Figure 14.6. Comparison of experimental data and theoretical results on a bifurcation diagram of the nondimensional shell amplitude at $x = L/2$ versus the nondimensional water-flow velocity for an aluminium shell with internal water flow; mode with $n = 6$ and transmural pressure at $x = L/2\Delta P_{tm} = 5.8$kPa; —, stable solutions; – –, unstable solutions; •, experimental data points. Simply supported model with 18 dofs and clamped-clamped model with 7 dofs. Courtesy of Dr. K. Karagiozis.

coordinates: $A_{1,5}$, $B_{1,5}$, $A_{2,5}$, $B_{2,5}$, $A_{1,0}$, $A_{3,0}$ and $A_{5,0}$. This is the smallest model giving reasonably accurate results. In Figure 14.7, the amplitude of oscillation of the generalized coordinate $A_{1,5}$ is shown for several values of the fluid velocity, without companion mode participation, that is, $B_{1,5} = 0$. The dynamical behavior shows softening-type nonlinearity for the entire flow velocity range explored. It is interesting to note that the maximum amplitude of oscillation varies with the axial flow velocity V and reaches a minimum around $V = 1.35$, which practically corresponds to the appearance of the bifurcated equilibrium position ($V = 1.31$ with the 7 dofs model and 1.15 with the 18 dofs model used in Section 14.3). Thus, increasing the flow velocity from $V = 0$ to $V = 1.35$, the vibration amplitude decreases. After this value, the amplitude suddenly increases a great deal with the flow velocity. Therefore, it seems that the effect of the flow is to reduce the amplitude, the nonlinearity and the frequency of the system response up to the velocity corresponding to the instability threshold under disturbance; after this velocity, the frequency of the response is still reduced, but amplitude and nonlinearity of the response increase with an easy jump to the bifurcated solution, giving divergence instability. It must be observed that vibration modes are not normal anymore but complex; therefore, mode shapes change during the vibration period.

14.5.2 Unsteady and Chaotic Motion

The tool used here to investigate parametrically the response of the shell with flow is the Poincaré map and the corresponding bifurcation diagrams for the 7 dofs model. These have been calculated by direct integration of the equations of motion by using the IMSL DIVPAG package, which uses the Adams-Gear integration scheme. At any increment of the bifurcation parameter, which is the excitation frequency in this

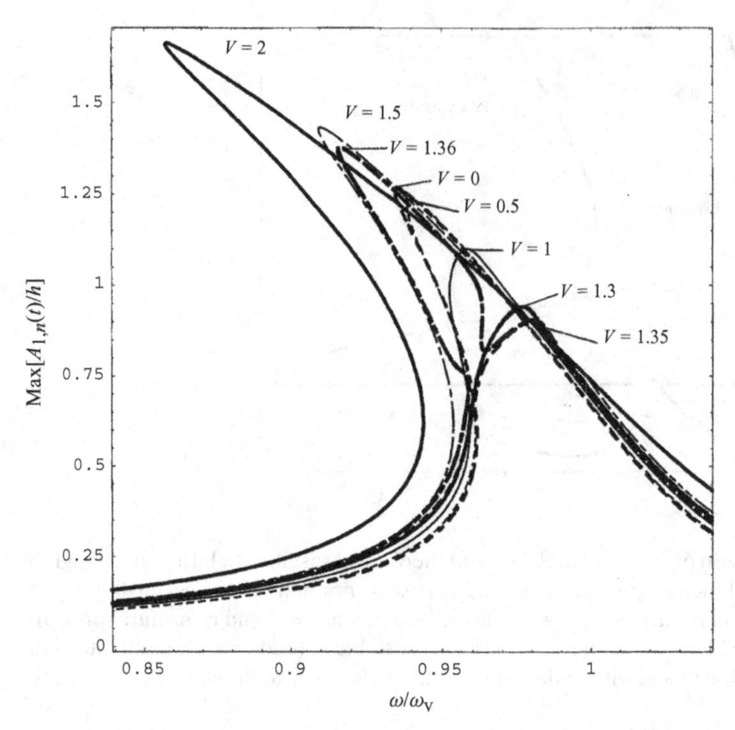

Figure 14.7. Frequency-response curves for the driven mode without companion mode participation at different flow velocities V; $f_{1,5} = 0.03\, h^2 \rho \omega_{1,5}^2$; 7 dofs model.

case, the solution is restarted by using the solution at the previous point as the initial condition. The initial point of the solution is obtained by using the software AUTO 97.

The existence of (i) simple periodic response (P), (ii) quasi-periodic, amplitude-modulated response (M), (iii) period-9 responses (periodic response with a period nine times the excitation period), and (iv) chaos (C) for sufficiently large excitation is confirmed by the bifurcation diagrams of the Poincaré map given in Figure 14.8, obtained by varying the excitation frequency ω (for fixed $\tilde{f} = 0.3$ and $V = 2$). The "blue sky catastrophe" phenomenon is observed, whereby a sudden, explosive change in the response occurs. It is obtained when the period-9 (9-T) periodic orbit loses stability at $\omega/\omega_V = 1.0643$ and this gives rise to a chaotic region (ω_V is the linear radian frequency for flow velocity V).

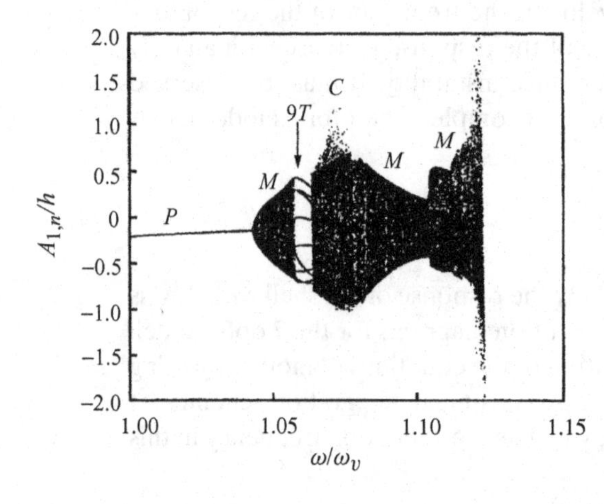

Figure 14.8. Bifurcation diagram for the driven mode $A_{1,5}(t)/h$ with companion mode participation for $f_{1,5} = 0.3\, h^2 \rho \omega_{1,5}^2$ and $V = 2$; 7 dofs model. P, simple periodic motion; M, modulated amplitude; 9T, periodic motion of multiple period; C, chaotic response.

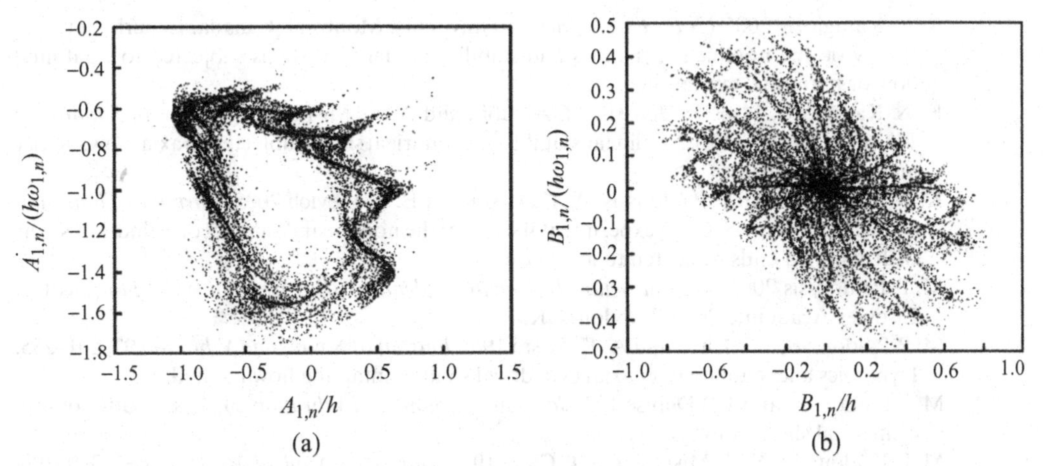

Figure 14.9. Poincaré maps for $f_{1,5} = 0.3\,h^2\rho\omega_{1,5}^2$, $\omega/\omega_V = 1.07446$ and $V = 2$. (a) $A_{1,5}(t)/h$. (b) $B_{1,5}(t)/h$.

In order to analyze the chaotic response immediately after the blue sky catastrophe, the system behavior is investigated for the following parameter values: $\omega/\omega_V = 1.07446$, $\tilde{f} = 0.3$ and $V = 2$. Figures 14.9(a,b) show the Poincaré maps projected on planes spanned by $(A_{1,5}, \dot{A}_{1,5})$ and $(B_{1,5}, \dot{B}_{1,5})$, where a strange attractor is found to be present. This chaotic trajectory is a perturbation of the unstable 9-T periodic orbit, as clarified by the power spectral analysis (Amabili et al. 2000b).

REFERENCES

M. Amabili and R. Garziera 2002a *Journal of Fluids and Structures* **16**, 31–51. Vibrations of circular cylindrical shells with nonuniform constraints, elastic bed and added mass. Part II: shells containing or immersed in axial flow.

M. Amabili and R. Garziera 2002b *Journal of Fluids and Structures* **16**, 795–809. Vibrations of circular cylindrical shells with nonuniform constraints, elastic bed and added mass. Part III: steady viscous effects on shells conveying fluid.

M. Amabili, F. Pellicano and M. P. Païdoussis 1999a *Journal of Sound and Vibration* **225**, 655–699. Non-linear dynamics and stability of circular cylindrical shells containing flowing fluid. Part I: stability.

M. Amabili, F. Pellicano and M. P. Païdoussis 1999b *Journal of Sound and Vibration* **228**, 1103–1124. Non-linear dynamics and stability of circular cylindrical shells containing flowing fluid. Part II: large-amplitude vibrations without flow.

M. Amabili, F. Pellicano and M. P. Païdoussis 2000a *Journal of Sound and Vibration* **237**, 617–640. Non-linear dynamics and stability of circular cylindrical shells containing flowing fluid. Part III: truncation effect without flow and experiments.

M. Amabili, F. Pellicano and M. P. Païdoussis 2000b *Journal of Sound and Vibration* **237**, 641–666. Non-linear dynamics and stability of circular cylindrical shells containing flowing fluid. Part IV: large-amplitude vibrations with flow.

M. Amabili, F. Pellicano and M. P. Païdoussis 2001 *Journal of Applied Mechanics* **68**, 827–834. Nonlinear stability of circular cylindrical shells in annular and unbounded axial flow.

M. Amabili, F. Pellicano and M. P. Païdoussis 2002 *Computers and Structures* **80**, 899–906. Non-linear dynamics and stability of circular cylindrical shells conveying flowing fluid.

M. Bathe and R. D. Kamm 2001 *Proceedings of the First MIT Conference on Computational Fluid and Solid Mechanics*, Cambridge, MA, USA, Vol. 2, 1068–1072. Elsevier, Amsterdam. Fluid-structure interaction analysis in biomechanics.

W. Horn, G. W. Barr, L. Carter and R. O. Stearman 1974 *AIAA Journal* **12**, 1481–1490. Recent contributions to experiments on cylindrical shell panel flutter.

K. N. Karagiozis 2005 *Ph.D. Thesis*, McGill University, Montreal, Canada. Experiments and theory on the nonlinear dynamics and stability of clamped shells subjected to axial fluid flow or harmonic excitation.

K. N. Karagiozis, M. P. Païdoussis, M. Amabili and A. K. Misra 2008 *Journal of Sound and Vibration* **309**, 637–676. Nonlinear stability of cylindrical shells subjected to axial flow: theory and experiments.

K. N. Karagiozis, M. P. Païdoussis, A. K. Misra and E. Grinevich 2005 *Journal of Fluids and Structures* **20**, 801–816. An experimental study of the nonlinear dynamics of cylindrical shells with clamped ends subjected to axial flow.

M. P. Païdoussis 2003 *Fluid-Structure Interactions: Slender Structures and Axial Flow*, Vol. 2. Elsevier Academic Press, London, UK.

M. P. Païdoussis, S. P. Chan and A. K. Misra 1984 *Journal of Sound and Vibration* **97**, 201–235. Dynamics and stability of coaxial cylindrical shells containing flowing fluid.

M. P. Païdoussis and J. P. Denise 1972 *Journal of Sound and Vibration* **20**, 9–26. Flutter of thin cylindrical shells conveying fluid.

M. P. Païdoussis, A. K. Misra and S. P. Chan 1985 *Journal of Applied Mechanics* **52**, 389–396. Dynamics and stability of coaxial cylindrical shells conveying viscous fluid.

D. S. Weaver and T. E. Unny 1973 *Journal of Applied Mechanics* **40**, 48–52. On the dynamic stability of fluid-conveying pipes.

15 Nonlinear Supersonic Flutter of Circular Cylindrical Shells with Imperfections

15.1 Introduction

Flutter instability is due to Hopf bifurcation (see Chapter 3) of static equilibrium solution and is observed as an harmonic (or quasi-periodic) oscillation in the absence of excitation. Plates and shells in airflow present flutter instability for high speed, so that the design of wing panels of aircraft must be verified for panel flutter. The natural modes of plates and shells become complex modes in the presence of flow, as discussed in Chapter 14, and their natural frequencies are changed. In particular, the frequency of the mode with one longitudinal half-wave can be raised by increasing the flow speed, whereas the frequency of the mode with two longitudinal half-waves is lowered. At a point, the frequencies of these two modes coincide with a coalescence of the two modes; this gives rise to coupled-mode flutter.

The first reported occurrence of flutter instability for circular cylindrical shells appears to have been on the V-2 rocket (see Figure 15.1). Since that time, the study of the aeroelastic stability of cylindrical shells in axial flow is fundamental in the skin panel design of aerospace vehicles, high-performance aircrafts and missiles. A fundamental contribution to the studies on this topic was due to the introduction of the piston theory by Ashley and Zartarian (1956).

The nonlinear stability of simply supported, circular cylindrical shells in supersonic axial external flow is investigated in this chapter by using an improved model with respect to the one developed by Amabili and Pellicano (2002). Therefore, new improved results, not previously published, are presented. In particular, both linear aerodynamics (first-order piston theory) and nonlinear aerodynamics (third-order piston theory) are used. Geometric nonlinearities, due to finite-amplitude shell deformations, are considered by using Donnell's nonlinear shallow-shell theory, and the effect of viscous structural damping is taken into account. Asymmetric and axisymmetric geometric imperfections of circular cylindrical shells and internal pressurization are taken into account. The system is discretized by using the Galerkin method.

Numerical calculations are performed for a copper circular shell, fabricated by electroplating and tested in the 8 × 7-ft supersonic wind tunnel, at fixed Mach number 3, at the NASA Ames Research Center in 1964 (Olson and Fung 1966; 1967). During the experiments, it was observed that the pertinent streamwise wavelengths of interest were very large with respect to the boundary layer thickness, suggesting that the influence of the boundary layer was probably negligible.

Figure 15.1. V-2 German rocket.

15.1.1 Literature Review

Many interesting studies have investigated the shell stability in supersonic flow by using a linear shell model. Among others, Dowell (1966; 1975), Olson and Fung (1966), Barr and Stearman (1970) and Ganapathi et al. (1994) predicted the onset of flutter instability. Experiments (Olson and Fung 1966; Horn et al. 1974) have indicated that the oscillation amplitude of flutter is of the same order of magnitude as the shell thickness; therefore, a nonlinear shell theory should be used in order to predict accurately the flutter amplitude. Extensive reviews of works on the aeroelasticity of plates and shells were written by Dowell (1970a; 1975) and Bismarck-Nasr (1992); a few nonlinear studies on shells and curved panels were included. A specific review on nonlinear panel flutter was written by Mei et al. (1999), including five studies on curved plates. Many experimental results on aeroelastic stability of circular cylindrical shells in axial airflow were collected by Horn et al. (1974).

Only a few researchers used a nonlinear shell model to investigate the aeroelastic stability of cylindrical shells and curved panels in axial supersonic and hypersonic flow. Librescu (1965; 1967) studied the stability of shallow panels and finite-length circular cylindrical shells by using Donnell's nonlinear shallow-shell theory and a simple mode expansion without considering the companion mode (a second standing-wave mode described angularly by $\sin(n\theta)$, the orientation of which is at $\pi/(2n)$ with respect to the original one, described by $\cos(n\theta)$, n being the number of nodal diameters) nor the interaction with the axisymmetric modes. The absence of the companion mode does not permit traveling-wave flutter. Expansions neglecting the axisymmetric modes are not able to capture the correct nonlinear response of circular shells and are only suitable for curved panels. The theory developed by Librescu (1965; 1967; 1975) is also suitable for composite shells, and nonlinear terms in the supersonic flow pressure calculated by using the third-order piston theory were included. No results on limit cycle amplitudes were given.

Olson and Fung (1967) modeled simply supported shells using a simplified form of Donnell's nonlinear shallow-shell theory and a simple two-mode expansion without considering the companion mode, but including an axisymmetric term. In their study, the supersonic flow was modeled by using the linear piston theory. In subsequent studies, Evensen and Olson (1967; 1968) added the companion mode, employing a 4 degrees of freedom (dofs) mode expansion. This expansion allows the study of traveling-wave flutter, where nodal lines are traveling circumferentially around the shell; this phenomenon is similar to traveling waves predicted and measured for large-amplitude forced vibrations of circular cylindrical shells (see Chapter 5). However, these expansions are not moment-free at the ends of the shell, as they should be for classical simply supported shells, and the homogeneous solution for the stress function is neglected. Evensen and Olson (1967; 1968) investigated periodic solutions by using the harmonic balance method and solved the nonlinear algebraic equations only for some special cases. The results obtained are different, from the qualitative point of view, from those of Olson and Fung (1967). This is due to the different order of the perturbation approach used. Olsson (1978) added to the problem the effect of a particular temperature field on the material properties by using a simple two-mode expansion.

Carter and Stearman (1968) and Barr and Stearman (1969; 1970) wrote a series of theoretical and experimental studies on the supersonic flutter of circular cylindrical shells by including geometric imperfections, but the theoretical analysis was linear. They also introduced an improved linear piston theory to describe the shell-flow interaction for Mach numbers $M > 1.6$.

Amabili and Pellicano (2001) studied the aeroelastic stability of simply supported, circular cylindrical shells without imperfections in supersonic flow. Nonlinearities due to large-amplitude shell motion were considered by using Donnell's nonlinear shallow-shell theory, and the effect of viscous structural damping was taken into account. Two different in-plane constraints were applied to the shell edges: (i) zero axial force and (ii) zero axial displacement. The linear piston theory was applied to describe the fluid-structure interaction by using two different formulations, taking into account or neglecting the curvature correction term. The system was discretized by the Galerkin method and was investigated by using a model involving 7 dofs, allowing for traveling-wave flutter of the shell and shell axisymmetric contraction. Modes with up to two streamwise half-waves were considered. Results show that the system loses stability by supercritical traveling-wave flutter. A good agreement between theoretical calculations and experimental data reported by Olson and Fung (1966) was found for flutter amplitudes. In a successive paper, Amabili and Pellicano (2002) improved their model by considering a 22 dofs model, including geometric imperfections of different shape, considering internal pressurization and comparing linear and third-order piston theory. Numerical results were in agreement with experimental data at Mach 3 obtained at the NASA Ames Research Center in 1964 (Olson and Fund 1966; 1967).

Bolotin (1963) treated the nonlinear flutter of curved plates in his book on nonconservative problems. Dowell (1969; 1970b) investigated the nonlinear flutter of curved plates of shallow curvature by using a modified Donnell's nonlinear shallow-shell theory. Both simply supported and clamped plates were considered. The linear piston theory was used to describe the fluid-structure interaction. Six modes, with different numbers of streamwise waves, were included in the mode expansion. Limit cycle amplitudes were calculated and the effect of an internal pressurization was investigated. The effect of the curvature in the flow direction was analyzed; results show that streamwise curvature is dramatically destabilizing for the onset of flutter. Nonlinear flutter of in-plane loaded flat plates was studied by several authors (see, e.g. Epureanu et al. (2004).

Vol'mir and Medvedeva (1973) investigated the nonlinear flutter of circular cylindrical panels with initial imperfection and axial loads in supersonic flow. They used Donnell's nonlinear shallow-shell theory to model the panel dynamics and linear piston theory for the fluid-structure interaction. The numerical solution was obtained by using the finite-difference method. A more recent study on the influence of curvature on supersonic flutter of simply supported panels is due to Krause and Dinkler (1998); in this study, the curvature of the panel is in the direction of the flow. Krause and Dinkler used the finite-element method to discretize the structure taking into account geometric nonlinearities (modeled with Donnell's theory); geometric imperfections were used to describe the curvature of the panel. The third-order piston theory was used to model the fluid-structure interaction. They found that the flutter boundary is much lower for largely curved panels than for slightly curved panels. They also predicted chaotic flutter motion.

Hypersonic flutter of simply supported, orthotropic curved panels was studied by Bein et al. (1993) and Nydick et al. (1995) by using Donnell's nonlinear shallow-shell theory, the Galerkin method and direct integration of the equations of motion. Expansion of flexural displacement involving modes up to eight longitudinal half-waves and one circumferential half-wave showed convergence of the solution. First-order piston theory, third-order piston theory, Euler equations and Navier-Stokes equations were used to describe the fluid-structure interaction with hypersonic flow. Significant differences were found by using the four different models. Nonsimple harmonic motion with modulations of amplitude was observed for sufficiently high postcritical dynamic pressure. An extensive experimental study on supersonic flutter of flat and slightly curved panels at Mach number 2.81 was performed by Anderson (1962).

15.2 Theoretical Approach

15.2.1 Linear and Third-Order Piston Theory

The fluid-structure interaction used in this study is based on the piston theory (Ashley and Zartarian 1956). It is assumed that the pertinent streamwise wavelengths of interest are very large with respect to the boundary layer thickness, so that the influence of the boundary layer is negligible. The circular cylindrical shell is assumed to be subjected to an external supersonic flow in the axial direction.

According to the piston theory, the radial aerodynamic pressure p_a applied to the external surface of a shell can be obtained by analogy with the instantaneous isentropic pressure on the face of a piston moving with velocity Z into a perfect gas that is confined in a one-dimensional channel. This pressure is given by

$$p_a = p_\infty \left(1 + \frac{\gamma - 1}{2}\frac{Z}{a_\infty}\right)^{2\gamma/(\gamma-1)}, \tag{15.1}$$

where γ is the adiabatic exponent, p_∞ is the free-stream static pressure and a_∞ is the free-stream speed of sound. In the analogy, the piston velocity Z is replaced by $Z = V_\infty\, \partial(w + w_0)/\partial x + \partial w/\partial t$, V_∞ being the free-stream velocity, in order to obtain the radial aerodynamic pressure p_a applied to the surface of the shell as a consequence of the external supersonic flow. Here, w is, as usual, the radial displacement of the shell and w_0 is the geometric imperfection in the radial direction. Equation (15.1) can be expanded into Taylor series for the variable Z/a_∞ close to zero; the third-order expansion, neglecting higher-order terms, gives (Ashley and Zartarian 1956; Librescu 1965; 1975)

$$p_a = -\gamma p_\infty \left\{ \left[M \frac{\partial(w + w_0)}{\partial x} + \frac{1}{a_\infty} \frac{\partial w}{\partial t} \right] + \frac{\gamma + 1}{4} \left[M \frac{\partial(w + w_0)}{\partial x} + \frac{1}{a_\infty} \frac{\partial w}{\partial t} \right]^2 + \frac{\gamma + 1}{12} \left[M \frac{\partial(w + w_0)}{\partial x} + \frac{1}{a_\infty} \frac{\partial w}{\partial t} \right]^3 \right\}. \tag{15.2}$$

Neglecting nonlinear terms, equation (15.2) can be transformed into

$$p_a = -\gamma p_\infty \left[M \frac{\partial(w + w_0)}{\partial x} + \frac{1}{a_\infty} \frac{\partial w}{\partial t} \right]. \tag{15.3a}$$

Equation (15.3a) can be substituted with a more accurate expression obtained by linearized potential flow theory (Barr and Stearman 1970; Krumhaar 1963):

$$p_a = -\frac{\gamma p_\infty M^2}{(M^2 - 1)^{1/2}} \left\{ \frac{\partial(w + w_0)}{\partial x} + \frac{1}{M a_\infty} \left[\frac{M^2 - 2}{M^2 - 1} \right] \frac{\partial w}{\partial t} - \frac{w + w_0}{2(M^2 - 1)^{1/2} R} \right\}. \tag{15.3b}$$

In equation (15.3b), the last term is the curvature correction term, which is neglected in some studies of shell stability based on the piston theory. Except for the curvature correction term, equation (15.3b) reduces to equation (15.3a) for sufficiently high Mach numbers; equation (15.3b) is more accurate for low supersonic speed and can be used for $M > 1.6$.

In this study, equation (15.3b) is used for linear aerodynamics (referred as linear piston theory) and equation (15.2), with the linear terms modified according to equation (15.3b), for nonlinear aerodynamics (third-order piston theory).

15.2.2 Structural Model

Donnell's nonlinear shallow-shell theory, equations (1.162) and (1.163), is used to study the system stability, where the fluid pressure p in equation (1.162) is

$$p = p_m - p_a. \tag{15.4}$$

In equation (15.4), p_m is the pressure differential across the shell skin (positive outward) and p_a is the aerodynamic pressure given by the piston theory (positive in the same direction of w, i.e. inward).

In this study, attention is focused on a finite, simply supported (with zero axial load), circumferentially closed circular cylindrical shell of length L. The boundary conditions are given in equations (5.48) and (5.49).

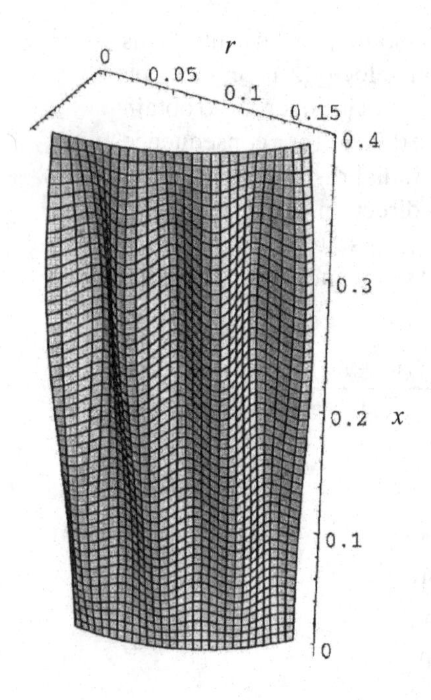

Figure 15.2. Mode shape of flutter for $p_\infty = 4500$ Pa in a scale with radial displacement augmented 1000 times. Only a part of 3/23 of the shell circumference is shown. A twist of the shell (skew flutter mode) and a larger movement on the downstream part of the shell are visible.

The radial displacement w is expanded by using the linear shell eigenmodes for zero flow, equation (5.50). In the numerical calculations, different expansions have been used and compared. The maximum number of degrees of freedom used in the numerical calculations for equation (5.50) is 22. It is observed, for reasons of symmetry, that the nonlinear interaction among linear modes of the chosen base involves only the asymmetric modes ($n > 0$) having a given n value, the asymmetric modes having a multiple value $k \times n$ of circumferential waves, where k is an integer, and axisymmetric modes ($n = 0$). The form of the radial displacement used in the numerical calculation is

$$w(x, \theta, t) = \sum_{m=1}^{4 \text{ or } 6} \sum_{k=1}^{1 \text{ or } 2} [A_{m,kn}(t) \cos(kn\theta) + B_{m,kn}(t) \sin(kn\theta)] \sin(\lambda_m x) + \sum_{m=1}^{6} A_{(2m-1),0}(t) \sin(\lambda_{(2m-1)} x). \tag{15.5}$$

Smaller expansions have been used for comparison purposes. It was observed in Chapter 14 that natural modes of shells are deformed by a flowing fluid and become complex modes; therefore, it is necessary to use in equation (15.5) modes with some longitudinal half-waves in order to reach good accuracy. The presence of couples of modes having the same shape but different angular orientations, the first one described by $\cos(n\theta)$ and the other by $\sin(n\theta)$, in the periodic response of the shell leads to the appearance of traveling-wave flutter around the shell in the angular direction. Moreover, the presence of $\cos(kn\theta)$ and $\sin(kn\theta)$ terms multiplied (i) by different time functions with a phase shift, and (ii) by $\sin(\lambda_m x)$ with different longitudinal wavenumber m, give rise to skew modes, which are flutter shapes with nodal lines no more parallel to the x axis, as shown in Figure 15.2, taken from Amabili and Pellicano (2001). Six axisymmetric modes are necessary to describe with accuracy the shell deformation due to pressurization.

The initial radial imperfection w_0 is expanded with equation (5.52). Nine terms in the expansion of imperfections are considered in the numerical calculations: (i) asymmetric imperfection having the same shape of the fluttering mode with one $(\tilde{A}_{1,n},\tilde{B}_{1,n})$, two $(\tilde{A}_{2,n}, \tilde{B}_{2,n})$ and three $(\tilde{A}_{3,n}, \tilde{B}_{3,n})$ longitudinal half-waves; (ii) axisymmetric imperfection with one, three and five longitudinal half-waves $(\tilde{A}_{1,0}, \tilde{A}_{3,0}, \tilde{A}_{5,0})$. Additional terms can be inserted, but the ones considered have the greatest effect on the shell dynamics.

The equations of motion are obtained by the Galerkin method.

15.3 Numerical Results

Numerical results have been obtained for a case studied experimentally by Olson and Fung (1966) and theoretically investigated by Olson and Fung (1967), Evensen and Olson (1967; 1968), Carter and Stearman (1968), Barr and Stearman (1969) and Amabili and Pellicano (2001; 2002). The shell and the airflow have the following characteristics: $R = 0.2032$ m, $L = 0.39116$ m, $R/h = 2000$, $E = 110.32 \times 10^9$ Pa, $\rho = 8905.37\,\text{kg/m}^3$, $\nu = 0.35$, $\gamma = 1.4$, $a_\infty = 213.36\,\text{m/s}$ and $M = 3$; the free-stream stagnation temperature is 48.9°C. A structural modal damping coefficient $\zeta_{1,n} = 0.0005$, which is compatible with the test shell (Evensen and Olson 1967; 1968), is assumed; for other modes, $\zeta_x = \zeta_{1,n}\omega_{1,n}/\omega_x$. The test shell was extremely thin, fabricated with copper by electroplating, and was tested in the 8×7-ft supersonic wind tunnel at the NASA Ames Research Center. The experimental boundary conditions at the shell edges were quite complex (Olson and Fung 1966). In particular, the test shell was soldered to two copper end rings, mounted over O-ring seals to allow thermal expansion. In the present calculations, they have been simulated with simply supported edges; actual boundary conditions, were between simply supported and clamped edges.

15.3.1 Linear Results

Eigenvalues associated with the first and second longitudinal modes with $n = 24$ circumferential waves and damping coefficient $\zeta_{1,n} = 0.0005$ are given in Figure 15.3 versus the value of the free-stream static pressure p_∞ by evaluating the aerodynamic pressure with equation (15.3b) for $p_m = 0$. The imaginary part of the eigenvalues λ gives the oscillation frequency, and the real part, being the exponent of the real exponential function, gives damping for negative values and instability for positive values. Flutter starts when the real part of λ crosses the zero value, in this case for $p_\infty = 3781$ Pa; this value gives the critical free-stream static pressure. In Figure 15.3(a) the lower and higher of the two branches give the oscillation frequencies of the first and second modes (with $n = 24$), respectively. When they coalesce, a single mode is obtained and its oscillation frequency slowly increases with p_∞ because the stiffness of the system is increased with p_∞ by the curvature correction term in equation (15.3b). This mode originates flutter when the real part of the eigenvalue becomes positive in Figure 15.3(b). The corresponding value of the free-stream static pressure is the critical p_∞ originating flutter.

Figure 15.3 has been obtained by using a model with 18 dofs, with the options $m = 1, \ldots, 6$ and $k = 1$ in equation (15.5). The convergence of the critical p_∞ originating flutter with the number of degrees of freedom is shown in Table 15.1. In particular, the effect of modes with $2n$ circumferential waves and modes with angular

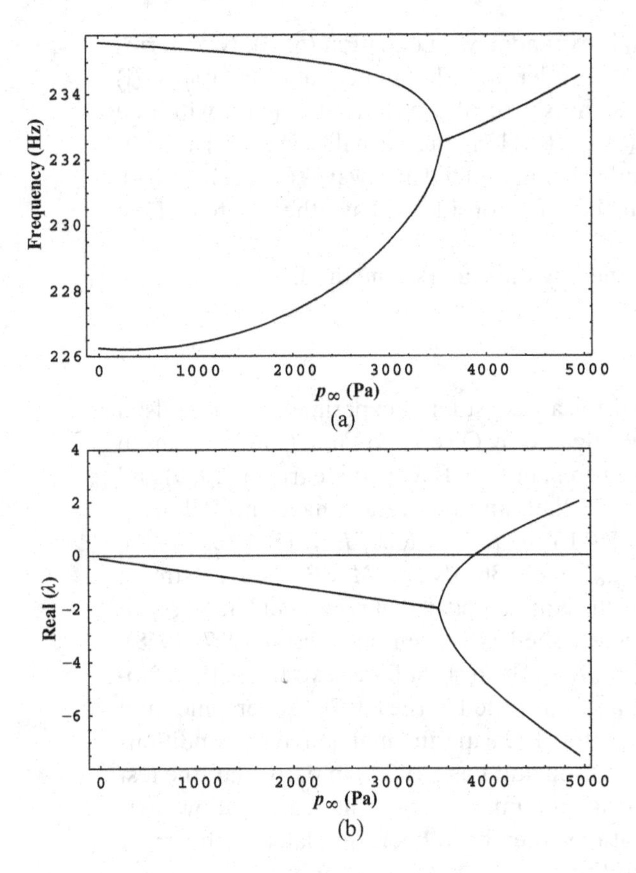

Figure 15.3. Eigenvalues of the system versus the free-stream static pressure p_∞; $n = 24$, $p_m = 0$, model with 18 dofs. (a) Imaginary part of the eigenvalues λ, i.e. frequency. (b) Real part of the eigenvalues λ, i.e. damping.

function $\sin(n\theta)$ is negligible on the onset of flutter. However, the contribution of the latter ones to the nonlinear flutter response is fundamental.

The critical free-stream static pressure p_∞ is reported in Figure 15.4 for different numbers of circumferential waves for $p_m = 0$. The value of n associated with the lowest critical free-stream static pressure is $n = 25$ for $p_m = 0$. For a different value of p_m, the shape of the curve is changed, as shown in Table 15.2.

Internal pressure p_m has a stabilizing effect on the system, as shown in Table 15.2. In particular, internal pressure gives rise to an initial axisymmetric deformation outward, which has a stabilizing effect. In most previous studies, this deformation was

Table 15.1. *Critical free-stream static pressure giving onset of flutter calculated for different expansions of the radial displacement w, equation (15.5); $n = 24$, $p_m = 0$. Case (i) with 22 modes: expansion given in equation (15.5) with the options of a maximum of 4 longitudinal half-waves and $k = 1, 2$. Case (ii) with 18 modes: $A_{m,n}$ and $B_{m,n}$ with $m = 1, \ldots, 6$, plus the six axisymmetric modes of equation (15.5). Case (iii) with 10 modes: elimination from case (i) of all the sine modes and modes with* $\cos(2n\theta)$. *Case (iv) with 5 modes: expansion retaining only $A_{1,n}$, $A_{2,n}$, $A_{1,0}$, $A_{3,0}$, $A_{5,0}$*

	Case (i)		Case (ii)	Case (iii)	Case (iv)
Number of modes	22	22	18	10	5
Piston theory	3rd order	linear	linear	linear	linear
Critical p_∞ (Pa)	3644	3644	3781	3644	2520

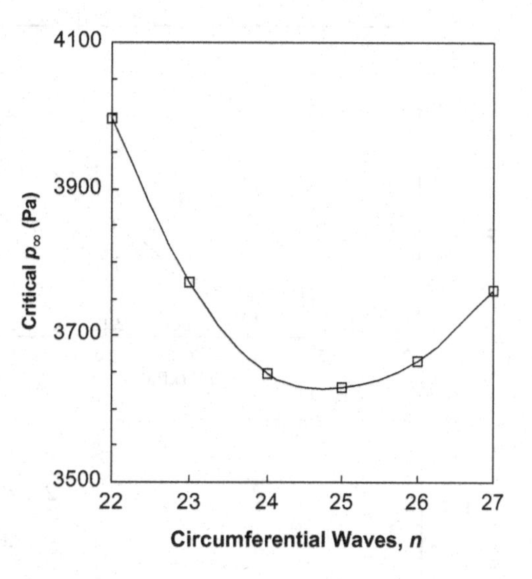

Figure 15.4. Critical free-stream static pressure p_∞ versus the number n of circumferential waves for $p_m = 0$. A model with 10 dofs for each value of n has been used.

neglected, because it should be evaluated by using a nonlinear deformation theory. Table 15.2 shows that this deformation has a large effect.

15.3.2 Nonlinear Results without Geometric Imperfections

In this subsection, the shell is considered without geometric imperfections. Calculations have been performed for a number $n = 24$ of circumferential waves and a pressure differential across the shell skin $p_m = 0$. Even if $n = 25$ is the most critical value for $p_m = 0$ (but no more for a different p_m), as shown by Figure 15.4, $n = 24$ is extremely close and the behavior of the system for these two values is almost the same. The free-stream static pressure p_∞ has been used as the bifurcation parameter instead of the Mach number M. In fact, experimental data available for comparison from the supersonic wind tunnel tests (Olson and Fung 1966; Barr and Stearman 1969) were collected by varying p_∞ and keeping M constant. The flight velocity is $U = Ma_\infty$ and $a_\infty = \sqrt{\gamma p_\infty/\rho_\infty}$; therefore, the flight velocity U can be easily related to the free-stream static pressure p_∞.

Solutions of the nonlinear equations of motion have been obtained numerically by using (i) the AUTO 97 computer program and (ii) direct integration of the equations of motion. AUTO is not able to detect surfaces coming out from a bifurcation point, but it can detect branches. For the axisymmetry, the system does not possess a

Table 15.2. *Critical free-stream static pressure p_∞ (Pa) giving onset of flutter calculated for different numbers of circumferential waves n and different internal pressurizations p_m neglecting and including the shell deflection due to pressurization. Model with 22 dofs*

	$n = 22$	$n = 23$	$n = 24$	$n = 25$	$n = 26$	$n = 27$
$p_m = 0$	3996	3773	3644	3628	3665	3763
$p_m = 3447.5$ Pa without pressure deflection	4673	4565	4548	4633	4730	4853
$p_m = 3447.5$ Pa with pressure deflection	6715	6722	6810	7083	6959	6481

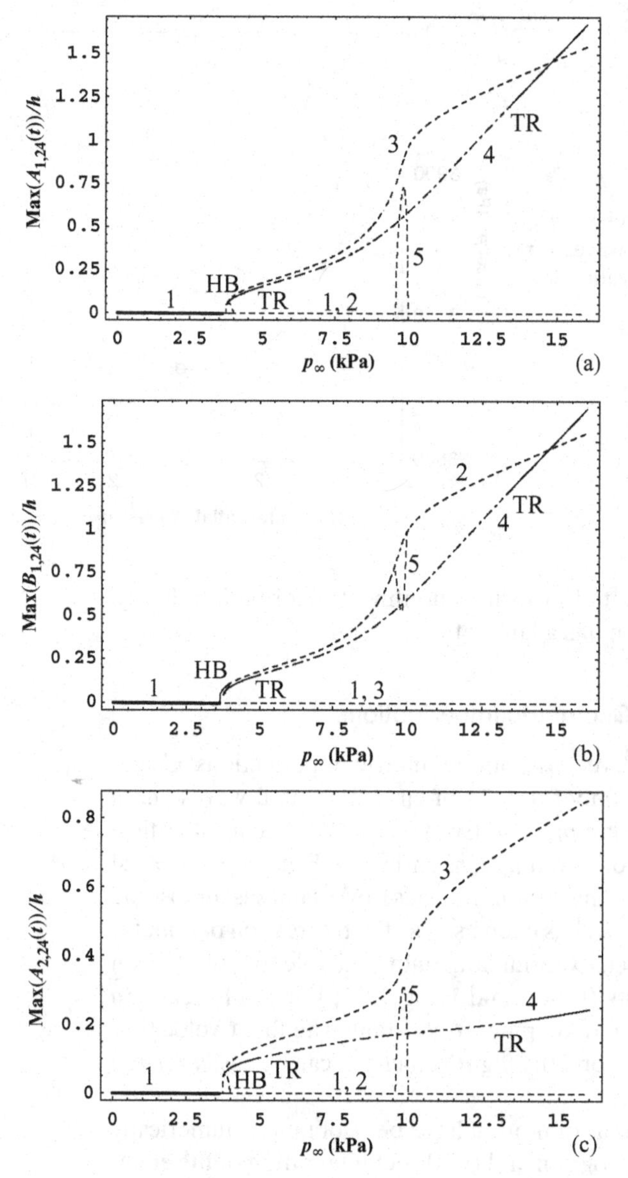

Figure 15.5. Amplitude of oscillatory solutions versus the free-stream static pressure; $n = 24$, linear piston theory, 22 dofs model. —, Stable solutions; –·–, stable quasi-periodic solution; – –, unstable solutions; HB, Hopf bifurcation; TR, Neimark-Sacker bifurcation. (a) Maximum amplitude of the first longitudinal mode $A_{1,24}(t)/h$. (b) Maximum amplitude of the first longitudinal mode $B_{1,24}(t)/h$. (c) Maximum amplitude of the second longitudinal mode $A_{2,24}(t)/h$. (d) Maximum amplitude of the second longitudinal mode $B_{2,24}(t)/h$. (e) Maximum amplitude of the third longitudinal mode $A_{3,24}(t)/h$. (f) Maximum amplitude of the first axisymmetric mode $A_{1,0}(t)/h$.

preferential angular coordinate θ to locate the deformation; in the present case, surfaces come out from bifurcation points. In order to use AUTO, a bifurcation analysis was performed introducing a small perturbation to the linear part of the system. This approach is analogous to having a very small difference in the stiffness of the system for the couple of modes described by the generalized coordinates $A_{1,n}$ and $B_{1,n}$. This perturbation allows normal bifurcation analysis, because line branches now emerge from bifurcation points instead of surfaces. A perturbation of 0.2% to the linear frequency of the mode corresponding to $B_{1,n}$ has been used in the present case, so that differences with respect to the actual systems are almost negligible.

The bifurcation curves for the most important modal coordinates versus the free-stream static pressure p_∞ are shown in Figure 15.5 for the aerodynamic pressure given by linear piston theory. In this case, 22 modes have been used. The expansion is given in equation (15.5) with the options of a maximum of 4 longitudinal half-waves and

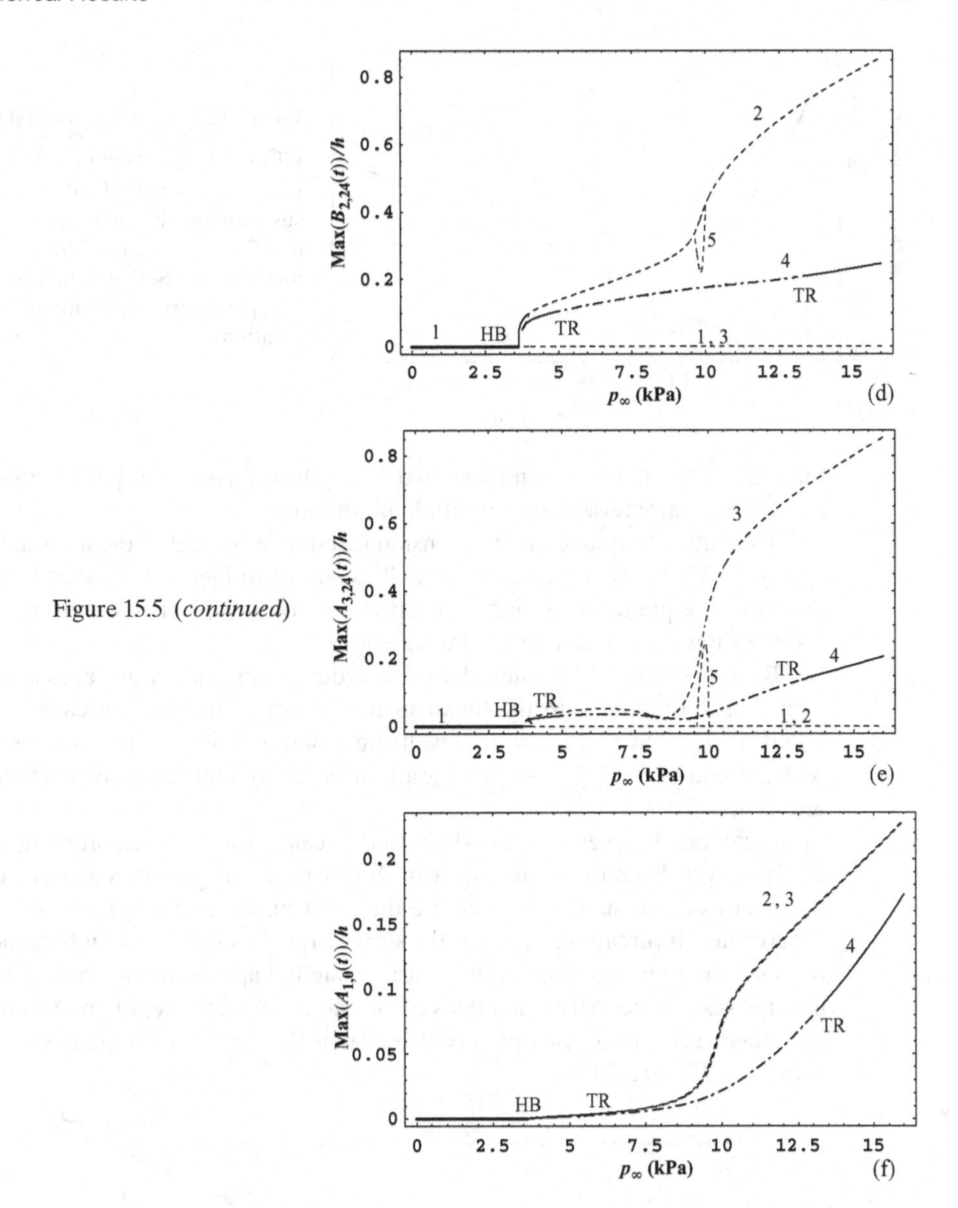

Figure 15.5 (*continued*)

$k = 1, 2$. In Figure 15.5, the curves correspond to the flutter amplitudes of the shell (excluding branch 1, which is relative to the trivial equilibrium position). Results show that the perfect shell lose stability for $p_\infty = 3644$ Pa through supercritical Hopf bifurcation on branch 1. Branches 2 and 3 correspond to standing-wave flutter and lose stability very soon through bifurcation. Branch 4, which is the attractive solution, represents a traveling-wave flutter around the shell circumference, with phase shift between $A_{1,n}(t)$ and $B_{1,n}(t)$ of $\vartheta_2 - \vartheta_1 = \pi/2$. Moreover, the couple of modes with equal longitudinal wavenumber m has the same amplitude, giving pure traveling-wave flutter. Note that, excluding a very small range of p_∞ after the onset of instability due to the perturbation introduced, all the stable flutter is a traveling wave around the circumference, as observed in the experiments by Olson and Fung (1966). Branch 4 in Figure 15.5 presents a Neimark-Sacker (torus) bifurcation for $p_\infty = 4863$ Pa and a second Neimark-Sacker bifurcation at $p_\infty = 13{,}318$ Pa. In

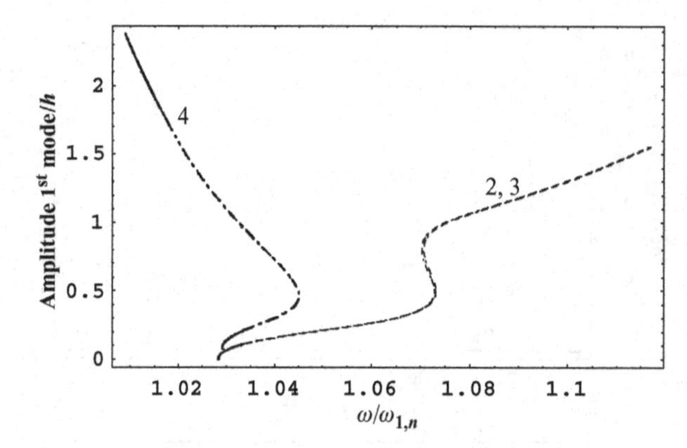

Figure 15.6. Nondimensional flutter amplitude $\sqrt{(A_{1,24}^2(t) + B_{1,24}^2(t))}/h$ of modes with one longitudinal half-wave versus nondimensional flutter frequency; $n = 24$, $\omega_{1,n} = 2\pi \times 226.3$ rad/s, 22 dofs model. —, Stable solutions; –·–, stable quasi-periodic solution; – –, unstable solutions.

the range confined between these two bifurcations, there is a quasi-periodic flutter oscillation characterized by amplitude modulations.

The flutter frequency, nondimensionalized with respect to the natural frequency $\omega_{1,n}$ $(n = 24)$ of the unpressurized shell, is shown in Figure 15.6. Results show that the flutter frequency of branch 4, the attractive solution, is almost constant; branches 2 and 3 show a mild supercritical behavior.

Results obtained by using the third-order piston theory are almost coincident with those obtained with the linear piston theory in the present case $(M = 3)$. In Figure 15.7, a comparison of the results obtained by using the two theories is shown with a zoom of the bifurcation diagram in order to appreciate the extremely small difference.

It can be observed that the shell used in calculations has a softening nonlinearity. However, because at the onset of flutter there is a coalescence of the natural frequency of two shell modes with a different number of longitudinal half-waves, supercritical bifurcations arise for the stable branches 2, 3 and 4. In fact, the passage from softening to hardening nonlinearity usually happens in the correspondence of internal resonances between modes of the shells, for example, when two modes have an integer ratio (one, two or three) between their natural frequency as shown in Chapter 5 (Figure 5.13).

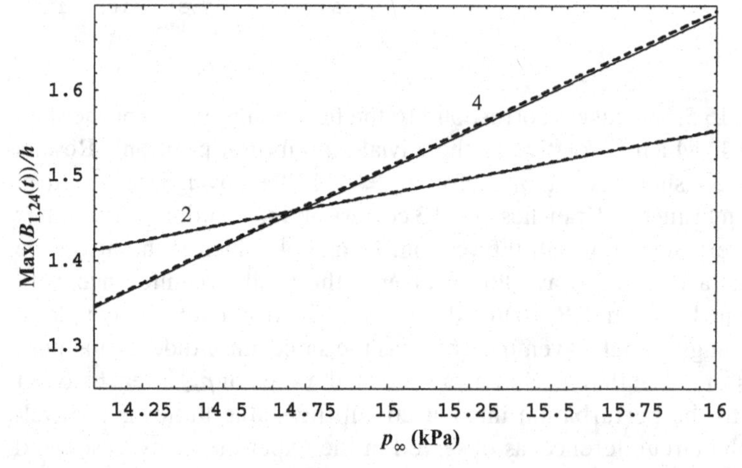

Figure 15.7. Amplitude of oscillatory solutions versus the free-stream static pressure; $n = 24$, 22 dofs model. – –, third-order piston theory; —, linear piston theory.

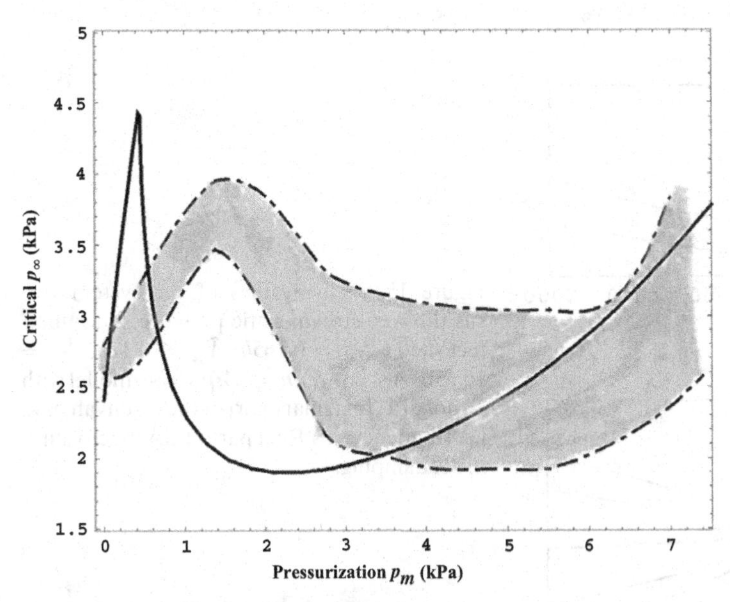

Figure 15.8. Critical free-stream static pressure p_∞ versus the pressure differential across the shell skin p_m. —, Theoretical results for imperfect shell, $\tilde{B}_{1,24} = 0.165h$, $\tilde{A}_{3,24} = 0.12h$, $\tilde{A}_{1,0} = 10.65h$, $\tilde{A}_{3,0} = -h$, 16 dofs model, $n = 24$; the gray area delimited by –·– represents the NASA experimental data (Barr and Stearman 1969).

15.3.3 Nonlinear Results with Geometric Imperfections

Results show that onset of flutter is very sensitive to small initial imperfections. In particular, asymmetric imperfections are ironed out by the pressurization of the shell, whereas the axisymmetric imperfections are not. This is in agreement with what Hutchinson (1965) predicted for the buckling of circular cylindrical shells. However, asymmetric imperfections change the natural frequency of the asymmetric modes, which are the fundamental ones to predict flutter boundary, more significantly than axisymmetric imperfections.

Calculations have been performed with different combinations of asymmetric and axisymmetric imperfections. However, for the sake of brevity, all the results reported in this section are relative to the following geometric imperfections: $\tilde{B}_{1,24} = 0.165h$, $\tilde{A}_{3,24} = 0.12h$, $\tilde{A}_{1,0} = 10.65h$ and $\tilde{A}_{3,0} = -h$; all the other coefficients in equation (5.52) are zero. Considering that h is about 0.1 mm, the asymmetric imperfections are almost imperceptible and the axisymmetric imperfections, which play a much smaller role, are compatible with the soldered connection of the shell to end rings, giving deflection inward. Calculations have been performed by linear piston theory and 16 modes have been used in the expansion of w. The expansion is the one in equation (15.5) with the option of a maximum of 4 longitudinal half-waves for $k = 1$ and only the first longitudinal mode for $k = 2$. Calculations have been performed for $n = 24$ by using linear piston theory. More accurate results could be obtained by merging results from models considering different numbers of circumferential waves; in fact, $n = 24$ can be not the most critical value by varying the pressurization p_m.

Figure 15.8 shows the flutter boundary versus the pressure differential across the shell skin for the imperfect shell. Experimental results obtained at the NASA

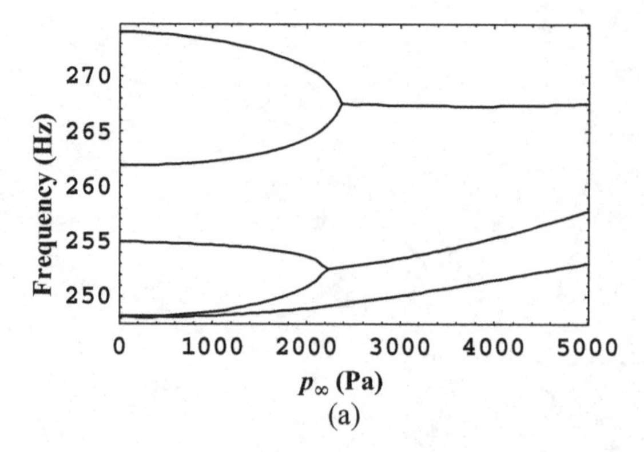

(a)

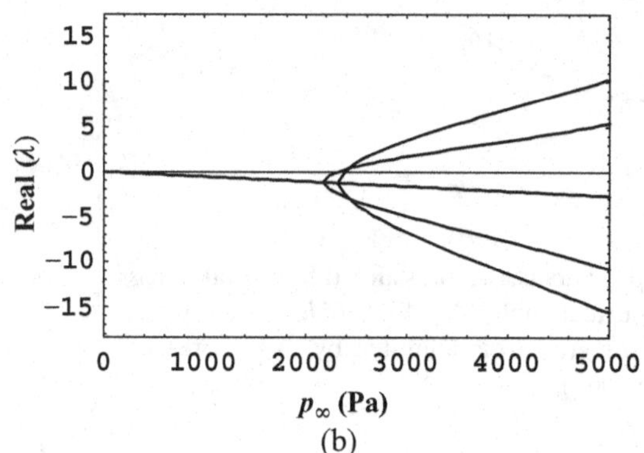

(b)

Figure 15.9. Eigenvalues of the system versus the free-stream static pressure p_∞; imperfect shell ($\tilde{B}_{1,24} = 0.165h$, $\tilde{A}_{3,24} = 0.12h$, $\tilde{A}_{1,0} = 10.65h$, $\tilde{A}_{3,0} = -h$), $n = 24$, $p_m = 0$, model with 16 dofs. (a) Imaginary part of the eigenvalues λ, i.e. frequency. (b) Real part of the eigenvalues λ, i.e. damping.

Ames Research Center in 1964 (Olson and Fung 1966; Barr and Stearman 1969) are also shown for comparison. The computed results are in satisfactory agreement with the experiments. In particular, the onset of flutter initially increases quickly with the pressure differential p_m up to a maximum. This part of the curve is associated with a Hopf bifurcation arising from merging of the frequencies of modes with two and three longitudinal half-waves as shown in Figure 15.9. The second part of the curve, to the right of the maximum, is associated with a Hopf bifurcation arising from merging of the frequencies of modes with one and two longitudinal half-waves. For increased pressure p_m, these modes are the first to merge, as for the perfect shell.

It can be observed that the maximum of the computed curve in Figure 15.8 is moved to the left with respect to experiments. Imperfections should be studied on a statistical basis because data on the specimens used in the experiments are not available. Moreover, imperfections having a different number of circumferential waves with respect to the fluttering mode should be considered. However, the imperfections introduced reproduce the experimental results of Olson and Fung (1966), who observed that: (i) small internal pressurization was very stabilizing; (ii) moderate pressurization reduced stability to the unpressurized level; (iii) high internal pressure completely stabilized the shell. In particular, observation (ii) was very surprising at that time. The observed behavior is well explained by asymmetric imperfections that are ironed out by the moderate pressurization of the shell.

In Figure 15.10 the curves correspond to maximum flutter amplitudes of the imperfect shell studied in Figure 15.8 for $p_m = 4750$ Pa; for brevity, only the generalized coordinates $A_{1,n}$ and $B_{1,n}$ are shown. All branches (excluding branch 1) correspond to traveling-wave flutter. The shell loses stability by Hopf bifurcation at

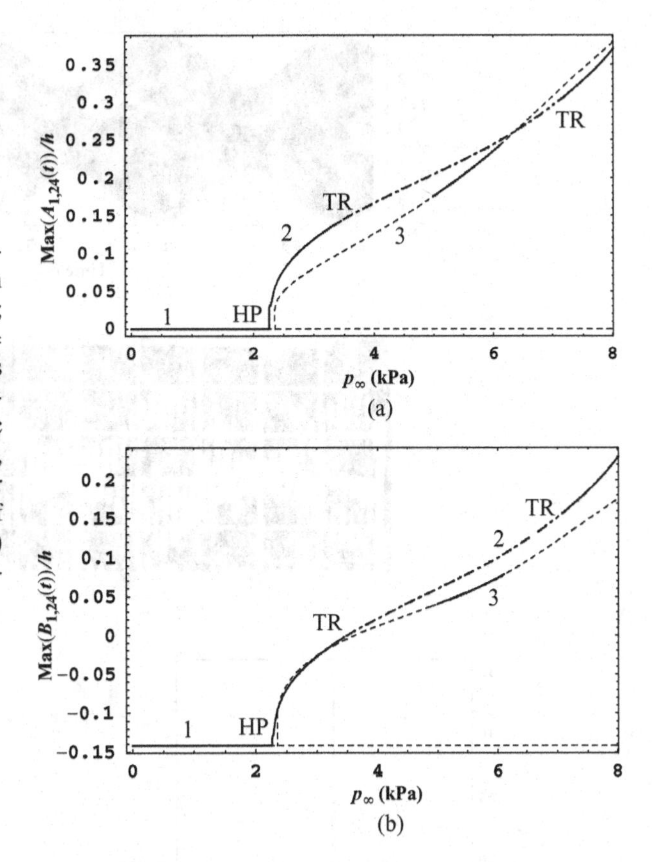

Figure 15.10. Amplitude of oscillatory solutions versus the free-stream static pressure; $n = 24$, $p_m = 4750$ Pa; imperfect shell ($\tilde{B}_{1,24} = 0.165h$, $\tilde{A}_{3,24} = 0.12h$, $\tilde{A}_{1,0} = 10.65h$, $\tilde{A}_{3,0} = -h$), 16 dofs model, linear piston theory. ——, Stable solutions; –·–, stable quasi-periodic solution; – –, unstable solutions; HB, Hopf bifurcation; TR, Neimark-Sacker bifurcation. (a) Maximum amplitude of the first longitudinal mode $A_{1,n}(t)/h$. (b) Maximum amplitude of the first longitudinal mode $B_{1,n}(t)/h$.

$p_\infty = 2263$ Pa; a second Hopf bifurcation arises at $p_\infty = 2356$ Pa. The flutter amplitudes obtained in Figure 15.10 are in very good agreement with experimental results described by Olson and Fung (1966) where a flutter amplitude of about $0.5h$ rms is reported at $x/L = 0.72$, corresponding to about $0.7h$ for simple harmonic oscillations. The attractive solution is branch 2 (even if branch 3 also presents a stable region) and corresponds to traveling-wave flutter with and without amplitude modulations. In particular, for $3445 < p_\infty < 7113$ Pa, that is, between two Neimark-Sacker bifurcations, traveling-wave flutter with amplitude modulations arises. The flutter frequency has only small variations, at about 665 Hz, as shown in Figure 15.11. This value is also in very good agreement with experimental results (Olson and Fung 1966).

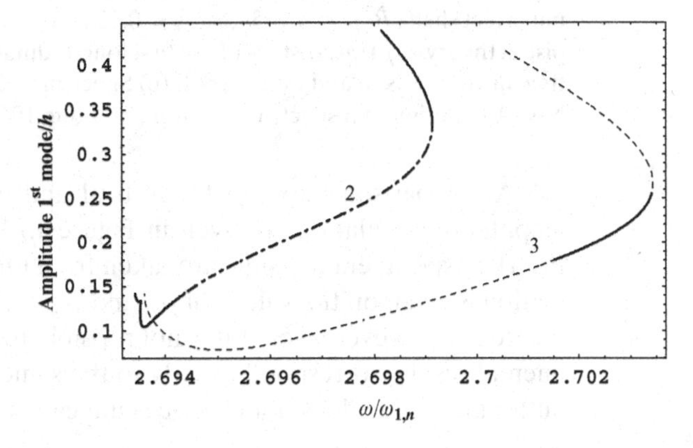

Figure 15.11. Nondimensional flutter amplitude $\sqrt{(A_{1,24}^2(t) + B_{1,24}^2(t))}/h$ of modes with one longitudinal half-wave versus nondimensional flutter frequency; $n = 24$, $p_m = 4750$ Pa, imperfect shell ($\tilde{B}_{1,24} = 0.165h$, $\tilde{A}_{3,24} = 0.12h$, $\tilde{A}_{1,0} = 10.65h$, $\tilde{A}_{3,0} = -h$), 16 dofs model, $\omega_{1,n} = 2\pi \times 248$ rad/s. ——, Stable solutions; –·–, stable quasi-periodic solution; – –, unstable solutions.

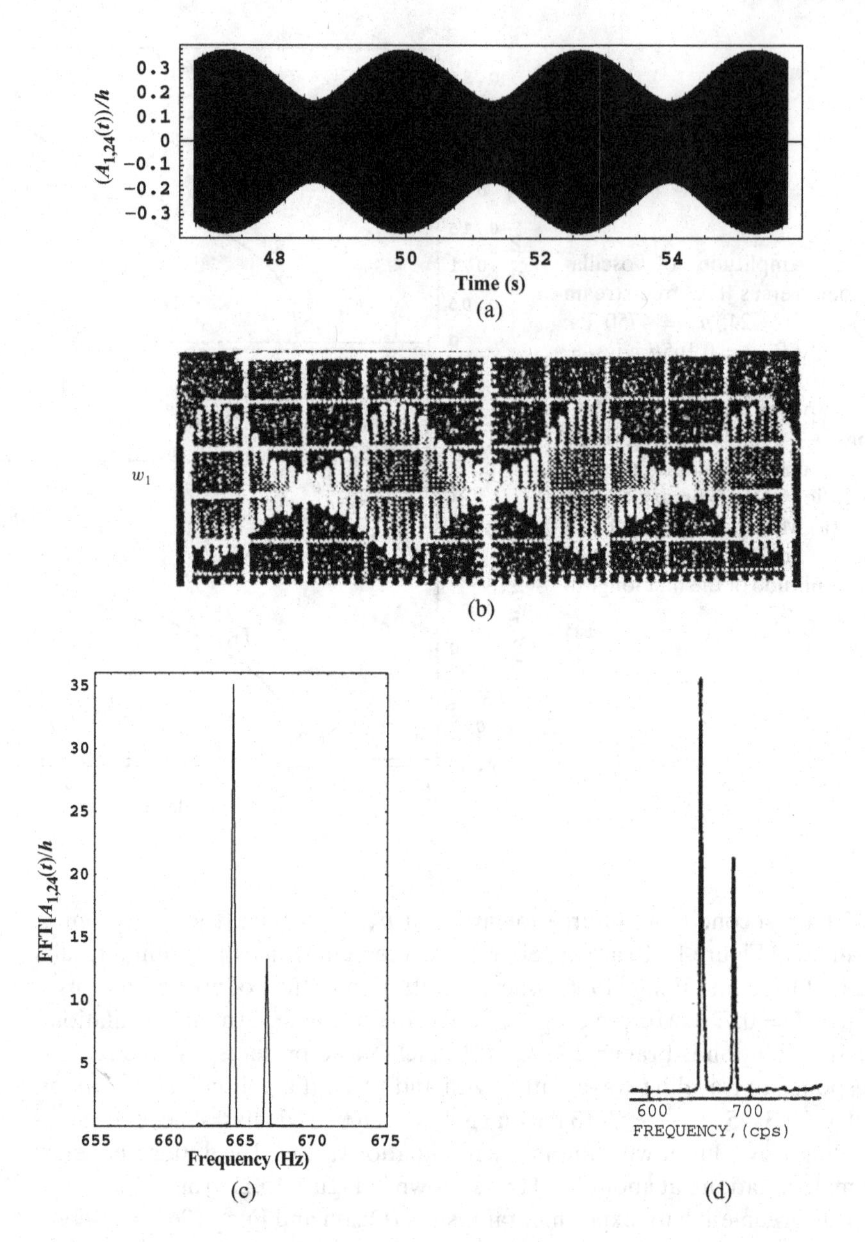

Figure 15.12. Flutter response of the imperfect shell for $p_\infty = 6700$ Pa; $n = 24$, $p_m = 4750$ Pa, imperfect shell ($\tilde{B}_{1,24} = 0.165h$, $\tilde{A}_{3,24} = 0.12h$, $\tilde{A}_{1,0} = 10.65h$, $\tilde{A}_{3,0} = -h$), 16 dofs model, linear piston theory. (a) Time history of the first longitudinal mode $A_{1,n}(t)/h$. (b) NASA experimental time history (Olson and Fung 1966). (c) Spectrum of the first longitudinal mode $A_{1,n}(t)/h$. (d) NASA experimental spectrum (Olson and Fung 1966).

A comparison between theoretical and experimental results for flutter with amplitude modulations is given in Figure 15.12 for $p_m = 4750$ Pa and $p_\infty = 6700$ Pa. The experimental results are taken from Olson and Fung (1966) where no information is given on the values of p_m and p_∞ for which these experimental data were recorded. However, even if it is not possible to say that these theoretical and experimental results correspond exactly to the same conditions, it was observed that the flutter amplitude does not change significantly with p_m and p_∞. Figure 15.12 shows that calculations and experiments are in good agreement. It seems that the numerical model is capable of reproducing quantitatively the experimental results obtained at the NASA Ames Research Center in 1964.

It is very important to observe that the nature of the Hopf bifurcation is supercritical for all the calculations performed for Mach number $M = 3$. This is also in agreement with the experiments available that show mild supersonic flutter with amplitude gradually increasing with the free-stream static pressure (i.e. with the flight speed). Differently, for subsonic incompressible flow, highly catastrophic subcritical divergence was always observed in Chapter 14.

REFERENCES

M. Amabili and F. Pellicano 2001 *AIAA Journal* **39**, 564–573. Nonlinear supersonic flutter of circular cylindrical shells.

M. Amabili and F. Pellicano 2002 *Journal of Applied Mechanics* **69**, 117–129. Multimode approach to nonlinear supersonic flutter of imperfect circular cylindrical shells.

W. J. Anderson 1962 Experiments on the flutter of flat and slightly curved panels at Mach number 2.81. *AFOSR 2996*, Graduate Aeronautical Laboratories, California Institute of Technology, Pasadena, CA, USA.

H. Ashley and G. Zartarian 1956 *Journal of Aeronautical Science* **23**, 1109–1118. Piston theory – a new aerodynamic tool for the aeroelastician.

G. W. Barr and R. O. Stearman 1969 *AIAA Journal* **7**, 912–919. Aeroelastic stability characteristics of cylindrical shells considering imperfections and edge constraint.

G. W. Barr and R. O. Stearman 1970 *AIAA Journal* **8**, 993–1000. Influence of a supersonic flowfield on the elastic stability of cylindrical shells.

T. Bein, P. P. Friedmann, X. Zhong and I. Nydick 1993 Hypersonic flutter of a curved shallow panel with aerodynamic heating. *Proceedings of the 34th AIAA/ASME/ASCE/AHS/ASC Structures, Structural Dynamics and Materials Conference*, 19–22 April, La Jolla, California, pp. 1–15 (AIAA paper 93–1318).

M. N. Bismarck-Nasr 1992 *Applied Mechanics Reviews* **45**, 461–482. Finite element analysis of aeroelasticity of plates and shells.

V. V. Bolotin 1963 *Nonconservative Problems of the Theory of Elastic Stability*. MacMillan, New York.

L. L. Carter and R. O. Stearman 1968 *AIAA Journal* **6**, 37–43. Some aspects of cylindrical shell panel flutter.

E. H. Dowell 1966 *AIAA Journal* **4**, 1510–1518. Flutter of infinitely long plates and shells. Part II: cylindrical shell.

E. H. Dowell 1969 *AIAA Journal* **7**, 424–431. Nonlinear flutter of curved plates.

E. H. Dowell 1970a *AIAA Journal* **8**, 385–399. Panel flutter: a review of the aeroelastic stability of plates and shells.

E. H. Dowell 1970b *AIAA Journal* **8**, 259–261. Nonlinear flutter of curved plates, II.

E. H. Dowell 1975 *Aeroelasticity of Plates and Shells*. Noordhoff International, Leyden, The Netherlands.

B. I. Epureanu, L. S. Tang and M. P. Païdoussis 2004 *International Journal of Non-Linear Mechanics* **39**, 977–991. Coherent structures and their influence on the dynamics of aeroelastic panels.

D. A. Evensen and M. D. Olson 1967 *Nonlinear Flutter of a Circular Cylindrical Shell in Supersonic Flow*. NASA TN D-4265, U.S. Government Printing Office, Washington, DC, USA.

D. A. Evensen and M. D. Olson 1968 *AIAA Journal* **6**, 1522–1527. Circumferentially travelling wave flutter of a circular cylindrical shell.

M. Ganapathi, T. K. Varadan and J. Jijen 1994 *Journal of Sound and Vibration* **171**, 509–527. Field-consistent element applied to flutter analysis of circular cylindrical shells.

W. Horn, G. W. Barr, L. Carter and R. O. Stearman 1974 *AIAA Journal* **12**, 1481–1490. Recent contributions to experiments on cylindrical shell panel flutter.

J. Hutchinson 1965 *AIAA Journal* **3**, 1461–1466. Axial buckling of pressurized imperfect cylindrical shells.

H. Krause and D. Dinkler 1998 The influence of curvature on supersonic panel flutter. *Proceedings of the 39th AIAA/ASME/ASCE/AHS/ASC Structures, Structural Dynamics and Materials Conference*, Long Beach, California, pp. 1234–1240 (AIAA paper 98–1841).

H. Krumhaar 1963 *AIAA Journal* **1**, 1448–1449. The accuracy of linear piston theory when applied to cylindrical shells.

L. Librescu 1965 *Journal de Mécanique* **4**, 51–76. Aeroelastic stability of orthotropic heterogeneous thin panels in the vicinity of the flutter critical boundary. Part I: simply supported panels.

L. Librescu 1967 *Journal de Mécanique* **6**, 133–152. Aeroelastic stability of orthotropic heterogeneous thin panels in the vicinity of the flutter critical boundary. Part II.

L. Librescu 1975 *Elastostatics and Kinetics of Anisotropic and Heterogeneous Shell-Type Structures*. Noordhoff, Leiden, The Netherlands.

C. Mei, K. Abdel-Motagaly and R. Chen 1999 *Applied Mechanics Reviews* **52**, 321–332. Review of nonlinear panel flutter at supersonic and hypersonic speeds.

I. Nydick, P. P. Friedmann and X. Zhong 1995 Hypersonic panel flutter studies on curved panels. *Proceedings of the 36th AIAA/ASME/ASCE/AHS/ ASC Structures, Structural Dynamics and Materials Conference*, New Orleans, Vol. 5, pp. 2995–3011 (AIAA paper 95–1485-CP).

M. D. Olson and Y. C. Fung 1966 *AIAA Journal* **4**, 858–864. Supersonic flutter of circular cylindrical shells subjected to internal pressure and axial compression.

M. D. Olson and Y. C. Fung 1967 *AIAA Journal* **5**, 1849–1856. Comparing theory and experiment for the supersonic flutter of circular cylindrical shells.

U. Olsson 1978 *AIAA Journal* **16**, 360–362. Supersonic flutter of heated circular cylindrical shells with temperature-dependent material properties.

S. Vol'mir and S. V. Medvedeva 1973 *Soviet Physics – Doklady* **17**, 1213–1214. Investigation of the flutter of cylindrical panels in a supersonic gas flow.

Index

Printed in the United States
by Baker & Taylor Publisher Services